# Lecture Notes in Computer Science 4315

*Commenced Publication in 1973*
Founding and Former Series Editors:
Gerhard Goos, Juris Hartmanis, and Jan van Leeuwen

## Editorial Board

David Hutchison
  *Lancaster University, UK*
Takeo Kanade
  *Carnegie Mellon University, Pittsburgh, PA, USA*
Josef Kittler
  *University of Surrey, Guildford, UK*
Jon M. Kleinberg
  *Cornell University, Ithaca, NY, USA*
Alfred Kobsa
  *University of California, Irvine, CA, USA*
Friedemann Mattern
  *ETH Zurich, Switzerland*
John C. Mitchell
  *Stanford University, CA, USA*
Moni Naor
  *Weizmann Institute of Science, Rehovot, Israel*
Oscar Nierstrasz
  *University of Bern, Switzerland*
C. Pandu Rangan
  *Indian Institute of Technology, Madras, India*
Bernhard Steffen
  *University of Dortmund, Germany*
Madhu Sudan
  *Massachusetts Institute of Technology, MA, USA*
Demetri Terzopoulos
  *University of California, Los Angeles, CA, USA*
Doug Tygar
  *University of California, Berkeley, CA, USA*
Gerhard Weikum
  *Max-Planck Institute of Computer Science, Saarbruecken, Germany*

Matthias S. Mueller   Barbara M. Chapman
Bronis R. de Supinski   Allen D. Malony
Michael Voss (Eds.)

# OpenMP Shared Memory Parallel Programming

International Workshops, IWOMP 2005 and IWOMP 2006
Eugene, OR, USA, June 1-4, 2005
Reims, France, June 12-15, 2006
Proceedings

Volume Editors

Matthias S. Mueller
TU Dresden, Zentrum für Informationsdienste und Hochleistungsrechnen
01062 Dresden, Germany
E-mail: matthias.mueller@tu-dresden.de

Barbara M. Chapman
University of Houston, Dept. of Computer Science
Houston, TX, 77204-3010, USA
E-mail: chapman@cs.uh.edu

Bronis R. de Supinski
Lawrence Livermore National Laboratory, Center for Applied Scientific Computing
Livermore, CA, 94551-0808, USA
bronis@llnl.gov

Allen D. Malony
University of Oregon, Dept. of Computer and Information Science
Eugene, OR, 97403-1202, USA
E-mail: malony@cs.uoregon.edu

Michael Voss
Intel Corporation
Champaign, IL 61820, USA
E-mail: MichaelJ.Voss@intel.com


Library of Congress Control Number: 2008927737

CR Subject Classification (1998): D.1.3, D.1, D.2, F.2, G.1-4, J.2, I.6

LNCS Sublibrary: SL 1 – Theoretical Computer Science and General Issues

ISSN      0302-9743
ISBN-10   3-540-68554-5 Springer Berlin Heidelberg New York
ISBN-13   978-3-540-68554-8 Springer Berlin Heidelberg New York

This work is subject to copyright. All rights are reserved, whether the whole or part of the material is
concerned, specifically the rights of translation, reprinting, re-use of illustrations, recitation, broadcasting,
reproduction on microfilms or in any other way, and storage in data banks. Duplication of this publication
or parts thereof is permitted only under the provisions of the German Copyright Law of September 9, 1965,
in its current version, and permission for use must always be obtained from Springer. Violations are liable
to prosecution under the German Copyright Law.

Springer is a part of Springer Science+Business Media

springer.com

© Springer-Verlag Berlin Heidelberg 2008
Printed in Germany

Typesetting: Camera-ready by author, data conversion by Scientific Publishing Services, Chennai, India
Printed on acid-free paper    SPIN: 12274802     06/3180      5 4 3 2 1 0

# Preface

OpenMP is an application programming interface (API) that is widely accepted as a standard for high-level shared-memory parallel programming. It is a portable, scalable programming model that provides a simple and flexible interface for developing shared-memory parallel applications in Fortran, C, and C++. Since its introduction in 1997, OpenMP has gained support from the majority of high-performance compiler and hardware vendors. Under the direction of the OpenMP Architecture Review Board (ARB), the OpenMP standard is being further improved. Active research in OpenMP compilers, runtime systems, tools, and environments continues to drive its evolution. To provide a forum for the dissemination and exchange of information about and experiences with OpenMP, the community of OpenMP researchers and developers in academia and industry is organized under cOMPunity (www.compunity.org).

Workshops on OpenMP have taken place at a variety of venues around the world since 1999: the European Workshop on OpenMP (EWOMP), the North American Workshop on OpenMP Applications and Tools (WOMPAT), and the Asian Workshop on OpenMP Experiences and Implementation (WOMPEI) were each held annually and attracted an audience from both academia and industry. The intended purpose of the new International Workshop on OpenMP (IWOMP) was to consolidate these three OpenMP workshops into a single, yearly international conference. The first IWOMP meeting was held during June 1–4, 2005, in Eugene, Oregon, USA. The second meeting took place during June 12–15, in Reims, France. Each event drew over 60 participants from research and industry throughout the world. In keeping with the objectives and format of the prior workshops, IWOMP includes technical papers and panels, tutorials, and a hands-on laboratory (OMPlab), where OpenMP users and developers worked together to test compilers, tune applications, and experiment with OpenMP tools. The first of these workshops was organized under the auspices of cOMPunity. In the meantime, a Steering Committee has been established to oversee the organization of these events and to guide the further development of the workshop series.

The first two IWOMP meetings were successful in every respect. To a large extent, this success was due to the generous support received from the IWOMP sponsors. Intel Corporation, Sun Microsystems, Hewlett Packard, STMicroelectronics, PathScale, Microsoft, the University and City of Reims, the Region Champagne-Ardenne, and the ARB all gave financial support to these conferences. Fujitsu Systems Europe LTD, Microway, the Technical University of Denmark, the Centre Informatique National de l'Enseignement Supérieur, Reims Universtiy, RWTH Aachen University, and Technische Universität Dresden provided access to system platforms for the OMPlab. The level of support given demonstrates a strong interest in the success of OpenMP in both industry and research.

The cOMPunity webpage (see http://www.compunity.org) provides access to the talks given at the meetings and to photos of the activities. The IWOMP webpage (see http://www.iwomp.org) provides information on the latest event. This book contains the proceedings of the first two IWOMP workshops. In total, 35 papers were accepted for the technical program sections.

It was a pleasure to help ignite the IWOMP workshop series. We look forward to a bright future for both OpenMP and this workshop.

February 2008

Matthias S. Müller
Barbara Chapman
Bronis R. de Supinski
Allen D. Malony
Michael Voss

# Organization

## Committee of IWOMP 2005

**General Chair**

Allen D. Malony　　　　　University of Oregon, USA

**Local Chair**

Sameer S. Shende　　　　University of Oregon, USA

**Chair of Program Committee**

Barbara Chapman　　　　University of Oregon, USA

**Program Committee**

| | |
|---|---|
| Dieter an Mey | RWTH Aachen University, Germany |
| Eduard Ayguade | CIRI, UPC, Spain |
| Mark Bull | EPCC, University of Edinburgh, UK |
| Luiz DeRose | Cray Inc., USA |
| Bronis R. de Supinski | LLNL, USA |
| Rudolf Eigenmann | Purdue University, USA |
| Lawrence Meadows | Intel, USA |
| Bernd Mohr | Research Centre Juelich, ZAM, Germany |
| Matthias S. Müller | University of Stuttgart, Germany |
| Mitsuhisa Sato | University of Tsukuba, Japan |
| Michael Voss | University of Toronto, Canada |
| Michael Wolfe | STMicroelectronics, Inc. |

## IWOMP 2006 Committee

**Organization Committee**

Chair: Michaël Krajecki　　　University of Reims, France

**Program Committee**

Chair: Matthias S. Müller　　University of Dresden, ZIH, Germany

## Program Committee

| | |
|---|---|
| Dieter an Mey | RWTH Aachen University, Germany |
| Eduard Ayguade | CEPBA-IBM Research Institute (CIRI), UPC, Spain |
| Luiz DeRose | Cray Inc., USA |
| Bronis R. de Supinkski | LLNL, USA |
| Rudolf Eigenmann | Purdue University, USA |
| Guang Gao | University of Delaware, USA |
| Ricky A. Kendall | ORNL, USA |
| Myungho Lee | MyongJi University, Korea |
| Federico Massaioli | CASPUR, Roma, Italy |
| Lawrence Meadows | Intel, USA |
| Bernd Mohr | Research Centre Juelich, ZAM, Germany |
| Mitsuhisa Sato | University of Tsukuba, Japan |
| Yoshiki Seo | NEC, Japan |

## External Reviewers

David R. Jefferson (LLNL), David Lowenthal (University of Georgia), Daniel J. Quinlan (LLNL), Markus Schordan (TU Vienna), Xavier Martorell (UPC), Toni Corte (UPC), Alex Durans (UPC)

## Steering Committee

| | |
|---|---|
| Chair: Bronis R. de Supinski | NNSA ASC, LLNL, USA |

## Steering Committee

| | |
|---|---|
| Dieter an Mey | CCC, RWTH Aachen University, Germany |
| Eduard Ayguade | Barcelona Supercomputing Center (BSC), Spain |
| Mark Bull | EPCC, UK |
| Barbara Chapman | CEO of cOMPunity, Houston, USA |
| Sanjiv Shah | Intel, OpenMP CEO |
| Christophe Jaillet | University of Reims, France |
| Ricky Kendall | ORNL, USA |
| Michaël Krajecki | University of Reims, France |
| Rick Kufrin | NCSA, USA |
| Federico Massaioli | CASPUR, Rome, Italy |
| Lawrence Meadows | KSL Intel, USA |
| Matthias S. Müller | University of Dresden, ZIH, Germany |
| Florent Nolot | University of Reims, France |
| Mitsuhisa Sato | University of Tsukuba, Japan |
| Ruud van der Pas | Sun Microsystems, Geneva, Switzerland |
| Matthijs van Waveren | Fujitsu, France |

# Table of Contents

## First International Workshop on OpenMP IWOMP 2005

### Performance Tools

Performance Analysis of Large-Scale OpenMP and Hybrid
MPI/OpenMP Applications with Vampir NG .......................... 5
  *Holger Brunst and Bernd Mohr*

ompP: A Profiling Tool for OpenMP ................................. 15
  *Karl Fürlinger and Michael Gerndt*

On the Interaction of Tiling and Automatic Parallelization ............ 24
  *Zhelong Pan, Brian Armstrong, Hansang Bae, and
  Rudolf Eigenmann*

Static Nonconcurrency Analysis of OpenMP Programs ................ 36
  *Yuan Lin*

CCRG OpenMP Compiler: Experiments and Improvements ........... 51
  *Huang Chun and Yang Xuejun*

### Compiler Technology

Implementing an OpenMP Execution Environment on InfiniBand
Clusters ....................................................... 65
  *Jie Tao, Wolfgang Karl, and Carsten Trinitis*

An Introduction to Balder—An OpenMP Run-time Library for Clusters
of SMPs ....................................................... 78
  *Sven Karlsson*

### Run-Time Environment

Experiences with the OpenMP Parallelization of DROPS, a
Navier-Stokes Solver Written in C++ .............................. 95
  *Christian Terboven, Alexander Spiegel, Dieter an Mey,
  Sven Gross, and Volker Reichelt*

A Parallel Structured Ecological Model for High End Shared Memory
Computers ................................................. 107
  *Dali Wang, Michael W. Berry, and Louis J. Gross*

Multi-cluster, Mixed-Mode Computational Modeling of Human Head
Conductivity .............................................. 119
  *Adnan Salman, Sergei Turovets, Allen D. Malony, and Vasily Volkov*

## Application I

An Evaluation of OpenMP on Current and Emerging
Multithreaded/Multicore Processors .......................... 133
  *Matthew Curtis-Maury, Xiaoning Ding,
  Christos D. Antonopoulos, and Dimitrios S. Nikolopoulos*

SPEC OpenMP Benchmarks on Four Generations of NEC SX Parallel
Vector Systems ............................................ 145
  *Matthias S. Müller*

Performance Evaluation of Parallel Sparse Matrix–Vector Products on
SGI Altix3700 ............................................. 153
  *Hisashi Kotakemori, Hidehiko Hasegawa, Tamito Kajiyama,
  Akira Nukada, Reiji Suda, and Akira Nishida*

## The OpenMP Language and Its Evaluation

The OpenMP Memory Model .................................... 167
  *Jay P. Hoeflinger and Bronis R. de Supinski*

Evaluating OpenMP on Chip MultiThreading Platforms ............ 178
  *Chunhua Liao, Zhenying Liu, Lei Huang, and Barbara Chapman*

Experiences Parallelizing a Web Server with OpenMP ............ 191
  *Jairo Balart, Alejandro Duran, Marc Gonzàlez, Xavier Martorell,
  Eduard Ayguadé, and Jesús Labarta*

## Second International Workshop on OpenMP IWOMP 2006

## Advanced Performance Tuning

Automatic Granularity Selection and OpenMP Directive Generation
Via Extended Machine Descriptors in the PROMIS Parallelizing
Compiler .................................................. 207
  *Walden Ko and Constantine D. Polychronopoulos*

Nested Parallelization of the Flow Solver TFS Using the ParaWise
Parallelization Environment ........................................ 217
    *Steve Johnson, Peter Leggett, Constantinos Ierotheou,*
    *Alexander Spiegel, Dieter an Mey, and Ingolf Hörschler*

Performance Characteristics of OpenMP Language Constructs on a
Many-core-on-a-chip Architecture ................................... 230
    *Weirong Zhu, Juan del Cuvillo, and Guang R. Gao*

Improving Performance of OpenMP for SMP Clusters Through
Overlapped Page Migrations......................................... 242
    *Woo-Chul Jeun, Yang-Suk Kee, and Soonhoi Ha*

## Aspects of Code Development

Adding New Dimensions to Performance Analysis Through
User-Defined Objects ............................................... 255
    *Gabriele Jost, Oleg Mazurov, and Dieter an Mey*

Performance Instrumentation and Compiler Optimizations for
MPI/OpenMP Applications .......................................... 267
    *Oscar Hernandez, Fengguang Song, Barbara Chapman,*
    *Jack Dongarra, Bernd Mohr, Shirley Moore, and Felix Wolf*

Supporting Nested OpenMP Parallelism in the TAU Performance
System ............................................................. 279
    *Alan Morris, Allen D. Malony, and Sameer S. Shende*

Parallelization of a Hierarchical Data Clustering Algorithm Using
OpenMP............................................................. 289
    *Panagiotis E. Hadjidoukas and Laurent Amsaleg*

OpenMP and C++ ................................................... 300
    *Christian Terboven and Dieter an Mey*

Common Mistakes in OpenMP and How to Avoid Them: A Collection
of Best Practices.................................................... 312
    *Michael Süß and Claudia Leopold*

Formal Specification of the OpenMP Memory Model ................. 324
    *Greg Bronevetsky and Bronis R. de Supinski*

## Applications II

Performance and Programmability Comparison Between OpenMP and
MPI Implementations of a Molecular Modeling Application ........... 349
    *Russell Brown and Ilya Sharapov*

OpenMP Implementation of SPICE3 Circuit Simulator .............. 361
  *Tien-Hsiung Weng, Ruey-Kuen Perng, and Barbara Chapman*

Automatic Generation of Parallel Code for Hessian Computations ...... 372
  *H. Martin Bücker, Arno Rasch, and Andre Vehreschild*

Geographical Locality and Dynamic Data Migration for OpenMP
Implementations of Adaptive PDE Solvers ........................ 382
  *Markus Nordén, Henrik Löf, Jarmo Rantakokko, and
  Sverker Holmgren*

## Proposed Extensions to OpenMP

A Comparison of Task Pool Variants in OpenMP and a Proposal for a
Solution to the Busy Waiting Problem ............................ 397
  *Alexander Wirz, Michael Süß, and Claudia Leopold*

A Proposal for OpenMP for Java ................................. 409
  *Michael Klemm, Ronald Veldema, Matthias Bezold, and
  Michael Philippsen*

A Proposal for Error Handling in OpenMP ........................ 422
  *Alejandro Duran, Roger Ferrer, Juan José Costa, Marc Gonzàlez,
  Xavier Martorell, Eduard Ayguadé, and Jesús Labarta*

Extending the OpenMP Standard for Thread Mapping and Grouping ... 435
  *Guansong Zhang*

**Author Index** .................................................. 447

# First International Workshop on OpenMP IWOMP 2005, June 1–4, Eugene, Oregon, USA

# Performance Tools

# Performance Analysis of Large-Scale OpenMP and Hybrid MPI/OpenMP Applications with Vampir NG

Holger Brunst[1] and Bernd Mohr[2]

[1] Center for High Performance Computing
Dresden University of Technology
Dresden, Germany
brunst@zhr.tu-dresden.de

[2] Forschungszentrum Jülich, ZAM
Jülich, Germany
b.mohr@fz-juelich.de

**Abstract.** This paper presents a tool setup for comprehensive event-based performance analysis of large-scale OpenMP and hybrid OpenMP/MPI applications. The KOJAK framework is used for portable code instrumentation and automatic analysis while the new VAMPIR NG infrastructure serves as generic visualization engine for both OpenMP and MPI performance properties. The tools share the same data base which enables a smooth transition from bottleneck auto-detection to manual in-depth visualization and analysis. With VAMPIR NG being a distributed data-parallel architecture, large problems on very large scale systems can be addressed.

**Keywords:** Parallel Computing, OpenMP, Program Analysis, Instrumentation.

## 1 Introduction

OpenMP is probably the most commonly used communication standard for shared-memory based parallel computing. The same applies to MPI when talking about parallel computing on distributed-memory architectures. Both approaches have widely accepted characteristics and qualities. OpenMP stands for an incremental approach to parallel computing which can be easily adapted to existing sequential software. MPI has a very good reputation with respect to performance and scalability on large problem and system sizes. Yet, it typically requires a thorough (re-) design of a parallel application. So far, most parallel applications are either native OpenMP or native MPI applications. With the emergence of large clusters of SMPs, this situation is changing. Clearly, hybrid applications that make use of both programming paradigms are one way to go. OpenMP has proven to work effectively on shared memory systems. MPI on the other hand can be used to bridge the gap between multiple SMP nodes. In a sense, this strategy follows the original idea of OpenMP which is to incrementally parallelize a given code.

In a hybrid scenario only minor changes (i. e. adding OpenMP directives) are required to achieve a moderate performance improvement while going beyond the memory boundaries of an SMP node requires more sophisticated techniques like message passing. Quite natural, a program that combines multiple programming paradigms is not easy to develop, maintain, and optimize. Portable tools for program analysis and debugging are almost essential in this respect. Yet, existing tools [1,2,3] typically concentrate on either MPI or OpenMP or exist for dedicated platforms only [4,5]. It is therefore difficult to get on overall picture of a hybrid large-scale application. This paper presents a portable, distributed analysis infrastructure which enables a comprehensive support of hybrid OpenMP applications. The paper is organized as follows. The next section deals with collecting, mapping, and automatic classification of OpenMP/MPI performance data. Based hereon, Section 3 goes a step further and presents an architecture for in-depth analysis of large hybrid OpenMP applications. In Section 4 mixed mode analysis examples are given. Finally, Section 5 concludes the joint tool initiative.

## 2 The KOJAK Measurement System

The KOJAK performance-analysis tool environment provides a complete tracing-based solution for automatic performance analysis of MPI, OpenMP, or hybrid applications running on parallel computers. KOJAK describes performance problems using a high level of abstraction in terms of execution patterns that result from an inefficient use of the underlying programming model(s). KOJAK's overall architecture is depicted in Figure 1. The different components are represented as rectangles and their inputs and outputs are represented as boxes with rounded corners. The arrows illustrate the whole performance-analysis process from instrumentation to result presentation.

The KOJAK analysis process is composed of two parts: a semi-automatic multi-level instrumentation of the user application followed by an automatic analysis of the generated performance data. The first part is considered semi-automatic because it requires the user to slightly modify the makefile.

To begin the process, the user supplies the application's source code, written in either C, C++, or Fortran, to OPARI, which is a source-to-source translation tool. OPARI performs automatic instrumentation of OpenMP constructs and redirection of OpenMP-library calls to instrumented wrapper functions on the source-code level based on the POMP OpenMP monitoring API [6,7]. This is done to capture OpenMP events relevant to performance, such as entering a parallel region. Since OpenMP defines only the semantics of directives, not their implementation, there is no equally portable way of capturing those events on a different level.

Instrumentation of user functions is done either during compilation by a compiler-supplied instrumentation interface or on the source-code level using TAU [8]. TAU is able to automatically instrument the source code of C, C++, and Fortran programs using a preprocessor based on the PDT toolkit [9].

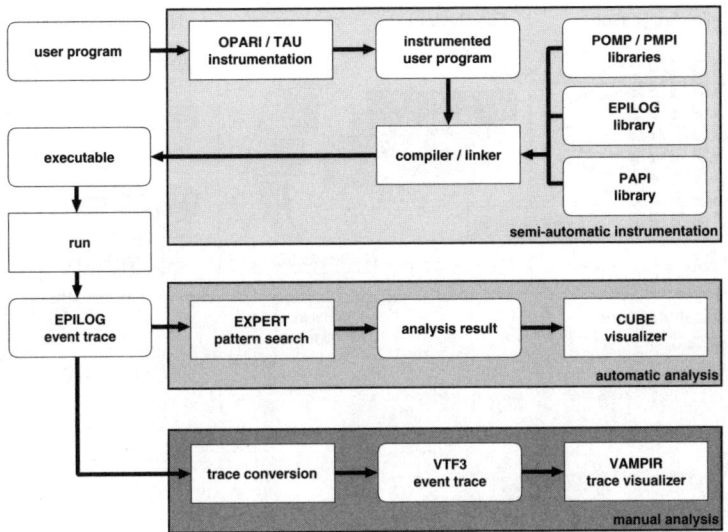

**Fig. 1.** KOJAK overall architecture

Instrumentation for MPI events is accomplished with a wrapper library based on the PMPI profiling interface, which generates MPI-specific events by intercepting calls to MPI functions. All MPI, OpenMP, and user-function instrumentation calls the EPILOG run-time library, which provides mechanisms for buffering and trace-file creation. The application can also be linked to the PAPI library [10] for collection of hardware counter metrics as part of the trace file. At the end of the instrumentation process, the user has a fully instrumented executable.

Running this executable generates a trace file in the EPILOG format. After program termination, the trace file is fed into the EXPERT analyzer. (See [11] for details of the automatic analysis, which is outside of the scope of this paper.) In addition, the automatic analysis can be combined with a manual analysis using VAMPIR [12] or VAMPIR NG [13], which allows the user to investigate the patterns identified by EXPERT in a time-line display via a utility that converts the EPILOG trace file into the VAMPIR format.

## 3 The Distributed VAMPIR NG Program Analysis System

The distributed architecture of the parallel performance analysis tool VAMPIR NG [13] outlined in this section has been newly designed based on the experience gained from the development of the performance analysis tool VAMPIR. The new architecture uses a distributed approach consisting of a parallel analysis server running on a segment of a parallel production environment and a visualization client running on a potentially remote graphics workstation. Both components interact with each other over the Internet through a socket based network connection.

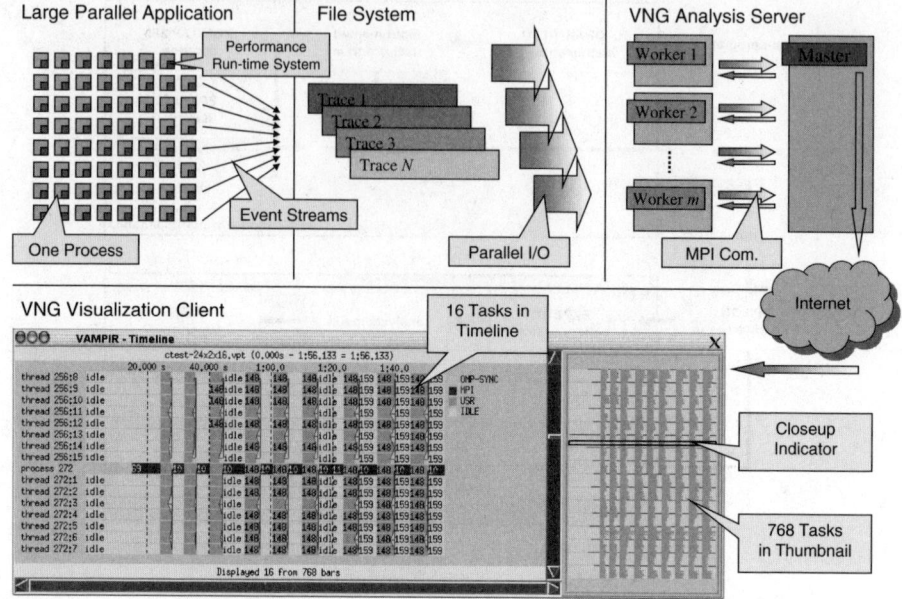

**Fig. 2.** VAMPIR NG Architecture Overview

The major goals of the distributed parallel approach are:

1. Keep event trace data close to the location where they were created.
2. Analyze event data in parallel to achieve increased scalability
   (# of events $\sim 1,000,000,000$ and # of streams (processes) $\sim 10,000$).
3. Provide fast and easy to use remote performance analysis on end-user platforms.

VAMPIR NG consists of two major components: an analysis server (**vngd**) and a visualization client (**vng**). Each is supposed to run on a different machine. Figure 2 shows a high-level view of the overall software architecture. Boxes represent modules of the components whereas arrows indicate the interfaces between the different modules. The thickness of the arrows gives a rough measure of the data volume to be transferred over an interface, whereas the length of an arrow represents the expected latency for that particular link.

In the top right corner of Figure 2 we can see the analysis server, which runs on a small interactive segment of a parallel machine. The reason for this is twofold. Firstly, it allows the analysis server to have closer access to the trace data generated by an application being traced. Secondly, it allows the server to execute in parallel. Indeed, the server is a heterogeneous parallel program, implemented using MPI and pthreads, which uses a master/worker approach. The workers are responsible for storage and analysis of trace data. Each of them holds a part of the overall data to be analyzed. The master is responsible for the communication to the remote clients. He decides how to distribute analysis requests among the workers. Once the analysis requests are completed, the master merges the results into a single response package that is subsequently sent to the client.

The bottom half of Figure 2 depicts a snapshot of the VAMPIR NG visualization client which illustrates the timeline of an application run with 768 independent tasks. The idea is that the client is not supposed to do any time consuming calculations. It is a straightforward sequential GUI implementation with a look-and-feel very similar to performance analysis tools like Jumpshot [1], Paraver [4], VAMPIR [12], Paje [3], etc. For visualization purposes, it communicates with the analysis server according to the user's preferences and inputs. Multiple clients can connect to the analysis server at the same time, allowing simultaneous viewing of trace results.

As mentioned above, the shape of the arrows indicates the quality of the communication links with respect to throughput and latency. Knowing this, we can deduce that the client-to-server communication was designed to not require high bandwidths. In addition, the system should operate efficiently with only moderate latencies in both directions. This is basically due to the fact that only control information and condensed analysis results are to be transmitted over this link. Following this approach we comply with the goal of keeping the analysis on a centralized platform and doing the visualization remotely.

The big arrows connecting the program traces with the worker processes indicate high bandwidth. The major goal is to get fast access to whatever segment of the trace data the user is interested in. High bandwidth is basically achieved by reading data in parallel by the worker processes. To support multiple client sessions, the server makes use of multi-threading on the boss and worker processes.

## 4 In-Depth Analysis of Large-Scale OpenMP Programs

The KOJAK analysis infrastructure primarily addresses automatic problem detection. Previously collected trace data is searched for pre-defined problems [14]. The results are displayed in a hierarchical navigator tool which provides links to the respective source code locations. This approach is very effective as it does not require complicated user interactions or expert knowledge. Yet, it is limited to known problems and sometimes the real cause of a phenomenon remains obscure.

With the help of the collected trace data it is even possible to go into further detail. The measurement system in KOJAK supports the generation of VAMPIR NG compatible traces which can be examined according to the hints made by the EXPERT tool.

Having access to the same central data base, VAMPIR NG offers a rich set of scalable remote visualization options for arbitrary program phases. In the following, the sPPM benchmark code [15] will serve as example application demonstrating combined OpenMP and MPI capabilities of VAMPIR NG. The code has been equipped with OpenMP directives and was executed on 128 MPI tasks with eight OpenMP threads each. The test platform was a Power4-based, 30-way SMP cluster system. Altogether, 1024 independent event data streams had to be handled.

## 4.1 Custom Profiles

VAMPIR NG supports a grouping concept for flat profile charts à la gprof. The summarized information reflects either the entire program run or a time interval specified by the user. The information provided is not limited to functions. Depending on the application, OpenMP and MPI related information like message sizes, counter values etc. can be summarized additionally.

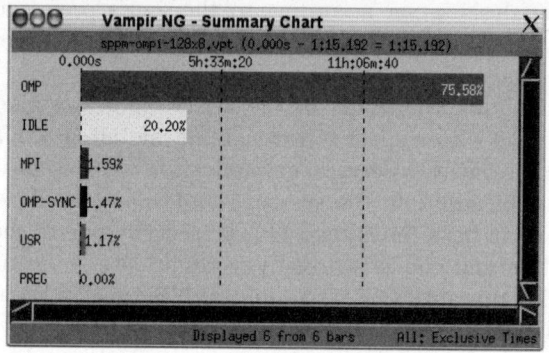

**Fig. 3.** Summary profile of a sPPM run on 1024 processes/threads

Figure 3 depicts a summary profile of the full program run which lasted 1:15 minutes. Exclusive timing information is shown as percentages relative to the overall accumulated run-time. The KOJAK OpenMP instrumentation creates the following six default sub-groups of program states:

1. *USR:* Code regions which are not parallelized with OpenMP
2. *OMP:* OpenMP parallel execution
3. *OMP-SYNC:* OpenMP implicit and explicit barrier synchronization
4. *PREG:* OpenMP thread startup and termination
5. *MPI:* MPI communication and synchronization
6. *IDLE:* Idle OpenMP threads

Quite obviously, the application spends too much time (20%) doing nothing (*IDLE*). Its cause is unknown. We will come to this phenomenon in the next section. 75% percent of the runtime is spent in OpenMP parallel code. The remaining five percent are spent in MPI and OpenMP synchronization code.

The same display can be used to further analyze the six sub-groups of program states. Figures 4(a) to 4(d) depict summary profiles for the states in *OMP, OMP-SYNC*, and *MPI* respectively. From Figure 4(a) we can read that our application has twelve major OpenMP do-loops from which six contribute with more than 8.5 seconds each (per process). Only these loops should be considered for further optimization. In Figure 4(b), OpenMP synchronization overhead is depicted. The first two barrier constructs are interesting candidates to be analyzed in further detail. Their active phases during run-time can be located with a navigator

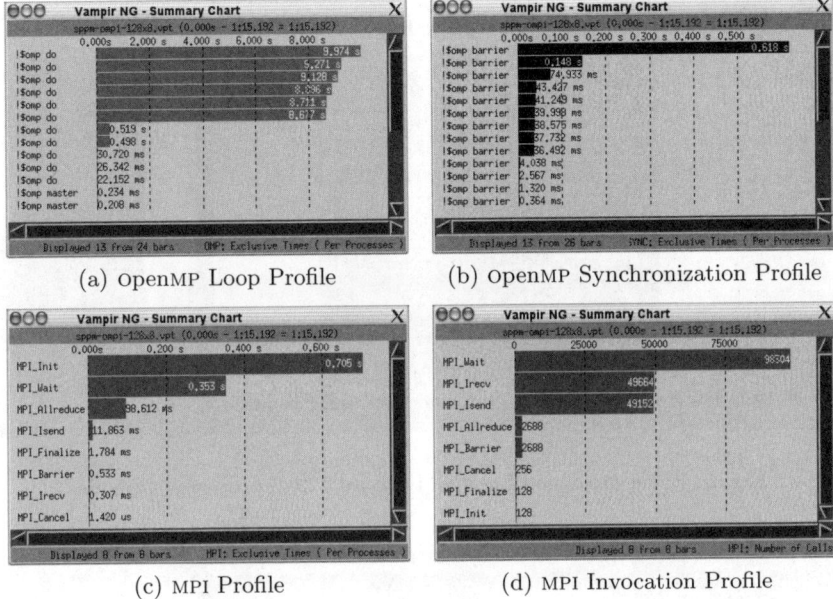

**Fig. 4.** Adaptive VAMPIR NG Profiles

display similar to traditional timelines. Depending on user defined queries, the "navigator" (not depicted) highlights selected states only. Figure 4(c) provides information on how MPI is used in this code. Synchronization is the dominant part. Last not least, the number of MPI function calls as depicted in Figure 4(d) tells us that approximately 100,000 messages are exchanged altogether. Considering the 128 MPI processes involved and the short amount of time spent in MPI, this information is more interesting for code debugging than for optimization.

### 4.2 Hierarchical Timelines

Sometimes, adaptive profiles are not sufficient for understanding an application's inner working. An event timeline as depicted in Figure 5 is very useful to obtain a better understanding. The event timeline visualizes the behavior of individual processes over time. Here, the horizontal axis reflects time, while the vertical axis identifies the process. Colors are used to represent the already mentioned sub-groups of program states. Apparently, navigating on the data of 1024 independent processing entities is a rather complex task. Therefore, an overview of the full set of processes and threads is depicted on the right hand side. The rectangular marking identifies the selection of the trace that is depicted in full detail on the left hand side (process 48 to process 64).

Having access to this kind of application overview, it quickly becomes evident where the 20% idle-time in the profile comes from. Due to the large number of processes and OpenMP threads, the execution platform needs a substantial time (approximately 17 seconds) to spawn the full application. Having a closer look

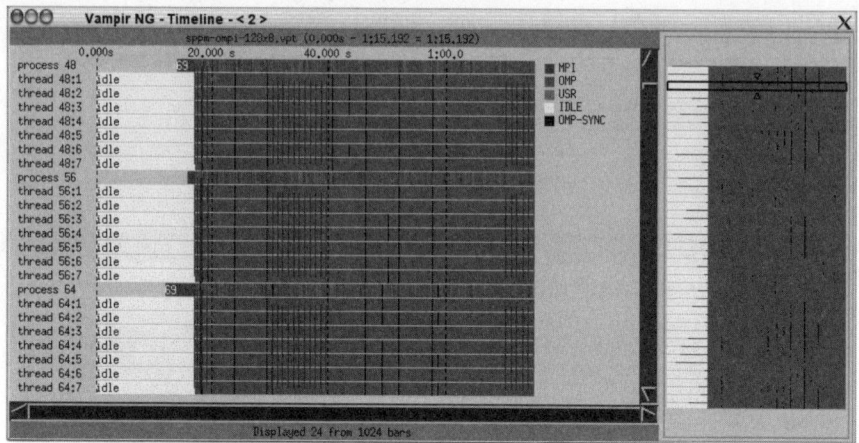

**Fig. 5.** Event timeline of a sPPM run on 128x8 processes/threads

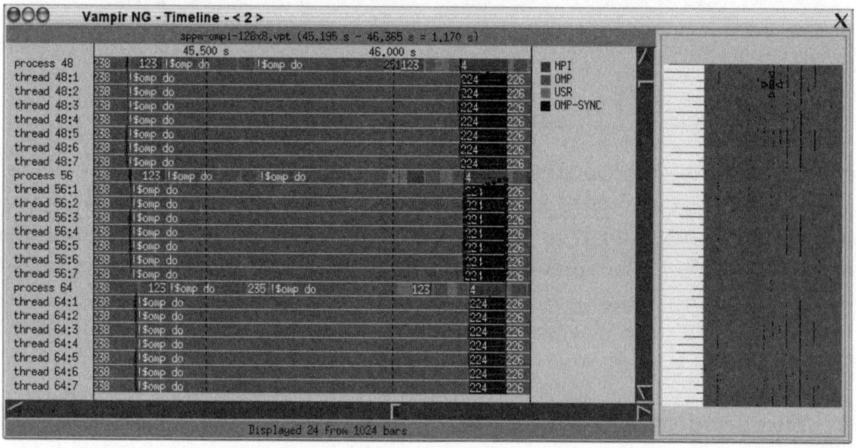

**Fig. 6.** Synchronization of sPPM run on 128x8 processes/threads

at the startup phase (light/beige section) reveals that the spawning of the MPI processes (MPI_Init) is varying a lot in time. Apparently, MPI_Init has to wait until all processes are up and running before it lets the processes start their individual tasks.

We will now take a closer look at the locations where OpenMP and MPI synchronization takes place. Figure 6 illustrates a section which includes the barrier that has been mentioned earlier in Section 4.1. The program highlights the selected OpenMP barrier with bold dotted lines. From this kind of display we can learn many things, one of which is that MPI and OpenMP synchronization have to work in close cooperation in mixed codes. This particular example shows how MPI collective communication is carried out on the master threads only (which

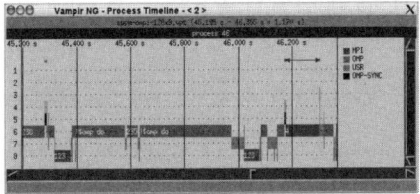 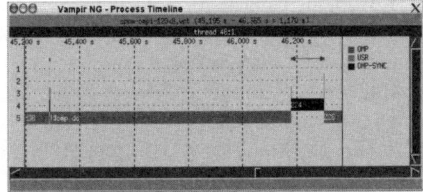

(a) Single Timeline – MPI Process     (b) Single Timeline – OpenMP Thread

**Fig. 7.** Hybrid MPI/OpenMP Synchronization

is a common MPI constraint) while OpenMP barriers guarantee that the thread parallelism is not continuing to process inconsistent data. Figure 7 illustrates the differences between an MPI communication process and a respective OpenMP thread by means of a single-task-timeline showing the detailed function call-path.

## 5 Conclusion

Data distribution and synchronization in large-scale OpenMP and hybrid MPI/OpenMP applications can lead to critical performance bottlenecks. Profiling alone can hardly help to identify the real cause of problems that fall into this category. Event-based approaches on the other hand are known to generate large volumes of data. In this difficult situation, automatic event-based performance analysis has the potential to quickly detect most known synchronization problems. When dealing with uncommon features or for detailed examination of already detected problems, manual analysis has certain advantages due to human intuition and pattern recognition capabilities. Therefore, an incremental approach with profiling and automatic techniques forming a solid starting point and event-based analysis being used for more detailed questions is advisable. PAPI [10] counter support in both tools completes the detailed performance examination. Finally, our work has shown that both approaches can be effectively combined in a portable way.

## References

1. Zaki, O., Lusk, E., Gropp, W., Swider, D.: Toward scalable performance visualization with Jumpshot. High Performance Computing Applications 13, 277–288 (1999)
2. Rose, L.D., Zhang, Y., Reed, D.A.: Svpablo: A multi-language performance analysis system. In: 10th International Conference on Computer Performance Evaluation - Modelling Techniques and Tools - Performance Tools 1998, Palma de Mallorca, Spain, pp. 352–355 (1998)
3. de Kergommeaux, J.C., de Oliveira Stein, B., Bernard, P.: Pajè, an interactive visualization tool for tuning multi-threaded parallel applications. Parallel Computing 26, 1253–1274 (2000)

4. European Center for Parallelism of Barcelona (CEPBA): Paraver - Parallel Program Visualization and Analysis Tool - Reference Manual (2000), http://www.cepba.upc.es/paraver
5. Intel: Intel thread checker (2005), http://www.intel.com/software/products/threading/tcwin
6. Mohr, B., Mallony, A., Hoppe, H.C., Schlimbach, F., Haab, G., Shah, S.: A Performance Monitoring Interface for OpenMP. In: Proceedings of the fourth European Workshop on OpenMP - EWOMP 2002 (September 2002)
7. Mohr, B., Malony, A., Shende, S., Wolf, F.: Design and Prototype of a Performance Tool Interface for OpenMP. The Journal of Supercomputing 23, 105–128 (2002)
8. Bell, R., Malony, A.D., Shende, S.: A Portable, Extensible, and Scalable Tool for Parallel Performance Profile Analysis. In: Kosch, H., Böszörményi, L., Hellwagner, H. (eds.) Euro-Par 2003. LNCS, vol. 2790, pp. 17–26. Springer, Heidelberg (2003)
9. Lindlan, K.A., Cuny, J., Malony, A.D., Shende, S., Mohr, B., Rivenburgh, R., Rasmussen, C.: A Tool Framework for Static and Dynamic Analysis of Object-Oriented Software with Templates. In: Proceedings of Supercomputing 2000 (November 2000)
10. Browne, S., Dongarra, J., Garner, N., Ho, G., Mucci, P.: A Portable Programming Interface for Performance Evaluation on Modern Processors. The International Journal of High Performance Computing Applications 14, 189–204 (2000)
11. Wolf, F., Mohr, B.: Automatic Performance Analysis of Hybrid MPI/OpenMP Applications. Journal of Systems Architecture, Special Issue 'Evolutions in parallel distributed and network-based processing' 49, 421–439 (2003)
12. Nagel, W., Arnold, A., Weber, M., Hoppe, H.C., Solchenbach, K.: Vampir: Visualization and Analysis of MPI Resources. Supercomputer 12, 69–80 (1996)
13. Brunst, H., Nagel, W.E., Malony, A.D.: A distributed performance analysis architecture for clusters. In: IEEE International Conference on Cluster Computing, Cluster 2003, Hong Kong, China, pp. 73–81. IEEE Computer Society, Los Alamitos (2003)
14. Fahringer, T., Gerndt, M., Riley, G., Träff, J.L.: Formalizing OpenMP performance properties with ASL. In: Valero, M., Joe, K., Kitsuregawa, M., Tanaka, H. (eds.) ISHPC 2000. LNCS, vol. 1940, pp. 428–439. Springer, Heidelberg (2000)
15. Lawrence Livermode National Laboratory: the sPPM Benchmark Code (2002), http://www.llnl.gov/asci/purple/benchmarks/limited/sppm/

# ompP: A Profiling Tool for OpenMP*

Karl Fürlinger and Michael Gerndt

Institut für Informatik,
Lehrstuhl für Rechnertechnik und Rechnerorganisation
Technische Universität München
{Karl.Fuerlinger, Michael.Gerndt}@in.tum.de

**Abstract.** In this paper we present a simple but useful profiling tool for OpenMP applications similar in spirit to the MPI profiler mpiP [16]. We describe the implementation of our tool and demonstrate its functionality on a number of test applications.

## 1 Introduction

For developers of scientific and commercial applications it is essential to understand the performance characteristics of their codes in order to take most advantage of the available computing resources. This is especially true for parallel programs, where a programmer additionally has to take issues such as load balancing, synchronization and communication into consideration. Accordingly, a number of tools with varying complexity and power have been developed for the major parallel programming languages and systems.

Generally, tools collect performance data either in the form of traces or profiles. Tracing allows a more detailed analysis as temporal characteristics of the execution is preserved, but it is usually more intrusive and the analysis of the recorded traces can be involved and time-consuming. Profiling, on the other hand, has the advantage of giving a concise overview where time is spent while causing less intrusion.

The best-known tracing solution for MPI is Vampir [12] (now Intel Trace Analyzer [6]) while mpiP [16] is a compact and easy to use MPI profiler. Both Vampir and mpiP rely on the MPI profiling interface that allows the interception and replacement of MPI routines by simply re-linking the user-application with the tracing or profiling library. Unfortunately no similar standardized profiling or performance analysis interface exists for OpenMP yet, making OpenMP performance analysis dependant on platform- and compiler specific mechanisms.

Fortunately, a proposal for a profiling interface for OpenMP is available in the form of the POMP specification and an instrumenter called Opari [10] has been developed that inserts POMP calls around instrumented OpenMP constructs. The authors of POMP and Opari also provide a tracing library, while we have implemented a straightforward POMP-based profiler that is similar in spirit to mpiP and which accordingly we call ompP [13].

---

* This work was partially funded by the Deutsche Forschungsgemeinschaft (DFG) under contract GE1635/1-1.

The rest of the paper is organized as follows: In Sect. 2 we describe the design and implementation of our tool and in Sect. 3 we demonstrate its functionality on some example programs. Finally in Sect. 4 we review related work, we conclude and present ideas for future work in Sect. 5.

## 2 Tool Design and Implementation

In this section we present the design and implementation of our profiling tool ompP.

### 2.1 Instrumentation

Opari [10] is an OpenMP source-to-source instrumenter for C, C++ and Fortran developed by Mohr et al. that inserts calls to a POMP compliant monitoring library around OpenMP constructs. For each instrumented OpenMP construct Opari creates a *region descriptor* structure that contains information such as the name of the construct, the source file and the begin and end line numbers. Each POMP_* call passes a pointer to the descriptor of the region being affected. In the example shown in Fig. 1, Opari creates one region descriptor for the parallel region and this descriptor is used for the POMP_Parallel_[fork,join,begin,end] and also for the POMP_Barrier_[Enter,Exit] calls. The barrier is added by Opari in order to measure the load imbalance in the parallel region, similar *implicit* barriers are added to OpenMP worksharing constructs.

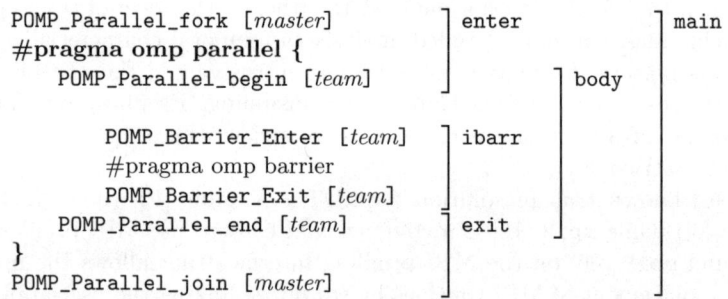

**Fig. 1.** Instrumentation added by Opari for the OpenMP parallel construct. The original code is shown in boldface, the square brackets denote the threads that execute a particular POMP_* call. The right part shows the pseudo region nesting used by ompP.

### 2.2 Performance Data Collection

Our profiler keeps track of counts and inclusive times for the instrumented OpenMP constructs. In order to simplify performance data bookkeeping (the same region descriptor can be used in a multitude of POMP_* calls), each Opari region is broken down into smaller conceptual "pseudo" regions and performance data (i.e, timestamps and execution counts) are recorded on the basis of these pseudo regions. In the example shown in Fig. 1, the pseudo regions are main, body, enter, exit and ibarr.

Instead of keeping track of all possible POMP_* calls for the individual Opari regions there are only two events for a pseudo region, namely enter and exit. For an enter event we record the enter timestamp (wall-clock time) that is later used in the exit event to increment the summed execution time by the elapsed time. Additionally a counter is incremented to count the number of executed instances of the pseudo region.

|  | seq | main | body | ibarr | enter | exit |
|---|---|---|---|---|---|---|
| MASTER | × |  |  |  |  |  |
| ATOMIC |  | × |  |  |  |  |
| BARRIER |  | × |  |  |  |  |
| FLUSH |  | × |  |  |  |  |
| USER_REGION |  | × |  |  |  |  |
| LOOP |  | × |  | × |  |  |
| SECTIONS |  | × | × | × |  |  |
| SINGLE |  | × | × | × |  |  |
| CRITICAL |  | × | × |  | × | × |
| WORKSHARE |  | × |  | × |  |  |
| PARALLEL | × |  | × | × | × | × |
| PARALLEL_LOOP | × |  | × | × | × | × |
| PARALLEL_SECTIONS | × | × | × | × | × | × |
| PARALLEL_WORKSHARE | × |  | × | × | × | × |

**Fig. 2.** List of pseudo regions for the different OpenMP constructs

A list of pseudo regions for the different Opari regions is shown in Fig. 2, the first column gives the name of the corresponding OpenMP construct as reported by ompP (LOOP refers to the for construct in C and the do construct in Fortran). The pseudo regions have the following semantic meaning:

main  Corresponds to the main region of the construct (unless the region is executed by one thread only then this role is taken by seq), if the construct has nested sub-regions, this refers to the "outer" part of a construct. An example is a sections construct that contains one or more section blocks.
body  Corresponds to the "inner" part of a construct, for example a section region inside a sections directive.
ibarr Corresponds to the implicit barrier added by Opari to worksharing constructs (unless a nowait clause is present) to measure load imbalance.
enter Allows the measurement of the time required to enter a construct. For critical sections this is the waiting time at the entry of the critical section. For parallel sections (and combined parallel worksharing regions) this measures the thread startup overhead.

**exit** Measures the time required to leave a construct. For critical sections this is the time required for leaving the critical section.[1] For parallel sections and combined parallel worksharing constructs this corresponds to the thread teardown overhead.

**seq** Measures the sequential execution time for a construct, i.e., the time spent by the master thread in a `master` construct or the time passing between `POMP_Fork` and `POMP_Join` in a `parallel` construct or combined parallel worksharing constructs.

Performance data is collected on a *region stack* basis. That is, similar to a call-path profile [8,3] where performance data is attributed not to a function itself (that would be a flat profile) but rather to the call-path that leads to a function, a stack of entered Opari regions is maintained and data is attributed to the stack that leads to a certain region. This region stack is currently maintained for POMP regions only, i.e., only automatically instrumented `OpenMP` constructs and user-instrumented regions[2] are placed on the stack, general functions or procedures are not handled unless manually instrumented by the user.

Including all called functions in our region stack would certainly be useful. However, this requires us to either perform a stackwalk (as `mpiP` does) or make use of compiler-supplied function instrumentation (i.e., the `-f instrument-functions` for the GNU compiler collection). Note that in either approach unwanted exposure to the compiler's implementation of the OpenMP standard (e.g., compiler outlining of parallel regions) has to be expected.

### 2.3 Performance Data Presentation

The performance data collected by `ompP` is kept in memory and written to a report file when the program finishes. The report file has the following sections:

- A header containing general information such as date and time of the program run.
- A list of all identified Opari regions with their type (`PARALLEL`, `ATOMIC`, `BARRIER`, ...) source file and line number information.
- A region summary list where performance data is summarized over the threads in the parallel execution. This list is sorted according to summed execution time and is intended to enable the developer to quickly identify the most time-consuming regions (and thus the most promising optimization targets).
- A detailed region summary for each identified region and for a specific region stack. This information allows the identification of load imbalances in the execution time and many other causes of inefficient or incorrect behavior.

---

[1] Usually one doesn't expect much waiting time at the end of a critical section. However, a thread might incur some overhead for signaling the critical section as "free" to other waiting threads.

[2] Users can instrument arbitrary regions by using the `pomp inst begin(`*name*`)` and `pomp inst end(`*name*`)` pragmas.

- A region summary for each region, where data is summed over all different region stacks that lead to the particular region (i.e., the flat profile for the region).

To produce a useful and concise profiling report, data are not reported as times and counts for each individual pseudo regions but specific semantic names are given according to the underlying Opari region. The following times and counts are reported:

- execT and execC count the number of executions and the total inclusive time spent for each thread (this is derived from the main or body pseudo region depending on the particular OpenMP construct).
- exitBarT and exitBarC are derived from the ibarr pseudo region and correspond to time spent in the implicit "exit barrier" in worksharing constructs or parallel regions. Analyzing the distribution of this time reveals load imbalances.
- startupT and startupC are defined for the OpenMP parallel construct and for the combined parallel work-sharing constructs (parallel for and parallel sections and parallel workshare), the data is derived from the enter pseudo region. If large fraction of time is spent in startupT and startupC is high, this indicates that a parallel region was repeatedly executed (maybe inside a loop) causing high overhead for thread creation and destruction.
- shutdownT and shutdownC are defined for the OpenMP parallel construct and for the combined parallel work-sharing constructs, the data is derived from the exit pseudo region. Its interpretation is similar to startupT and startupC.
- singleBodyT and singleBodyC are reported for single regions and report the time and execution counts spent inside the single region for each thread, the data is derived from the body pseudo region.
- sectionT and sectionC are reported for a sections construct and give the time and counts spent inside a section construct for each thread. The data is derived from the body pseudo region.
- enterT, enterC, exitT and exitC give the counts and times for entering and exiting critical sections, the data is derived from the enter and exit pseudo regions.

## 3 Application Examples

We report on a number of experiments that we have performed with ompP, all measurements have been performed on a single 4-way Itanium-2 SMP systems (1.3 GHz, 3 MB third level cache and 8 GB main memory), the Intel compiler version 8.0 was used.

## 3.1 APART Test Suite (ATS)

The ATS [11] is a set of test applications (MPI and OpenMP) developed within the APART [3] working group to test the functionality of automated and manual performance analysis tools. The framework is based on functions that generate a sequential amount of work for a process or thread and on a specification of the distribution of work among processes or threads. Building on this basis, individual programs are generated that exhibit a certain pattern of inefficient behavior, for example "imbalance in parallel region".

Previous work already tested existing OpenMP performance analysis tools with respect to their ability to detect the performance problems in the ATS framework [2]. With Expert [17], also a POMP-based tool was tested and generally with ompP a developer is able to detect the same set of OpenMP related problems as Expert (although with Expert the process is somewhat more automated).

The ompP output below is from a profiling run for the ATS program that demonstrates the "imbalance in parallel loop" performance problem. Notice the exitBarT column and the uneven distribution of time with respect to threads {0,1} and {2,3}. This example is typical for a number of load imbalance problems that are easily spottable by analyzing the exit barrier.

```
R00003    LOOP            pattern.omp.imbalance_in_parallel_loop.c (15--18)
   001:   [R0001]   imbalance_in_parallel_loop.c (17--34)
   002:   [R0002]   pattern.omp.imbalance_in_parallel_loop.c (11--20)
   003:   [R0003]   pattern.omp.imbalance_in_parallel_loop.c (15--18)

TID      execT      execC    exitBarT   exitBarC
 0        6.32        1        2.03        1
 1        6.32        1        2.02        1
 2        6.32        1        0.00        1
 3        6.32        1        0.00        1
 *       25.29        4        4.05        4
```

## 3.2 Quicksort

Süß and Leopold compare several parallel implementations of the Quicksort algorithm with respect to their efficiency in representing its recursive divide-and-conquer nature [15]. The code is now part of the OpenMP source code repository [1] and we have analyzed a version with a global work stack (called sort_omp_1.0 in [15]) with ompP. In this version there is a single stack of work elements (sub-sequences of the vector to be sorted) that are placed on or taken from the stack by the threads. Access to the stack is protected by critical section. The ompP output below shows the two critical sections in the code and it clearly indicates that a considerable amount of time is spent due to critical section contention. The total execution time of the program (summed over threads) was 61.02 seconds so the 9.53 and 6.27 seconds represent a considerable amount.

---

[3] Automated Performance Analysis: Real Tools.

```
R00002      CRITICAL         cpp_qsomp1.cpp (156--177)
  001:      [R0001]  cpp_qsomp1.cpp (307--321)
  002:      [R0002]  cpp_qsomp1.cpp (156--177)
  TID       execT     execC    enterT     enterC     exitT      exitC
  0          1.61    251780      0.87     251780      0.31     251780
  1          2.79    404056      1.54     404056      0.54     404056
  2          2.57    388107      1.38     388107      0.51     388107
  3          2.56    362630      1.39     362630      0.49     362630
  *          9.53   1406573      5.17    1406573      1.84    1406573

R00003      CRITICAL         cpp_qsomp1.cpp (211--215)
  001:      [R0001]  cpp_qsomp1.cpp (307--321)
  002:      [R0003]  cpp_qsomp1.cpp (211--215)
  TID       execT     execC    enterT     enterC     exitT      exitC
  0          1.60    251863      0.85     251863      0.32     251863
  1          1.57    247820      0.83     247820      0.31     247820
  2          1.55    229011      0.81     229011      0.31     229011
  3          1.56    242587      0.81     242587      0.31     242587
  *          6.27    971281      3.31     971281      1.25     971281
```

To improve the performance of the code, Süß and Leopold implemented a second version using thread-local stacks to reduce the contention for the global stack. We also analyzed the second version with ompP and the timing result for the two critical sections appears below.

In this version the overhead with respect to critical sections is clearly smaller than the first one (enterT and exitT have been improved by about 25 percent) The overall summed runtime reduces to 53.44 seconds, an improvement of about 12 percent, which is in line with the results reported in [15]. While this result demonstrates a nice performance gain with relatively little effort, our analysis clearly indicates room for further improvement; an idea would be to use lock-free data structures.

```
R00002      CRITICAL         cpp_qsomp2.cpp (175--196)
  001:      [R0001]  cpp_qsomp2.cpp (342--358)
  002:      [R0002]  cpp_qsomp2.cpp (175--196)
  TID       execT     execC    enterT     enterC     exitT      exitC
  0          0.67    122296      0.34     122296      0.16     122296
  1          2.47    360702      1.36     360702      0.54     360702
  2          2.41    369585      1.31     369585      0.53     369585
  3          1.68    246299      0.93     246299      0.37     246299
  *          7.23   1098882      3.94    1098882      1.61    1098882

R00003      CRITICAL         cpp_qsomp2.cpp (233--243)
  001:      [R0001]  cpp_qsomp2.cpp (342--358)
  002:      [R0003]  cpp_qsomp2.cpp (233--243)
```

| TID | execT | execC | enterT | enterC | exitT | exitC |
|-----|-------|--------|--------|--------|-------|--------|
| 0   | 1.22  | 255371 | 0.55   | 255371 | 0.31  | 255371 |
| 1   | 1.16  | 242924 | 0.53   | 242924 | 0.30  | 242924 |
| 2   | 1.32  | 278241 | 0.59   | 278241 | 0.34  | 278241 |
| 3   | 0.98  | 194745 | 0.45   | 194745 | 0.24  | 194745 |
| *   | 4.67  | 971281 | 2.13   | 971281 | 1.19  | 971281 |

## 4 Related Work

A number of performance analysis tools for OpenMP exist. Vendor specific tools such as Intel Thread Profiler [5] and Sun Studio [14] are usually limited to the respective platform but can make use of details of the compiler's OpenMP implementation.

Expert [17] is a tool based on POMP that performs tracing of hybrid MPI and OpenMP applications. After a program run traces are analyzed by Expert which performs an automatic search for patterns of inefficient behavior. Another POMP-based profiler called PompProf is mentioned in [4] but no further details are given.

TAU [7] is also able to profile OpenMP applications by utilizing the Opari instrumenter. TAU additionally profiles user functions, provides support for hardware counters and includes a visualizer for performance results. ompP differs in the way performance data is presented. We believe that due to its simplicity, limited purpose and scope, ompP might be easier to use for programmers wanting to get an overview of the behavior of their OpenMP codes that the more complex and powerful TAU tool set.

## 5 Conclusion and Future Work

We have presented our OpenMP profiler ompP. The tool can be used to quickly identify regions of inefficient behavior. In fact by analyzing execution counts the tool is also useful for correctness debugging in certain cases (for example to verify that a critical section is actually entered a certain, known number of times for given input data).

An important benefit is the immediate availability of the textual profiling report after the program run, as no further post-processing step is required. Furthermore the tool is naturally very portable and can be used on virtually any platform making it straightforward to compare the performance (and the performance problems) on a number of different platforms.

For the future we are considering the inclusion of hardware performance counters in the data gathering step. Additionally we are investigating to use Tool Gear [9] to be able to related the profiling data to the user's source code in a nice graphical representation.

# References

1. Dorta, A.J., Rodríguez, C., de Sande, F., Gonzáles-Escribano, A.: The OpenMP source code repository. In: Proceedings of the 13th Euromicro Conference on Parallel, Distributed and Network-Based Processing (PDP 2005), February 2005, pp. 244–250 (2005)
2. Gerndt, M., Mohr, B., Träff, J.L.: Evaluating OpenMP performance analysis tools with the APART test suite. In: Danelutto, M., Vanneschi, M., Laforenza, D. (eds.) Euro-Par 2004. LNCS, vol. 3149, pp. 155–162. Springer, Heidelberg (2004)
3. Graham, S.L., Kessler, P.B., McKusick, M.K.: gprof: A call graph execution profiler. SIGPLAN Not. 17(6), 120–126 (1982)
4. IBM HPC Toolkit, http://www.spscicomp.org/ScicomP10/Presentations/Austin_Klepacki.pdf
5. Intel Thread Profiler, http://www.intel.com/software/products/threading/tp/
6. Intel Trace Analyzer, http://www.intel.com/software/products/cluster/tanalyzer/
7. Malony, A.D., Shende, S.S.: Performance technology for complex parallel and distributed systems, pp. 37–46 (2000)
8. Malony, A.D., Shende, S.S.: Overhead Compensation in Performance Profiling. In: Danelutto, M., Vanneschi, M., Laforenza, D. (eds.) Euro-Par 2004. LNCS, vol. 3149, pp. 119–132. Springer, Heidelberg (2004)
9. May, J., Gyllenhaal, J.: Tool Gear: Infrastructure for parallel tools. In: Proceedings of the 2003 International Conference on Parallel and Distributed Processing Techniques and Applications (PDPTA 2003), pp. 231–240 (2003)
10. Mohr, B., Malony, A.D., Shende, S.S., Wolf, F.: Towards a performance tool interface for OpenMP: An approach based on directive rewriting. In: Proceedings of the Third Workshop on OpenMP (EWOMP 2001) (September 2001)
11. Mohr, B., Träff, J.L.: Initial design of a test suite for automatic performance analysis tools. In: Eighth International Workshop on High-Level Parallel Programming Models and Supportive Environments (HIPS 2003), pp. 77–86 (2003)
12. Nagel, W.E., Arnold, A., Weber, M., Hoppe, H.-C., Solchenbach, K.: VAMPIR: Visualization and analysis of MPI resources. Supercomputer 12(1), 69–90 (1996)
13. ompp webpage, http://www.ompp-tool.com
14. Sun Studio, http://developers.sun.com/prodtech/cc/hptc_index.html
15. Süß, M., Leopold, C.: A user's experience with parallel sorting and OpenMP. In: Proceedings of the Sixth Workshop on OpenMP (EWOMP 2004) (October 2004)
16. Vetter, J.S., Mueller, F.: Communication characteristics of large-scale scientific applications for contemporary cluster architectures. J. Parallel Distrib. Comput. 63(9), 853–865 (2003)
17. Wolf, F., Mohr, B.: Automatic performance analysis of hybrid MPI/OpenMP applications. In: Proceedings of the 11th Euromicro Conference on Parallel, Distributed and Network-Based Processing (PDP 2003), February 2003, pp. 13–22. IEEE Computer Society, Los Alamitos (2003)

# On the Interaction of Tiling and Automatic Parallelization*

Zhelong Pan, Brian Armstrong, Hansang Bae, and Rudolf Eigenmann

Purdue University, School of ECE, West Lafayette, IN, 47907
{zpan, barmstro, baeh, eigenman}@purdue.edu

**Abstract.** Iteration space tiling is a well-explored programming and compiler technique to enhance program locality. Its performance benefit appears obvious, as the ratio of processor versus memory speed increases continuously. In an effort to include a tiling pass into an advanced parallelizing compiler, we have found that the interaction of tiling and parallelization raises unexplored issues. Applying existing, sequential tiling techniques, followed by parallelization, leads to performance degradation in many programs. Applying tiling *after* parallelization without considering parallel execution semantics may lead to incorrect programs. Doing so conservatively, also introduces overhead in some of the measured programs. In this paper, we present an algorithm that applies tiling *in concert with* parallelization. The algorithm avoids the above negative effects. Our paper also presents the first comprehensive evaluation of tiling techniques on compiler-parallelized programs. Our tiling algorithm improves the SPEC CPU95 floating-point programs by up to 21% over non-tiled versions (4.9% on average) and the SPEC CPU2000 Fortran 77 programs up to 49% (11% on average). Notably, in about half of the benchmarks, tiling does not have a significant effect.

## 1 Introduction and Motivation

With processor speeds increasing faster than memory speeds, many compiler techniques have been developed to improve cache performance. Among them, iteration space tiling is a well known technique, used to reduce capacity misses [15,?]. Tiling combines stripmining and loop-permutation to partition a loop's iteration space into smaller chunks, so as to help the data stay in the cache until it is reused. Several contributions have improved the initial tiling algorithms, by tiling imperfectly-nested loops [1,14], carefully selecting the tile size, and avoiding conflict misses by copying and/or padding [7,8,?,?,?].

Enhancing locality is an important optimization technique to gain better performance, not only on a single processor, but also on a parallel machine. Tiling has been applied in parallelizing compilers, based on the sequential tiling algorithms [3,9]. It has also been used in distributed memory machines [13]. The

* This work was supported in part by the National Science Foundation under Grants 0103582-EIA, and 0429535-CCF.

present paper was motivated by an effort to include a tiling technique into our Polaris parallelizing compiler [5,12] for shared memory machines. (Polaris translates sequential Fortran 77 programs into parallel OpenMP form. The transformed parallel program will be compiled by the OpenMP backend compiler.) We have found existing tiling techniques to be insufficient for this purpose, as they are defined on a sequential program. Although a performance memory model has been presented that trades off parallelism against locality [9], it has not been discussed in the context of tiling. Also, a tiling technique for parallel programs was introduced in [3], however the interaction of the technique with other compiler passes has not been considered.

Without considering the interaction of tiling and parallelization, two approaches are open: *pre-parallelization tiling* and *post-parallelization tiling*. The *pre-parallelization tiling* algorithm performs tiling on the sequential program, followed by the parallelization pass. We have measured that this approach causes substantial performance degradation, primarily due to load imbalance of small loops. The *post-parallelization tiling* algorithm performs tiling after parallelization. To avoid incorrect results, this transformation needs to be conservative, also causing overheads. In Section 5, we will use these two tiling options as reference points and discuss their overheads in more detail.

The goal of this paper is to present an algorithm for *tiling in concert with parallelization*. First, the algorithm selects the candidate loop nests for tiling, based on data dependence and reuse information. Next, it trades off parallelism versus locality and performs the actual tiling transformation through loop stripmining and permutation. It factors parallelism information into tile sizes and the load balancing scheme. It also interacts with other parallelization passes by properly updating the list of private and reduction variable attributes.

Our algorithm outperforms both *pre-parallelization* and *post-parallelization* tiling. It improves the SPEC CPU95 floating point benchmarks by up to 21% (4.9% on average) over the parallel codes without tiling. The SPEC CPU 2000 Fortran 77 benchmarks are improved by up to 49% (11% on average). Our measurements confirm that tiling can have a significant performance impact on individual programs. However, they also show that, on today's architectures, about half of the programs benefit insignificantly.

The specific contributions of this paper are as follows:

1. We show how tiling affects the parallelism attributes of a loop nest. We prove these properties from data dependence information.
2. We introduce a new parallelism-aware tiling algorithm and show that it performs significantly better than existing techniques.
3. We discuss tiling-related issues in a parallelizing compiler: load balancing, tile size, and the trade-off between parallelism and locality.
4. We compare the performance of our algorithm with best alternatives. We discuss the measurements relative to an upper limit that tiling may achieve.

In the next section, we review some basic concepts of tiling, data reuse analysis, and data dependence directions. Section 3 analyzes the parallelism of tiled loops. Section 4 presents the algorithm for tiling in concert with parallelism and discusses related issues arising in a parallelizing compiler. Section 5 shows experimental results using the SPEC benchmarks and compares our new tiling algorithm to the pre-parallelization and post-parallelization tiling algorithms.

## 2 Background

### 2.1 Tiling Algorithm

Tiling techniques combine stripmining and loop permutation in order to reduce the volume of data accessed between two references to the same array element. Thus, it increases the chances that cached data can be reused. It has often been shown that tiling can significantly improve the performance of matrix multiplication and related linear algebra algorithms [8,9,10].

```
        (a) Matrix Multiply                (b) Tiled Matrix Multiply
                                    DO K2 = 1, M, B
                                    DO J2 = 1, M, B
     DO I = 1, M                    DO I = 1, M
       DO K = 1, M                    DO K1 = K2, MIN(K2+B-1,M)
         DO J = 1, M                    DO J1 = J2, MIN(J2+B-1,M)
           Z(J,I) = Z(J,I) + X(K,I) * Y(J,K)    Z(J1,I) = Z(J1,I) + X(K1,I) * Y(J1,K1)
```

**Fig. 1.** Tiling of a matrix multiplication code

Figure 1 shows a simple example of the original and the tiled versions of a matrix multiplication code. Loop K in the original version is stripmined into two loops, K1 and K2. Loop K1 in the tiled version iterates through a strip of B; we call it the *in-strip* loop. Loop K2 in the tiled version iterates across different strips; we call this the *cross-strip* loop. B is called the tile size.

### 2.2 Data Reuse Analysis

Data reuse analysis [10] identifies program data that is accessed repeatedly, and it quantifies the amount of data touched between consecutive accesses. To improve data locality, one attempts to permute loop nests so as to place the loop that carries the most reuse in the innermost position. If there are multiple accesses to the same memory location, we say that there is temporal reuse. If there are multiple accesses to a nearby memory location that share the same cache line, we say that there is spatial reuse. Both types of reuse may result from a single array reference, which we call self reuse, or from multiple references, which we call group reuse [11,15].

### 2.3 Direction Vectors

In this paper, we use data dependence direction vectors to determine parallelism and the legality of a loop permutation. The direction "<" denotes a forward cross-iteration dependence. The direction ">" denotes a backward cross-iteration dependence. We refer to [2,4,18,17] for a thorough description of direction vectors. We make use of the following lemmas, given in these papers.

**Lemma 1. Reordering:** *Permuting the loops of a nest reorders the elements of the direction vector in the same way.*

**Lemma 2. Permutability:** *A loop permutation is legal as long as it does not produce an illegal direction vector. In a legal direction vector, the leftmost non-equal direction must be "<" (i.e., it cannot be ">").*

**Lemma 3. Parallelism:** *Given a direction vector, its leftmost "<" direction makes the corresponding loop serial. Furthermore, serializing this loop covers the given dependence on all inner loops. That is, w.r.t. this dependence, all inner loops are parallel.*

**Lemma 4.** *After stripmining the loop $L$ into $(L', L'')$, its direction vector changes from $[d]$ to $[d', d'']$, as follows: $[=] \rightarrow [=, =]$; $[<] \rightarrow [=, <]$ or $[<, *]$; $[>] \rightarrow [=, >]$ or $[>, *]$. That is, either the cross-strip direction is "=" and the in-strip loop takes on the direction of the original loop, or the cross-strip loop takes on the original direction ("<" or ">") and the in-strip direction becomes unknown ("*").*

Direction vectors of the original (pre-tiling) loops can be used to determine the direction vectors of the tiled loops [19]. Lemma 1 and 4 aid in deriving those new direction vectors. Lemma 2 aids in finding all legally tiled versions of a loop nest. Lemma 3 decides parallelism of the tiled loops.

## 3 Parallelism of Tiled Loops

**Theorem 1.** *After tiling, the in-strip loops have the same parallelism as the original ones. The cross-strip loop $L'_i$ is serial, if the corresponding original loop $L_i$ is serial. But the cross-strip loop may become serial, even if the corresponding original is parallel.*

```
        (a) Original loop                    (b) Tiled loop
                                    DO J1 = 1, M, B            (serial)
   DO I = 1, N          (serial)    DO I = 1, N                (serial)
   DO J = 1, M          (parallel)  DO J = J1, MIN(J1+B-1,M)   (parallel)
   A(J,I) = A(J+1,I+1)               A(J,I) = A(J+1,I+1)
```

**Fig. 2.** Reduced parallelism as a result of tiling

This theorem can be strictly proved based on the previous lemmas. In this paper, limited by space, we only explain the rationales. The dependence inside one tile is essentially the same as the dependence in the pre-tiling loops. So, parallelism of the in-strip loops is not changed. According to Lemma 4, tiling introduces new dependence across the tiles. So, a cross-strip loop may become serial. Figure 2 shows an example. In the original loop, loop I is serial and loop J is parallel. After tiling, the cross-strip loop J1 is serial, loop I is serial, and loop J is parallel.

Theorem 1 shows that a parallelizing compiler cannot simply apply tiling after parallelization. The compiler needs to analyze the parallelism of the cross-strip loops, unless it only chooses to parallelize the in-strip loops, whose parallelism does not change. The following section develops the tiling algorithm considering the interaction of tiling and parallelization.

## 4 Parallelism-Aware Tiling

### 4.1 Algorithm

Our tiling in concert with parallelization algorithm uses the direction vectors of the original loop nest to determine the parallelism of the tiled loop nest, as discussed in Section 2 and Section 3. It then trades off parallelism and locality and determines a balanced tile size. Figure 3 shows the pseudo code.

**Subroutine** ParallelTiling(LoopNest $L$)
1  $P$ = the number of processors;
2  $DVs$ = the set of all direction vectors;
3  Perform data reuse analysis;
4  For each possible tiled version $V$ of $L$
5    Decide parallelism of $V$ based on the $DVs$;
6    $C$ = the cost of $V$ based on its parallelism and reuse information;
7  $X$ = the tiled version with the least cost;
8  $T$ = raw tile size computed by LRW [10], considering
      loop parallelism and cache configuration;
9  $S$ = BalancedTileSize($X,T,P$);
10 Substitute the tile size $S$ into the tiled version $X$;
11 Update reduction/private variable attributes;
12 Generate two versions if iteration number unknown;
   //$L$ is called when not enough iterations; Otherwise, $X$ is called.

**Fig. 3.** Parallelism-aware tiling algorithm

Our algorithm considers all legally tiled loop nest versions and selects the one with the least cost. It is worth noting that the order of the cross-strip loops may be different from the order of the in-strip loops. Enumerating all possible tiled versions is feasible, because most loops are nested with two or three levels. Step 5 follows Section 2 and Section 3. Step 6 uses a simple model that assumes that placing a parallel loop in an outer position is preferable over increased reuse, which will be discussed in Section 4.2. (This model suffices for our machine environment; more advanced schemes can be used without change of the algorithm). Steps 8 and 9 follow Sections 4.3 and 4.4, respectively. Step 12 is important for reducing potential tiling overheads. If the number of loop iterations is unknown at compile time, a two-version loop is created that selects between the tiled and non-tiled variants at runtime.

Our compiler pass also deals with imperfectly nested loops. It transforms such loops into perfect nests through loop fusion, loop distribution and code sinking [16]. Inner loops with fixed small number of iterations are unrolled. Then, tiling is applied to the perfectly nested loops. We have verified that this approach generates comparable results to the methods proposed in [1,?] for the SPEC CPU benchmarks, except in TOMCATV and SWIM. (In SWIM, the higher performance was achieved through manual source modifications; TOMCATV is an obsolete benchmark.)

## 4.2 Trading Off Parallelism and Locality

Locality enhancement and parallelization may have conflicting performance goals. Per Theorem 1, although the parallelism of the in-strip loops is the same as that of the original loops, the parallelism of the cross-strip loops can be different. For example, in Figure 2, loop $I$ is serial in both versions, while loop $J$ is parallel in both versions. However, after tiling, the cross-strip loop, $J1$, becomes serial. Thus, the tiled nest invokes the parallel loop more times than the original loop nest, causing higher fork-join overhead.

Table 1. Effect of tiling on fork-join overhead

| Original parallelism | Parallelism after tiling | Fork-join overhead |
|---|---|---|
| $[S, P]$ | $[S, S, P]$ | increased |
| $[S, P]$ | $[P, S, P]$ | decreased |
| $[P, S]$ | $[S, P, S]$ | increased |
| $[S, S]$ | $[S, S, S]$ | not changed |
| $[P, P]$ | $[P, P, P]$ | not changed |

Five different scenarios may occur after tiling a doubly nested loop, depending on the parallelism. We list these cases in Table 1. $S$ indicates that the corresponding loop is serial; $P$ indicates the loop is parallel. (If there are more than two loops, the change in fork-join overhead can be determined by similar analysis.) For example, in the first and second rows of Table 1, the original outer loop is serial and the original inner loop is parallel. Per Section 4, after tiling, the cross-strip loop is serial (in Row 1) or parallel (in Row 2), which results in an increase or decrease of fork-join overhead.

In summary, the parallelism of the cross-strip loops determines if tiling will increase or reduce fork-join overhead. Thus, tiling a parallel program can result in either higher or reduced parallel loop execution cost. In an advanced performance model, both the fork-join overhead and the benefit of increased locality need to be considered.

## 4.3 Tile Size Selection

The tile size is a critical parameter for tiling. We use the LRW algorithm [10] to compute the raw tile size, which fits in cache. In addition, for distributed caches, if the parallel loop is an in-strip loop, a tile needs to fit in multiple caches; for a shared cache, if the parallel loop is a cross-strip loop, the cache needs to hold multiple tiles. So, the computation of the raw tile size depends on the cache configurations and parallelism of the loops. This computed raw tile size is tuned further to balance the loads among processors, which will be described in Section 4.4.

### 4.4 Load Balancing

If the cross-strip loop is executed in parallel, load balancing can become an important issue. Tiling splits the number of iterations of the parallel loop into chunks. If the split is uneven, load imbalance results. This effect is more pronounced for programs or program sections that operate on small data sets relative to the available cache size. Hence, for a given data size, this issue tends to increase with newer generations of processors.

For example, in Figure 4, all loops are parallel and the tile size is 80. Suppose we run the program on a four processor machine. Before tiling, loop $I$ has 512 iterations and each processor executes 128 iterations. But after tiling, the cross-strip loop $J1$ has $512/80 + 1 = 7$ iterations, which cannot be evenly divided among the processors, causing load imbalance.

```
    (a) Before tiling (balanced)      (b) After tiling (not balanced)
                                      DO J1 = 1, 512, 80
    DO I = 1, 512                     DO I = 1, 512
      DO J = 1, 512                     DO J = 1, MIN(J1+79,512)
      ...                                 ...
```

**Fig. 4.** Load imbalance after tiling

Tiling sequential loops does not require balanced strip-mining. The tile size is obtained by computing the number of memory references that fit in the cache. However, the parallelizing compiler needs to tune the tile size, so that each processor will execute nearly the same number of iterations. For the previous example, the compiler can set the tile size to be 64. Then, after tiling the cross-strip loop $J1$ has $512/64 = 8$ iterations. Each processor will get 2 iterations with the same load. A more general rule is to find the largest size that is less than the original tile size and that creates a balanced load:

Suppose that the un-tuned, raw tile size is $T$, the number of iterations is $I$, and the number of processors is $P$. We choose a tile size $S$ such that $S * P$ is divisible by $I$ and $S$ is as close to $T$ as possible, based on the following formula.

$$S = \frac{I}{\lceil I/(P*T) \rceil * P}$$

This formula applies to parallel cross-strip loops (cases 2, 3, and 5 in Table 1). If the in-strip loop is parallel (case 1), it suffices to make the tile size a multiple of the number of processors.

## 5 Experiments

### 5.1 Reference Points: Tiling Independent of Parallelization

In order to verify the effectiveness of our tiling algorithm, we compare it with two algorithms that apply tiling independent of parallelization: *pre-parallelization* tiling and *post-parallelization tiling*. To our knowledge, they represent the best that can be realized with tiling techniques for sequential programs, as proposed in related work.

*Pre-parallelization tiling* determines the tiled loop shape and tile size before the parallelization passes. In most cases, the chosen parallel loop is a cross-strip loop. Load balancing, as discussed in Section 4.4 is not applied. Sequential tiling semantics is fully valid in this case, as parallelization has not yet been applied.

*Post-parallelization tiling* would generate incorrect code, if the tiling algorithm simply propagated parallel loop attributes from an original loop to its stripmined pair. So, conservatively, the cross-strip loop is *always* serialized. To further increase the fairness of our comparison, we have added an optimization to reduce fork-join overheads, when the scheduled parallel loop does not carry cross-processor dependences. In that case, this optimization moves the parallel region to the outer loops and reduces the number of barrier synchronizations by using the OpenMP *"nowait"* clause on the parallel loop.

## 5.2 Experimental Environment

We implemented the tiling algorithm presented in Section 4 in the Polaris [6,12] parallelizing compiler. The experiments were done on an Ultra SPARC II machine with four 250 MHZ processors. Each processor is equipped with a $16K$ direct-mapped L1 cache and a $1M$ direct-mapped L2 cache. Both caches are distributed. We measured the performance of all compiler-parallelized SPEC CPU95 floating point benchmarks with and without tiling. In addition, in order to evaluate how increasing data sets impacts the performance of tiling, we measured all of the six SPEC CPU2000 Fortran 77 benchmarks.

## 5.3 Experimental Results

The baseline of our experimentation is parallelization without tiling. In Figure 5 and Figure 6, the first two bars for every benchmark show the performance of the reference points. The third bar shows the performance of our new algorithm for tiling in concert with parallelization.

In most benchmarks of the SPEC CPU95 suite, pre-parallelization tiling does not improve performance over parallelization without tiling. Two main effects degrade the performance of APSI, HYDRO2D, MGRID, SWIM and TOMCATV significantly. First, the data size is small relative to the available cache size, so that the important loops contain very few tiles, causing load imbalance. Second, since no information about parallel loops is known, the tiling algorithm does not permute the most beneficial loops to outermost positions. In the SPEC CPU2000 codes, the data size is much larger, reducing the load imbalance effect. For the measured SPEC CPU2000 benchmarks, APSI, APPLU and SWIM show improvements over parallelization alone. SWIM is improved by 49%, most of which is due to the fact that tiling yields many stride-one access patterns.

In the SPEC CPU95 suite, post-parallelization tiling also degrades the performance over parallelization without tiling for APSI, HYDRO2D, MGRID

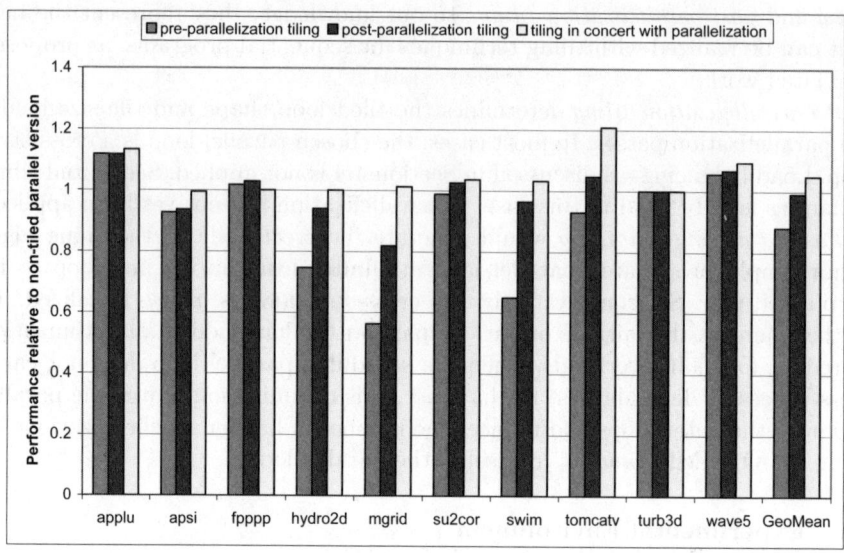

**Fig. 5.** Performance of tiling relative to non-tiled parallel codes for SPEC95

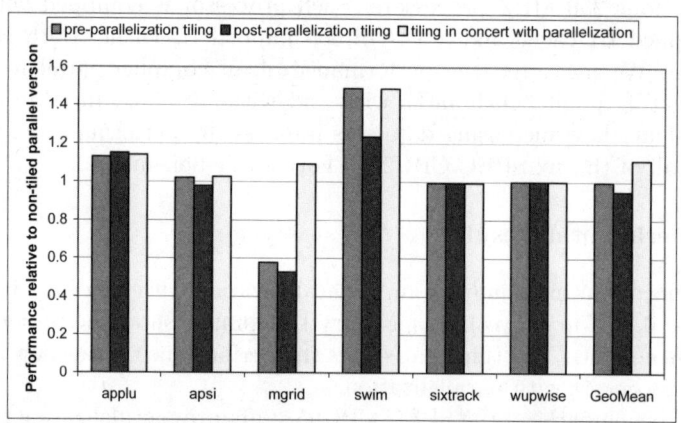

**Fig. 6.** Performance of tiling relative to non-tiled parallel codes for selected SPEC2000 benchmarks

and SWIM. First, post-parallelization tiling may cause load imbalance, if it parallelizes the in-strip loop and the tile size is small. Second, in post-parallelization tiling, the chosen parallel loop tends to have finer granularity than in tiling in concert with parallelization. Although the synchronization optimization reduces the fork-join overhead, some of this overhead remains. Another observation is that, in general, post-parallelization tiling performs better than pre-parallelization tiling for SPEC CPU95 benchmarks, but not for SPEC CPU2000.

The reason is that a large data size reduces load imbalance for pre-parallelization tiling, but not for post-parallelization tiling.

Our experiments show that tiling in concert with parallelization performs significantly better than tiling independent of parallelization. Our new algorithm never degrades performance. Five out of the ten benchmarks in the SPEC CPU95 suite show improvements over parallelization alone. The largest improvement is 21% in TOMCATV (TOMCATV is a small kernel benchmark, which is now considered obsolete). Tiling can add some control overhead, offsetting parallel performance. We have found this to be the reason for very minor performance degradation to APSI and HYDRO2D. In FPPPP, post-parallelization tiling performs slightly better than our algorithm. It is a rare case where the cost of computing a balanced tile size at runtime is noticeable. Most SPEC CPU2000 benchmarks show improvements. SWIM is improved by 49%. It is important to note, that although matrix multiply (a code frequently used to demonstrate tiling) is very important in WUPWISE, all matrices are small and do not benefit from tiling.

As expected, our measurements show only small improvements on the SPEC CPU95 codes, whose data sets mostly fit in cache. Tiling improves more significantly the SPEC CPU2000 codes, which have larger data sets and, consequently, increased cache misses in the original programs.

### 5.4 On Performance Bounds for Tiling

The fact that half of our program suite does not benefit from tiling raises the question of how much better a further improved algorithm could perform. In order to find an upper bound on the performance achievable by tiling, we measured the percentage of tilable loops in the SPEC CPU95 benchmarks based on reuse analysis. A tilable loop nest must satisfy two conditions. First, at least two loops in the nest carry reuses, otherwise loop interchanging would suffice. Second, it does not contain subroutine calls or I/O operations. The last column in Table 2 shows the execution time percentage of the loop nests satisfying both conditions. This percentage gives us an upper bound on tilable loops.

**Table 2.** Percentage of tilable loops based on reuse analysis. Each column shows, respectively, the numbers of loops, loops carrying reuses, loop nests with at least two loops carrying reuses, and those loop nests without subroutine calls or I/O operations. The data in the parentheses are the execution time percentage of the loop nests with more than two loops carrying reuses and without subroutine calls.

| Benchmark | Total | Reuse | Nested | w/o Call |
|---|---|---|---|---|
| APPLU | 149 | 125 | 55 | 54 (97.60%) |
| APSI | 388 | 310 | 111 | 59 (19.50%) |
| FPPPP | 49 | 37 | 15 | 8 ( 5.80%) |
| HYDRO2D | 170 | 117 | 21 | 21 (53.70%) |
| MGRID | 38 | 24 | 8 | 8 (86.40%) |
| SU2COR | 208 | 177 | 37 | 22 (14.90%) |
| SWIM | 24 | 15 | 3 | 3 (60.10%) |
| TOMCATV | 16 | 14 | 5 | 5 (95.90%) |
| TURB3D | 64 | 43 | 12 | 11 (22.20%) |
| WAVE5 | 362 | 274 | 59 | 57 (19.70%) |

For all benchmarks other than HYDRO2D, the result in Table 2 is consistent with that in Figure 5. The benchmarks gaining significant performance from our tiling algorithm spend a large percentage of execution time in tilable loop nests, and vice versa. In HYDRO2D, although 53.7% of the execution time is spent in tilable loops, each loop nest only refers to a small amount of memory, which can fit into cache. Therefore, tiling does not reduce cache misses.

Our results also show that, while tiling can be an important locality enhancement technique for individual programs, especially for stencil operations, its performance benefit is not as broad as commonly assumed. Tiling does not gain significant performance in half of the benchmarks. The major reason is limited data reuse that is amenable to tiling.

## 6 Conclusions

We have presented a new tiling algorithm that works in concert with other parallelization passes. We have shown that applying existing tiling techniques, designed for sequential programs before or after parallelization, would lead to significant performance degradation or incorrect programs. Our algorithm avoids these negative effects, hence it represents new technology, relevant to any parallelizing compiler.

Furthermore, in evaluating tiling techniques comprehensively, we have found that the benefit is less than commonly assumed. Tiling – along with other locality enhancement techniques – is believed to be very important, as the memory-to-processor speed ratio in new computer architectures keeps decreasing. However, this technique has often been demonstrated on simple linear algebra kernels. Although our measurements confirm improvements on stencil computations, tiling has only limited effect on other programs, which is due to limited data reuse, amenable to tiling. Increasing cache sizes and increasing data sets are two opposite trends that will impact the performance of tiling techniques on future computer systems.

## References

1. Ahmed, N., Mateev, N., Pingali, K.: Tiling imperfectly-nested loop nests. In: Proceedings of the 2000 ACM/IEEE conference on Supercomputing (CDROM), p. 31 (2000)
2. Allen, R., Kennedy, K.: Dependence: Theory and Practice. Optimizing compilers for modern architectures, pp. 45–55. Morgan Kaufman Publishers, San Francisco (2002)
3. Anderson, J.M., Lam, M.S.: Global optimizations for parallelism and locality on scalable parallel machines. In: Proceedings of the conference on Programming language design and implementation, pp. 112–125 (1993)
4. Banerjee, U., Eigenmann, R., Nicolau, A., Padua, D.A.: Automatic program parallelization. Proceedings of the IEEE 81(2), 211–243 (1993)
5. Blume, W., Doallo, R., Eigenmann, R., Grout, J., Hoeflinger, J., Lawrence, T., Lee, J., Padua, D., Paek, Y., Pottenger, B., Rauchwerger, L., Tu, P.: Advanced program restructuring for high-performance computers with polaris (1996)

6. Blume, W., Doallo, R., Eigenmann, R., Grout, J., Hoeflinger, J., Lawrence, T., Lee, J., Padua, D., Paek, Y., Pottenger, B., Rauchwerger, L., Tu, P.: Parallel programming with Polaris. IEEE Computer 29(12), 78–82 (1996)
7. Chame, J., Moon, S.: A tile selection algorithm for data locality and cache interference. In: Proceedings of the 13th international conference on Supercomputing, pp. 492–499 (1999)
8. Coleman, S., McKinley, K.S.: Tile size selection using cache organization and data layout. In: Proceedings of the conference on Programming language design and implementation, pp. 279–290 (1995)
9. Kennedy, K., McKinley, K.S.: Optimizing for parallelism and data locality. In: Proceedings of the 6th international conference on Supercomputing, pp. 323–334 (1992)
10. Lam, M.D., Rothberg, E.E., Wolf, M.E.: The cache performance and optimizations of blocked algorithms. In: Proceedings of the fourth international conference on Architectural support for programming languages and operating systems, pp. 63–74 (1991)
11. McKinley, K.S., Carr, S., Tseng, C.-W.: Improving data locality with loop transformations. ACM Transactions on Programming Languages and Systems 18(4), 424–453 (1996)
12. Min, S.J., Kim, S.W., Voss, M., Lee, S.I., Eigenmann, R.: Portable compilers for OpenMP. In: Eigenmann, R., Voss, M.J. (eds.) WOMPAT 2001. LNCS, vol. 2104, pp. 11–19. Springer, Heidelberg (2001)
13. Ramanujam, J., Sadayappan, P.: Tiling multidimensional iteration spaces for multicomputers. Journal of Parallel and Distributed Computing 16(2), 108–230 (1992)
14. Song, Y., Li, Z.: A compiler framework for tiling imperfectly-nested loops. In: Languages and Compilers for Parallel Computing, pp. 185–200 (1999)
15. Wolf, M.E., Lam, M.S.: A data locality optimizing algorithm. In: Proceedings of the conference on Programming language design and implementation, pp. 30–44 (1991)
16. Wolf, M.E., Maydan, D.E., Chen, D.-K.: Combining loop transformations considering caches and scheduling. In: Proceedings of the 29th annual ACM/IEEE international symposium on Microarchitecture, pp. 274–286 (1996)
17. Wolfe, M.J.: Optimizing supercompilers for supercomputers. PhD thesis (1982)
18. Wolfe, M., Banerjee, U.: Data dependence and its application to parallel processing. Int. J. Parallel Program. 16(2), 137–178 (1987)
19. Xue, J.: On tiling as a loop transformation. Parallel Processing Letters 7(4), 409–424 (1997)

# Static Nonconcurrency Analysis of OpenMP Programs

Yuan Lin

Sun Microsystems, Inc.
yuan.lin@sun.com

## 1 Introduction

Writing correct and efficient parallel programs is more difficult than doing so for sequential programs. One of the challenges comes from the nature of concurrent execution of a parallel program by different threads.[1] Determining exact concurrency is NP-hard[10], and is impossible for real-world programs at compile time.

OpenMP provides an easy and incremental way to write parallel programs. The well-structured OpenMP constructs and well-defined semantics of OpenMP directives make compiler analyses more effective on OpenMP programs than on more loosely structured parallel programs that are solely based on runtime libraries, such as MPI and Pthreads.

In this paper, we present a static nonconcurrency analysis technique that detects, at compile time, whether two statements in an OpenMP program will not be executed concurrently by different threads in a team. Similar to the method presented in [5], ours is a close underestimation of the real nonconcurrency in a program. When our method determines that the executions of two statements are nonconcurrent, these two statements will not be executed concurrently. When the method fails, the two statements may, but need not, execute concurrently.

Our nonconcurrency analysis models and uses the semantics of OpenMP directives. For example, in the following codes,

```
1.   !$omp parallel
2.
3.     a = ...
4.
5.     !$omp single
6.       b = ...
7.       c = ...
8.     !$omp end single
9.
```

---

[1] Concurrency is where the execution order of different threads is not enforced, and thus synchronization must be used to control shared resources. Parallelism is where different threads actually execute in parallel. Parallelism is an instance of concurrency. Parallel execution is concurrent, but concurrent execution is not necessarily parallel.

```
10.    !$omp do
11.      do i=1, 100
12.        c(i) = ...
13.      end do
14.    !$omp end do nowait
15.
16. !$omp end parallel
```

there is one implicit barrier at line 8, which partitions the statements inside the parallel region (lines 3-14) into two phases. Phase one contains statements 3 through 8, and phase two contains statements 10 through 14. No two statements from different phases (such as statements 3 and 12) will ever be executed concurrently, while statements within the same phase (such as statements 3 and 6, or two instances of statement 3) may. In addition, the **single** directive mandates only one thread can execute statements 6 and 7. Therefore statements 6 and 7, though in the same phase, will never execute concurrently. Our analysis is able to recognize the OpenMP directives and use them to derive the nonconcurrency information.

This paper makes the following contributions,

- It gives a graph representation (OpenMP control flow graph) to model the control flow in parallel OpenMP programs, and a tree representation (OpenMP region tree) to model the hierarchical structure of loops and OpenMP constructs. Similar to the control flow graph and loop tree representations for sequential programs, these two representations serve as the base for further compiler analysis of parallel OpenMP programs.
- It presents an efficient static nonconcurrency analysis for OpenMP programs which are more tractable than general parallel programs. The phase partitioning algorithm has a complexity that is linear to the size of the program being analyzed for most real-life applications.
- It shows the usefulness of the nonconcurrency analysis by building a compile-time data race detection technique upon it.

The rest of the paper is organized as follows. Section 2 describes the OpenMP control flow graph and OpenMP region tree. Section 3 presents the phase partition algorithm. Section 4 gives the static nonconcurrency analysis. Section 5 uses data race detection to illustrate the use of nonconcurrency analysis. Section 6 compares related work and section 7 concludes the paper.

To be concise, we use Fortran as the base language, while our technique is not specific to Fortran.

## 2 OpenMP Control Flow Graph and OpenMP Region Tree

### 2.1 Program Model

The techniques in this paper work on OpenMP standard compliant programs[1]. Nested parallelism and orphaned directives are allowed, recognized and handled

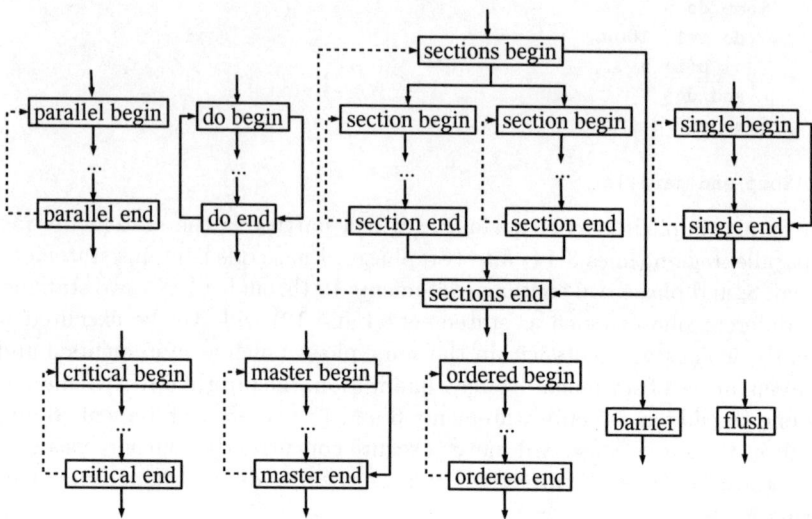

**Fig. 1.** Directive nodes in OpenMP control flow graph (solid lines are flow edges and dotted lines are construct edges)

accordingly. Our techniques also recognize and use the properties of all OpenMP synchronization constructs and directives (such as **barrier, master, critical** and **ordered**).

We assume 1) all parallel regions can be active and none is serialized; 2) there is an infinite number of threads available; and 3) the exact number of threads that execute any particular parallel region is unspecified. These assumptions not only simplify the problem but also make the result of our nonconcurrency analysis independent of runtime environment. We ignore calls to OpenMP runtime lock routines, and make no attempt to recognize 'roll-your-own' synchronizations, such as busy-waiting. Knowledge of this information could add to the nonconcurrency result, but could never invalidate a nonconcurrency relationship between statements that our method finds.

### 2.2 OpenMP Control Flow Graph

An OpenMP control flow graph (OMPCFG) models the transfer of control flow in a subroutine of an OpenMP program.

The statements in an OpenMP subroutine are partitioned into basic blocks and each OpenMP directive is put into an individual block. Each block becomes a node in OMPCFG. The nodes representing basic blocks are called basic nodes, and the nodes representing directive blocks are called directive nodes. A single *Entry* node and a single *Exit* node are created for an OMPCFG.

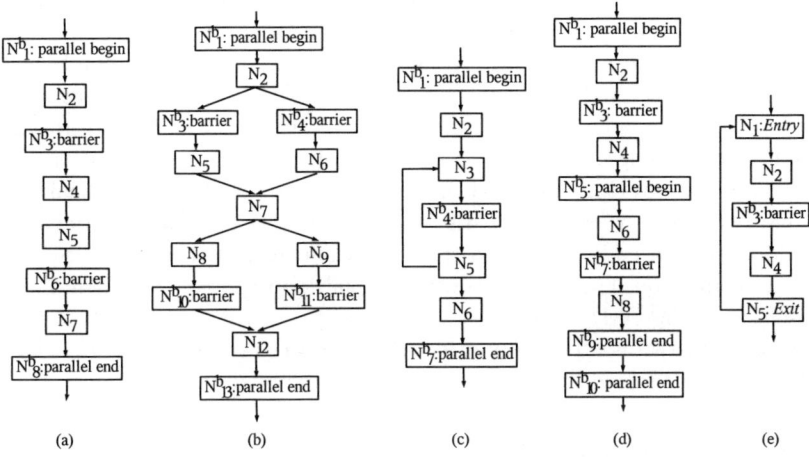

**Fig. 2.** (a) simple phases in one parallel region. (b) phases when there is a branch (c) phases in a loop. (d) phases in nested parallel regions. (e) orphan phases.

**Table 1.** Notations: attributes of different types OpenMP constructs

| | |
|---|---|
| $ORC(N)$ | the immediately enclosing OpenMP construct for node $N$. |
| $ORC(N).type$ | the type of $ORC(N)$, i.e. **root, do, sections, section, critical, single, master, ordered**. |
| $ORC(N).crit.name$ | the name for $ORC(N)$ whose $ORC(N).type$ is **critical**. |
| $ORC(N).ordered.bound$ | the binding worksharing **do** loop for $ORC(N)$ whose $ORC(N).type$ is **ordered** |
| $ORC(N).parent$ | the parent OpenMP construct of $ORC(N)$ in the OpenMP region tree |
| $ORC(N).pregion$ | the parallel region that encloses $ORC(N)$ |

In an OMPCFG, to make compiler analysis easier, implicit barriers are made explicit[2], and each combined parallel work-sharing construct (such as **parallel do** and **parallel sections**) is separated into a **nowait** work-sharing construct nested in a parallel region. **parallel begin** directive nodes and **parallel end** directive nodes are considered as barrier nodes in the parallel region defined by the two directive nodes. Fortran specific 'WORKSHARE' construct can be converted into a set of other OpenMP constructs, therefore it is not presented directly in an OMPCFG.

An edge in OMPCFG represents a possible transfer of control flow executed by a thread. Edges between basic nodes are created in a way similar to that in

---

[2] To model the dataflow in an OpenMP program, it would be better to make all implicit flushes explicit. To be concise, we do not do so in this paper because our nonconcurrency analysis does not depend on inter-thread dataflow information.

sequential programs. Edges between basic nodes and directive nodes and edges between directive nodes are created according to OpenMP semantics.

Statements inside an OpenMP construct form a single-entry/single-exit region. For each OpenMP construct, an edge is created from the directive begin node to the single entry node of the region for the construct, and an edge is created from the single exit node of the region to the directive end node. Edges to and from the **barrier** and **flush** nodes are created as if they are basic nodes. An edge is created from a **sections begin** node to each binding **section begin** node. And an edge is created from each **section end** node to its binding **sections end** node. For the **do** construct, the loop control statements are not represented in the OMPCFG.

Figure 1 illustrates all the directive nodes and the corresponding edges. The construct edges (dotted lines) are explained in the next section.

## 2.3 OpenMP Region Tree

In OpenMP programs, we use a region tree to model both the hierarchical loop structure and the hierarchical OpenMP construct structure in a subroutine.

In OMPCFG, for each OpenMP construct (except for **do** constructs), we add an edge from the end construct directive node to the begin construct directive node. We call this edge a construct edge and represent it using a dotted line in an OMPCFG. A construct edge does not reflect any control flow. It is inserted so that an OpenMP construct forms a cycle in the OMPCFG. Therefore, the normal loop tree detection algorithm for sequential programs can be used to find both loops and OpenMP construct regions in an OMPCFG.

Because the statements in an OpenMP construct form a single-entry/single-exit region, the OpenMP constructs in a subroutine are properly nested. If we treat the whole subroutine as a **root** construct, then all the OpenMP constructs form a tree structure. The OpenMP constructs are also properly nested with loops in the subroutine. When we combine the loop tree with the OpenMP construct tree, we get the OpenMP region tree. Each node in an OpenMP region tree represents either a loop or an OpenMP construct.

For a node $N$ in an OMPCFG, we use $ORC(N)$ to represent the immediately enclosing OpenMP construct for node $N$ in the OpenMP region tree. Table 1 lists the notations used to represent the attributes of different types of OpenMP construct.

## 3 Phase Partitioning

### 3.1 Phases in a Parallel Region

Barrier is the most frequently used synchronization method in OpenMP. Barriers can be inserted by using the BARRIER directive, and are also implied at the end of worksharing constructs or parallel constructs.

In addition, OpenMP standard requires[1]

> BARRIER directives must be encountered by all threads in a team or by none at all, and they must be encountered in the same order by all threads in a team.

The restriction the OpenMP standard imposes on the use of barriers essentially partitions the execution of a parallel region into a set of distinct, non-overlapping run-time phases. No two statement instances in two different run-time phases will ever be executed concurrently by different threads in a team. For example, the barriers in Fig. 2(a) (barrier nodes are marked with a superscript b) put the non-barrier nodes into three phases - one phase with node $N_2$, another phase with node $N_4$ and node $N_5$, and yet another with node $N_7$. Statements in $N_2$ and statements in $N_4$ will not be executed concurrently by different threads in a team. We should also note that the restriction the OpenMP language imposes on barriers does not apply to threads in different teams.

### 3.2 Static Phases

In this section, we give an algorithm that computes the static phases in an OpenMP subroutine at compile-time. Our algorithm works on basic blocks instead of statements. All construct edges in OMPCFG are ignored since they do not represent any control flow. A special edge from the *Exit* node to the *Entry* node is added to help analysis of subroutines that contain orphaned OpenMP directives.

A static phase $\langle N_i^b, N_j^b \rangle$ consists of a sequence of nodes along all barrier free paths in the OMPCFG that start at one barrier node $N_i^b$ and end at another (possibly the same) barrier node $N_j^b$ in the same parallel region. Table 2 lists the phases for each OMPCFG in Fig. 2.

Note that in Fig. 2(e), node $N_2$ and node $N_4$ are in the same phase. Node $N_3^b$ is an orphaned barrier, and there is no lexically visible parallel region in the subroutine. It is possible that the call-site of the subroutine is inside a loop, therefore there might be a barrier free path from node $N_4$ to node $N_2$ at runtime. Without interprocedural analysis, we have to assume such a loop exists. That's the reason why a special edge from the *Exit* node to the *Entry* node is inserted.

Also note that a node may belong to different static phases. For example, in Fig. 2(b), node $N_5$ belongs to two both phase $\langle N_3^b, N_{10}^b \rangle$ and phase $\langle N_3^b, N_{11}^b \rangle$.

Each static phase has its owner parallel region, which is its immediate enclosing parallel region. A static phase is not considered as a static phase in a parallel region that is not its owner parallel region. For example, in Fig. 2(d), the owner parallel region of static phase $\langle N_5^b, N_7^b \rangle$ is the inner parallel region, and it is not a static phase in the outer parallel region. For a static phase that starts and ends at orphaned barriers, its owner parallel region is **root**.

### 3.3 Algorithm to Compute Static Phases

The algorithm to partition an OMPCFG into phases is shown in Fig. 3. In the following text, when we say 'phase', we mean 'static phase'.

**Table 2.** Phases and nodes in each phase for each OMPCFG in Fig. 2

| OMPCFG | Phase | Nodes in Phase |
|---|---|---|
| (a) | $\langle N_1^b, N_3^b \rangle$ | $N_2$ |
|  | $\langle N_3^b, N_6^b \rangle$ | $N_4, N_5$ |
|  | $\langle N_6^b, N_8^b \rangle$ | $N_7$ |
| (b) | $\langle N_1^b, N_3^b \rangle$ | $N_2$ |
|  | $\langle N_1^b, N_4^b \rangle$ | $N_2$ |
|  | $\langle N_3^b, N_{10}^b \rangle$ | $N_5, N_7, N_8$ |
|  | $\langle N_3^b, N_{11}^b \rangle$ | $N_5, N_7, N_9$ |
|  | $\langle N_4^b, N_{10}^b \rangle$ | $N_6, N_7, N_8$ |
|  | $\langle N_4^b, N_{11}^b \rangle$ | $N_6, N_7, N_9$ |
|  | $\langle N_{10}^b, N_{13}^b \rangle$ | $N_{12}$ |
|  | $\langle N_{11}^b, N_{13}^b \rangle$ | $N_{12}$ |
| (c) | $\langle N_1^b, N_4^b \rangle$ | $N_2, N_3$ |
|  | $\langle N_4^b, N_4^b \rangle$ | $N_5, N_3$ |
|  | $\langle N_4^b, N_7^b \rangle$ | $N_5, N_6$ |
| (d) | $\langle N_1^b, N_3^b \rangle$ | $N_2$ |
|  | $\langle N_3^b, N_{10}^b \rangle$ | $N_4, N_5^b, N_6, N_7^b, N_8, N_9^b$ |
|  | $\langle N_5^b, N_7^b \rangle$ | $N_6$ |
|  | $\langle N_7^b, N_9^b \rangle$ | $N_8$ |
| (e) | $\langle N_3^b, N_3^b \rangle$ | $N_1, N_2, N_4, N_5$ |

We use the following notations in the algorithm:

- $phase(N_i^b, N_j^b)$
  the set of nodes that belong to phase $\langle N_i^b, N_j^b \rangle$.
- $in\_phase(N)$
  the set of phases that node $N$ belongs to.
- $p\_start(N)$
  the set of starting barriers of phases that node $N$ belongs to, i.e. $\{N_i^b | \langle N_i^b, N_j^b \rangle \in in\_phase(N)\}$.
- $p\_end(N)$
  the set of ending barriers of phases that node $N$ belongs to, i.e. $\{N_j^b | \langle N_i^b, N_j^b \rangle \in in\_phase(N)\}$.

The algorithm does a forward depth-first-search, and a backward depth-first-search from each barrier node (including pseudo barrier nodes, i.e. *Entry*, *Exit*, parallel-begin, and parallel-end). During each search, if a barrier node in the same parallel region is encountered, the search does not continue with successors or predecessors of the barrier node. In a forward search from a barrier $N^b$, we put $N^b$ in $p\_start(N)$ of each node $N$ reached. In a backward search from a barrier $N^b$, we put $N^b$ in $p\_end(N)$ of each node $N$ reached. After all searches finish, for each non-barrier node $N$, we compute $in\_phase(N)$ as $\{\langle N_i^b, N_j^b \rangle \mid N_i^b \in p\_start(N), N_j^b \in p\_end(N), ORC(N_i^b).pregion = ORC(N_j^b).pregion\}$.

In general, the complexity of this algorithm is $O(\sum_{i=1}^{K} \sum_{j=1}^{M_i} node(i,j)\ nbar(i,j))$. Basically, if the OMPCFG for a parallel region is disconnected at each barrier node, then the OMPCFG is separated into several disconnected sub-graphs. Here, $K$ is the number of parallel regions; $M_i$ is the number of the disconnected sub-graphs for parallel region $i$; $node(i,j)$ is the number of nodes in sub-graph $j$ of parallel region $i$; and $nbar(i,j)$ is the number of barrier nodes that separates sub-graph $j$ from other sub-graphs in parallel region $i$. Nested-parallel regions are rarely used and each sub-graph of a parallel region contains only two barrier nodes (one starting barrier node and one ending barrier). Therefore, in most cases, the complexity of the algorithm is $O(n)$, where $n$ is the number of nodes in OMPCFG.

## 4 Nonconcurrency Analysis

In this section, we describe our static nonconcurrency analysis. Given two statements in a subroutine and an OpenMP parallel region, the analysis detects at compile time whether these two statements can be executed concurrently by different threads in the team for the parallel region.

The algorithm in Section 3 partitions a subroutine into phases. Depending on whether the two statements belong to the same phase or not, we use two different methods to check the nonconcurrency. Because two statements are executed concurrently if and only if their basic blocks are executed concurrently, we will work on basic blocks instead of statements.

### 4.1 Two Nodes in Different Phases

If two nodes in a parallel region do not share any static phase, then the runtime instances of these two nodes will be in different runtime phases. Therefore these two nodes will not be executed concurrently by different threads in the team that executes the parallel region.

For example, in Fig. 2(a), node $N_2$ and node $N_4$ will never be executed concurrently. However, node $N_4$ and node $N_5$ may be executed concurrently, because $N_4 \in \langle N_3^b, N_6^b \rangle$, and $N_5 \in \langle N_3^b, N_6^b \rangle$.

In Fig. 2(b), node $N_5$ and node $N_6$ will never be executed concurrently. Node $N_5$ and node $N_9$ may be executed concurrently.

In Fig. 2(c), node $N_2$ and node $N_5$ will never be executed concurrently. Node $N_3$ and node $N_5$ may be executed concurrently.

In Fig. 2(d), node $N_6$ and node $N_8$ will never be executed concurrently by different threads in the team that executes the inner parallel region. However, these two nodes may be executed concurrently by different threads in the team that executes the outer parallel region.

In Fig. 2(e), node $N_2$ and node $N_4$ may be executed concurrently.

In summary, given two nodes $N_1$ and $N_2$ whose immediate common enclosing parallel region is $PR$ (could be **root**), if there does not exist a phase in $in\_phase(N_1) \cap in\_phase(N_2)$ whose owner parallel region is $PR$, then $N_1$ and $N_2$ will not be executed concurrently by different threads in the team that executes $PR$.

```
foreach barrier $N_i^b$ in OMPCFG
    foreach successor $N_j$ of $N_i^b$
        $forward\_mark(N_j, N_i^b)$ ;
    foreach predecessor $N_k$ of $N_i^b$
        $backward\_mark(N_k, N_i^b)$ ;

foreach non-barrier node $N$ in OMPCFG
    foreach $N_i^b$ in $p\_start(N)$
        foreach $N_j^b$ in $p\_end(N)$ that $ORC(N_i^b).pregion = ORC(N_j^b).pregion$
            $phase(N_i^b, N_j^b) := phase(N_i^b, N_j^b) \cup \{N\}$ ;
            $in\_phase(N) := in\_phase(N) \cup \{\langle N_i^b, N_j^b \rangle\}$ ;


$forward\_mark(N, N^b)$
{
    if ($N$ is a barrier node and $ORC(N).pregion = ORC(N^b).pregion$)
        return ;

    $p\_start(N) := p\_start(N) \cup \{N^b\}$ ;

    foreach successor $N_j$ of $N$
        $forward\_mark(N_j, N^b)$ ;
}


$backward\_mark(N, N^b)$
{
    if ($N$ is a barrier node and $ORC(N).pregion = ORC(N^b).pregion$)
        return ;

    $p\_end(N) := p\_end(N) \cup \{N^b\}$ ;

    foreach predecessor $N_k$ of $N$
        $backward\_mark(N_k, N^b)$ ;
}
```

**Fig. 3.** Algorithm: phase partitioning

### 4.2 Two Nodes in the Same Phase

The semantics of OpenMP constructs also prohibits some statements within the same phase to be executed concurrently, e.g. statement 6 and statement 7 in the example at the beginning of this paper.

Given two basic blocks $N_1$ and $N_2$ (possibly the same) that $\langle N_i^b, N_j^b \rangle \in in\_phase(N_1) \cap in\_phase(N_2)$, and the owner parallel region of $\langle N_i^b, N_j^b \rangle$ is $PR$, the two blocks $N_1$ and $N_2$ will not be executed concurrently by different threads in a team that executes $PR$ in the following situations.

1. **master**
    Both $N_1$ and $N_2$ are in **master** constructs that belong to $PR$.
    $$ORC(N_1).type = ORC(N_2).type = master$$
    $$ORC(N_1).pregion = ORC(N_2).pregion = PR$$
2. **ordered**
    Both $N_1$ and $N_2$ are in **ordered** constructs in $PR$ and are bound to the same **do** construct.
    $$ORC(N_1).type = ordered,\ ORC(N_2).type = ordered$$
    $$ORC(N_1).pregion = ORC(N_2).pregion = PR$$
    $$ORC(N_1).ordered.bound = ORC(N_2).ordered.bound$$
3. **single**
    Both $N_1$ and $N_2$ are in the same **single** construct in $PR$
    $$ORC(N_1) = ORC(N_2),\ ORC(N_1).type = ORC(N_2).pregion = single$$
    $$ORC(N_1).pregion = ORC(N_2).pregion = PR$$
    and one of the following is true.
    - the **single** construct is not in any loop within the parallel region $PR$.
    - the **single** construct is in a loop within the parallel region $PR$, and there is no barrier-free path from the **single** end directive node to the header of the immediately enclosing loop.
    - the **single** construct is in a loop within the parallel region, and there is no barrier-free path from the header of the immediately enclosing loop to the **single begin** directive node.

OpenMP requires a **single** construct to be executed by only one thread in a team. However, it does not specify which thread. If the **single** construct is inside a loop, then two different threads may each execute one instance of the **single** construct in different iterations. If there is no barrier, then the two threads may execute the construct concurrently.

Also note that we do not check for critical sections. A critical section enforces serial execution, but does not enforce synchronization. Different instances of the statements in a critical section cannot be executed in parallel, but can be executed concurrently.

## 5 Application: Static Race Detection

Static nonconcurrency analysis can help many useful analyses and optimizations, such as race detection, lock/barrier removal, synchronization optimization, etc. A static nonconcurrency analysis similar to the above has been implemented in Sun Studio$^{TM}$ 9 compilers. It serves as one of the analysis engines for the OpenMP autoscoping feature, which automatically detects the data sharing attributes of variables in an OpenMP application[6]. It also serves as an engine for the static OpenMP error detection feature provided in Sun compilers. Here, we show how to build a static race detection algorithm upon the static nonconcurrency analysis.

## 5.1 The Method

There are two different types of races, *synchronization races* and *data races*, which are collectively called *general races* [8]. A general race happens when the order of two accesses (at least one is write) to the same memory location is not enforced by synchronizations. A data race happens when a general race happens and the access to the memory is not guarded by a critical section. A general race that is not a data race is called synchronization race. A correct OpenMP program may contain synchronization races, but is usually expected to be free of data race. For example, in a producer/consumer code, the producer and the consumer may execute asynchronously, but they should not corrupt the shared data. Many OpenMP programs are parallelized from serial codes and their behavior is usually deterministic. Such programs should be free of both synchronization races and data races.

If any two accesses to the same memory location cannot be executed concurrently, then these two accesses must be ordered and a general race is impossible. If the two accesses can be executed concurrently and the accesses are guarded by critical sections, then a synchronization race may happen while a data race is impossible. Based on the above logic and our nonconcurrency analysis, we can develop a static race detection method for OpenMP programs.

Given two statements $s_1$ and $s_2$ that access the same shared memory location (at least one of them writes to the location) and a parallel region $PR$, the following steps detect whether the two statements may cause a race in $PR$.

1. Find the basic block $N_1$ for $s_1$ and the basic block $N_2$ for $s_2$.
2. Use the method in Section 4 to check the nonconcurrency relationship between $N_1$ and $N_2$ in parallel region $PR$.
3. If $N_1$ and $N_2$ will not be executed concurrently, then the two statements will not cause a race in $PR$.
4. Otherwise, if both $N_1$ and $N_2$ are in **critical** constructs that have the same name or both are unnamed.

$$ORC(N_1).type = critical, \ ORC(N_2).type = critical$$

$$ORC(N_1).crit.name = ORC(N_2).crit.name$$

then the two statement may cause a synchronization race, but will not cause a data race in $PR$.
5. Otherwise, the two statements may cause a data race in $PR$.

## 5.2 Example

```
1.         function foo (n, x, y)
2.         integer      n, i
3.         real         x(*), y(*)
4.         real         w, mm, m, foo
5.
6.         w = 0.0
7.
```

```
 8. c$omp parallel private(i,mm,t), firstprivate(n),
 9. c$omp+ shared(m,x,y), reduction(+:w)
10.
11. c$omp single
12.         m = 0.0
13. c$omp end single nowait
14.
15.         mm = 0.0
16.
17. c$omp do
18.         do i = 1, n
19.            t = x(i)
20.            y(i) = t
21.            if (t .gt. mm) then
22.               w = w + t
23.               mm = t
24.            end if
25.         end do
26. c$omp end do nowait
27.
28. c$omp critical
29.         if ( m .le. mm ) then
30.            m = mm
31.         end if
32. c$omp end critical
33.
34. c$omp end parallel
35.
36.         foo = w - m
37.
38.         return
39.         end
```

Function foo contains a parallel region (line 8-34), whose purpose is to copy array x() to array y(), set the maximum value of all positive elements of x() to a scalar variable m, and compute the sum w of some elements of x(). Scalar m is a shared variable. A **single** construct (line 11-13) is used to initialize m. Each thread uses a private variable mm to store the maximum value the thread gets in the worksharing **do** loop (line 17-26). At the end of the parallel region, the shared variable m is updated by all threads in a critical section (line 28-32) which is used to avoid data race.

In an attempt to speed up the execution of the parallel region, two **nowait** clauses (line 13 and line 26) are inserted to remove the implicit barriers. Threads not executing the **single** can go ahead to work on the worksharing **do** without having to wait for the thread who is initializing m. And threads that have finished their share of the work in the **do** can continue to update the shared variable m, and don't have to wait for other threads.

However, the program may not deliver the expected result because of the use of these two **nowait** clauses. Our nonconcurrency analysis will find that statements

11 through 32 are all in one static phase, and statements 12 and 29, as well as statements 12 and 30 may be executed concurrently. Because statement 12 is not guarded by a critical section, either case will cause a data race and lead to a nondeterministic execution result.

Sun Studio 9 Fortran compiler will give the following warning when the above code is compiled with the parallel error checking option -vpara.

```
>f90 -xopenmp -vpara -xO3 -c t.f
''t.f'', line 8: Warning: inappropriate scoping
        variable 'm' may be scoped inappropriately as 'SHARED'
      . write at line 30 and write at line 12 may cause data race
```

If the **single** construct is changed to a **critical** construct, then our race detection method will find that the code may cause a synchronization race and the execution result may still be nondeterministic.

If either one of the two **nowait** clauses is not there, our nonconcurrency analysis will find that the parallel region is partitioned into two nonconcurrent phases - one contains statement 12 and the other contains statement 29 and statement 30. Therefore statement 12 will never be executed concurrently with either statement 29 or statement 30. According to the our race detection method, the code has no race conditions now.

## 6  Related Work

Many researchers [4][3][2][7] have proposed different methods to detect race conditions and non-determinacy in parallel programs that use low-level event variable synchronization, such as post/wait and locks. Our technique is different from theirs because ours uses high-level semantic information exposed by OpenMP directives and constructs. Our method is simpler and more efficient for analyzing OpenMP programs. It is not clear how to represent OpenMP semantics using event variables. Nevertheless, since our method does not handle OpenMP lock API calls, their techniques can be incorporated into our method to refine analysis results.

Jeremisassen and Eggers [5] present a compile-time nonconcurrency analysis using barriers. Their method assumes the SPMD model and is similar to our method in Section 4.1 as it also divides the program into a set of phases separated by barriers. They assume a general SPMD model that is not OpenMP specific. Therefore they cannot take advantage of restrictions that OpenMP has on the use of barriers. For example, their method will say node $N_5$ and node $N_6$ in Fig. 2(b) may be executed concurrently while ours does not. Their method is purely based on barriers and does not detect nonconcurrency within one phase.

In [9], Satoh et. al. describe a 'parallel flow graph' that is similar to our OMPCFG. They connect flush operations with special edges that present the ordering constraints between the flushes. They do not have construct edges as they do not build a hierarchical structure like our OpenMP region tree. The two representations are different because they serve different purposes. Theirs is more data flow oriented, while ours is more control flow oriented. It is possible to combine these two graphs together.

## 7 Conclusion

We have presented a method for compile-time detection of nonconcurrency information in OpenMP programs. The analysis uses the semantics of OpenMP directives and takes advantage of the fact that standard compliant OpenMP programs are well-structured. The analysis has a complexity that is linear to the size of the program in most applications, and can handle nested parallelism and orphaned OpenMP constructs. The OpenMP control flow graph and the OpenMP region tree developed in this work can be used for other compiler analyses/optimizations of OpenMP programs as well. We have also demonstrated the use of the nonconcurrency information by building a compile-time race detection algorithm upon it.

**Acknowledgement.** The author would like to thank Nawal Copty and Eric Duncan for reviewing the paper and providing insightful comments.

## References

1. OpenMP Fortran Application Program Interface, Version 2.0 (November 2000), http://www.openmp.org/specs
2. Balasundaram, V., Kennedy, K.: Compile-time detection of race conditions in a parallel program. In: Conference Proceedings, 1989 International Conference on Supercomputing, Crete, Greece, June 5– 1989, ACM SIGARCH, pp. 175–185 (1989)
3. Callahan, D., Kennedy, K., Subhlok, J.: Analysis of event synchronization in a parallel programming tool. In: Proceedings of the Second ACM SIGPLAN Symposium on Principles and Practice of Parallel Programming, Seattle WA, March 1990, pp. 21–30 (1990)
4. Emrath, P., Padua, D.: Automatic detection of nondeterminacy in parallel programs. In: Proceedings of the ACM SIGPLAN and SIGOPS Workshop on Parallel and Distributed Debugging, January 1989, vol. 24(1), pp. 89–99. ACM Press, New York (1989)
5. Jeremiassen, T.E., Eggers, S.J.: Static analysis of barrier synchronization in explicitly parallel programs. In: International Conference on Parallel Architectures and Compilation Techniques, August 1994, pp. 171–180 (1994)
6. Lin, Y., Terboven, C., an Mey, D., Copty, N.: Automatic scoping of variables in parallel regions of an openmp program. In: Proceedings of the 2004 Workshop on OpenMP Applications and Tools, Houston, TX (May 2004)
7. Netzer, R.H.B., Ghosh, S.: Efficient race condition detection for shared-memory programs with post/wait synchronization. In: Shin, K.G. (ed.) Proceedings of the 1992 International Conference on Parallel Processing, Software, Ann Arbor, MI, August 1993, vol. 2, pp. 242–246. CRC Press, Boca Raton (1993)

8. Netzer, R.H.B., Miller, B.P.: What are race conditions? Some issues and formalizations. ACM Letters on Programming Languages and Systems 1(1), 74–88 (1992)
9. Satoh, S., Kusano, K., Sato, M.: Compiler optimization techniques for openMP programs. Scientific Programming 9(2-3), 131–142 (2001)
10. Taylor, R.N.: Complexity of analyzing the synchronization structure of concurrent programs. Acta Informatica 19, 57–84 (1983)

# CCRG OpenMP Compiler: Experiments and Improvements

Huang Chun and Yang Xuejun

National Laboratory for Parallel and Distributed Processing, P.R. China
chunhuang73@hotmail.com

**Abstract.** In this paper, we present the design and experiments of a practical OpenMP compiler for SMP, called CCRG OpenMP Compiler, with the focus on its performance comparison with commercial Intel Fortran Compiler 8.0 using SPEC OMPM2001 benchmarks. The preliminary experiments showed that CCRG OpenMP is a quite robust and efficient compiler for most of the benchmarks except *mgrid* and *wupwise*. Then, further performance analysis of *mgrid* and *wupwise* are provided through *gprof* tool and Intel optimization report respectively. Based on the performance analysis, we present the optimized static schedule implementation and inter-procedural constant propagation techniques to improve the performance of CCRG OpenMP Compiler. After optimization, all of the SPEC OMPM2001 Fortran benchmarks can be executed on SMP systems efficiently as expected.

## 1 Introduction

The OpenMP[1] has gained momentum in both industry and academy, and has become the de-facto standard for parallel programming on shared memory multiprocessors. The open source compilers and runtime infrastructures promote the development and acceptance of OpenMP effectively. There have been several recent attempts, such as NanosCompiler[2] and PCOMP[3] for Fortran77, Omni[4] and Intone[5] for Fortran77 and C, OdinMP[6] for C/C++, and Nanos Mercurium[7] on top of Open64 compilers. All of them are source-to-source translators that transform the code into the equivalent version with calls to the associated runtime libraries.

CCRG OpenMP Compiler[1] (CCRG, for short) aims to create a freely available, fully functional and portable set of implementations of the OpenMP Fortran specification for a variety of different platforms, such as Symmetric Multiprocessor (SMP) as well as Software Distributed Shared Memory (SDSM) system. As the above compilers, CCRG also uses the approach of the source-to-source translation and runtime support to implement OpenMP. CCRG has the following features.

---

[1] Both the compiler and runtime library will be available from the CCRG Fortran95 Compiler project web site at http://cf95.cosoft.org.cn

- Generate only one external subroutine for each subprogram which may specify one or several parallel regions, and the parallel regions in the subprogram are implemented using ENTRY statements. Therefore, the size of code generated by the source-to-source translator has been reduced significantly.
- Fully support Fortran90/95 programming languages except the type declaration statements whose kind-selector involves the intrinsic procedure.
- Be robust enough to enable testing with real benchmarks.
- Support multiple target processors and platforms, including Digital Alpha[9], Intel Itanium and Pentium, SDSM - JIAJIA[10] and SMP.

In this paper, we present the design and experiments of CCRG for SMP, with the focus on performance comparison with commercial Intel OpenMP Compiler 8.0[11] using SPEC OMPM2001 benchmarks[12]. The preliminary experiments showed that most of the Fortran benchmarks with CCRG OpenMP are executed as fast as with Intel OpenMP Compiler on SMP, except *mgrid* and *wupwise*. Based on performance analysis for *mgrid* and *wupwise*, we present the optimized static schedule implementation and inter-procedural optimization(IPO) which improve the performance of *mgrid* and *wupwise* as desired.

In the next section we briefly outline the design of the CCRG OpenMP Compiler. Section 3 describes the experiments and performance analysis using SPEC OMPM2001 in detail. Section 4 presents the optimization techniques based on the result of section 3 and reports the performance improvements. Conclusion and future work are given in section 5.

## 2 CCRG OpenMP Compiler

CCRG OpenMP Compiler has fully implemented OpenMP 1.0 and partial features of OpenMP 2.0 Fortran API on the POSIX thread interface. As most of the open source OpenMP compilers[2,3,4,5,6,7], it includes a source-to-source translator to transform OpenMP applications into the equivalent Fortran programs with the runtime library calls. The source-to-source translator is based on Sage++[14] and consists of two parts, a Fortran OpenMP syntax parser and a translator which converts the internal representation into the parallel execution model of the underlying machine. Sage++ is an object-oriented compiler preprocessor toolkit for building program transformation systems for Fortran 77, Fortran 90, C and C++ languages. Though many features of Fortran 90/95 are not supported in Sage++, it is not very difficult to add new elements to the system because of its well-structured architecture. In the syntax parser, the OpenMP syntax description is added for supporting OpenMP directives as well as the new Fortran90/95 languages elements, as shown in Fig.1.

The parser recognizes the OpenMP directives and represents their semantics in a machine independent binary internal form. A *.dep* file is produced to store the internal representation for each OpenMP source file. The translator reads the *.dep* file and exports the normal Fortran program with calls to the runtime library. In [2,3,4,5,6,7], a subroutine is generated for each parallel region by the translators.

```
omp_directive:
    omp_parallel
    | omp_paralleldo
    | omp_parallelsections
    | omp_parallelworkshare
    | omp_single
    | omp_master
    | ......;

omp_parallel:
    PARALLEL end_spec needkeyword omp_clause_opt keywordoff
    {
       omp_binding_rules (OMP_PARALLEL_NODE);
       $$ = get_bfnd (fi, OMP_PARALLEL_NODE, SMNULL, $4, LLNULL, LLNULL);
    }
```

**Fig. 1.** OpenMP Syntax Description

Two separate subroutines are needed to implement the two parallel regions in Fig.2, which means that many same declare statements are included. The internal subroutine can be used to reduce the size of code generated by the source-to-source translator. Some commercial OpenMP compilers use this strategy to implement OpenMP parallel region, such as IBM XLF compiler[13]. But the special support of compilers is needed because the Fortran standard specifies some constraints on using an internal subroutine. Therefore, we use an alternative approach by using ENTRY statement to eliminate these same statements in CCRG OpenMP. If a subroutine contains one or more ENTRY statements, it defines a procedure for each ENTRY statement and permits this procedure reference to begin with a particular executable statement within the subroutine in which the ENTRY statement appears. Therefore, ENTRY name can be used to guide all the threads to execute parallel regions correctly, such as test_$1 and test_$2 shown in Fig.3.

The source-to-source translator encapsulates all parallel regions of a main program or subprogram into one external subroutine. The ENTRY procedures are generated to implement parallel regions, as shown in Fig.3. So, only one external subroutine test_$0 is generated for the OpenMP example in Fig.2,

```
      SUBROUTINE test(a)
         DIMENSION a(100)
     !$OMP PARALLEL DO PRIVATE(K)
         DO 100 k = 1, 100
     100     a(k) = 0.9
            .........
     !$OMP PARALLEL NUM_THREADS(4)
            .........
     !$OMP END PARALLEL
         END
```

**Fig. 2.** An OpenMP Example

```
      SUBROUTINE test_$0(a)
        DIMENSION a(100)
        INTEGER lc_k
        INTEGER _omp_dolo, _omp_dohi, comp_static_more
      !The first parallel region
        ENTRY test_$1 ()
          CALL comp_static_setdo (1, 100, 1, 0)
          DO WHILE (comp_static_more(_omp_dolo, _omp_dohi, 1).eq.1)
            DO 100 lc_k = _omp_dolo, _omp_dohi, 1
  100         a(lc_k) = 0.9
          END DO
          CALL comp_barrier()
        RETURN
      !The second parallel region
        ENTRY test_$2()
          .........
          CALL comp_barrier()
        RETURN
      END

      SUBROUTINE test(a)
        DIMENSION a(100)
        CALL comp_runtime_init ()
        CALL comp_parallel (test_$1, 0, 1, a)
        .........
        CALL comp_parallel (test_$2, 4, 1, a)
        CALL comp_exit ()
      END
```

**Fig. 3.** Fortran Program using ENTRY Statement after Transformation

which contains two ENTRY procedures test_$1 and test_$2. The procedures defined by ENTRY statements share the specification parts. Therefore, the size of code generated by the translator is reduced largely.

The CCRG OpenMP runtime library for SMP has been implemented based on the standard POSIX thread interface. The library is platform-independent except few functions, such as comp_parallel, comp_barrier and comp_flush. It focuses on three tasks: thread management, task schedule, and implementation of OpenMP library routines and environment variables. The "comp_" functions shown in Fig.3 are main functions for thread management and task schedule. comp_runtime_init initializes the runtime system and reads the associated environment variables. comp_exit terminates all the slaves in the thread pool and releases memory. Function comp_static_setdo and comp_static_more implement the static schedule in OpenMP. comp_barrier synchronizes all the threads in the current thread team. comp_parallel is the most complex function in the library, which creates slave threads when necessary and starts the slave threads in the thread pool to execute the parallel region procedures. It has following form.

```
comp_parallel (parallel_region_procedure_name,
               num_threads, num_parameter, param1, param2,...)
```

If there is no NUM_THREADS clause in a OpenMP parallel region directive, the value of num_threads is 0, as shown in the first parallel region in Fig.3. comp_parallel decides the number of the threads in the team according to the environment variable or library calls, or the default value which is equal to the number of physical processors of the underlying target.

## 3 Experiments

To evaluate CCRG OpenMP Compiler, SPEC OMPM2001[12] Fortran benchmarks are compiled and executed. The host platform for the experiments is a HP server rx2600 with four Itanuim2 processors (1.5GHz) and Linux IA-1 2.4.18-e.12smp.

### 3.1 Results

The backend compiler of CCRG can be any compilers executed over the target machines, including commercial compilers(Intel, PGI, etc.) and GNU compiler. To compare CCRG with commercial Intel compiler exactly, Intel Fortran Compiler 8.0 is used as the backend compiler of CCRG. Fig.4 and Fig.5 show the Base Ratios of SPEC OMPM2001 Fortran benchmarks of CCRG and Intel OpenMP Compiler 8.0[2] with four OpenMP threads . "-O3" and "-O3 -ipo" options are used respectively. "-ipo" option enables inter-procedural optimization(IPO) across files.

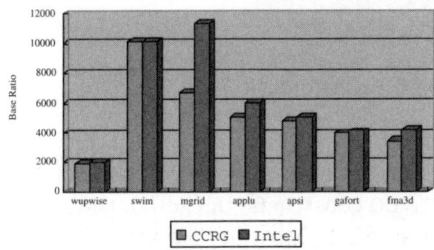 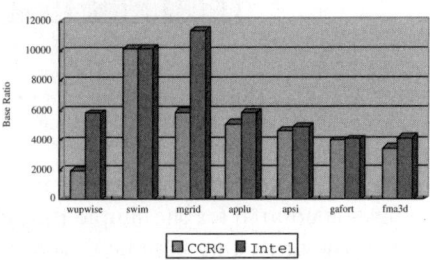

**Fig. 4.** Base Ratios of CCRG and Intel without IPO

**Fig. 5.** Base Ratios of CCRG and Intel with IPO

The performance data in Fig.4 and Fig.5 suggest that CCRG indeed makes good use of the multiprocessing capabilities offered by the underlying platform as Intel OpenMP Compiler with two exceptions: *mgrid* and *wupwise*. The Base Ratio of *mgrid* with CCRG is only half of that with Intel whether inter-procedure optimization option is used or not. When "-ipo" option is used, the performance of *wupwise* with Intel can be improved greatly, while CCRG seems to block some further optimization.

---

[2] 318.galgel can not execute correctly using Intel OpenMP Compiler 8.0 on our server.

## 3.2 Performance Analysis

For most of the SPEC OMPM2001 Fortran benchmarks, CCRG OpenMP Compiler results in almost exactly the same Base Ratios as Intel OpenMP Compiler. But the performance of *mgrid* and *wupwise* with CCRG are much worse than that with Intel OpenMP Compiler. In this section, we analyze and explain why *mgrid* and *wupwise* perform poorly in detail. HP server rx2600 with 2 Itanuim2 processors (1.0GHz) is used for performance analysis here.

***mgrid.*** Fig.6 shows the execution time of the top six pocedures[3] in *mgrid* with "TRAIN" input sets. "_p1" and "_p2" denote the procedures generated for the first and second parallel regions in one procedure respectively. For example, the column of resid_p1 in Fig.6 denotes the execution time of the first parallel region of resid. The procedures resid_p1, psinv_p1, rprj3_p1, interp_p1 and interp_p2 cause the different execution time between CCRG and Intel.

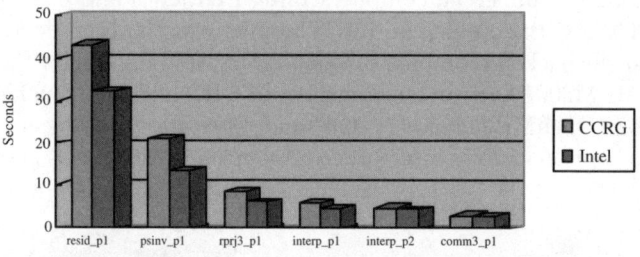

**Fig. 6.** Execution Time of Top 6 Procedures in *mgrid*

resid_p1, psinv_p1, rprj3_p1, interp_p1 and interp_p2 are the procedures generated for the simple PARALLEL DO constructs. In CCRG, the structure of the procedure is same as that shown in Fig.3. Most of the execution time of the procedures are spent to execute the nested DO-loop. Comparing with Intel OpenMP Compiler, CCRG introduces an additional loop level to implement the schedule types in OpenMP.

```
DO WHILE (comp_static_more(_omp_dolo,_omp_dohi,_omp_doin).eq.1)
    ........
END DO
```

This additional loop is a while loop whose control condition is a logical expression containing a function call. It encloses the original loops and becomes the most outer loop. Therefore, the performance of the whole procedure degrades significantly.

---

[3] In Intel OpenMP Compiler, the procedure should be the "T-region".

**Table 1.** Execution Time of Top 4 procedures in *wupwise*

|   | CCRG | | Intel | |
|---|---|---|---|---|
|   | Subroutine | Execution time(sec) | Subroutine | Execution time(sec) |
| 1 | zgemm | 82.10 | dlaran | 9.77 |
| 2 | gammul | 10.77 | zaxpy | 8.57 |
| 3 | zaxpy | 7.74 | zgemm | 7.91 |
| 4 | dlaran | 7.35 | lsame | 1.87 |

***wupwise.*** Unlike *mgrid*, the performance of *wupwise* is affected by IPO largely. Table 1 shows the execution time of the top four procedures in *wupwise* with "TRAIN" input sets. The time of subroutine zgemm with CCRG is 82.10 seconds, while the time with Intel is only 7.91 seconds.

The Intel compiler provides the extensive support for inter-procedural analysis and optimization, such as points-to analysis and mod/ref analysis required by many other optimizations. However, only the equivalent version transformed by the source-to-source translator of CCRG can be seen by the backend compiler Intel Fortran Compiler. Because the parallel region procedures are called by function comp_parallel as actual parameters, the source-to-source translator dose not keep the information about the caller-callee relationship between the original procedures. So, the backend compiler can not process the further inter-procedural analysis and optimization.

Though zgemm is only called in su3mul with several constants which are used to control the loops of zgemm, these constants have not been propagated to zgemm in CCRG. From the above two optimization reports, it is obvious that inter-procedural constant propagation has been applied to zgemm when

```
High Level Optimizer Report for: zgemm_
Block, Unroll, Jam Report:
(loop line numbers, unroll factors and type of transformation)
Loop at line 2194 unrolled with remainder by 6
Loop at line 2177 unrolled with remainder by 6
Loop at line 2158 unrolled with remainder by 6
.........
                    (a) For Program Transformed by CCRG
```

```
High Level Optimizer Report for: zgemm_
Block, Unroll, Jam Report:
(loop line numbers, unroll factors and type of transformation)
Loop at line 2377 completely unrolled by 3
Loop at line 2379 completely unrolled by 3
Loop at line 2360 completely unrolled by 3
.........
                    (b) For Source OpenMP Program
```

**Fig. 7.** Optimization Report Generated by Intel Fortran Compiler 8.0

using Intel OpenMP Compiler. Many loops in zgemm are completely unrolled according to the value of actual parameter.

## 4 Optimization

Section 3.1 describes the key factors which influence on the performance of *mgrid* and *wupwise* programs. In this section, the optimized static schedule implementation and inter-procedural optimization are presented to CCRG OpenMP Compiler for the improvement of performance of these programs.

### 4.1 Optimized Static Schedule

In the following three cases, function comp_static_more is .TRUE. only once for each thread when executing a parallel region procedure in CCRG.
- Absence of the SCHEDULE clause.
- Static schedule without chunk, i,e. SCHEDULE(STATIC) is specified.
- Static schedule, and both chunk size and number of iteration are known during compile time, and (chunk size × number of threads) $\leq$ number of iteration.

Therefore, the parallel region subroutine code in Fig.3 can be replaced with the codes in Fig.8. comp_static_once is called only once to implement the PARALLEL DO directive in Fig.2.

```
SUBROUTINE test_$0(a)
   DIMENSION a(100)
   INTEGER lc_k
 ENTRY test_$1 ()
   CALL comp_static_setdo(1,100,1,0)
   CALL comp_static_once(_omp_dolo,_omp_dohi,1)
   DO 100 lc_k=_omp_dolo,_omp_dohi,1
100   a(lc_k)=0.9
   CALL comp_barrier()
   RETURN
END
```

**Fig. 8.** Implementation of STATIC Schedule without Chunk Size

After optimization, the execution time of all procedures in Fig.6 has been reduced significantly. The middle columns in Fig.9 are the execution time using the optimized static schedule implementation. Obviously, the performance of whole *mgrid* has been improved largely too.

SCHEDULE clause is not specified in most of OpenMP programs, we can just use comp_static_once instead of comp_static_more with introducing an additional outer loop.

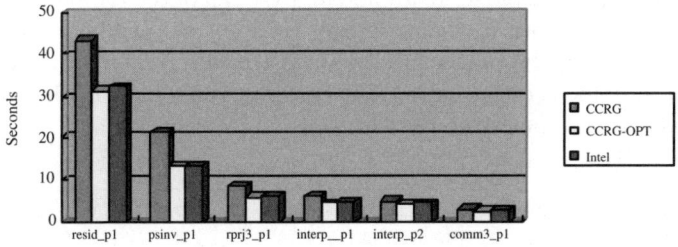

**Fig. 9.** Execution time of Main Procedures after Optimization

## 4.2 Inter-Procedural Constant Propagation

Because CCRG uses the source-to-source approach, some inter-procedural optimizations may no longer be applicable for some OpenMP programs. For example, in *wupwise*, subroutine zgemm is called only once in subroutine su3mul where the third, fourth and fifth actual parameters are integer constants. But these constants have not been propagated to zgemm.

This is a native problem of source-to-source OpenMP compilers. We present an approach to solving it by adding inter-procedural optimization in source-to-source translator. Inter-procedural optimization contains two-pass compilation, as shown in Fig.10.

In the first pass, the parser scans all the project files to record the information about procedure calls, such as procedure name, formal parameters, procedure name and actual parameters called by the procedures in the files. Temporary file *tmp_filename.i* is generated for each file in the project. For example, *tmp_su3mul.i* for *su3mul.f* in *wupwise* contains the information as follows.

```
{SUBROUTINE "SU3MUL"
   (FORMAL ("U" COMPLEX*16 DIMENSION(2 3 *))
           ("TRANSU" CHARACTER*1 SCALAR)
           ("X" COMPLEX*16 DIMENSION(1 *))
           ("RESULT" COMPLEX*16 DIMENSION(1 *)))
   (SUBROUTINE "ZGEMM"
     (ACTUAL (TRANSU, 'NO TRANSPOSE',3,4,3, ONE,U,3,X,3,ZERO,RESULT,3)
}
```

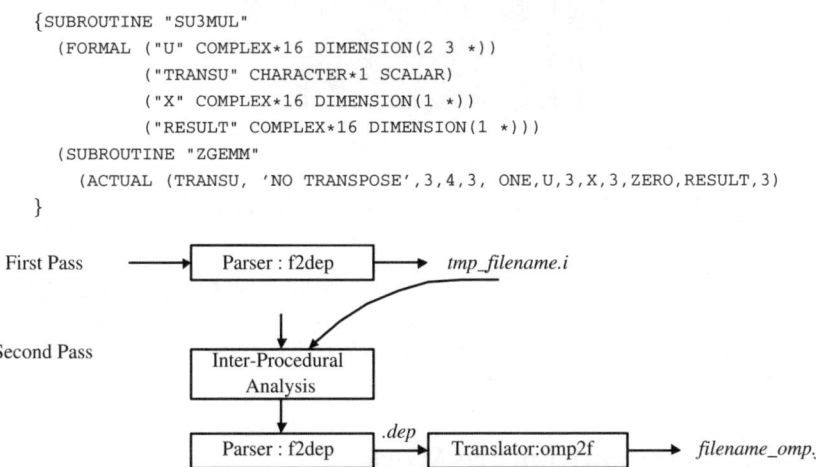

**Fig. 10.** Inter-Procedural Optimization

In the second pass, the parser reads and analyzes all of the temporary files firstly. If the callers always use the same integer constant as certain actual parameter to call a procedure, the parser inserts an assignment statement before the first executable statement in the callee. The constant is assigned to the parameter in the assignment statement(see Fig.11).

```
    SUBROUTINE ZGEMM ( TRANSA, TRANSB, M, N, K, ALPHA, A, LDA, B, LDB,
$                     BETA, C, LDC )
    ! Variables Declaration Statements.........
    ! Assignment to Formal parameters
      M = 3
      N = 4
      K = 3
    !Other Executable Statements
    END
```

**Fig. 11.** Inserts the Assignment Statements in the Callee

Therefore, the constant propagation is implemented through assignments to formal parameters. That is, the source-to-source translator only provides the initial values of the formal parameters after inter-procedural analysis, the backend compiler utilizes the information to make further optimization. After optimization, the execution time of *wupwise* has been reduced significantly as shown in Fig.12, whose middle columns are the execution time after inter-procedural optimization.

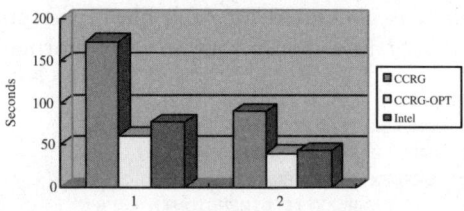

**Fig. 12.** Execution Time of *wupwise* after Optimization

Complete inter-procedure constant propagation needs to be supported by other optimizations, such as source-level data flow analysis and constant propagation within a procedure. At present only integer constant actual parameters can be propagated cross procedures.

## 5 Conclusion and Future Work

The CCRG OpenMP Compiler is a mature source-to-source compiler for OpenMP. All SPEC OMPM2001 Fortran benchmarks have been compiled and executed on SMP system efficiently. CCRG supports OpenMP Fortran90/95 programming, and all of the Fortran benchmarks achieve the comparable Base Ratios as Intel OpenMP Compiler.

In the paper, we show our experience for performance improvement of CCRG. We analyze two benchmarks *mgrid* and *wupwise* whose execution time with CCRG were much longer than that with Intel. Two optimization techniques, namely, optimized static schedule and inter-procedural constant propagation, are presented to resolve the performance problems in these two programs. After optimization, the performances of *mgrid* and *wupwise* are improved significantly.

In the future, we plan to design and implement more source-to-source optimization strategies and complete inter-procedural optimization framework. This framework will be applicable for not only IPO but also for profile-guided optimization aimed at OpenMP. In addition, the performance of CCRG will be evaluated on the large SMP systems.

## Acknowledgements

This work was supported by National 863 Hi-Tech Programme of China under grant No. 2002AA1Z2101 and 2004AA1Z2210.

## References

1. The OpenMP Forum. OpenMP Fortran Application Program Interface, Version 2.0, (November 2000), http://www.OpenMP.org
2. Gonzalez, M., Ayguade, E., Labarta, J., Martorell, X., Navarro, N., Oliver, J.: NanosCompiler: A Research Platform for OpenMP Extensions. In: Proc. of the 1st European Workshop on OpenMP (EWOMP 1999), Lund, Sweden (October 1999)
3. Min, S.J., Kim, S.W., Voss, M., Lee, S.I., Eigenmann, R.: Portable Compilers for OpenMP. In: Eigenmann, R., Voss, M.J. (eds.) WOMPAT 2001. LNCS, vol. 2104, pp. 11–19. Springer, Heidelberg (2001)
4. Omni OpenMP Compiler Project., http://phase.hpcc.jp/Omni/
5. INTONE: Innovative Tools for Non Experts, IST/FET project (IST-, 1999-20252), http://www.cepba.upc.es/intone/
6. Brunschen, C., Brorsson, M.: OdinMP/CCp - A Portable Implementation of OpenMP for C. In: Proc. of the 1st European Workshop on OpenMP(EWOMP 1999), Lund, Sweden (October 1999)
7. Balart, J., Duran, A., Gonz'alez, M., Martorell, X., Ayguade, E., Labarta, J.: Nanos Mercurium: a Research Compiler for OpenMP. In: Proc. of the 6th European Workshop on OpenMP (EWOMP 2004), Stockholm, Sweden (October 2004)
8. Open64 Compiler and Tools, http://sourceforge.net/projects/open64
9. Chun, H., Xuejun, Y.: Performance Analysis and improvement of OpenMP on Software Distributed Shared Memory System. In: Proc. of the 5th European Workshop on OpenMP (EWOMP 2003), Aachen, Germany (September 2003)
10. Hu, W., Shi, W., Tang, Z.: JIAJIA: An SVM System Based on A New Cache Coherence Protocol. In: Sloot, P.M.A., Hoekstra, A.G., Bubak, M., Hertzberger, B. (eds.) HPCN-Europe 1999. LNCS, vol. 1593, pp. 463–472. Springer, Heidelberg (1999)

11. Intel Corporation. Intel Fortran Compilers for Linux Application Development (2003), http://www.intel.com/software/products/compilers/linux
12. Aslot, V., Domeika, M., Eigenmann, R., Gaertner, G., Jones, W.B., Parady, B.: SPEComp: A New Benchmark Suite for Measuring Parallel Computer Performance. In: Eigenmann, R., Voss, M.J. (eds.) WOMPAT 2001. LNCS, vol. 2104, pp. 1–10. Springer, Heidelberg (2001)
13. Grassl, C.: Shared Memory Programming: Pthreads and OpenMP (October 2003), http://www.csit.fsu.edu/~burkardt/fsu/7.OpenMP.pdf
14. Sage++ Users Guide, http://www.extreme.indiana.edu/sage/

# Compiler Technology

# Implementing an OpenMP Execution Environment on InfiniBand Clusters

Jie Tao[1], Wolfgang Karl[1], and Carsten Trinitis[2]

[1] Institut für Technische Informatik
Universität Karlsruhe (TH)
76128 Karlsruhe, Germany
{tao,karl}@ira.uka.de

[2] Lehrstuhl für Rechnertechnik und Rechnerorganisation
Technische Universität München
Boltzmannstr.3, 85748 Garching, Germany
trinitic@cs.tum.edu

**Abstract.** Cluster systems interconnected via fast interconnection networks have been successfully applied to various research fields for parallel execution of large applications. Next to MPI, the conventional programming model, OpenMP is increasingly used for parallelizing sequential codes. Due to its easy programming interface and similar semantics with traditional programming languages, OpenMP is especially appropriate for non-professional users.

For exploiting scalable parallel computation, we have established a PC cluster using InfiniBand, a high-performance, de facto standard interconnection technology. In order to support the users with a simple parallel programming model, we have implemented an OpenMP execution environment on top of this cluster. As a global memory abstraction is needed for shared data, we first built a software distributed shared memory implementing a kind of Home-based Lazy Release Consistency protocol. We then modified an existing OpenMP source-to-source compiler for mapping shared data on this DSM and for handling issues with respect to process/thread activities and task distribution. Experimental results based on a set of different OpenMP applications show a speedup of up to 5.22 on systems with 6 processor nodes.

## 1 Motivation

Clusters are regarded as adequate platforms for exploring high performance computing. In contrast to tightly-coupled multiprocessor systems, like SMPs, clusters have the advantage of scalability and cost-effectiveness. Therefore, they are generally deployed in a variety of both research and commercial areas for performing parallel computation.

As a consequence, we have also established a cluster system using modern processors. More specifically, this cluster is connected via InfiniBand [7], a high-performance interconnect technology. Besides its low latency and high bandwidth, InfiniBand supports Remote Data Memory Access (RDMA), allowing access to remote memory locations via the network without any involvement of the receiver. This feature allows InfiniBand to achieve higher bandwidth for inter-node communication, in comparison with other interconnect technologies such as Giga-Ethernet and Myrinet.

As the first step towards cluster computing, we have built an MPI environment on top of this cluster. However, we note that increasingly, users have no special knowledge about parallel computing and they usually bring OpenMP codes. Since it offers an easier programming interface with semantics similar to that of sequential codes, OpenMP is preferred by non-professional users to develop parallel programs. In order to support these users, we established the OpenMP execution environment on top of our InfiniBand clusters.

As a global memory abstraction is the basis for any shared memory programming model, we first developed ViSMI (Virtual Shared Memory for InfiniBand clusters), a software-based distributed shared memory. ViSMI implements a kind of home-based lazy release consistency model and provides annotations for dealing with issues with respect to parallel execution, such as process creation, data allocation, and synchronization. We then developed Omni/Infini, a source-to-source OpenMP compiler using ViSMI as the supporting interface. Omni/Infini is actually an extended version of the Omni compiler. We have modified Omni in order to replace the thread interface with ViSMI interface, to map shared data on the distributed shared memory, and to coordinate the work of different processes.

The established OpenMP execution environment has been verified using both applications from standard benchmark suites, like NAS and SPLASH-II, and several small kernels. Experimental results show different behavior with applications. However, for most applications, scalable speedup has been achieved.

The remainder of this paper is organized as follows. Section 2 gives an introduction to the InfiniBand cluster and the established software DSM. This is followed by a brief description of Omni/Infini in Section 3. In Section 4 first experimental results are illustrated. The paper concludes with a short summary and some future directions in Section 5.

## 2 The InfiniBand Cluster and the Software DSM

InfiniBand [7] is a point-to-point, switched I/O interconnect architecture with low latency and high bandwidth. For communications, InfiniBand provides both channel and memory semantics. While the former refers to traditional send/receive operations, the latter allows the user to directly read or write data elements from or to the virtual memory space of a remote node without involving the remote host processor. This scenario is referred to as Remote Direct Memory Access (RDMA).

The original configuration of our InfiniBand cluster included 6 Xeon nodes and 4 Itanium 2 (Madison) nodes. Recently we have added 36 Opteron nodes into the cluster. The Xeon nodes are used partly for interactive tasks and partly for computation, while the others are purely used for computation. These processor nodes are connected through switches with a theoretical peak bandwidth of 10 Gbps.

As the first step towards an infrastructure for shared memory programming, we implemented ViSMI [16], a software-based distributed shared memory system.

The basic idea behind software distributed shared memory is to provide the programmers with a virtually global address space on cluster architectures. This idea is first proposed by Kai Li [13] and implemented in IVY [14]. As the memories are actually distributed across the cluster, the required data could be located on a remote node and also multiple copies of shared data could exist. The latter leads to consistency issues, where a write operation on shared data has to be seen by other processors. For tackling this problem, software DSMs usually rely on the page fault handler of the operating system to implement invalidation-based consistency models.

The concept of memory consistency models is to precisely characterize the behavior of the respective memory system by clearly defining the order in which memory operations are performed. Depending on the concrete requirement, this order can be strict or less strict, hence leading to various consistency models.

The most strict one is sequential consistency [12], which forces a multiprocessor system to achieve the same result of any execution as if the operations of all the processors were executed in some sequential order and the operations of each individual processor appear in the order specified by its programmers. This provides an intuitive and easy-to-follow memory behavior, however, the strict ordering requires the memory system to propagate updates early and prohibits optimizations in both hardware and compilers. Hence, other models have been proposed to relax the constraints of sequential consistency with the goal of improving the overall performance.

Relaxed consistency models [3,6,9,10] define a memory model for programmers to use explicit synchronization. Synchronizing memory accesses are divided into *Acquires* and *Releases*, where an *Aquire* allows the access to shared data and ensures that the data is up-to-date, while *Release* relinquishes this access right and ensures that all memory updates have been properly propagated. By separating the synchronization in this way invalidations are only performed by a synchronization operation, therefore reducing the unnecessary invalidations caused by an early coherence operation.

A well-known relaxed consistency model is Lazy Release Consistency (LRC) [10]. Within this model, invalidations are propagated at the acquire time. This allows the system to delay communication of write updates until the data is actually needed. To reduce the communications caused by false sharing, where multiple unrelated shared data locate on the same page, LRC protocols usually support a multiple-writer scheme. Within this scheme, multiple writable copies of the same page are allowed and a clean copy is generated after an invalidation. Home-based Lazy Release Consistency (HLRC) [17], for example, implements such a multiple-writer scheme by specifying a home for each page. All updates to a page are propagated to the home node at synchronization points, such as lock release and barrier. Hence the page copy on home is up-to-date.

**Table 1.** ViSMI annotations for shared memory execution

| Annotation | Description |
| --- | --- |
| HLRC_Malloc | allocating memory in shared space |
| HLRC_Myself | querying the ID of the calling process |
| HLRC_InitParallel | initialization of the parallel phase |
| HLRC_Barrier | establishing synchronization over processes |
| HLRC_Acquire | acquiring the specified lock |
| HLRC_Release | releasing the specified lock |
| HLRC_End | releasing all resources and terminating |

ViSMI implements such a Home-based Lazy Release Consistency protocol. For each shared page a default home is specified during the initialization phase and then the node first accessing the page becomes its home. Each processor can maintain a copy of the shared page, but by a synchronization operation all copies are invalidated. Also at this point, an up-to-date version of the page is created on the home node. For this, the updates of all processors holding a copy must be aggregated. ViSMI uses a *diff*-based mechanism, where the difference (*diffs*) between each dirty copy and the clean copy is computed. This is similar to that used by the Myrias parallel do mechanism [2]. To propagate the updates, ViSMI takes advantage of the hardware-based multicast provided by InfiniBand to minimize the overheads for interconnection traffic. The *diffs* are then applied to the clean copy and the up-to-date version of the page is generated. For further computation page fault signals are issued on other processors and the missing page is fetched from the home node. To handle the incoming communication, each node maintains an additional thread, besides the application thread. This communication thread is only active when a communication request occurs. We use the event notification scheme of InfiniBand to achieve this.

For parallel execution, ViSMI establishes a programming interface for developing shared memory applications. This interface is primarily composed of a set of annotations that handle issues with respect to parallelism. The most important annotations and a short description about them are listed in Table 1.

## 3 Omni/Infini: Towards OpenMP Execution on Clusters

OpenMP is actually initially introduced for parallel multiprocessor systems with physically global shared memory. Recently, compiler developers have been extending the existing compilers to enable the OpenMP execution on cluster systems, often using a software distributed shared memory as the basis. Well-known examples are the Nanos Compiler [5,15], the Polaris parallelizing compiler [1], and the Omni/SCASH compiler [18].

Based on ViSMI and its programming interface, we similarly implemented an OpenMP compiler for the InfiniBand cluster. This compiler, called Omni/Infini, is actually a modification and extension of the original Omni compiler for SMPs [11]. The major work has been done with coordination of processes, allocation of shared data, and a new runtime library for parallelization, synchronization, and task scheduling.

**Process structure vs. thread structure.** The Omni compiler, like most others, uses a thread structure for parallel execution. It maintains a master thread and a number of slave threads. Slave threads are created at the initialization phase, but they are idle until a parallel region is encountered. This indicates that sequential regions are implicitly executed by the master thread, without any special task assignment. The ViSMI interface, on the other hand, uses a kind of process structure, where processes are forked at the initialization phase. These processes execute the same code, including both sequential regions and parallel parts. Clearly, this structure burdens processors with unnecessary work. In order to maintain the conventional OpenMP semantics with parallelism and also to save the CPU resources, we have designed special mechanisms to clearly specify which process does what job. For code regions needed to be executed on a single processor, for example, only the process on the host node is assigned with tasks.

**Shared data allocation and initialization.** ViSMI maintains a shared virtual space visible to all processor nodes. This space is reserved at the initialization phase and consists of memory spaces from each processor's physical main memory. Hence, all shared data in an OpenMP code must be allocated into this virtual space in order to be globally accessible and consistent. This is an additional work for a cluster-OpenMP compiler. We extended Omni for detecting shared variables and further changing them to data objects which will be allocated to the shared virtual space at runtime.

Another issue concerns the initialization of shared variables. Within a traditional OpenMP implementation, this is usually done by a single thread. Hence, an OpenMP directive SINGLE is often applied in case that such initialization occurs in a parallel region. This causes problems when running the applications on top of ViSMI. ViSMI allocates memory spaces for shared data structures on all processor nodes [1]. These data structures must be initialized before further use for parallel computation. Hence, the initialization has to be performed on all nodes. Currently, we rely on an implicit barrier operation inserted to SINGLE to tackle this problem. With this barrier, updates to shared data are forced to aggregate on the host node and a clean copy is created. This causes performance lost because a barrier operation is not essential for all SINGLE operations. For the next version of Omni/Infini, we intend to enable compiler-level automatic distinction between different SINGLE directives.

**Runtime library.** Omni contains a set of functions to deal with runtime issues like task scheduling, lock and barrier, reduction operations, environment variables, and specific code regions such as MASTER, CRITICAL, and SINGLE. These functions require information, like thread ID number and number of threads, to perform correct actions for different threads. This information

---

[1] ViSMI allocates on each processor a memory space for shared variables. Each processor uses the local copy of shared data for computation.

is stored within data structures for threads, which are not available in ViSMI. Hence, we modified all related functions in order to remove the interface to thread structure of Omni and to build the connection to the ViSMI programming interface. In this way, we created a new OpenMP runtime library that is linked to the applications for handling OpenMP runtime issues.

## 4  Initial Experimental Results

Based on the extension and modification with both sides, the Omni compiler and ViSMI, we have developed this OpenMP execution environment for InfiniBand clusters. In order to verify the established environment, various measurements have been done using our InfiniBand cluster. Since ViSMI is currently based on a 32-bit address space and the Opteron nodes are available to users quite recently, most of the experiments were carried out on the six Xeon nodes. Only the last experiment for examining scalability was performed with Opteron processors.

We use a variety of applications for examining different behavior. Four of them are chosen from the NAS parallel benchmark suite [4,8] and the OpenMP version of the SPLASH-2 Benchmark suite [20]. Two Fortran programs are selected from the benchmark suite developed for an SMP programming course [19]. In addition, two self-coded small kernels are also examined. A short description, the working set size, and the required shared memory size of these applications are shown in Table 2.

**Table 2.** Description of benchmark applications

| Application | Description | Working set size | Shared memory size | Benchmark |
|---|---|---|---|---|
| LU | LU-decomposition for dense matrices | 2048×2048 matrix | 34MB | SPLASH-2 |
| Radix | Integer radix sort | 18.7M keys | 67MB | SPLASH-2 |
| FT | Fast Fourier Transformations | 64×64×64 | 51MB | NAS |
| CG | Grid computation and communication | 1400 | 3MB | NAS |
| Matmul | Dense matrix multiplication | 2048×2048 | 34MB | SMP course |
| Sparse | Sparse matrix multiplication | 1024×1024 | 8MB | SMP course |
| SOR | Successive Over Relaxation | 2671×2671 | 34MB | self-coded |
| Gauss | Gaussian Elimination | 200×200 | 1MB | self-coded |

First, we measured the speedup of the parallel execution using different number of processors. Figure 1 shows the experimental results. It can be seen that applications behave quite differently. LU achieves the best performance with a scalable speedup of as high as 5.52 on a 6-node cluster system. Similarly, Matmul and SOR also show a scalable speedup with close parallel efficiency on different systems (efficiency is calculated with the speedup divided by the number of processors and reflects the scalability of a system). Sparse and Gauss behave poorly, with either no speedup or running even slower on multiprocessor systems. This is caused by the smaller working set size of both codes. Actually, we selected this size in order to examine how system overhead influences the parallel performance; and we see that due to the large percentage of overhead

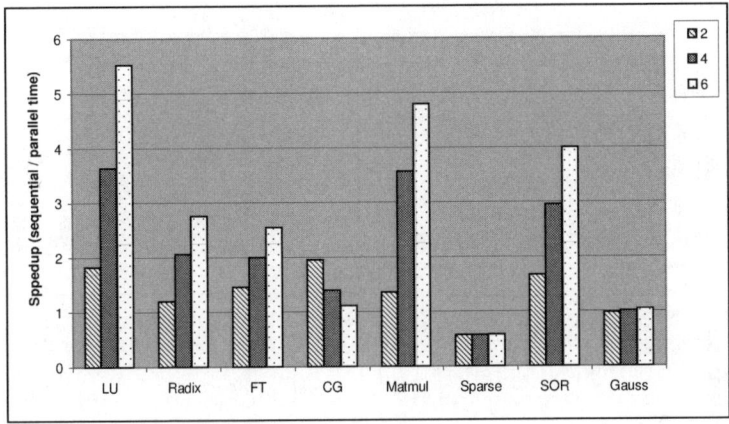

**Fig. 1.** Speedup on systems with different number of processors

in the overall execution time, applications with smaller data size can not gain speedup on systems with software-based distributed shared memory.

CG behaves surprisingly with a decreasing speedup as the number of processors increases. For detecting the reasons, we measured the time needed for different activities when running the applications on 6-node systems. Figure 2 shows the experimental results.

In Figure 2, *exec.* denotes the overall execution time, while *comput.* specifies the time for actually executing the application, *page* is the time for fetching pages, *barrier* denotes the time for barrier operations [2], *lock* is the time for performing locks, *handler* is the time needed by the communication thread, and *overhead* is the time for other protocol activities. The sum of all partial times equals to the total *exec.* time.

It can be seen that LU, Matmul, and SOR show a rather high proportion in computation time, and hence achieve better speedup than other applications. Radix and FT behaves worse than them, but most of the time is used for calculation. For Sparse and Gauss roughly only the half time is spent for computation and hence nearly no speedup can be observed. The worst case is with CG, where only 33% of the overall time is used for running the program and more time is spent on other activities like inter-node communication and synchronization. As it is a fact that each processor introduces such overhead, slowdown can be caused with more processors running the code. CG has shown this behavior.

In order to further verify this, we measured the time for different activities with CG also on 2-node and 4-node systems. Figure 3 shows the experimental results. It can be seen that while the time with *handler* and *overhead* is close on different systems, *page* and *barrier* show a drastic increase with more processors on the system. As a result, a decreasing speedup has been observed.

---

[2] This includes the time for synchronization and that for transferring *diff*s.

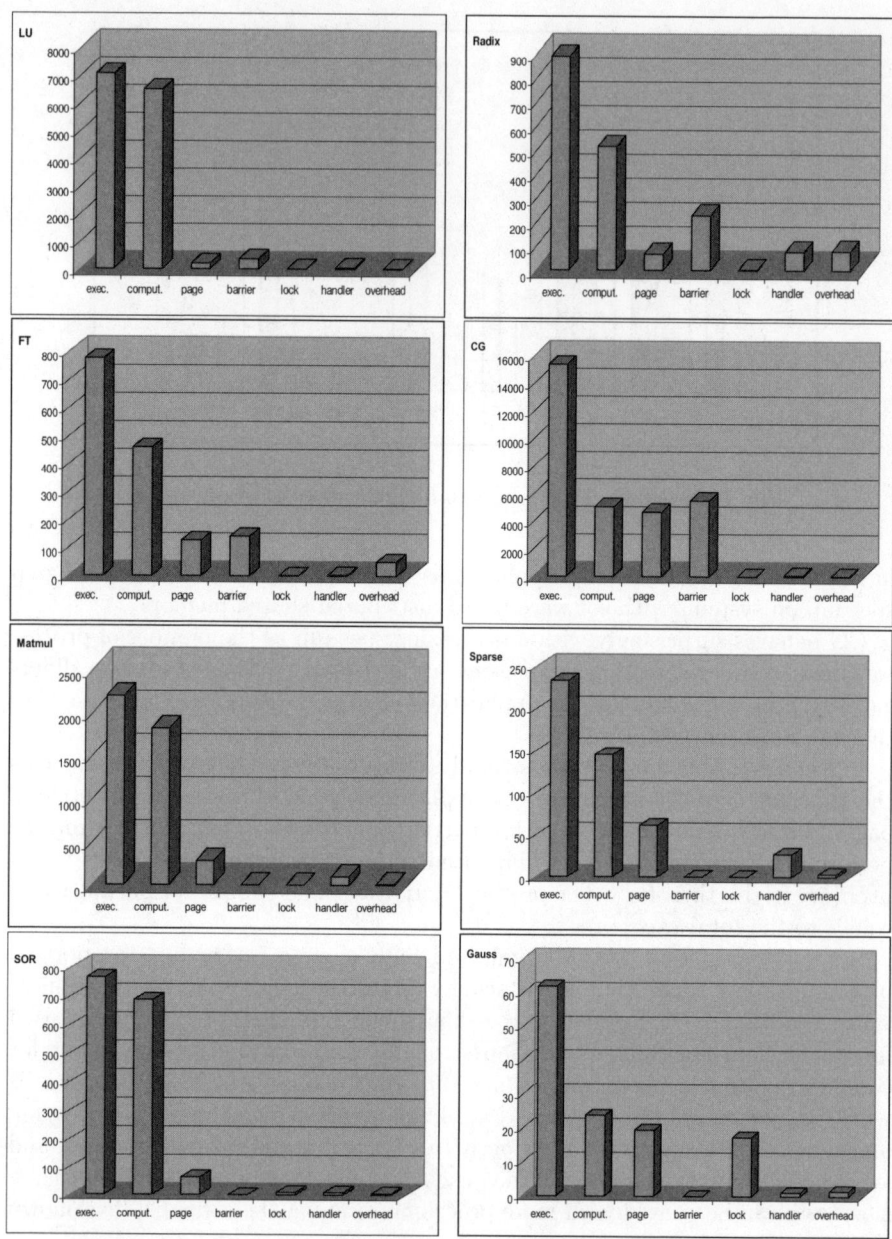

**Fig. 2.** Normalized execution time breakdown of tested applications

In addition, Figure 2 also shows a general case where page fetching and barrier operations introduce the most overhead. In order to further examine these critical issues, we measured the concrete number of page faults, barriers, and locks. Table 3 depicts the experimental results.

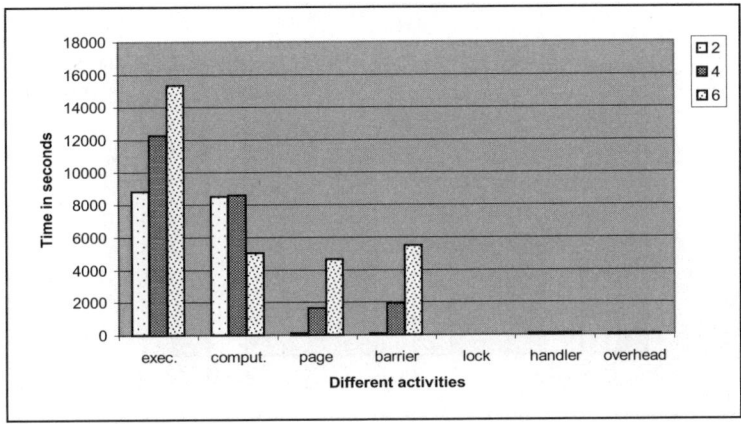

**Fig. 3.** Execution time of different activities on different systems (CG code)

This table shows the number of page fault, barrier operation, and locks. It also gives the information about data and barrier traffic over processors. All applications, except Gauss with smaller working set, show high number of page fault, and hence the large amount of resulted data transfer. In contrast, only fewer barriers have been measured, except the CG code. However, a barrier operation could introduce significant overhead, since all processors have to send updates to the home node, which introduces transfer overhead, and to wait for a reply, which causes synchronization overhead. In addition, the computation can continue only after a clean copy has been created. Therefore, even though only a few of barriers are performed, still large proportion of the overall time is spent on barrier operations, as having been illustrated in Figure 2. However, LU is an exception, where few time is needed for barriers. This can be explained by the fact that with LU rather small amount of *diff*s are created and hence overhead for data transfer at barriers is small.

In order to reduce the overhead with page fault and barrier, we propose adaptive approaches. Actually, page fault occurs when a page copy has to be

**Table 3.** Value of several performance metrics

|        | page fault | barriers | lock acquired | data traffic | barrier traffic |
|--------|-----------|----------|---------------|--------------|-----------------|
| LU     | 5512      | 257      | 0             | 14.4MB       | 0.015M          |
| Radix  | 6084      | 10       | 18            | 13.9MB       | 0.12M           |
| FT     | 3581      | 16       | 49            | 21MB         | 0.03M           |
| CG     | 8385      | 2496     | 0             | 8.3MB        | 0.03M           |
| Matmul | 4106      | 0        | 0             | 17.4MB       | 0               |
| Sparse | 1033      | 0        | 0             | 1.4MB        | 0               |
| SOR    | 2739      | 12       | 0             | 4.7MB        | 0.006M          |
| Gauss  | 437       | 201      | 0             | 0.17MB       | 0.02M           |

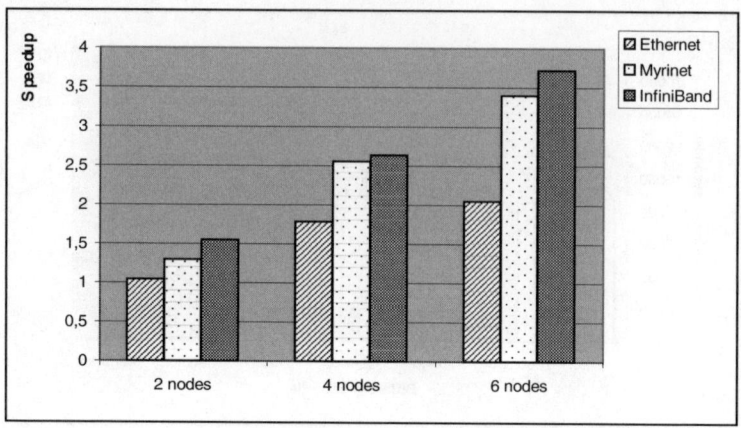

**Fig. 4.** Speedup comparison with other SDSM-based OpenMP implementation

invalidated, while a clean copy of this shared page is created. The current HLRC implementation uses an approach, where all processors send the *diff*s to the home node. The home node then applies the *diff*s to the page and generates a clean copy. After that, other processors can fetch this copy for further computation. A possible improvement to this approach is to multicast the *diff*s to all processors and apply them directly on all dirty copies. In this way, the number of page fault and the resulted data transport could be significantly reduced, while at the same time the overhead for multicasting is much smaller due to the special hardware-level support of InfiniBand. This optimization can also reduce the overhead for barriers, because in this case synchronization is not necessary; rather a processor can go on with the computation, as soon as the page copy on it is updated. We will implement such optimizations in the next step of this research work.

The following experiment was done with the laplace code provided by the Omni compiler. The goal is to compare the performance of OpenMP execution on InfiniBand clusters with that of software DSM based OpenMP implementation on clusters using other interconnection technologies. For the latter, we apply the data measured by the Omni/SCASH researchers on both Ethernet and Myrinet clusters. Figure 4 gives the speedup on 2, 4, and 6 node systems.

It can be seen that our system provides the best performance, with an average improvement of 62% to Ethernet and 10% to Myrinet. This improvement shall be contributed by the specific properties of InfiniBand.

With the last experiment we intended to examine the scalability of the established OpenMP platform on top of a software DSM. For this, we ran the LU, Matmul, and SOR codes, which achieved the best performance as shown in Figure 1, on the Opteron processors and measured speedup on systems with different number of nodes. Figure 5 illustrates the experimental results.

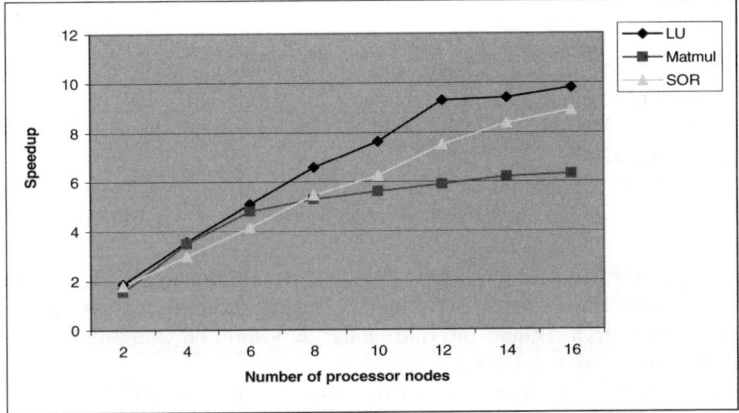

**Fig. 5.** Speedup of LU, Matmul, and SOR on Opteron Processors

As shown in this figure, the scalability with these codes is rather different. While LU scales up to 12 processors and SOR up to 16, Matmul achieves only a slight speedup increase on systems with more than 6 processor nodes. This also indicates that larger applications and computation intensive codes benefit more from the parallel execution on a system relying on software DSM. Other codes, however, could not achieve expected performance on large systems due to the high overhead for maintaining the shared space.

## 5 Conclusions

OpenMP is traditionally designed for shared memory multiprocessors with physically global shared memory. Due to the architectural restriction, however, such machines suffer from scalability. On the other hand, cluster systems are widely used for parallel computing, raising the need for establishing OpenMP environments on top of them.

In this paper, we introduce an approach for building such an environment on InfiniBand clusters. First, a software DSM is developed, which creates a shared virtual memory space visible to all processor nodes on the cluster. We then modified and extended the Omni OpenMP compiler in order to deal with issues like data mapping, task scheduling, and the runtime. Experimental results based on a variety of applications show that the parallel performance depends on applications and their working set size. Overall, a speedup of up to 5.22 on 6 nodes has been achieved.

Besides the optimization with page fault and barrier, we also intend to apply more features of InfiniBand to further reduce the system overhead. In addition, the software DSM will be extended to 64-bit address space, allowing a full use of the whole cluster for OpenMP execution.

# References

1. Basumallik, A., Min, S.-J., Eigenmann, R.: Towards OpenMP Execution on Software Distributed Shared Memory Systems. In: Zima, H.P., Joe, K., Sato, M., Seo, Y., Shimasaki, M. (eds.) ISHPC 2002. LNCS, vol. 2327, pp. 457–468. Springer, Heidelberg (2002)
2. Beltrametti, M., Bobey, K., Zorbas, J.R.: The Control Mechanism for the Myrias Parallel Computer System. ACM SIGARCH Computer Architecture News 16(4), 21–30 (1988)
3. Cox, A.L., Dwarkadas, S., Keleher, P.J., Lu, H., Rajamony, R., Zwaenepoel, W.: Software Versus Hardware Shared-Memory Implementation: A Case Study. In: Proceedings of the 21th Annual International Symposium on Computer Architecture, April 1994, pp. 106–117 (1994)
4. Bailey, D., et al.: The NAS Parallel Benchmarks. Technical Report RNR-94-007, Department of Mathematics and Computer Science, Emory University (March 1994)
5. Gonzàlez, M., Ayguadé, E., Martorell, X., Labarta, J., Navarro, N., Oliver, J.: NanosCompiler: Supporting Flexible Multilevel Parallelism in OpenMP. Concurrency: Practice and Experience 12(12), 1205–1218 (2000)
6. Iftode, L., Singh, J.P.: Shared Virtual Memory: Progress and Challenges. Proceedings of the IEEE, Special Issue on Distributed Shared Memory, 87, 498–507 (1999)
7. InfiniBand Trade Association. InfiniBand Architecture Specification, vol. 1 (November 2002)
8. Jin, H., Frumkin, M., Yan, J.: The OpenMP Implementation of NAS Parallel Benchmarks and Its Performance. Technical Report NAS-99-011, NASA Ames Research Center (October 1999)
9. Keleher, P., Dwarkadas, S., Cox, A., Zwaenepoel, W.: TreadMarks: Distributed Shared Memory On Standard Workstations and Operating Systems. In: Proceedings of the 1994 Winter Usenix Conference, January 1994, pp. 115–131 (1994)
10. Keleher, P.J.: Lazy Release Consistency for Distributed Shared Memory. PhD thesis, Department of Computer Science, Rice University (January 1995)
11. Kusano, K., Satoh, S., Sato, M.: Performance Evaluation of the Omni OpenMP Compiler. In: Valero, M., Joe, K., Kitsuregawa, M., Tanaka, H. (eds.) ISHPC 2000. LNCS, vol. 1940, pp. 403–414. Springer, Heidelberg (2000)
12. Lamport, L.: How to Make a Multiprocessor That Correctly Executes Multiprocess Programs. IEEE Transactions on Computers 28(9), 241–248 (1979)
13. Li, K.: Shared Virtual Memory on Loosely Coupled Multiprocessors. PhD thesis, Yale University (September 1986)
14. Li, K.: IVY: A Shared Virtual Memory System for Parallel Computing. In: Proceedings of the International Conference on Parallel Processing, Software, vol. II, pp. 94–101 (1988)
15. Martorell, X., Ayguadé, E., Navarro, N., Corbalán, J., González, M., Labarta, J.: Thread Fork/Join Techniques for Multi-Level Parallelism Exploitation in NUMA Multiprocessors. In: Proceedings of the 1999 International Conference on Supercomputing, Rhodes, Greece, June 1999, pp. 294–301 (1999)
16. Osendorfer, C., Tao, J., Trinitis, C., Mairandres, M.: ViSMI: Software Distributed Shared Memory for InfiniBand Clusters. In: Proceedings of the 3rd IEEE International Symposium on Network Computing and Applications (IEEE NCA 2004), September 2004, pp. 185–191 (2004)

17. Rangarajan, M., Iftode, L.: Software Distributed Shared Memory over Virtual Interface Architecture: Implementation and Performance. In: Proceedings of the 4th Annual Linux Showcase, Extreme Linux Workshop, Atlanta, USA, October 2000, pp. 341–352 (2000)
18. Sato, M., Harada, H., Hasegawa, A.: Cluster-enabled OpenMP: An OpenMP compiler for the SCASH software distributed shared memory system. Scientific Programming 9(2-3), 123–130 (2001)
19. Standish, R.K.: SMP vs Vector: A Head-to-head Comparison. In: Proceedings of the HPCAsia 2001 (September 2001)
20. Woo, S.C., Ohara, M., Torrie, E., Singh, J.P., Gupta, A.: The SPLASH-2 programs: characterization and methodological considerations. In: Proceedings of the 22nd Annual International Symposium on Computer Architecture, June 1995, pp. 24–36 (1995)

# An Introduction to Balder — An OpenMP Run-time Library for Clusters of SMPs

Sven Karlsson

Department of Microelectronics and Information Technology,
Royal Institute of Technology, KTH, Sweden
`Sven.Karlsson@sven.karlsson.name`

**Abstract.** In this paper a run-time library, called Balder, for OpenMP 2.0 is presented. OpenMP 2.0 is an industry standard for programming shared memory machines. The run-time library presented can be used on SMPs and clusters of SMPs and it will provide a shared address space on a cluster. The functionality and design of the library is discussed as well as some features that are being worked on. The performance of the library is evaluated and is shown to be competitive when compared to a commercial compiler from Intel.

## 1 Introduction

OpenMP has during the last few years gained considerable acceptance as the shared memory programming model of choice. OpenMP is an industry standard and utilizes a fork-join programming model based on compiler directives [1,2].

The directives are used by the programmer to instruct an OpenMP aware compiler to transform the program into a parallel program. In addition to the directives, OpenMP also specifies a number of run-time library functions.

To use OpenMP, an OpenMP aware compilation system is thus needed. The compilation system generally consists of a compiler and an OpenMP run-time library. The library is not only used to handle the run-time library functions as defined by the OpenMP specification but also to aid the compiler with a number of functions that efficiently spawn threads, synchronize threads, and help share work between threads.

In this paper, an open source OpenMP run-time library, called *Balder*, is presented which is capable of fully handling OpenMP 2.0 including nested parallelism [2]. The library supports not only single SMPs efficiently but also clusters of SMPs making it possible to do research on extensions to the OpenMP specification in the areas of SMP centric and cluster centric extensions. A compiler, called OdinMP, targeting the library is already readily available [3,4]. A more detailed description of the OpenMP transformations is available on OdinMPs homepage [4]. I will not discuss the API of Balder in detail in this paper and instead I refer the interested reader to the aforementioned website. The source code of the library is also available from the same website.

The library is highly portable and there are currently ports available to several processor architectures such as IA32 and ARM and to several different operating systems including Unix variants and Windows versions. Balder is currently at version 1.0.1.

The rest of the paper is organized as follows. Section 2 provides a brief overview of the library while section 3 provides information of the sub-libraries that Balder builds on. Section 4 describes the OpenMP run-time itself and presents some details on the implementation of important primitives while section 5 discusses work-in-progress and future features for the Balder run-time. Experimental results are presented in section 6. The paper is summarized in section 7.

## 2 Overview of the Run-Time Library

The run-time library is implemented in ANSI C [5] and provides the following functionality:

- Full implementation of all OpenMP 2.0 intrinsic functions, i.e., the run-time library functions described by the specification.
- Efficient handling of threads including thread creation, and thread synchronization.
- Parallel for-loop primitives to aid the compiler when transforming work sharing constructs.
- Support for OpenMPs threadprivate variables.
- Support for OpenMPs copyprivate clause.
- A built-in software distributed shared memory system, *software DSM system*, to achieve a shared address space on a cluster.
- Memory management for the shared address space.
- Support for shared stack storage.

The library builds on previous experience [3,6]. It is designed to be highly portable and to achieve portability it is designed as a layered system. Most of the functionality outlined in a previous paper is implemented [7].

The library utilizes three sub libraries: *Balder Threads*, *Balder·Oslib*, and *Balder Messages* as can be seen in figure 1. These libraries provide thread services, operating system services, and cluster messaging services respectively.

The software DSM system and OpenMP run-time library is then implemented on top of the sub-libraries making it possible to implement the run-time in ordinary ANSI C without any dependencies on processor architecture or operating system features. In essence, this means that porting Balder is a matter of porting the well-defined primitives in the three sub-libraries.

### 2.1 Related Work

Balder is complete re-write of the run-time described in a previous paper which only supported uni-processor nodes [6]. No code has been borrowed or taken from that prototype or other OpenMP run-time libraries. Balder differs from previous efforts in that it is designed from the ground up to be both efficient

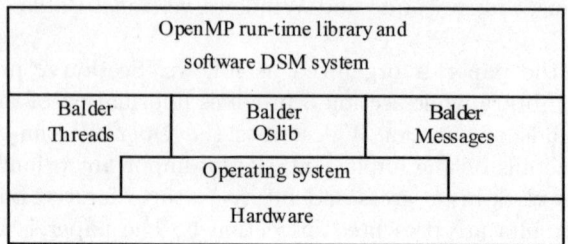

**Fig. 1.** Overview of the design of the Balder library

and portable. In a previous paper, the Balder run-time was mentioned with the OdinMP compiler [3]. In that paper the focus was on the compiler and the design of Balder was not discussed.

In the next section, I will continue with a description of the run-time sub-libraries.

## 3 The Balder Sub-libraries

The three sub-libraries are instrumental in achieving a high degree of functionality and portability. A layered design approach has, whenever possible, been used even in the sub-libraries to aid porting efforts.

### 3.1 Balder Messages

Balder Messages is a packet-based, network technology independent messaging library with support for prioritized communication. The library is described in previous work [7,8] and I will here only provide a very brief introduction. The transmission and reception primitives in the library use a data format based on linked lists for the packets. This facilitates scatter-gather I/O. It is also possible to allocate memory for packets in network hardware buffers so as to achieve zero-copy communication. Finally, there exist primitives for both synchronous and asynchronous communication as well as active message communication [9].

The library is divided into several parts. One part handles the construction of packets as linked lists, another handles flow control and queuing of packets and the last is the network back-end. There exist network back-ends for UDP and MPI [10]. Back-ends with partial support for Myrinet, LAPI, and System Vs shared memory primitives exist [11].

### 3.2 Balder Oslib

Balder relies heavily on the virtual memory management system of the operating system. The virtual memory systems of operating systems differ from each other in both APIs and provided services. The Balder Oslib sub-library acts as a virtualization layer and provides an operating system independent API. There is support for:

- Creation of a virtual memory region which is guaranteed to be at the same address range on all cluster nodes.
- Setting of access permissions on individual virtual memory pages.
- Catching of page-faults and similar exceptions.
- Timers and timing management.

In addition, Balder Oslib implements a registry under which arbitrary data structures can be filed using strings as keys. This registry is used to store global system options and data structures and is instrumental in making it possible to modularize Balder properly without being hampered by artificial cross-module dependencies.

The Oslib is the sub-library which is the least layered. This is largely due to the virtual memory systems differing so much between operating systems. The Windows port, for instance, requires a completely different implementation than Unixes and cannot share any code with other ports.

## 3.3 Balder Threads

Balder Threads provides an efficient processor architecture independent multi-threading API [12]. Balder Threads has primitives for:

- Thread creation and destruction.
- Thread synchronization using monitors, monitor signals, and barriers.
- Work queues.
- Stack frame creation so that arbitrary functions can be called.

Balder Threads uses a system of assembly macros that describes the processor architecture. A port to a new processor architecture is in most cases a matter of adapting the macros of an existing port.

The assembly macros describe how a function stack frame is organized, how functions are called, e.g. which parameters are passed on the stack and which are passed in registers, and provide implementations for important low-level primitives. Balder Thread uses POSIX threads as underlying thread library but implements efficient synchronization primitives using the mentioned low-level primitives [13]. The primitives used by Balder Threads are test-and-set, fetch-and-add, and a memory fence operation. Most modern processor architectures require that memory fence operations are embedded into lock and barrier operations or they will otherwise not work. This is not the only reason for providing memory fences as a low-level primitive. The memory fence operations are also used to implement the support for the OpenMP flush directive. In the absence of a required primitive, in the form of macros, Balder Threads will first revert to an implementation which only uses test-and-set and then, if no test-and-set primitive is provided, to POSIX threads. The memory fence operation is a no-operation on processor architectures that do not require memory fences.

Using this scheme of portable synchronization primitives based on low-level assembly macros, Balder Threads is able to achieve a synchronization overhead close to an order of magnitude less than POSIX threads. The synchronization primitives are generally based on test-test-and-set with a time-out so as to avoid excessive busy wait.

Balder Threads only provides primitives which can be used by threads running on the same cluster node. The OpenMP run-time layer provides primitives which can be used across cluster nodes.

## 4 Software DSM System and OpenMP Run-Time Library

The OpenMP run-time library is built on-top of the sub-libraries and can thus be written completely architecture independent. It provides support for a shared address space on clusters via a software DSM system, and it also provides support for high-level OpenMP primitives as outlined in section 2.

The software DSM system uses the virtual memory management system to provide, on a cluster, a shared address space which acts as a shared memory. The system does not rely on any particular hardware except a decent virtual memory system and a network interconnect system.

In Balder, the software DSM system is built on-top of Balder Oslib and Balder Messages. The software DSM system uses home-based lazy release consistency, HLRC [14]. The reason for using HLRC is HLRCs robustness and relative simplicity. Some special features of the HLRC variant used in Balder will be mentioned when discussing the support functions for OpenMP.

As mentioned in section 2, the OpenMP run-time library provides a number of primitives in different areas of functionality. I will now go through these areas and provide an overview of the functionality provided and how the implementation is done.

### 4.1 Parallel Regions

OpenMP is a fork-join programming model and the basis of parallelism are parallel regions as defined by compiler directives. The compiler will transform each parallel region into a function which is executed in parallel by a number of threads. The compiler then inserts a call to a run-time library function, which handles creation of threads and execution of the function, into the program. The run-time library function is the basis for parallelism and is called in__tone_spawnparallel.

Balder uses a pool of threads to avoid excessive and time consuming thread creation and destruction operations. Threads are never destroyed and are only created when no threads are left in the pool. All threads in the thread pool wait on a work queue as provided by Balder Threads.

Internally the implementation of in__tone_spawnparallel is straight-forward. It merely performs book-keeping, adds work to the thread pools work queue, the calling thread executes the parallel region, and then waits for any spawned threads to finish executing the parallel region. After finishing, the spawned threads returns to the thread pool and can be reused in other parallel regions.

Each cluster node has its own thread pool and so message passing is used to hand out work when running on a cluster. The cluster code also makes use of a limitation to simplify the message passing and the memory coherence protocol. Balder does not currently allow nested parallelism when running on a cluster although different approaches to lift the limitation are being investigated. All nested parallel regions are serialized yielding one level of parallelism. This is allowed by the OpenMP specification and does not break OpenMP compliance. Balder does, however, fully support nested parallelism when running on one single SMP node.

## 4.2 Lock Functions

The OpenMP specification describes a number of lock functions as run-time library functions. To handle these function, a set of distributed lock primitives have been implemented on top of Balder Threads and Balder Messages. The lock primitives are unfair in the sense that if a lock is held on a cluster node, the threads on that node get precedence over other threads when setting the lock. This reduces the network activity and the latency in case of lock contention [15].

Intra-node synchronization is performed with Balder Thread primitives while inter-node synchronization is performed with message passing.

If the system is running on one single node, the lock primitives revert back to the efficient primitives in the Balder Thread sub-library, thus avoiding the overhead of the distributed lock algorithm.

## 4.3 Barriers

The threads executing a specific parallel region form a thread team. The threads in a thread team can be synchronized with barriers and the OpenMP run-time provides a function for barrier synchronization.

Inter-node barriers are performed in two phases. First there is an intra-node barrier as provided by Balder Threads. Next, when all threads internal to a node are synchronized, message passing takes place which synchronizes all the nodes to each other using a centralized barrier algorithm. No threads are allowed to proceed until the second phase is finished.

The barrier operation is an operation with particularly high overhead on clusters. The overhead, is however, overlapped with memory coherence operations so as to not waste CPU resources. No message passing or memory coherence activities are performed when only one cluster node is used.

## 4.4 Worksharing Primitives

The worksharing constructs in OpenMP can all be mapped onto parallel for-loops and so Balder only provides support for such loops. The primitives essentially split a range of iterations into smaller pieces which then are executed in parallel by different threads.

It is, when running on one single cluster node, straight-forward to implement such primitives and in Balder the fetch-and-add primitive as provided by the Balder Thread assembly macros is used, if available, to minimize thread synchronization.

A cluster implementation is, however, much more complicated and easily leads to excessive message passing. A trick is used to reduce the message passing. The entire iteration space is first divided statically among the cluster nodes so that each cluster node receives a piece of the iteration space proportional to the number of threads, taking part of the parallel for-loop, that are executing on the node. These smaller pieces are then divided and handed out to the threads. This way no message passing is needed to implement parallel for-loops at the expense of potentially worse load imbalance. Load balancing algorithms, to reduce any load imbalance, are planned but not implemented.

This trick is OpenMP compliant as the static assignment of iterations performed follows all rules for the scheduling of for-loops stated in the OpenMP specification.

### 4.5 Advanced Data Clauses

The OpenMP specification defines several data clauses which describe how different variables are handled. One such data clause is the *threadprivate* data clause which defines a variable to be private to a thread and to keep its value between parallel regions. Such variables are called *threadprivate variables*. This essentially means that the threadprivate variables cannot be stored on the stack but must be associated with the threads themselves.

The run-time provides, for this, a thread-local storage space and two primitives to access the storage space. One primitive is used to allocate and initialize space and another is used to retrieve a pointer to the space.

The compiler generates code which at startup defines the thread-local storage. The storage space is, however, not allocated or initialized until it is used.

All accesses to threadprivate variables are changed by the compiler so that the pointer to the storage space is retrieved using the second primitive and all accesses are performed relative to that pointer. Allocation and initialization of the storage space will be performed in the primitive before the pointer is returned if no space has been allocated for the thread.

Another data clause is the *copyprivate* data clause. It makes it possible to broadcast the value of a variable private to a thread to the other threads in the same thread team. The copyprivate clause has been added in the second revision of the OpenMP specification to aid programming of applications utilizing nested parallelism.

A set of primitives has been added to the run-time library to implement the copyprivate data clause. These primitives make it possible for the broadcasting thread to send a variable via the run-time library to receiving threads and then wait for the receiving threads to properly receive the variable. For efficiency, the primitives are devised so that several variables can be sent and received after each other.

Naturally, message passing is used for inter cluster node broadcasts.

## 4.6 Handling of Shared Memory

The software DSM system is providing a shared address space across the cluster which can be accessed as a shared memory. To manage this address space, the run-time library provides memory allocation and de-allocation functions similar to ANSI Cs malloc and free [5]. These are used instead of malloc and free when running on a cluster. The software DSM system is inactive when running on a single node, i.e., a single SMP, and so the normal heap as provided by the operation system is used.

A few advanced code transformations are required to run OpenMP applications on clusters. The transformations involve the handling of shared global variables and shared stack variables. The OdinMP compiler has a command line option which if enabled forces the compiler to perform these transformations. The default is to not perform the transformations as they are not needed when generating code for SMPs.

Shared global variables must be allocated in the shared address space when running on a cluster. The OdinMP compiler can emit code to do the allocation and it also transforms declarations of, and accesses to, shared variables so that shared variables are accessed through pointers pointing to allocated memory regions in the shared address space.

In OpenMP, variables located on a threads stack can also be shared. This can occur if a parallel region is started. The thread that starts the parallel region, i.e., executes the in_tone_spawnparallel primitive, is called the master thread. Variables on the master threads stack can be shared among the threads in the spawned thread team.

Balder handles this by implementing one single shared stack located in the shared address space. One stack is enough as, on clusters, only one level of parallelism is allowed which means that only the thread that executes the serial portions of the OpenMP application can spawn parallelism.

The OdinMP compiler can transform the application code to make use of the stack. It changes how variables are allocated so that the potentially shared variables are allocated on the shared stack and it inserts code that builds stack frames on, and removes them from, the shared stack upon entry and exit from functions respectively.

One shared variable is in the example in figure 2. The transformed code is presented in figure 3. The examples are simplified for brevity.

The code inserted by OdinMP is very much similar to what a compiler emits at function entry and exit to build and remove stack frames. One thing differing is, however, the checks to see if the function is being called from a serial or

```
void f(void) {
  int shared_variable;    /* The shared variable. */

  shared_variable=5;      /* The shared variable is accessed. */
}
```

Fig. 2. An example with a shared variable

parallel region of the code. This is necessary as the shared stack must only be used from a serial portion of the code. A frame pointer is inserted as a local variable and used for accessing the shared variable.

In OpenMP, the flush directive is used to enforce consistency. The flush directive is a memory fence operation that controls when memory updates are conveyed and when memory operations are performed.

The application programmer must, essentially, insert a flush directive before an access or update of any shared variable that could have been updated by another thread and a flush directive must be inserted after an update for the update to be conveyed to other threads. Most OpenMP directives such as the barrier directives have implied flush directives and this reduces the need to insert extra flush directives. The OpenMP directives are defined so that extra flush directives are only needed in some very special cases, e.g. when implementing thread synchronization not using the OpenMP synchronization primitives.

The semantics of the flush directive is implemented in Balder just like in an earlier prototype and a more in-depth description is available in [6]. A special distributed lock is used to implement the flush directive. The lock is not accessible from the OpenMP application but is internally handled by the run-time library just like any other OpenMP lock. A flush directive consists of a set of the lock followed by a release. Information on memory updates are piggy-backed onto the lock. When setting the lock, the received information is used to invalidate memory contents so as to drive coherency. The received update information is merged with information on locally performed memory updates and then sent with the lock when the lock is released to a remote node.

The implied flush directives can be and are in most cases optimized. The barrier directive, as an example, has an implied flush directive but the coherency information is piggy-backed on the barrier message passing, thus removing the use of the special lock mentioned above.

### 4.7  OpenMP Intrinsic Functions

All the OpenMP 2.0 intrinsic functions are implemented in Balder. The implementation is rather straight-forward and mainly involves inquiring or updating the internal state of the run-time library.

```
void f(void ) {
struct in__tone_c_dt0
{
 int shared_variable;
};        /* The declaration of the stack frame which is used on the
             shared stack. */

          /* The shared stack must only be used in the serial portions of
             the code. The declaration below makes space on the threads
             stack to be used in parallel regions. */
struct in__tone_c_dt0 in__tone_c_dt0;
          /* The variable below is true when executing in a parallel
             region. in__tone_in_parallel() is a run-time library call
             that returns 1 iff the calling thread is executing in a
             parallel region. */
const int in__tone_sdsm_in_parallel=in__tone_in_parallel();
          /* The declaration below is for the frame pointer. The frame
             pointer is set to either the shared stack or the stack frame
             on the threads stack. */
struct in__tone_c_dt0 * const in__tone_sdsm_i_framep
 =in__tone_sdsm_in_parallel ? &in__tone_c_dt0:
          /* The shared stack pointer is called in__tone_sdsm_stackptr.
             It is below subtracted to make space for a new frame. This
             piece of code will only be executed if executing in a serial
             portion of the code.*/
 (struct in__tone_c_dt0 * const ) (in__tone_sdsm_stackptr=
  in__tone_sdsm_stackptr-sizeof(struct in__tone_c_dt0 ));
          /* Below is the access to the shared_variable. It is now done
             through the frame pointer. */
in__tone_sdsm_i_framep->in__tone_c_dt0.shared_variable=5;
          /* When leaving the function, either at the end of the function
             or with return, the shared stack pointer must be updated so
             as to remove the frame if executing in a serial portion of
             the code. The code below does that. */
if (!in__tone_sdsm_in_parallel)
  in__tone_sdsm_stackptr=in__tone_sdsm_stackptr+
                    sizeof(struct in__tone_c_dt0 );
}
```

Fig. 3. Transformed example with shared variable

## 5  Advanced Features

Balder has a few planned features currently under implementation. Apart from prefetch and producer-push [16,17], support for a fine-grain Software DSM system is under implementation as outlined in the next section.

## 5.1 A Compiler Supported Hybrid Fine-Grain/Coarse-Grain Software DSM System

The OdinMP compiler can gather information from the source code of OpenMP applications which can be used to further optimize the performance of said applications on Balder. This can be used to insert coherency checks into the compiler output so as to reduce the granularity of the coherency while still being able to fall back on a page-based system.

In short this means that whenever a shared variable can potentially be accessed the compiler needs to make sure that there is code inserted, prior to the access, which assures the shared variable is cached locally. The key point here is that only the shared variable itself, and not the virtual memory pages on which the variable is located, has to be locally cached. This reduces the latency of remotely requesting shared data and also reduces the average memory access latency of shared variables thus increasing the performance, i.e., reducing the execution time, of applications running on the Software DSM system provided the overhead of the inserted coherency checks is small enough.

OdinMP is being augmented to insert the mentioned coherency checks and two primitives is being added to Balder. The coherency checks are based on a check-in/check-out scheme where a piece of shared address space is requested and then later returned. Prototypes for the two primitives are:

```
void* in__tone_sdsm_checkout_memory(
                        void*           shared_address,
                        unsigned int    length
                        int             write_permissions);
void in__tone_sdsm_commit_memory(
                        void*           shared_address,
                        void*           local_address,
                        unsigned int    length
                        int             data_written);
```

The `in__tone_sdsm_checkout_memory` function is used to request an up-to-date version of a memory region. The memory region is defined by the parameters `shared_address` which is the shared address to the region and `length` which is the length in bytes of the region. The `write_permissions` parameter is 1 if a copy with write permissions is requested. The primitive returns a pointer to a copy of the memory region.

The `in__tone_sdsm_commit_memory` primitive is used to hand a previously requested copy back to the system and commit changes. It takes as parameters the shared address in `shared_address`, the pointer to the copy of the region in `local_address`, and the length, in bytes, of the region in `length`. The `data_written` parameter is 1 if the copy has been written to and has thus been updated.

## 6 Experimental Results

Some experiments have been conducted so as to evaluate the performance of Balder. These experiments are not exhaustive but they do give an indication of Balders performance. The cluster parts of Balder are not evaluated as they are being tested and are not ready to be evaluated yet.

A dual Pentium-III workstation running Linux version 2.4.25 was used as experimental platform. The processors were running at a clock rate of 1 GHz.

The EPCC micro-benchmark suite was used in the experiments [18]. The EPCC micro-benchmark suite is a set of benchmarks which measure the overhead of individual OpenMP constructs and thus also the run-time library. The benchmarks were compiled with OdinMP version 0.284.1 and GCC 3.3.4. The Balder library version 1.0.1 was used and was also compiled with GCC 3.3.4.

For comparison, the same set of benchmarks were compiled with the Intel C/C++ compiler version 8.0 and run on the experimental platform. The Intel compiler supports C/C++. The highest possible optimization level was used in both compilation systems. The Balder library cannot currently be compiled with the Intel compiler.

The overheads in microseconds for the most common OpenMP constructs as reported by the EPCC micro-benchmarks are summarized in table 1. The overheads are presented with their 95% confidence interval.

**Table 1.** Overheads in microseconds of common OpenMP constructs

| OpenMP Construct | Intel compiler | Balder with OdinMP |
|---|---|---|
| Parallel construct | $1.43 \pm 0.11$ | $2.91 \pm 0.15$ |
| For construct | $0.79 \pm 0.17$ | $2.93 \pm 0.30$ |
| Barrier construct | $0.48 \pm 0.19$ | $0.49 \pm 0.12$ |
| Lock and unlock primitives | $0.48 \pm 0.33$ | $0.47 \pm 0.12$ |

The overheads in the first three rows are for one single parallel region, parallel for-loop, and barrier respectively. The lock and unlock primitives row is the overhead of setting and then releasing a single lock once.

The overheads of the primitives in the Balder library are very competitive for barrier and lock synchronization. The overhead of parallel for loops are much higher for the Balder run-time. The reason for this is the fact that OdinMP is a source-to-source compiler and cannot do aggressive optimizations of the parallel for-loops as the Intel compiler can as it is compiling to object code.

The overhead of parallel regions are also higher for the Balder run-time. One contributing factor is the transformation of parallel regions as outlined in section 4.1. The functions created out of the parallel regions by the OdinMP compiler can take arguments. These arguments are managed by the in__tone_spawnparallel run-time primitive. The arguments are copied once more than actually needed on an SMP during the handling of the arguments and the creation of stack frames performed in the run-time library. The extra copying performed is, however, necessary when running on clusters. I'm investigating to see if this can be improved in future versions of Balder.

The differences in overheads between code generated by the Intel compiler and the OdinMP/Balder combination are very small. The overheads are in the same range as found in a previous study [3]. It was found in the same study that small differences in overheads are unlikely to influence end performance of OpenMP applications. The overheads measured thus suggests that Balder paired with OdinMP should be very competitive to commercial compilation systems.

## 7 Summary

This paper provides an overview of the current status of Balder, an OpenMP run-time library. The organization of the library is presented and the functionality and design of the different modules are described. Some selected parts of the implementation are discussed. Planned future and work in progress for the Balder library and the OdinMP compiler is presented. Some experimental data is presented which shows the Balder library to be competitive when compared to a commercial OpenMP compiler.

## Acknowledgements

The research in this thesis has been in part financially supported by the Swedish Research Council for Engineering Sciences under contract number, TFR 1999-376, and by the Swedish National Board for Industrial and Technical Development (NUTEK) under project number P855. It was also partially financed by the European Commission under contract number IST-1999-20252.

Nguyen-Thai Nguyen-Phan has implemented parts of the work-sharing, memory allocation and distributed lock primitives. He has been instrumental in testing the library. The API used by Balder is a super-set of the API for OpenMP run-time libraries developed during the Intone project by the Intone project partners. Eduard Ayguadé, Marc González, and Xavier Martorell at UPC in Spain were particularly helpful in that effort.

## References

1. OpenMP Architecture Review Board: OpenMP specification, C/C++ version 1.0 (October 1998)
2. OpenMP Architecture Review Board: OpenMP specification, C/C++ version 2.0 (March 2002)
3. Karlsson, S., Brorsson, M.: A Free OpenMP Compiler and Run-Time Library Infrastructure for Research on Shared Memory Parallel Computing. In: Proceedings of The 16th IASTED International Conference on Parallel and Distributed Computing and Systems (PDCS 2004) (November 2004)

4. Karlsson, S.: OdinMP homepage (retrieved on May 1, 2005), http://www.odinmp.com
5. Information Technology Industry Council: ISO/IEC 9899:1999 Programming languages - C, 2nd edn. American National Standards Institute (1999)
6. Karlsson, S., Lee, S.-W., Brorsson, M.: A Fully Compliant OpenMP Implementation on Software Distributed Shared Memory. In: Sahni, S.K., Prasanna, V.K., Shukla, U. (eds.) HiPC 2002. LNCS, vol. 2552, pp. 195–206. Springer, Heidelberg (2002)
7. Karlsson, S., Brorsson, M.: An Infrastructure for Portable and Efficient Software DSM. In: Proceedings of 1st Workshop on Software Distributed Shared Memory (WSDSM 1999), Rhodes, Greece, June 25 (1999); Also available from Department of Information Technology, Lund University, P.O. Box 118, SE-221 00 Lund, Sweden
8. Karlsson, S., Brorsson, M.: Priority Based Messaging for Software Distributed Shared Memory. Journal on Cluster Computing 6(2), 161–169 (2003)
9. von Eicken, T., Culler, D.E., Goldstein, S.C., Schauser, K.E.: Active Messages: a Mechanism for Integrated Communication and Computation. In: Proceedings of the 19th International Symposium on Computer Architecture, Gold Coast, Qld., Australia, May 1992, pp. 256–266 (1992)
10. Message Passing Interface Forum: MPI: A Message-Passing Interface Standard, version 1.1. (June 12, 1995)
11. Boden, N.J., Cohen, D., Felderman, R.E., Kulawik, A.E., Seitz, C.L., Seizovic, J.N., Su, W.-k.: Myrinet: A gigabitpersecond local area network. IEEE Micro 15(1), 29–36 (1995)
12. Karlsson, S.: A portable and efficient thread library for OpenMP. In: Proceedings of EWOMP 2004 (October 2004)
13. IEEE: IEEE std 1003.1-1996 POSIX part 1: System Application Programming Interface (1996)
14. Zhou, Y., Iftode, L., Li, K.: Performance evaluation of two home-based lazy release consistency protocols for shared virtual memory systems. In: Proceedings of the 2nd Operating Systems Design and Implementation Symposium (October 1996)
15. Radovic, Z., Hagersten, E.: Hierarchical Backoff Locks for Nonuniform Communication Architectures. In: Proceedings of the Ninth International Symposium on High Performance Computer Architecture (HPCA-9) (February 2003)
16. Karlsson, M., Stenström, P.: Evaluation of Dynamic Prefetching in Multiple-Writer Distributed Virtual Shared Memory Systems. Journal of Parallel and Distributed Computing 43(7), 79–93 (1997)
17. Karlsson, S., Brorsson, M.: Producer-Push — a Protocol Enhancement to Page-based Software Distributed Shared Memory Systems. In: Proceedings of the 1999 International Conference on Parallel Processing (ICPP 1999), September 1999, pp. 291–300 (1999)
18. Bull, J.M.: Measuring Synchronization and Scheduling Overheads in OpenMP. In: Proceedings of the First European Workshop on OpenMP, September 1999, pp. 99–105 (1999)

# Run-Time Environment

# Experiences with the OpenMP Parallelization of DROPS, a Navier-Stokes Solver Written in C++

Christian Terboven[1], Alexander Spiegel[1], Dieter an Mey[1], Sven Gross[2], and Volker Reichelt[2]

[1] Center for Computing and Communication, RWTH Aachen University, Germany
{Terboven,Spiegel,anMey}@rz.rwth-aachen.de
http://www.rz.rwth-aachen.de

[2] Institut für Geometrie und Praktische Mathematik, RWTH Aachen University, Germany
{Gross,Reichelt}@igpm.rwth-aachen.de
http://www.igpm.rwth-aachen.de

**Abstract.** In order to speed-up the Navier-Stokes solver DROPS, which is developed at the IGPM (Institut für Geometrie und Praktische Mathematik) at the RWTH Aachen University, the most compute intense parts have been tuned and parallelized using OpenMP. The combination of the employed template programming techniques of the C++ programming language and the OpenMP parallelization approach caused problems with many C++ compilers, and the performance of the parallel version did not meet the expectations.

## 1 Introduction

The Navier-Stokes solver DROPS [2] is developed at the IGPM (Institut für Geometrie und Praktische Mathematik) at the RWTH Aachen University, as part of an interdisciplinary project (SFB 540: Model-based Experimental Analysis of Kinetic Phenomena in Fluid Multi-phase Reactive Systems [1]) where complicated flow phenomena are investigated.

The object-oriented programming paradigm offers a high flexibility and elegance of the program code facilitating development and investigation of numerical algorithms. Template programming techniques and the C++ Standard Template Library (STL) are heavily used.

In cooperation with the Center for Computing and Communication of the RWTH Aachen University detailed runtime analysis of the code has been carried out and the computationally dominant program parts have been tuned and parallelized with OpenMP.

The UltraSPARC IV- and Opteron-based Sun Fire SMP-Clusters have been the prime target platforms, but other architectures have been investigated, too.

It turned out that the sophisticated usage of template programming in combination with OpenMP is quite demanding for many C++ compilers. We observed a high variation in performance and many compiler failures.

In chapter 2 the DROPS package is described briefly. In chapter 3 we take a look at the performance of the original and the tuned serial code versions. In chapter 4 we describe the OpenMP parallelization. The performance of the OpenMP version is discussed in chapter 5. Chapter 6 contains a summary of our findings.

## 2 The DROPS Multi-phase Navier-Stokes Solver

The aim of the ongoing development of the DROPS software package is to build an efficient software tool for the numerical simulation of three-dimensional incompressible multi-phase flows. More specifically, we want to support the modeling of complex physical phenomena like the behavior of the phase interface of liquid drops, mass transfer between drops and a surrounding fluid, or the coupling of fluid dynamics with heat transport in a laminar falling film by numerical simulation. Although quite a few packages in the field of CFD already exist, a black-box solver for such complicated flow problems is not yet available.

From the scientific computing point of view it is of interest to develop a code that combines the efficiency and robustness of modern numerical techniques, such as adaptive grids and iterative solvers, with the flexibility required for the modeling of complex physical phenomena.

For the simulation of two-phase flows we implemented a levelset technique for capturing the phase interface. The advantage of this method is that it mainly adds a scalar PDE to the Navier-Stokes system and therefore fits nicely into the CFD framework. But still, the coupling of the phase interface with the Navier-Stokes equations adds one layer of complexity.

The main building blocks of the solution method are the following:

*Grid generation and grid refinement.* Only tetrahedral grids without hanging nodes are used. The grids form a hierarchy of stable triangulations to enable the use of multi-grid solvers. The hierarchical approach also facilitates the coarsening of the grids.

*Time discretization.* For the stable time discretization of the instationary problems an implicit Fractional Step scheme is used.

*Spatial discretization.* The LBB-stable Taylor-Hood Finite Element pair ($P_2$-$P_1$) is used for the spatial discretization of the Navier-Stokes equations. For the level set equation the quadratic $P_2$ element is used.

*Iterative solution methods.* We decouple the Navier-Stokes-Level-Set system via a fixed point iteration which is also used to linearize the Navier-Stokes equations. The linearized equations which are of Stokes-type are treated by a Schur complement (inexact Uzawa) technique. The resulting convection-diffusion problems are solved by Krylov-subspace or multi-grid methods.

The several layers of nesting in the solvers (from the outer fixed point iteration down to the convection-diffusion-type solvers) induced by the structure of the mathematical models require fast inner-most solvers as well as fast discretization methods since many linear systems have to be regenerated in each time step.

Apart from the numerical building blocks, software engineering aspects such as the choice of suitable data structures in order to decouple the grid generation and finite element discretization (using a grid based data handling) as much as possible from the iterative solution methods (which use a sparse matrix format) are of main importance for performance reasons.

The code is programmed in C++ and uses several attractive facilities offered by this programming language.

## 3 Portability and Performance of the Serial Program Version

### 3.1 Platforms

The main development platform of the IGPM is a standard PC running Linux using the popular GNU C++ compiler [3]. Because this compiler does not support OpenMP, we had to look for adequate C++ compilers supporting OpenMP on our target platforms.

Table 1 lists compilers and platforms which we considered for our tuning and parallelization work. It also introduces abbreviations for each combination of hardware, operating system and compiler, which will be referred to in the remainder of the paper.

The programming techniques employed in the DROPS package (Templates, STL) caused quite some portability problems due to lacking standard conformance of the compilers (see table 3). The code had to be patched for most compilers.

From the early experiences gathered by benchmarking the original serial program and because of the good availability of the corresponding hardware we concentrated on the OPT+icc and USIV+guide platforms for the development of the OpenMP version. We used XEON+icc (running Windows) for verification of the OpenMP codes using the Intel ThreadChecker.

### 3.2 Runtime Profile

The runtime analysis (USIV+guide platform) shows that assembling the stiffness matrices (SETUP) costs about 52% of the total runtime, whereas the PCG-method including the sparse-matrix-vector-multiplication costs about 21% and the GMRES-method about 23%. Together with the utility routine LINCOMB these parts of the code account for 99% of the total runtime. All these parts have been considered for tuning and for parallelization with OpenMP.

It must be pointed out that the runtime profile heavily depends on the number of mesh refinements and on the current timesteps. In the beginning of a program run the PCG-algorithm and the matrix-vector-multiplication take about 65% of the runtime, but because the number of iterations for the solution of the linear equation systems shrinks over time, the assembly of the stiffness matrices is

**Table 1.** Compilers and platforms

| code | machine | processor | operating system | compiler |
|---|---|---|---|---|
| XEON+gcc333 | standard PC | 2x Intel Xeon 2.66 GHz | Fedora-Linux | GNU C++ V3.3.3 |
| XEON+gcc343 | | | | GNU C++ V3.4.3 |
| XEON+icc81 | standard PC | 2x Intel Xeon 2.66 GHz | Fedora-Linux and Windows 2003 | Intel C++ V8.1 |
| XEON+pgi60 | standard PC | 2x Intel Xeon 2.66 GHz | Fedora-Linux | PGI C++ V6.0-1 |
| XEON+vs2005 | standard PC | 2x Intel Xeon 2.66 GHz | Windows 2003 | MS Visual Studio 2005 beta 2 |
| OPT+gcc333 | Sun Fire V40z | 4x AMD Opteron 2.2 GHz | Fedora-Linux | GNU C++ V3.3.3 |
| OPT+gcc333X | | | | GNU C++ V3.3.3, 64bit |
| OPT+icc81 | Sun Fire V40z | 4x AMD Opteron 2.2 GHz | Fedora-Linux | Intel C++ V8.1 |
| OPT+icc81X | | | | Intel C++ V8.1, 64bit |
| OPT+pgi60 | Sun Fire V40z | 4x AMD Opteron 2.2 GHz | Fedora-Linux | PGI C++ V6.0-1 |
| OPT+pgi60X | | | | PGI C++ V6.0-1, 64bit |
| OPT+path20 | Sun Fire V40z | 4x AMD Opteron 2.2 GHz | Fedora-Linux | PathScale EKOpath 2.0 |
| OPT+path20X | | | | PathScale EKOpath 64bit |
| OPT+ss10 | Sun Fire V40z | 4x AMD Opteron 2.2 GHz | Solaris 10 | SunStudio C++ V10 |
| USIV+gcc331 | Sun Fire E2900 | 12x UltraSPARC IV 1.2 GHz, dual core | Solaris 9 | GNU C++ V3.3.1 |
| USIV+ss10 | Sun Fire E2900 | 12x UltraSPARC IV 1.2 GHz, dual core | Solaris 9 | Sun Studio C++ V10 |
| USIV+guide | Sun Fire E2900 | 12x UltraSPARC IV 1.2 GHz, dual core | Solaris 9 | Intel-KSL Guidec++ V4.0 + Sun Studio 9 |
| POW4+guide | IBM p690 | 16x Power4 1.7 GHz, dual core | AIX 5L V5.2 | Intel-KSL Guidec++ V4.0 |
| POW4+xlC60 | IBM p690 | 16x Power4 1.7 GHz, dual core | AIX 5L V5.2 | IBM Visual Age C++ V6.0 |
| POW4+gcc343 | IBM p690 | 16x Power4 1.7 GHz, dual core | AIX 5L V5.2 | GNU C++ V3.3.3 |
| IT2+icc81 | SGI Altix 3700 | 128x Itanium 2 1.3 GHz | SGI ProPack Linux | Intel C++ V8.1 |

getting more and more dominant. Therefore we restarted the program after 100 time steps and let it run for 10 time steps with 2 grid refinements for our comparisons.

### 3.3 Data Structures

In the DROPS package the Finite Element Method is implemented. This includes repeatedly setting up the stiffness matrices and then solving linear equation systems with PCG- and GMRES-methods.

Since the matrices arising from the discretization are sparse, an appropriate matrix storage format, the CRS (compressed row storage) format is used, in which only nonzero entries are stored. It contains an array *val* - which will be referred to later - for the values of the nonzero entries and two auxiliary integer arrays that define the position of the entries within the matrix.

The data structure is mainly a wrapper class around a `valarray<double>` object, a container of the C++ Standard Template Library (STL).

Unfortunately, the nice computational and storage properties of the CRS format are not for free. A disadvantage of this format is that insertion of a non-zero element into the matrix is rather expensive. Since this is unacceptable when building the matrix during the discretization step, a sparse matrix builder class has been designed with an intermediate storage format based on STL's `map` container that offers write access in logarithmic time for each element. After the assembly, the matrix is converted into the CRS format in the original version.

### 3.4 Serial Tuning Measures

On the Opteron systems the PCG-algorithm including a sparse-matrix-vector-multiplication and the preconditioner profits from manual prefetching. The performance gain of the matrix-vector-multiplication is 44% in average, and the speed-up of the preconditioner is 19% in average, depending on the addressing mode (64bit mode profits slightly more than 32bit mode).

As the setup of the stiffness matrix turned out to be quite expensive we reduced the usage of the `map` datatype. As long as the structure of the matrix does not change, we reuse the index vectors and only fill the matrix with new data values. This leads to a performance plus of 50% on the USIV+guide platform and about 57% on the OPT+icc platform. All other platforms benefit from this tuning measure as well.

Table 2 lists the results of performance measurements of the original serial version and the tuned serial version. Note that on the Opteron the 64bit addressing mode typically outperforms the 32bit mode, because in 64bit mode the Opteron offers more hardware registers and provides an ABI which allows for passing function parameters using these hardware registers. This outweights the fact that 64bit addresses take more cache space.

## 4 The OpenMP Approach

### 4.1 Assembly of the Stiffness Matrices

The matrix assembly could be completely parallelized, but it only scales well up to about 8 threads, because the overhead increases with the number of threads used (see table 4).

The routines for the assembly of the stiffness matrices typically contain loops like the following:

```
for (MultiGridCL::const_TriangTetraIteratorCL
    sit=_MG.GetTriangTetraBegin(lvl),
    send=_MG.GetTriangTetraEnd(lvl);
    sit != send; ++sit)
```

**Table 2.** Platforms, compiler options and serial runtime of the original and the tuned versions. Note that we didn't have exclusive access to the Power4 and Itanium2 based systems for timing measurements.

| code | compiler options | runtime [s] original version | runtime [s] tuned version |
|---|---|---|---|
| XEON+gcc333 | -O2 -march=pentium4 | 3694.9 | 1844.3 |
| XEON+gcc343 | -O2 -march=pentium4 | 2283.3 | 1780.7 |
| XEON+icc81 | -O3 -tpp7 -xN -ip | 2643.3 | 1722.9 |
| XEON+pgi60 | -fast -tp piv | 8680.1 | 5080.2 |
| XEON+vs2005 | compilation fails | n.a. | n.a. |
| OPT+gcc333 | -O2 -march=opteron -m32 | 2923.3 | 1580.3 |
| OPT+gcc333X | -O2 -march=opteron -m64 | 3090.9 | 1519.5 |
| OPT+icc81 | -O3 -ip -g | 2516.9 | 1760.7 |
| OPT+icc81X | -O3 -ip -g | 2951.3 | 1521.2 |
| OPT+pgi60 | -fast -tp k8-32 -fastsse | 6741.7 | 5372.9 |
| OPT+pgi60X | -fast -tp k8-64 -fastsse | 4755.1 | 3688.4 |
| OPT+path20 | -O3 -march=opteron -m32 | 2819.3 | 1673.1 |
| OPT+path20X | -O3 -march=opteron -m64 | 2634.5 | 1512.3 |
| OPT+ss10 | -fast -features=no%except -xtarget=opteron | 3657.8 | 2158.9 |
| USIV+gcc331 | -O2 | 9782.4 | 7845.4 |
| USIV+ss10 | -fast -xtarget=ultra3cu -xcache=64/32/4:8192/512/2 | 7749.9 | 5198.0 |
| USIV+guide | -fast +K3 -xipo=2 -xtarget=ultra4 -xcache=64/32/4:8192/128/2 -lmtmalloc | 7551.0 | 5335.0 |
| POW4+guide | +K3 -backend -qhot -backend -O3 -backend -g [-bmaxdata:0x80000000] | 5251.9 | 2819.4 |
| POW4+xlC60 | compilation fails | n.a. | n.a. |
| POW4+gcc343 | -O2 -maix64 -mpowerpc64 | 3193.7 | 2326.0 |
| IT2+icc81 | -O3 -ip -g | 9479.0 | 5182.8 |

Such a loop construct cannot be parallelized in OpenMP, because the loop iteration variable is not of type integer. Therefore the pointers of the iterators are stored in an array in an additional loop, so that afterwards a simpler loop running over the elements of this array can be parallelized.

Reducing the usage of the map STL datatype during the stiffness matrix setup as described in chapter 3 turned out to cause additional complexity and memory requirements in the parallel version. In the parallel version each thread fills a private temporary container consisting of one map per matrix row. The structure of the complete stiffness matrix has to be determined, which can be parallelized over the matrix rows. The master thread then allocates the valarray STL objects. Finally, the matrix rows are summed up in parallel.

If the structure of the stiffness matrix does not change, each thread fills a private temporary container consisting of one valarray of the same size as the array *val* of the final matrix.

This causes massive scalability problems for the guidec++-compiler. Its STL library obviously uses critical regions to be threadsafe. Furthermore the guidec++ employs an additional allocator for small objects which adds more overhead. Therefore we implemented a special allocator and linked to the Sun-specific memory allocation library *mtmalloc* which is tuned for multithreaded applications to overcome this problem.

### 4.2 The Linear Equation Solvers

In order to parallelize the PCG- and GMRES-method, matrix and vector operations, which beforehand had been implemented using operator overloading, had to be rewritten with C-style `for` loops with direct access to the structure elements. Thereby some synchronizations could be avoided and some parallelized `for`-loops could be merged.

The parallelized linear equation solvers including the sparse-matrix-vector-multiplication scale quite well, except for the intrinsic sequential structure of the Gauss-Seidel preconditioner which can only be partially parallelized. Rearranging the operations in a blocking scheme improves the scalability (`omp_block`) but still introduces additional organization and synchronization overhead.

A modified parallelizable preconditioner (`jac0`) was implemented which affects the numerical behavior. It leads to an increase in iterations to fulfill the convergence criterium. Nevertheless it leads to an overall improvement with four or more threads.

The straight-forward parallelization of the sparse matrix vector multiplication turned out to have a load imbalance. Obviously the nonzero elements are not equally distributed over the rows. The load balancing could be easily improved by setting the loop scheduling to `SCHEDULE(STATIC,128)`.

### 4.3 Compilers

Unfortunately not all of the available OpenMP-aware compilers were able to successfully compile the final OpenMP code version. Table 3 gives a survey of how successful the compilers have been.

Only the GNU C++ and the Pathscale C++ compilers were able to compile the DROPS code without any source modifications. Unfortunately the GNU C++ compiler does not support OpenMP, and the Pathscale C++ compiler currently does not support OpenMP in conjunction with some C++ constructs.

The Intel C++ compiler does not respect that a `valarray` is guaranteed to be filled with zero after construction. This is necessary for DROPS working correctly, so we changed the declaration by explicitly forcing a zero-filled construction.

In all cases marked with an (ok) modifications were necessary to get the serial DROPS code to compile and run.

Table 3. Compiler's successes

| code | DROPS serial | OpenMP support | DROPS parallel |
|---|---|---|---|
| XEON+gcc333 | ok | no | n.a. |
| XEON+gcc343 | ok | no | n.a. |
| XEON+icc81 | (ok) | yes | ok |
| XEON+pgi60 | (ok) | yes | compilation fails |
| XEON+vs2005 | compilation fails | yes | compilation fails |
| OPT+gcc333 | ok | no | n.a. |
| OPT+gcc333X | ok | no | n.a. |
| OPT+icc81 | (ok) | yes | ok |
| OPT+icc81X | (ok) | yes | compilation fails |
| OPT+pgi60 | (ok) | yes | compilation fails |
| OPT+pgi60X | (ok) | yes | compilation fails |
| OPT+path20 | ok | no | n.a. |
| OPT+path20X | ok | no | n.a. |
| OPT+ss10 | (ok) | yes | compilation fails |
| USIV+gcc331 | ok | no | n.a. |
| USIV+ss10 | (ok) | yes | ok |
| USIV+guide | (ok) | yes | ok |
| POW4+guide | (ok) | yes | ok |
| POW4+xlC60 | compilation fails | yes | compilation fails |
| POW4+gcc343 | ok | no | n.a. |
| IT2+icc81 | (ok) | yes | 1 thread only |

## 5 Performance of the OpenMP Version

OpenMP programs running on big server machines operating in multi-user mode suffer from a high variation in runtime. Thus it is hard to see clear trends concerning speed-up. This was particularly true for the SGI Altix. Exclusive access to the 24 core Sun Fire E2900 system helped a lot.

On the 4-way Opteron systems the *taskset* Linux command was helpful to get rid of negative process scheduling effects.

### 5.1 Assembly of the Stiffness Matrices

Setting up the stiffness matrices could be completely parallelized as described in the previous chapter. Nevertheless the scalability of the chosen approach is limited. The parallel algorithm executed with only one thread clearly performs worse than the tuned serial version, because the parallel algorithm contains the additional summation step as described above (see 4.1). It scales well up to about 8 threads, but then the overhead which is caused by a growing number of dynamic memory allocations and memory copy operations increases. On the USIV+ss10 platform there is still some speedup with more threads, but on the USIV+guide platform we had to limit the number of threads used for the SETUP routines to a maximum of eight in order to prevent a performance decrease for a higher thread count (table 7 and 8). Table 4 shows the runtime of the matrix setup routines on the USIV+guide platform.

Table 4. C++ + OpenMP: matrix setup

| code | serial original | serial tuned | parallel (jac0) | | | | |
|---|---|---|---|---|---|---|---|
| | | | 1 | 2 | 4 | 8 | 16 |
| XEON+icc81 | 1592 | 816 | 1106 | 733 | 577 | n.a. | n.a. |
| OPT+icc81 | 1368 | 778 | 1007 | 633 | 406 | n.a. | n.a. |
| USIV+guide | 4512 | 2246 | 2389 | 1308 | 745 | 450 | 460 |
| USIV+ss10 | 4604 | 2081 | 2658 | 1445 | 820 | 523 | 383 |
| POW4+guide | 4580 | 2119 | 2215 | 2285 | 3659 | 4726 | 5995 |

## 5.2 The Linear Equation Solvers

The linear equation solvers put quite some pressure on the memory system. This clearly reveals the memory bandwidth bottleneck of the dual processor Intel-based machines (XEON+icc).

The ccNUMA-architecture of the Opteron-based machines (OPT+icc) exhibits a high memory bandwidth if the data is properly allocated. But it turns out that the OpenMP version of DROPS suffers from the fact that most of the data is allocated by the master thread because of the usage of the STL datatypes.

As an experiment we implemented a modification of the stream benchmark using the STL datatype `valarray` on one hand and simple C-style arrays on the other hand. These arrays are allocated with *malloc* and initialized in a parallel region.

Table 5 lists the memory bandwidth in GB/s for the four stream kernel loops and a varying number of threads. It is obvious that the memory bandwidth does not scale when `valarrays` are used. The master thread allocates and initializes (after construction a `valarray` has to be filled with zeros by default) a contiguous memory range for the `valarray` and because of the first touch memory allocation policy, all memory pages are put close to the master thread's processor. Later on, all other threads have to access the master thread's memory in parallel regions thus causing a severe bottleneck.

The Linux operating system currently does not allow an explicit or automatic data migration. The Solaris operating system offers the Memory Placement Optimization feature (MPO), which can be used for an explicit data migration. In our experiment we measured the Stream kernels using `valarrays` after the data has been migrated by a "next-touch" mechanism using the *madvise* runtime function, which clearly improves parallel performance (see table 5).

This little test demonstrates how sensitive the Opteron architecture reacts to disadvantageous memory allocation and how a "next-touch" mechanism can be employed beneficially.

On the USIV+guide and USIV+ss10 platforms we were able to exploit the MPO feature of Solaris to improve the performance of DROPS, but currently there is no C++ compiler available for Solaris on Opteron capable of compiling the parallel version of DROPS.

**Table 5.** Stream benchmark, C++ (valarray) vs. C, memory bandwidth in GB/s on OPT+ss10

| Stream kernel | Data structure | Initialization method | 1 Thread | 2 Threads | 3 Threads | 4 Threads |
|---|---|---|---|---|---|---|
| assignment | valarray | implicit | 1.60 | 1.84 | 1.94 | 1.79 |
| | valarray | implicit+madvise | 1.60 | 3.19 | 4.78 | 6.36 |
| | C-array | explicit parallel | 1.69 | 3.35 | 5.00 | 6.64 |
| scaling | valarray | implicit | 1.51 | 1.81 | 1.93 | 1.78 |
| | valarray | implicit+madvise | 1.50 | 2.98 | 4.47 | 5.94 |
| | C-array | explicit parallel | 1.62 | 3.22 | 4.81 | 6.38 |
| summing | valarray | implicit | 2.12 | 2.16 | 2.16 | 2.03 |
| | valarray | implicit+madvise | 2.12 | 4.20 | 6.22 | 8.22 |
| | C-array | explicit parallel | 2.19 | 4.34 | 6.42 | 8.49 |
| saxpying | valarray | implicit | 2.11 | 2.16 | 2.15 | 2.03 |
| | valarray | implicit+madvise | 2.10 | 4.18 | 6.20 | 8.20 |
| | C-array | explicit parallel | 2.15 | 4.26 | 6.30 | 8.34 |

On the whole the linear equation solvers scale reasonably well given that frequent synchronizations in the CG-type linear equation solvers are inevitable. The modified preconditioner takes more time than the original recursive algorithm for few threads, but it pays off for at least four threads. Table 6 shows the runtime of the solvers.

**Table 6.** C++ + OpenMP: linear equation solvers

| code | serial original | serial tuned | parallel (omp_block) | | | | | parallel (jac0) | | | | |
|---|---|---|---|---|---|---|---|---|---|---|---|---|
| | | | 1 | 2 | 4 | 8 | 16 | 1 | 2 | 4 | 8 | 16 |
| XEON+icc81 | 939 | 894 | 746 | 593 | 780 | n.a. | n.a. | 837 | 750 | 975 | n.a. | n.a. |
| OPT+icc81 | 1007 | 839 | 823 | 590 | 496 | n.a. | n.a. | 699 | 526 | 466 | n.a. | n.a. |
| USIV+guide | 2682 | 2727 | 2702 | 1553 | 1091 | 957 | 878 | 1563 | 902 | 524 | 320 | 232 |
| USIV+ss10 | 2741 | 2724 | 2968 | 1672 | 1162 | 964 | 898 | 2567 | 1411 | 759 | 435 | 281 |
| POW4+guide | 398 | 428 | 815 | 417 | 333 | 1171 | 18930 | 747 | 267 | 308 | 12268 | 37142 |

### 5.3 Total Performance

Table 7 shows the total runtime of the DROPS code on all platforms for which a parallel OpenMP version could be built. Please note that we didn't have exclusive access to the POW4 platform. Table 8 shows the resulting total speedup.

**Table 7.** C++ + OpenMP: total runtime

| code | serial original | serial tuned | parallel (omp_block) 1 | 2 | 4 | 8 | 16 | parallel (jac0) 1 | 2 | 4 | 8 | 16 |
|---|---|---|---|---|---|---|---|---|---|---|---|---|
| XEON+icc81 | 2643 | 1723 | 2001 | 1374 | 1353 | n.a. | n.a. | 2022 | 1511 | 1539 | n.a. | n.a. |
| OPT+icc81 | 2517 | 1761 | 2081 | 1431 | 1093 | n.a. | n.a. | 1962 | 1382 | 1048 | n.a. | n.a. |
| USIV+guide | 7551 | 5335 | 5598 | 3374 | 2319 | 1890 | 1796 | 4389 | 2659 | 1746 | 1229 | 1134 |
| USIV+ss10 | 7750 | 5198 | 6177 | 3629 | 2488 | 2001 | 1782 | 5683 | 3324 | 2067 | 1457 | 1151 |
| POW4+guide | 5252 | 2819 | 3467 | 3310 | 4534 | 7073 | 26037 | 3290 | 2871 | 4338 | 17465 | 43745 |

**Table 8.** Speedup for the USIV+guide, USIV+ss10 and OPT+icc platforms

| Version | USIV+guide omp_block | jac0 | USIV+ss10 omp_block | jac0 | OPT+icc omp_block | jac0 |
|---|---|---|---|---|---|---|
| serial (original) | 1.00 | — | 1.00 | — | 1.00 | — |
| serial (tuned) | 1.42 | — | 1.49 | — | 1.43 | — |
| parallel (1 Thread) | 1.35 | 1.72 | 1.26 | 1.36 | 1.21 | 1.28 |
| parallel (2 Threads) | 2.24 | 2.84 | 2.14 | 2.33 | 1.76 | 1.82 |
| parallel (4 Threads) | 3.26 | 4.32 | 3.11 | 3.75 | 2.30 | 2.40 |
| parallel (8 Threads) | 3.99 | 6.14 | 3.87 | 5.32 | — | — |
| parallel (16 Threads) | 4.20 | 6.66 | 4.35 | 6.73 | — | — |

# 6 Summary

The compute intense program parts of the DROPS Navier-Stokes solver have been tuned and parallelized with OpenMP. The heavy usage of templates in this C++ program package is a challenge for many compilers. As not all C++ compilers support OpenMP, and some of those which do fail for the parallel version of DROPS, the number of suitable platforms turned out to be quite limited.

We ended up with using the guidec++ compiler from KAI (which is now part of Intel) and the Sun Studio 10 compilers on our UltraSPARC IV-based Sun Fire servers (platform USIV+guide) and the Intel compiler in 32 bit mode on our Opteron-based Linux cluster (OPT+icc).

The strategy which we used for the parallelization of the Finite Element Method implemented in DROPS was straight forward. Nevertheless the obstacles which we encountered were manifold, many of them are not new to OpenMP programmers.

Finally the USIV+guide and USIV+ss10 versions exhibit some scalability. The best effort OpenMP version runs 6.7 times faster with 16 threads than the original serial version on the same platform. But as we improved the serial version during the tuning and parallelization process the speed-up compared to the tuned serial version is only 4.7.

As an Opteron processor outperforms a single UltraSPARC IV processor core it only takes 3 threads on the Opteron-based machines to reach the same absolute speed. On the other hand Opteron processors are not available in large shared memory machines. So shorter elapsed times are not attainable.

As tuning is a never ending process, there still is room for improvement. Particularly the data locality has to be improved for the ccNUMA-architecture of the 4-way Opteron machines.

## Acknowledgements

The authors would like to thank Uwe Mordhorst at University of Kiel and Bernd Mohr at Research Center Jülich for granting access to and supporting the usage of their machines, an SGI Altix 3700 and an IBM p690.

## References

1. Reusken, A., Reichelt, V.: Multigrid Methods for the Numerical Simulation of Reactive Multiphase Fluid Flow Models (DROPS), http://www.sfb540.rwth-aachen.de/Projects/tpb4.php
2. Gross, S., Peters, J., Reichelt, V., Reusken, A.: The DROPS package for numerical simulations of incompressible flows using parallel adaptive multigrid techniques, ftp://ftp.igpm.rwth-aachen.de/pub/reports/pdf/IGPM211_N.pdf
3. GNU Compiler documentation, http://gcc.gnu.org/onlinedocs/
4. Intel C/C++ Compiler documentation, http://support.intel.com/support/performancetools/c/linux/manual.htm
5. Guide-Compiler of the KAP Pro/Toolset, http://support.rz.rwth-aachen.de/Manuals/KAI/KAP_Pro_Reference.pdf, http://developer.intel.com/software/products/kappro/
6. PGI-Compiler documentation, http://www.pgroup.com/resources/docs.htm
7. KCC-Compiler, component of guidec++, http://support.rz.rwth-aachen.de/Manuals/KAI/KCC_docs/index.html
8. Pathscale-Compiler, http://www.pathscale.com
9. Sun Analyzer of Sun Studio 9, http://developers.sun.com/tools/cc/documentation/ss9_docs/
10. Intel Threading Tools, http://www.intel.com/software/products/threading/
11. Karlsson, S., Brorsson, M.: OdinMP OpenMP C/C++ Compiler, http://odinmp.imit.kth.se/projects/odinmp

# A Parallel Structured Ecological Model for High End Shared Memory Computers

Dali Wang[1], Michael W. Berry[1], and Louis J. Gross[2]

[1] Department of Computer Science,
203 Claxton Complex,
University of Tennessee, Knoxville, TN 37996
{dwang,berry}@cs.utk.edu
[2] The Institute for Environmental Modeling,
569 Dabney Hall,
University of Tennessee, Knoxville, TN 37996
gross@tiem.utk.edu

**Abstract.** This paper presents a new approach to parallelize spatially-explicit structured ecological models. Previous investigations have mainly focused on the use of spatial decomposition for parallelization of these models. Here, we exploit the partitioning of species age structures (or layers) as part of an integrated ecosystem simulation on a high-end shared memory computer using OpenMP. As an example, we use a parallel spatially-explicit structured fish model (ALFISH) for regional ecosystem restoration to demonstrate the parallelization procedure and associated model performance evaluation. Identical simulation results, validated by a comparison with a sequential implementation, and impressive parallel model performance demonstrate that layer-wised partitioning offers advantages in parallelizing structured ecological models on high-end shared memory computers. The average execution time of the parallel ALFISH model, using 1 computational thread, is about 11 hours, while the execution of the parallel ALFISH model using 25 computational threads is about 39 minutes (the speedup factor being about 16).

## 1 Introduction

Recently, emphasis on integrated multi-component ecosystem modeling, involving complex interactions between some or all of the trophic layers, has increased, resulting in coupled ecosystem models with large numbers of state variables. Current efforts in integrated ecosystem modeling have led to the realization that model and software development utilizing only a single approach within a traditional computing framework can hinder innovation of complex highly-integrated model simulation involving diverse spatial, temporal and organismal scales. The Across Trophic Level System Simulation (ATLSS) [1] is an example of a new type of spatially-explicit ecosystem modeling package that utilizes different modeling approaches or ecological multimodeling [2]. In this paper, we focus on the parallelization of an age-structured population model, within the context of ATLSS, for freshwater fish functional groups (ALFISH) in South Florida.

ALFISH is an Intermediate Trophic Level Functional Groups model which includes two main subgroups (small planktivorous fish and large piscivirous fish), structured by size. In the complex integrated system of ATLSS, ALFISH is an important link to the higher-level landscape models, since it provides a food base for several wading bird models [3]. An objective of the ALFISH model is to compare, in a spatially explicit manner, the relative effects of alternative hydrologic scenarios on fresh-water fish densities across South Florida. Another objective is to provide a measure of dynamic, spatially-explicit food resources available to wading birds.

The ALFISH model has been developed in part to integrate with two wading bird models: a proxy individual-based wading bird (WB) model [4], and a Spatially-Explicit Species Index (SESI) wading bird model [5]. Major concerns associated with integrating ALFISH into the ATLSS architecture include its long runtime. The average runtime is 35 hours on a 400MHz (Ultra Sparc II-based) Sun Enterprise 4500 for a typical 31-year simulation. One objective of this parallelization is to speedup the ALFISH execution for its practical use within ATLSS by allowing real-time data generation and information sharing between multiple ATLSS models, including the individual-based wading bird model.

## 2 Parallelization Strategy

A successful approach to the parallelization of landscape based (spatially-explicit) fish models is spatial decomposition [6,7]. For these cases, each processor only simulates the ecological behaviors of fish on a partial landscape. This approach is efficient in stand-alone fish simulations because the low movement capability of fish does not force large data movement between processors [8]. However, in an integrated simulation with an individual-based wading bird model, intensive data immigration across all processors is inevitable, since a bird's flying distance may cover the whole landscape. In this paper, we present an age-structured decomposition (or layer-wised partition). As opposed to a spatial decomposition, this approach partitions the computational domain along the fish age-structure, so that each processor computes the dynamics of fish at certain ages on the whole landscape. This parallel strategy is suitable for shared-memory computational platforms, where the behind-the-scenes data exchanges between processors have been highly optimized by the hardware and underlying operating system through related library routines and directives.

## 3 Computational Platform and Parallel Programming Model

The computational platform used in this research is a 256-processor SGI Altix system (Ram) at the Center for Computational Sciences (CCS) of Oak Ridge National Laboratory. Ram is unique in the CCS in that it has a very large, shared memory. Ram is comprised of 256 Intel Itanium2 processors running at

1.5 GHz, each with 6 MB of L3 cache, 256K of L2 cache, and 32K of L1 cache. Ram also has 8 GB of memory per processor for a total of 2 Terabytes of total system memory. This system has a theoretical total peak performance of 1.5 TeraFLOP/s. The operating system on Ram is a 64-bit version of Linux. The parallel programming model in this research is supported by the multithreaded application programming interface referred to as OpenMP [9].

## 4 Functionality and Parallel Implementation of ALFISH Model

### 4.1 Model Structure

The study area for ALFISH modeling contains 26 regions as determined by the South Florida Water Management Model [10]. A complete list of these regions is provided in Fig. 1.

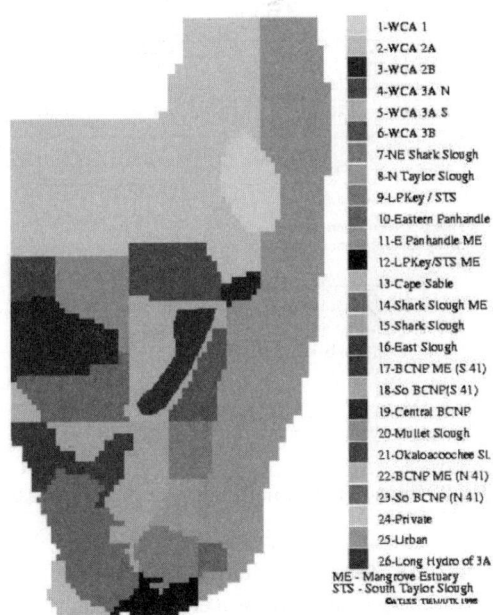

**Fig. 1.** Subregions used by the ALFISH model

The total area of Everglades modeled in ALFISH contains approximately 111,000 landscape cells, with each cell 500m on a side. Each landscape cell has two basic types of area: marsh and pond. The difference between marsh and pond areas is that the latter is always considered wet (contains standing water) regardless of any available water data. In the marsh area of each cell, there is a distribution of elevations based upon a hypsograph [3]. This hypsograph is used to determine the fraction of the marsh area that is still under water at a given water depth. A pond refers to permanently wet areas of small size, such as alligator holes, which are a maximum of 50 $m^2$ or 0.02% of the cell.

The fish population simulated by ALFISH is size-structured and is divided into two functional groups: small and large fishes. Both of these groups appear in each of the marsh and pond areas. Each functional group in each area is further divided into several fish categories according to age, referred to as ageClass, and each age class has 6 size classes, referred to as sizeClass. The fish population in each cell is summarized by the total fish density (biomass) within that cell. Each cell, as an element of the landscape matrices, contains an array of floating-point numbers representing individual fish density of various age classes. The length of the array corresponds to the number of age classes for that functional group. Normally, when a fish density is referenced, the value reflects the total fish densities of all the fish age classes combined. Fig. 2 shows the simple age and size categorization of the fish functional groups in four landscape matrices.

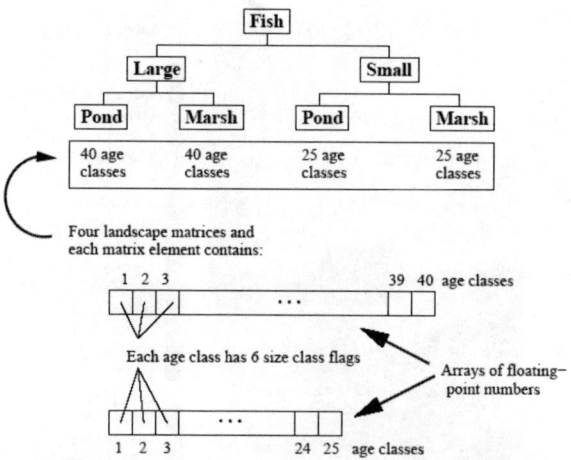

**Fig. 2.** Simple fish functional group categorization

In ALFISH, spatial and temporal fluctuations in fish populations are driven by a number of factors, especially the water level. Fluctuations in water depth, which affect all aspects of the trophic structure in the Everglades area, are provided through an input hydrology data file for each timestep throughout the execution of the model.

### 4.2 Layer-Wised Partition

To support the parallel computation of fish population dynamics in all age classes, two data structures (*groupInfo* and *layerInfo*) are introduced to support layer-wised partitioning.

```
Struct layerInfo        // information on fish function groups
  { int Ngroup;         // number of function group
    int size[5];        // size of each function group
  };

Struct groupInfo        // information on layer-wised partition
  { int Ngroup;         // number of fish function group on
                        //   this partition domain
    int group[2];       // size of each fish function group
                        //   on this domain
    int position[10]    // start and end layer position of
                        //   fish function group
  };
```

In our case, *layerInfo.Ngroup* is 2, and *layerInfo.size* is set as [25 40]. We partition all 65 age classes (including 25 classes for small fish and 40 classes for large fish) into all processors, and use the structure *groupInfo* to store appropriate information. Fig. 4 illustrates the initial layer-wised partition across 3 processors.

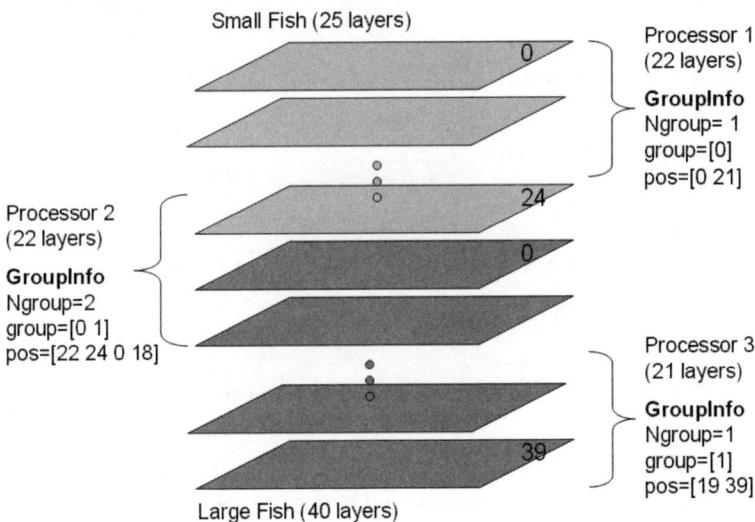

**Fig. 3.** Layer-wise partition across 3 processors

Initially, the value of position in *groupInfo* is equivalent to the value of age classes. However, every 30 days (6 timesteps of 5-days each), all of the fish are moved to the next age class. Thus, we introduce an *ageOffset* variable to eliminate the data movement involved in the sequential implementation (more details are provided in Sect. 4.3).

## 4.3 Fish Dynamics and Parallelization

The fish population model simulated by ALFISH is size-structured and is divided into two functional groups: small fish and large fish. Both of these groups are used in each of the marsh and pond areas. Each functional group in each area is further divided into several fish categories according to size. The fish population that occupies a cell area is represented as the fish density (*biomass*) within that cell. Basic behaviors of fish, including escape, diffusive movement, mortality, aging, reproduction and growth, are simulated in the model. Beside the fish dynamics, the model has to update the lower trophic level (food resources for the fish) data, hydrological data and execute a certain amount of I/O operations at each timestep. The parallel ALFISH deploys a master-slave communication model, which is illustrated in Fig. 4.

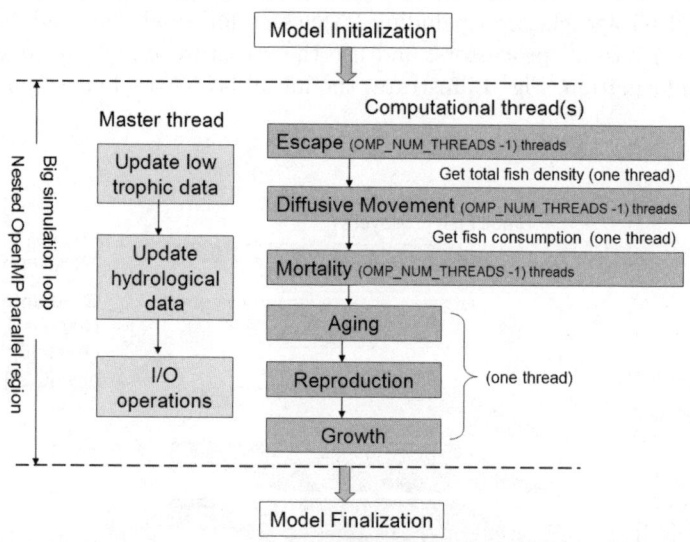

**Fig. 4.** Computational Model of Parallel ALFISH

There are some advantages associated with this implementation. From a performance aspect, the time-consuming I/O operations are separated from the much faster computational phases, which increase the efficiency of data access during the simulation. More importantly, the master thread provides a uniform, transparent interface for integration with other ATLSS ecological models. The main task of the master thread is to collect the data from computational threads, and write the results to the local disk. Therefore, the time-consuming output operations on the master processor are no longer a key component of the total execution time. All the computational threads are dedicated to simulate the fish dynamics. Since it is a multithreaded code, no explicit data exchange is needed during simulation. Data sharing in this implementation is supported by globally accessing the same memory location. The synchronizations required within each computational step are guaranteed by either explicit OMP_BARRIER or implicit barriers associated with the OpenMP parallel code sections. In this model, all the computational threads (OMP_NUM_THREADS-1) were used to simulate computational intensive fish behaviors (i.e., *Escape, Diffusive Movement* and *Mortality*), while only one thread was used to simulate other fish behaviors (i.e., *Aging, Reproduction* and *Growth*). Considering the efficiency of thread creation and join, those operations are deployed within the computational loop. More details on the nested OpenMP parallel region are provided in the Appendix.

**Escape.** This function is designed to simulate the fish movement between marsh and pond areas within each cell caused by drying out or reflooding. At the current timestep, if the water depth of the cell has increased, and the cell has a pond, an appropriate fraction of fish (of the sizes previously too large) is moved from pond areas into marsh areas. If the cell water has decreased, an appropriate fraction of those fish are moved from the marsh area into pond area, another fraction of those fish are moved to adjacent cells, and the remaining portion of those fish are eliminated, referred as dying-out mortality [11]. Following this, one thread collects the total fish density in both marsh and pond areas, which will be used in the following diffusive movements.

**Diffusive Movement.** This function is designed to simulate the movement of fish between adjacent cells mainly due to the relative differences in water depth and fish densities. The mathematical formula used to determine the number of fish to be moved is not presented here (see [3,11]). Since this kind of fish movement is density dependent, the fish landscape matrix is only updated after all movement calculations are complete, to remove any order-based bias. After that, one computational thread summarizes the amount of biomass needed (*req_biomass*) for every fish to grow from its current size class to the next. At this stage, different parameters were classified to describe the bio-consumption related to fish function group (small and large fish) and the fish size. The *req_biomass* in each landscape will be used to calculate the fish mortality.

**Mortality.** Besides the mortality caused by the drying out of a landscape cell (see Sect. 4.3), the ALFISH model also simulates background mortality (*AgeMortality*) and density-dependent mortality (*FoodMortality*) in both marsh and pond areas in each landscape cell. *AgeMortality* is dependent on the individual fish age class, but it is independent of population size. *FoodMortality* is due to starvation. As this density of available prey decreases, the mortality rate of that specific age class and functional group increases. It is assumed that the starvation affects all age classes equally. In ALFISH, these two types of mortality are compared and the greater one is applied to the population. The mathematical formula to determine those fish mortalities are presented in [3,11].

**Aging.** The age classes for the fish functional groups are defined by 30-day intervals. Every 30 days (6 timesteps of 5-days each), all of the fish are moved to the next age class. We do not use the mathematical formula defined in the sequential implementation, which would require extra date movement in parallelization. Rather, an ageOffset variable affiliated with each partition was introduced. The value *ageOffset* was initialized as 0, and every 30 days, *ageOffset* is increased by 1. Therefore, the value of the fish age can be derived using: $age = (position + ageOffset)$ % $nAge$, where *age* is the value of fish age classes, position refers to the value of *groupInfo.position* and *nAge* is the total number of age classes of a particular fish functional group. For small and large fish, the values of $nAge$ are 25 and 40, respectively.

**Reproduction.** For each functional group, if it is the appropriate time of year, the number of offspring is calculated. To prevent the population from producing too many new fish in a reproductive event, a constant maximum reproduction density is used. The new fish population from this stage is collected and used to update the fish population at position (layer) $p$, where the value of $[(p+ageOffset)$ % $nAge]$ equals 0.

**Growth,** Finally, fish that have survived are moved into the next size class since the size classes are equivalent to the 5-day timestep. All fish with an age class advance synchronously in size class, with the size class $s$ incremented using $s=(s$ % $6)+1$.

## 5 Selected Model Results and Performance

### 5.1 Scenarios

The ALFISH models are mainly used to determine the pattern of fish density on the landscape for a variety of hydrology scenarios. The motivation for the particular scenarios chosen was the Restudy process for the selection of a plan for Everglades restoration [12]. In this paper, one scenario referred to as F2050 was applied. F2050 is a standard base scenario, which uses water data based on a 31-year time series of historical rainfall from 1965 through 1995, as well as sea level, population level and socioeconomic conditions projected for the year 2050. It also includes all of the previously legislated structural changes for the water control measures. Therefore, the simulation time of both the sequential and parallel ALFISH models is 31 years, from 1965 to 1995 using a timestep of 5 days.

### 5.2 Comparison of Selected Outputs

To verify parallel model correctness, that is, its ability to produce results similar to those of the sequential model, we compared outputs of both the sequential and parallel models. We analyzed several outputs and selected one set of data for comparison (the 31-year mean fish density and distribution on October 1). Fig. 5 shows the mean fish density map comparison on October 1. The left graph represents the output from the parallel ALFISH model, and the right graph is the output from the sequential ALFISH model. There are no observable differences between the outputs of these two models.

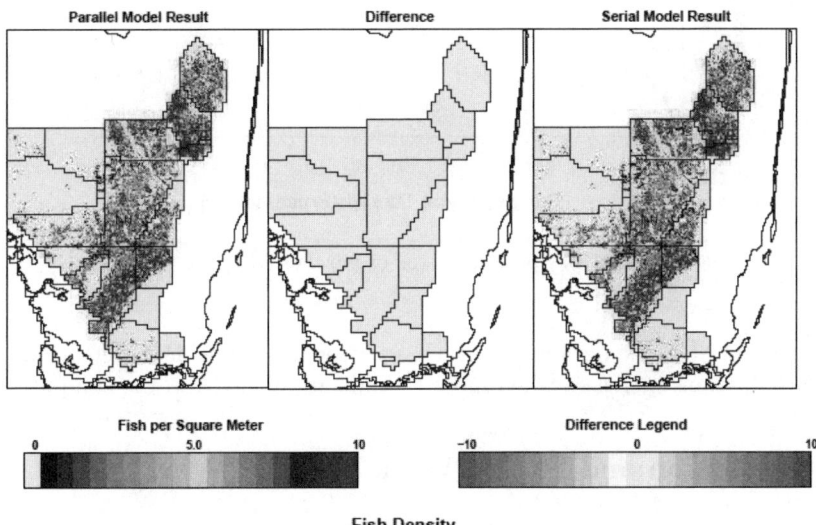

**Fig. 5.** Spatial 31-year average fish density map comparison in Everglades on Oct 1

### 5.3 Model Performance

In order to measure the scalability of the parallel ALFISH model, we deployed a series of parallel simulations using different numbers of threads (processors) (OMP_NUM_THREADS ranges from 2 to 66). Herein we define speedup ($S$) as

$$Speedup(S) = \frac{Model\ Execution\ Time(using\ 1\ computational\ thread)}{Model\ Execution\ Time(using\ N\ computational\ threads)} \quad (1)$$

Fig. 6 shows the speedup factor of the parallel ALFISH. Although static partitioning is applied in the model, the parallel ALFISH demonstrates satisfactory scalability when the number of computational threads is less than 25. The average execution time of the parallel ALFISH, using 1 computational thread, is about 11 hours, while the execution of the parallel ALFISH using 25 computational threads is about 39 minutes (the speedup factor being about 16). The speedup factor decreases when more than 25 computational threads were used, since ($i$) single thread execution portion within each computational loop is

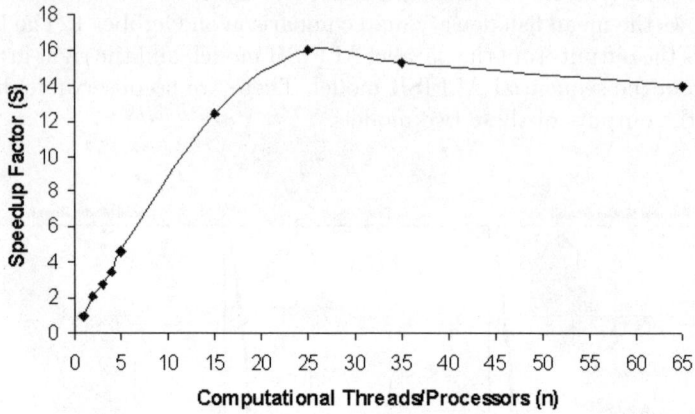

**Fig. 6.** Speedup factor of parallel ALFISH

constant, and (ii) the parallelism overhead associated with thread creations and synchronizations increases when more threads are involved in the computation.

## 6 Discussion and Future Work

The nearly identical outputs and excellent speed improvement (especially when the number of computational threads is less than 25) obtained from the parallel ALFISH model provide strong evidence that layer-wised partitioning can be highly effective for age- and size-structured spatially explicit landscape fish models on high-end shared memory computers using OpenMP. Our results indicate that even with simple static partitioning, the parallel ALFISH model demonstrates satisfactory scalability. In this paper, we adapted a one-dimensional layer-wised partitioning, which minimized the potential data sharing between computational threads, and allowed simple parallel implementation using OpenMP nested parallel sections. From the user's perspective, more OpenMP features related to thread management will be appreciated. We are now developing a hybrid, reconfigurable two-dimensional partitioning (using both landscape (spatial) decomposition and age structure decomposition (layer-wised partition)) using a hybrid MPI/OpenMP model. Further plans for ALFISH model development include integration with other ATLSS models, (including an individual-based wading bird model) for parallel multi-scale ecosystem simulation on high performance computational platforms via grid computing [13,14].

## Acknowledgements

This research has been supported by the National Science Foundation under grant No. DEB-0219269. The sequential implementation was developed with support from the U.S. Geological Survey, through Cooperative Agreement No. 1445-CA09-95-0094 with the University of Tennessee, and the Department of Interior's Critical Ecosystem Studies Initiative. Authors also appreciate the support via the Computational Science Initiatives through the Science Alliance at the University of Tennessee (UT) and Oak Ridge National Libratory (ORNL). This research used resources of the Center for Computational Sciences at Oak Ridge National Laboratory, which is supported by the Office of Science of the Department of Energy under Contract DE-AC05-00OR22725.

## References

1. ATLSS: Across Trophic Level System Simulation., http://www.atlss.org
2. Gross, L., DeAngelis, D.: Multimodeling: New Approaches for Linking Ecological Models. In: Scott, J.M., Heglund, P.J., Morrison, M.L. (eds.) Predicting Species Occurrences: Issues of Accuracy and Scale, pp. 467–474 (2002)
3. Gaff, H., DeAngelis, D., Gross, L., Salinas, R., Shorrosh, M.: A Dynamic Landscape Model for Fish in the Everglades and its Application to Restoration. Ecological Modeling 127, 33–53 (2000)
4. Wolff, W.F.: An Individual-Oriented Model of a Wading Bird Nesting Colony. Ecological Modelling 114, 72–75 (1994)
5. Curnutt, J.L., Comiskey, E.J., Nott, M.P., Gross, L.J.: Landscape-based spatially explicit species index models for Everglade restoration. Ecological Applications 10, 1849–1860 (2000)
6. Wang, D., Gross, L., Carr, E., Berry, M.: Design and Implementation of a Parallel Fish Model for South Florida. In: Proceedings of the 37th Annual Hawaii International Conference on System Sciences (HICSS 2004) (2004), http://csdl.computer.org/comp/proceedings/hicss/2004/2056/09/205690282c.pdf
7. Immanuel, A., Berry, M., Gross, L., Palmer, M., Wang, D.: A parallel Implementation of ALFISH: Compartmentalization Effects on Fish Dynamics in the Florida Everglades. Simulation Practice and Theory 13(1), 55–76 (2005)
8. Wang, D., Berry, M., Carr, E., Gross, L.: A Parallel Fish Landscape Model for Ecosystem Modeling. Simulation: Transactions of the Society of Modeling and Simulation International 82(7), 451–465 (2006)
9. OpenMP: Simple, Portable, Scalable SMP Programming, http://www.openmp.org
10. Fennema, R., Neidrauer, C., Johnson, R., MacVicar, T., Perkins, W.: A Computer Model to Simulate Everglades Hydrology. In: Davis, S.M., Ogden, J.C. (eds.) Everglades: The Ecosystem and Its Restoration, pp. 249–289. St. Lucie Press (1994)
11. Gaff, H., Chick, J., Trexler, J., DeAngelis, D., Gross, L., Salinas, R.: Evaluation of and insights from ALFISH: a spatial-explicit, landscape-level simulation of fish populations in the Everglades. Hydrobiologia vol. 520, pp. 73–87 (2004)
12. U.S. Army Corps of Engineers, C & SF Restudy Draft Feasibility Report (accessed, November 17, 1998), http://www.restudy.org/overview.pdf
13. Wang, D., Carr, E., Palmer, M., Berry, M., Gross, L.: A Grid Service Module for Natural Resource Managers. IEEE Internet Computing 9(1), 35–41 (2005)
14. Wang, D., Carr, E., Ecomiske, J., Plamer, M., Berry, M., Gross, L.: Grid Computing for Regional Ecosystem Restoration., http://www.sc-conference.org/sc2004-/schedule/index.php?module=Default\&action=ShowDetail\&eventid=131

# Appendix: Nested OpenMP Parallel Code Section

```
for (date=current_date; date<=end_date; date+=5) {
#pragma omp parallel private(rank, myInfo, group, start, end)
  { rank = omp_get_thread_num();
    if (rank != 0) {
        remapping(current_partition, layers, rank, myInfo);
        #pragma omp parallel {
            compute fish.escape functionality
            #pragma omp barrier // synchronize computational threads
            if (rank == 1)  getfishTotaldensity
            #pragma omp barrier // make sure to update shared data
            compute fish.move functionality
            #pragma omp barrier // synchronize computational threads
            if (rank == 1)  getfishConsumption
            #pragma omp barrier // make sure to update shared data
            compute fish.mortality functionality
            #pragma omp barrier // synchronize computational threads
            if (rank == 1)  compute fish.aging functionality
            #pragma omp barrier // make sure to update shared data
            if (it is the right time))
                compute fish.reproduction functionality
            #pragma omp barrier // synchronize computational threads
            if (rank == 1) {
                if (it is the right time)
                    compute fish.reproduction functionality
                compute fish.growth functionality
            }
        } // end of inner p-section (implicit synchronization)
    }
    #pragma omp barrier // block the master thread once
    if (rank ==0 ) {
        collect and save date at previous timestep
    }
} // end of external parallel section (implicit synchronization)
```

# Multi-cluster, Mixed-Mode Computational Modeling of Human Head Conductivity

Adnan Salman[1], Sergei Turovets[1], Allen D. Malony[1], and Vasily Volkov[2]

[1] NeuroInformatics Center, 5219 University of Oregon, Eugene, OR 97403, USA
(adnan,sergei,malony)@cs.uoregon.edu
[2] Institute of Mathematics, Academy of Sciences, 11 Surganov St, Minsk 220072, Belarus
volk@im.bas-net.by

**Abstract.** A multi-cluster computational environment with mixed-mode (MPI + OpenMP) parallelism for estimation of unknown regional electrical conductivities of the human head, based on realistic geometry from segmented MRI up to $256^3$ voxels resolution, is described. A finite difference multi-component alternating direction implicit (ADI) algorithm, parallelized using OpenMP, is used to solve the forward problem calculation describing the electrical field distribution throughout the head given known electrical sources. A simplex search in the multi-dimensional parameter space of tissue conductivities is conducted in parallel across a distributed system of heterogeneous computational resources. The theoretical and computational formulation of the problem is presented. Results from test studies based on the synthetic data are provided, comparing retrieved conductivities to known solutions from simulation. Performance statistics are also given showing both the scaling of the forward problem and the performance dynamics of the distributed search.

## 1 Introduction

The essence of most tomographic techniques is to determine unknown complex coefficients in PDEs governing the physics of the particular experimental modality. Such problems are typically non-linear and ill-poised. The first step in solving such an inverse problem is to find a numerical method to calculate the direct (*forward*) problem. When the physical model is three-dimensional and geometrically complex, the forward solution can be difficult to construct and compute. However, this is only the first stage of the tomographic solution. The second stage involves a search across a multi-dimensional parameter space of unknown (to be found) model properties. The search employs the forward problem with chosen parameter estimates and a function that determines the error of the forward calculation with an empirically measured result. As the error residuals of local inverse searches are minimized, the global search determines convergence to final property estimates based on its knowledge of how well the parameter space has been sampled.

Fundamental problems in neuroscience involving experimental modalities like electroencephalography (EEG) and magnetoencephalograpy (MEG) are naturally expressed as tomographic imaging problems. The difficult problems of *source localization* and *impedance imaging* require modeling and simulating the associated bioelectric fields.

Forward calculations are necessary in the computational formulation of these problems. Until recently, most practical research in this field has opted for analytical or semi-analytical models of a human head in the forward calculations [1,2]. This is in contrast to approaches that use realistic 3D head geometry for purposes of significantly improving the accuracy of the forward and inverse solutions. To do so, however, requires that the geometric information be available from MRI or CT scans. With such image data, the tissues of the head can be better segmented and more accurately represented in the computational model. Unfortunately, these realistic modeling techniques have intrinsic computational complexities that grow as the image resolution increases. This is the primary reason such techniques have not be used in the past.

In source localization we are interested in finding the electrical source generators for the potentials that might be measured by EEG electrodes on the scalp surface. Here, the inverse search is looking for those sources (their position and amplitude) on the cortex surface whose forward solution most accurately describes the electrical potentials observed. The computational formulation of the source localization problem assumes the forward calculation is without error. However, this assumption in turn assumes the conductivity values of the modeled head tissues are known. In general, for any individual, they are not known. Thus, the impedance imaging problem is actually a predecessor problem to source localization. In impedance imaging, the inverse search finds those tissue impedance values whose forward solution best matches measured scalp potentials when experimental stimuli are applied. In either problem, source localization or impedance imaging, solving the inverse search usually involves the large number of runs of the forward problem. Therefore, computational methods for the forward problem, which are stable, fast and eligible for parallelization, as well as intelligent strategies and techniques for multi-parameter search, are of paramount importance.

To deal with complex geometries, PDE solvers use finite element (FE) or finite difference (FD) methods [3,4]. The main computational idea behind these methods is to reduce a continuous problem with infinitely many unknown field values to a finite number of unknowns by discretizing the solution region into elements. Application of each of these approximation methods to the governing equations for the specific modality yields eventually a system of linear equations of the form $AX = b$, which must be solved to obtain the final solution. The solution techniques can be broadly categorized as direct and iterative solvers. The choice of the particular solution method is highly dependent upon the approximation technique employed to obtain the linear system, upon the size of the resulting system, and upon accessible computational resources.

Usually, for the geometry with the given complexity level, the FE methods are more economical in terms of the number of unknowns (the size of the stiffness matrix A, is smaller, as homogeneous segments do not need a dense mesh) and resulting computational cost. However, the FE mesh generation for a 3D, highly heterogeneous subject with irregular boundaries (e.g., the human brain) is a difficult task. The process involves a significant degree of preprocessing and smoothing of the initial geometry through manual means. A fully automated process of image segmentation and mesh generation is unavailable at present.

At the same time, the FD method with a regular cubed grid is generally the easiest method to code and implement. It is often chosen over FE methods for simplicity and the fact that MRI/CT segmentation map is also based on a cubed lattice of nodes. Therefore, meshes are relatively easy to construct (once segmentation is accomplished) as the cubic/rectangular elements can be "mapped" directly from the voxels of the medical images (3D MRI scans). Many anatomical details (e.g., olfactory perforations and internal auditory meatus) or structural defects in case of trauma (e.g., skull cracks and punctures) can be included as the computational load is based on the number of elements and not on the specifics of tissues differentiation. Thus, the model geometry accuracy can be the same as the resolution of MRI scans (e.g., $1 \times 1 \times 1mm$), while in the FEM approach, simplification of the geometry is unavoidable as a result of mesh generation. In addition, the multiscale (multigrid) strategy of calculations on a hierarchy of coarser grids (starting with $64 \times 64 \times 44$ and feeding the results into the next cycle of iterations on the finer grid) can be easily implemented in a FD forward solver. The FD grid can be made non-uniform and/or applied in the spherical coordinates to capture more details in the regions of interest.

In the present work we adopt a model based on FD methods and construct a heterogeneous distributed and mixed-mode parallel simulation environment for conductivity optimization through inverse simplex search. FE simulation [7] is used to solve for relatively simple phantom geometries that we then apply as "gold standards" for validation.

## 2 Mathematical Description of the Problem

The relevant frequency spectrum in EEG and MEG is typically below $1kHz$, and most studies deal with frequencies between 0.1 and $100Hz$. Therefore, the physics of EEG/MEG can be well described by the quasi-static approximation of Maxwell's equations, the Poisson equation. The electrical *forward problem* can be stated as follows: given the positions, orientations and magnitudes of current sources, as well as geometry and electrical conductivity of the head volume $\Omega$ calculate the distribution of the electrical potential on the surface of the head (scalp) $\Gamma_\Omega$. Mathematically, it means solving the linear Poisson equation:

$$\nabla \cdot \sigma(x,y,z)\nabla\phi(x,y,z) = S, \qquad (1)$$

in $\Omega$ with no-flux Neumann boundary conditions on the scalp:

$$\sigma(\nabla\phi) \cdot n = 0, \qquad (2)$$

on $\Gamma_\Omega$. Here $\sigma = \sigma_{ij}(x,y,z)$ is an inhomogeneous tensor of the head tissues conductivity and $S$ is the source current. Having computed potentials $\phi(x,y,z)$ and current densities $J = -\sigma(\nabla\phi)$, the magnetic field $B$ can be found through the Biot-Savart law. In this paper, we do not consider anisotropy or capacitance effects (the latter because the frequencies of interest are too small), but they can be included in a straightforward manner. Eq. (1) becomes complex-valued, and complex admittivity should be used.

We have built a finite difference forward problem solver for Eq. (1) and (2) based on the multi-component alternating directions implicit (ADI) algorithm [8,9]. It is a generalization of the classic ADI algorithm as described by Hielscher et al [6], but with improved stability in 3D (the multi-component FD ADI scheme is unconditionally

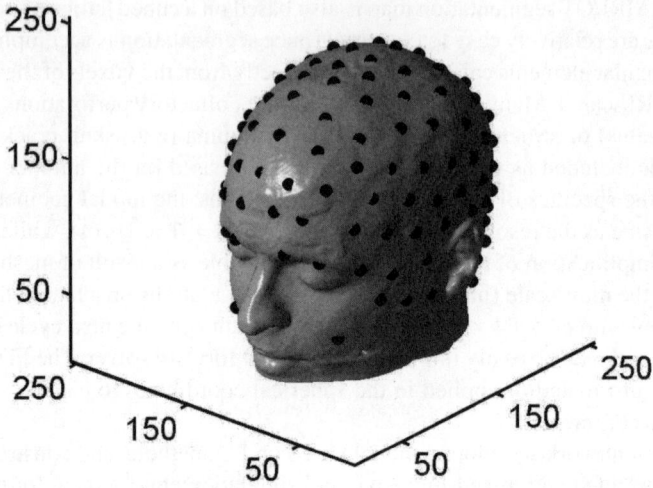

**Fig. 1.** A visualization of a 3D human head CT scan with the measuring electrodes

stable in 3D for any value of the time step [8,9]). The algorithm has been extended to accommodate anisotropic tissues parameters and sources. To describe the electrical conductivity in the heterogeneous biological media within arbitrary geometry, the method of the embedded boundaries has been used. Here an object of interest is embedded into a cubic computational domain with extremely low conductivity values in the external complimentary regions. This effectively guarantees there are no current flows out of the physical area (the Neuman boundary conditions, Eq. (2), is naturally satisfied). The idea of the iterative ADI method is to find the solution of Eq. (1) and (2) as a steady state of the appropriate evolution problem. At every iteration step the spatial operator is split into the sum of three 1D operators, which are evaluated alternatively at each sub-step. For example, the difference equations in $x$ direction is given as [9]

$$\frac{\phi_i^{n+1} - \frac{1}{3}(\phi_i^n + \phi_j^n + \phi_k^n)}{\tau} + \delta_x \phi_i^{n+1} + \delta_y \phi_j^n + \delta_z \phi_k^n = S, \qquad (3)$$

where $\tau$ is a time step and $\delta_{x,y,z}$ is a notation for the appropriate $1D$ second order spatial difference operator (for the problems with variable coefficients it is approximated on a "staggered" mesh). Such a scheme is accurate to $O(\tau + \Delta x^2 + \Delta y^2 + \Delta z^2)$. In contrast with the classic ADI method, the multi-component ADI does not require the operators to be commutative. In addition, it uses the regularization (averaging) for evaluation of the variable at the previous instant of time.

It is worth noting, that the multi-component ADI algorithm can be also easily adapted for solving PDEs describing other tomographic modalities. In particular, we have used it in other related studies, for example, in simulation of photon migration (diffusion) in a human head in near-infrared spectroscopy of brain injuries and hematomas.

The inverse problem for the electrical imaging modality has the general tomographic structure. From the assumed distribution of the head tissue conductivities, $\sigma_{ij}$, and the given injection current configuration, $S$, it is possible to predict the set of potential measurement values, $\phi^p$, given a forward model $F$ (Eq. (1), (2)), as the nonlinear functional [5,6]:

$$\phi^p = F(\sigma_{ij}(x,y,z)). \qquad (4)$$

Then an appropriate objective function is defined, which describes the difference between the measured, $V$, and predicted data, $\phi^p$, and a search for the global minimum is undertaken using advanced nonlinear optimization algorithms. In this paper, we used the simple least square error norm:

$$E = \left( \frac{1}{N} \sum_{i=1}^{N} (\phi_i^p - V_i)^2 \right)^{1/2}, \qquad (5)$$

where $N$ is a total number of the measuring electrodes (cl. Fig. 1). To solve the nonlinear optimization problem in Eq.(5), we employed the downhill simplex method of Nelder and Mead as implemented by Press et al[3]. In the strictest sense, this means finding the conductivity at each node of the discrete mesh. In simplified models with the constraints imposed by the segmented MRI data, one needs to know only the average regional conductivities of a few tissues, for example, scalp, skull, cerebrospinal fluid (CSF) and brain, which significantly reduces the demensionality of the parameter space in the inverse search, as well as the number of iterations in converging to a local minimum. To avoid the local minima, we used a statistical approach. The inverse procedure was repeated for hundreds sets of conductivity guesses from appropriate physiological intervals, and then the solutions closest to the global minimum solutions were selected using the simple critirea $E < E_{threshold}$.

## 3 Parallel Computional Design

The solution approach maps naturally to a multi-level computational design that can benefit from parallel execution both in the parametric search for conductivities and the forward problem calculations. Fig. 2 gives a schematic view of the approach we applied in a heterogeneous environment of parallel computing clusters. The *conductivity optimizer* (CO) is responsible for launching new inverse problems with guesses of conductivity values. Upon completion, the inverse solvers return conductivity solutions and error results to the master. Inverse solvers run on a separate computational server. The system design allows for the servers to be added dynamically and the number of processors per inverse solve to be decided at execution time, thus trading off inverse search parallelism versus forward problem speedup.

The CO interacts with each server using a TCP/IP-based interface. We use MPI to parallelize the inverse solvers as a master-worker computation. The *inverse master* (IM) manages multiple solvers at the same time. For each , the IM supplies new conductivity search values, lunches the simplex search and collects the results . The CO passes the initial seed to the IM to start simplex refinement for each new inverse worker. The IM sends a MPI message containing conductivity values to a free *inverse worker* (IW) to use in the forward calculation. The IM then waits to receives a solution from any IW, knowing which IW is working on what inverse solution. The *forward solver* (FS)

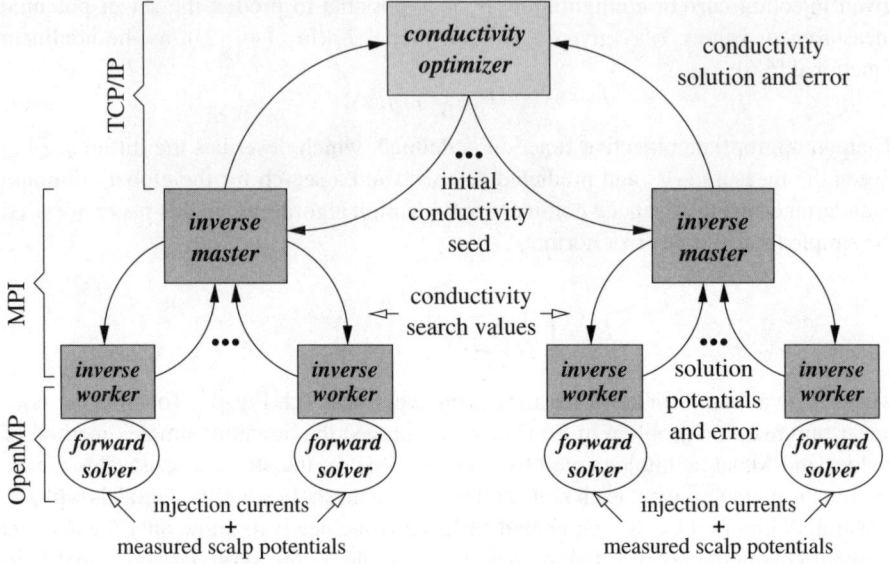

**Fig. 2.** Schematic view of the parallel computational system

is parallelized using OpenMP. It has been chosen over MPI as in the shared memory environment we avoid high data traffic naturally in solving PDE at 3D geometry. Parallelization of the ADI algorithm is straightforward, as it consists of nests of independent loops over "bars" of voxels for solving the effective 1D problem (Eq. (3)) at each iteration. These loops can be easily unrolled for efficient execution on a shared memory multiprocessor system.

The inverser solver MPI program executes as a mixed-mode parallel computation. Based on the number of cluster processors available and how the cluster is organized, we decide at runtime how many inverse workers to create and how many threads to assign to the forward calculation. In this manner, the program can be ported without change to both distributed memory and shared memory parallel clusters, and can naturally scale to meet available processing resources.

At the University of Oregon, we have access to a computational systems environment consisting of seven multiprocessor clusters. Of the shared memory clusters, three are 8-processor IBM Power4+ p655 machines, one is a 16-processor IBM Power4 p690 machine, and two (Phoenix and Optix) are 16-processor SGI Itanium-2 machines, an Altix and Prism machine. The one distributed memory cluster is a Dell 16x2-processor Pentium Xeon machine. All of the clusters run Linux and are connected by a high-speed gigabit network. The conductivity optimizer can run on any machine, including a workstation. In our experiments below, we show results only for the shared memory clusters. Also, the mixed-mode inverse solve program allocated four threads for the OpenMP forward calculation in each inverse worker.

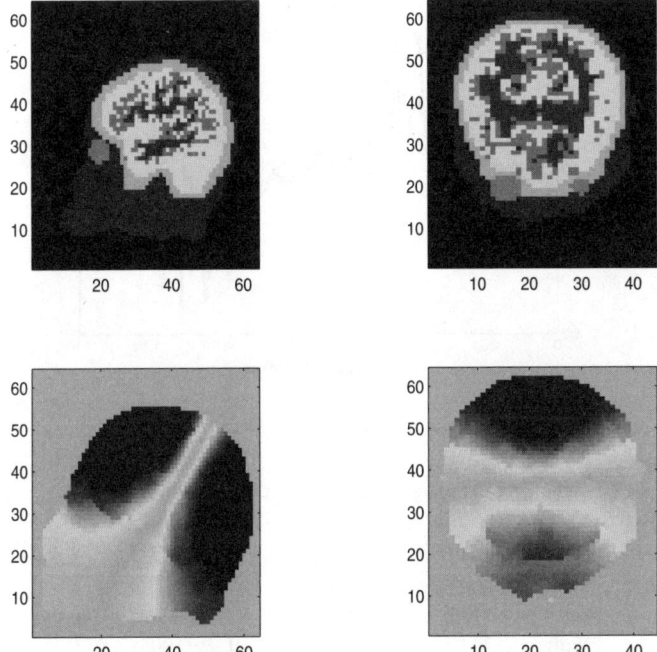

**Fig. 3.** Segmented MRI data (64x64x44 voxels resolution), top row, and calculated absolute value of potential, bottom row, for two points current injection (top and back of the head)

## 4 Computational Results

The forward solver was tested and validated against a 4-shell spherical phantom, and low ($64 \times 64 \times 44$) and high ($256 \times 256 \times 176$) voxels resolution human MRI data. For comparison purposes, the initial MRI data segmentation into ten tissues types as it is shown in the top row of Fig.3 was reduced to only four tissue types. Their values were set to those in the spherical model (cl. Table 1). We computed potentials at standard locations for the 129 electrodes configuration montage on the spherical phantom and compared the results with the analytical solution [2] available for a 4-shell spherical phantom in Fig. 4. One can we see very good agreement, save for some minor discrepancies caused by the mesh orientation effects (the cubic versa spherical symmetry).

Similarly, we found the good agreement for spherical phantoms between our results and the solution of the Poisson equation using the standard FEM packages such as FEMLAB [7]. Also, we have performed a series of computations for electric potentials and currents inside a human head with surgical or traumatic openings in the skull. We found that generally low resolution ($64 \times 64 \times 44$ voxels)like the one which is shown in the bottom row of Fig. 3 is not enough for accurate description of the current and potentials distribution through the head, as the coarse discretization creates artificial shunts for currents (mainly in the skull). With increased resolution ($128 \times 128 \times 88$ or $256 \times 256 \times 176$ voxels) our model has been shown to be capable to capture the fine

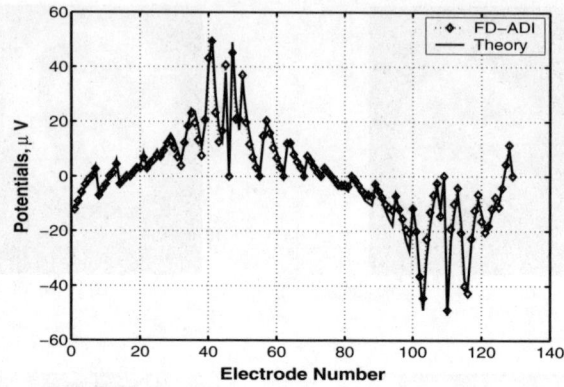

**Fig. 4.** Validation of the forward solver accuracy against analytics for a 4-shell spherical phantom

**Table 1.** Tissues parameters in 4-shell models[2]

| Tissue type | $\sigma(\Omega^{-1}m^{-1})$ | Radius(cm) | Reference |
|---|---|---|---|
| Brain | 0.25 | 8.0 | Geddes(1967) |
| Csf | 1.79 | 8.2 | Daumann(1997) |
| Skull | 0.018 | 8.7 | Law(1993) |
| Scalp | 0.44 | 9.2 | Burger(1943) |

details of current/potential redistribution caused by the structural perturbation. However, the computational requirements of the forward calculation increase significantly.

The forward solver was parallelized using OpenMP. The performance speedups (execution times) for $256 \times 256 \times 176$ sized problems on the IBM and SGI machines are shown in Fig. 5. While the performance is reasonable at present, we believe there are still optimizations that can be made, particularly on the SGI machines. The importance of understanding the speedup performance on the cluster compute servers is to allow flexible allocation of resources between inverse and forward processing.

To investigate the best balance of parallelism between inverse and forward processing, we conducted an experiment to optimize the numbers of MPI tasks and openMP threads at 12 processors of the 16-processors p690 machine. In this experiment we considered the total number of forward solutions performed by the cluster for several configurations in a fixed period of time. The number of iterations per a forward solution was fixed. The total number of forward solutions performed by a given cluster configuration was chosen as the figure of merit over the number of total inverse solutions due to the variation of the required number of forward computations in different inverse searches. The results are presented at Fig. 6. It can be seen that allocation of four threads per an inverse worker (3x4) gives the highest throughput for the total number of forward solutions.

In the inverse search the initial simplex was constructed randomly based upon the mean conductivity values (cl. Table 1) and their standard deviations as it is reported in the related biomedical literature. In the present test study we did not use the real

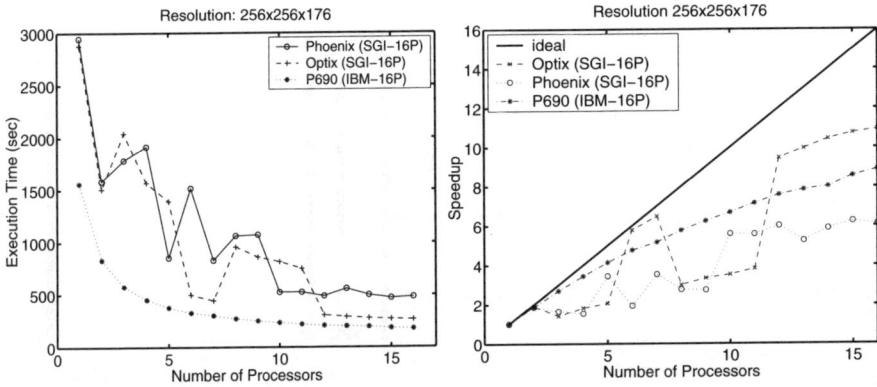

**Fig. 5.** Execution time (left) and speed-up (right) of the forward solver for grid size $256^3$ on SGI (Phoenix, Optix) and IBM (p690) machines

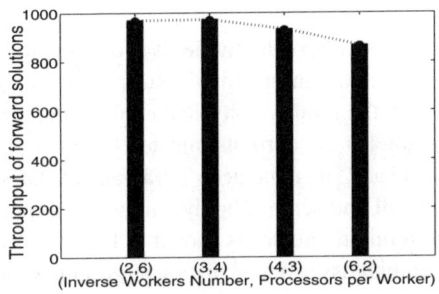

**Fig. 6.** Forward solutions throughput for different resource allocations between the forward and inverse problems. The total number of available processors is fixed to 12 in all configurations.

experimental human data, instead, we simulated the experimental set of the reference potentials $V$ in Eq. 5 using our forward solver with the mean conductivity values from Table 1, which had been assumed to be true, but not known a priory for a user running the inverse procedure. The search was stopped when one or two criteria were met. The first is when the decrease in the error function is fractionally smaller than some tolerance parameter. The second is when the number of steps of the simplex exceeds some maximum value. During the search, the conductivities were constrained to stay within their pre-defined plausible ranges. If the simplex algorithm attempted to step outside of the acceptable range, then the offending conductivity was reset to the nearest allowed value. Our procedure had the desired effect of guiding the search based on prior knowledge. Some number of solution sets included conductivities that were separated from the bulk of the distribution. These were rejected as outliers, based on the significant larger square error norm in Eq. (5) (i.e., the solution sets were filtered according to the criteria $E < E_{threshold}$). We have found empirically that setting $E_{threshold} = 1\mu V$ in most of our runs produced a fair percentage of solutions close to the global minimum.

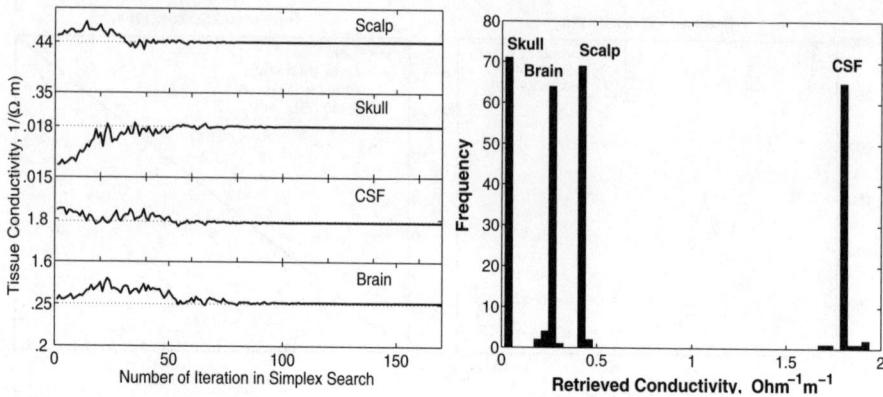

**Fig. 7.** Results of the inverse search. Dynamics of the individual search (left) and statistics of the retrieved conductivities for about 200 initial random guesses. The actual number of the solutions shown is 71, their error function is less than 1 microvolt.

The distribution of the retrieved conductivities is shown in Fig. 7 (right). The fact that the retrieved conductivities for the intracranial tissues (CSF and brain) have wider distributions is consistent with the intuitive physical explanation that the skull, as having the lowest conductivity, shields the currents injected by the scalp electrodes from the deep penetration into the head. Thus, the deep intracranial tissues are interrogated less in comparison with the skull and scalp. The dynamics of an individual inverse search convergence for a random initial guesses is shown in Fig. 7 (left). In general, the conductivities for the extra cranial tissue and skull converge somewhat faster than the brain tissues, due to the better interrogation by the injected current.

After filtering data according to the error norm magnitude, we fitted the individual conductivities to the normal distribution. The mean retrieved conductivities $\sigma(\Omega^{-1}m^{-1})$ and their standard deviations $\Delta\sigma$ are: Brain (0.24 / .01), CSF (1.79 / .03), Skull (0.0180 / .0002), and Scalp (0.4400 / .0002). It is interesting to compare these values to the "true" conductivities from Table 1. We can see excellent estimates for the scalp and skull conductivities and a little bit less accurate estimates for the intracranial tissues. We also have done some preliminary runs with the realistic noise included. These runs and the similar investigation in Ref. [2] for a spherical phantom suggest that noise leads to some deterioration of the distributions and more uncertainty in the results. In general, it still allows the retrieval of the unknown tissue parameters.

Finally, in Fig. 8 we present the dynamics of the performance of the inverse search in our distributed multi-cluster computational environment. Six curves with different markers show the dynamics of the inverse solution flux at the conductivity optimizer. The markers correspond to the instances of inverse solutions arrival to CO from a specific inverse master (cluster). The inverse solution rate varies between the clusters based on several factors: the number of processors available, the speed of the forward solve, and inverse search convergence rate. The markers seated at the "zero" error function line represent solutions that contribute to the final solution distribution, with the rest of the solutions rejected as outliers. In average, the throughput was 15 minutes per one

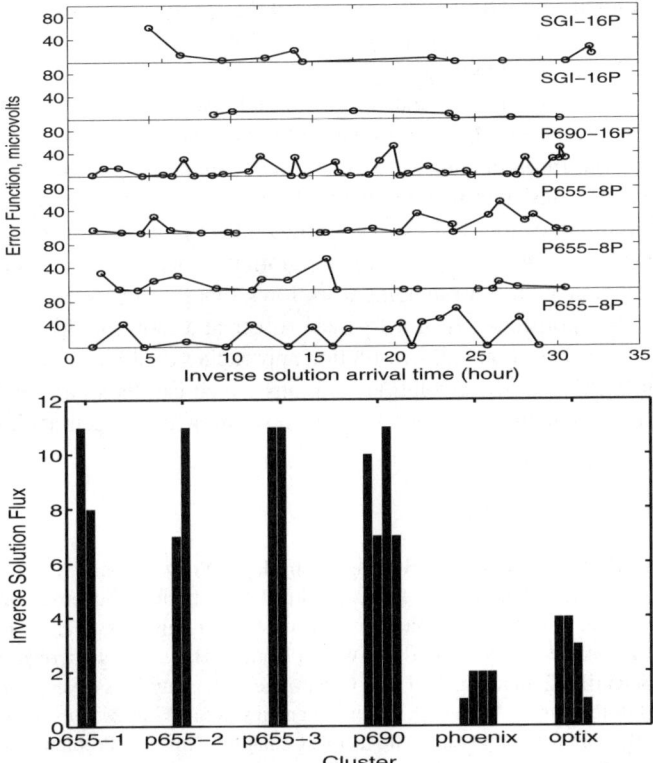

**Fig. 8.** Solutions flow at the conductivity optimizer (top). Inverse solutions arrival to the conductivity optimizer is marked. Number of inverse solutions per inverse worker (bottom).

inverse solution for the $128 \times 128 \times 88$ MRI resolution test case. The second graph shows the number of inverse solutions completed by the different clusters. Since we chose four threads to use in the OpenMP forward solve, the graph shows the number of inverse solutions completed per inverse worker.

## 5 Conclusion

We have built an accurate and robust 3D Poisson solver based on a finite difference multi-components ADI algorithm for modeling electrical and optical problems in heterogeneous biological tissues. We focus in particular on modeling the conductivity properties of the human head. The computational formulation utilizes realistic head geometry obtained from segmented MRI datasets. This is important to the effective use of impedance imaging and source localization in clinical neuroimaging applications where diagnostic accuracy depends significantly on the degree to which individual differences in head structure can be represented. The computational formulation of the problem is as a multi-cluster mixed-mode calculation suitable for parallel execution on a computational grid. Our results validate FDM approach for impedance imaging and provide a performance assessment of parallel computation on six clusters of the University of Oregon's ICONIC grid.

In the future, we will enhance the computational framework in several ways. Additional cluster resources will be used to naturally scale the performance of the conductivity optimization. In particular, we will add the 16-node, 2-processor per node Dell cluster to the mix. Consistent with the ICONIC grid, our intent is to evolve the present interprocess communication (IPC) socket-based code to one that uses grid middleware support, allowing the impedance imaging program to more easily access available resources and integrate with neuroimaging workflows. Finally, intrinsically parallel multi-component ADI algorithms [9] in a forward solver and more intelligent schemes of conductivity search based on multi-resolution approaches could be tried. The idea here is to first start with fast, low-resolution solutions which can then narrow the range of and guide initial conductivity guesses for high-resolution, more accurate investigation.

## References

1. Gulrajani, R.M.: Bioelectricity and Biomagnetism. John Wiley & Sons, New York (1998)
2. Ferree, T.C., Eriksen, K.J., Tucker, D.M.: Regional head tissue conductivity estimation for improved EEG analysis. IEEE Transactions on Biomedical Engineering 47, 1584–1592 (2000)
3. Press, W.H., Teukolsky, S.A., Vetterling, W.T., Flannery, B.P.: The Numerical Recipes in C: The art of Scientific Computing, 2nd edn. Cambridge University Press, New York (1992)
4. Jin, J.: The Finite Element Method in Electromagnetics. John Wiley & Sons, New York (1993)
5. Arridge, S.R.: Optical tomography in medical imaging. Inverse Problems, 15, R41–R93 (1999)
6. Hielscher, A.H., Klose, A.D., Hanson, K.M.: Gradient Based Iterative Image Reconstruction Scheme for Time-Resolved Optical Tomography. IEEE Transactions on Medical Imaging 18, 262–271 (1999)
7. http://www.comsol.com
8. Abrashin, V.N., Dzuba, I.A.: Economical Iterative Methods for solving multi- dimensional problems in Mathematical Physics. Differential Equations 30, 281–291 (1994)
9. Abrashin, V.N., Egorov, A.A., Zhadaeva, N.G.: On the Convergence Rate of Additive Iterative Methods. Differential Equations 37, 867–879 (2001)

# Application I

# An Evaluation of OpenMP on Current and Emerging Multithreaded/Multicore Processors

Matthew Curtis-Maury, Xiaoning Ding,
Christos D. Antonopoulos, and Dimitrios S. Nikolopoulos

Department of Computer Science
The College of William and Mary
McGlothlin–Street Hall, Williamsburg, VA 23187-8795

**Abstract.** Multiprocessors based on simultaneous multithreaded (SMT) or multicore (CMP) processors are continuing to gain a significant share in both high-performance and mainstream computing markets. In this paper we evaluate the performance of OpenMP applications on these two parallel architectures. We use detailed hardware metrics to identify architectural bottlenecks. We find that the high level of resource sharing in SMTs results in performance complications, should more than 1 thread be assigned on a single physical processor. CMPs, on the other hand, are an attractive alternative. Our results show that the exploitation of the multiple processor cores on each chip results in significant performance benefits. We evaluate an adaptive, run-time mechanism which provides limited performance improvements on SMTs, however the inherent bottlenecks remain difficult to overcome. We conclude that out-of-the-box OpenMP code scales better on CMPs than SMTs. To maximize the efficiency of OpenMP on SMTs, new capabilities are required by the runtime environment and/or the programming interface.

## 1 Introduction

As a shared-memory programming paradigm, OpenMP is suitable for parallelizing applications on simultaneous multithreaded (SMT) [17] and multicore (CMP) [16] processors. These processors appear to dominate both the high-end and mainstream computing markets. Products such as Intel's Hyperthreaded Pentium IV are already widely used for desktop and server computing, with similar products being marketed or in late stages of development by other vendors. At the same time, high-end, future microprocessors encompass aggressive multithreading and multicore technologies to form powerful computational building blocks for the next generation of supercomputers. All three vendors selected by the DARPA HPCS program (IBM, Cray and Sun) have adopted multithreaded and multicore processor designs, combined with different technological innovations such as streaming processor cores and proximity communication [5, 6, 15].

With the advent of multithreaded and multicore multiprocessors, a thorough evaluation of OpenMP on such architectures is a timely and necessary effort. In this paper we evaluate a comprehensive set of OpenMP codes, including complete parallel benchmarks and real-world applications, on both a real multi-SMT system, composed of

Intel's Hyperthreaded processors, and on a simulated multiprocessor with CMP processors. For the latter architectural class we use complete system simulation to factor in any operating system effects. Our evaluation uses detailed performance measurements and information from hardware performance counters to pinpoint architectural bottlenecks of SMT/CMP processors that hinder the scalability of OpenMP, as well as areas in which OpenMP implementations can be improved to better support execution on SMT/CMP processors.

We observe that the extensive resource sharing in SMTs often hinders scalability, should threads co-executing on the same physical processor have conflicting resource requirements. The significantly lower degree of resource sharing in CMPs, on the other hand, allows applications to effectively exploit the multiple execution cores of each physical processor. We quantitatively evaluate the effects of resource sharing on the L2 miss rate, the number of stall cycles and the number of data TLB misses. We then evaluate the effectiveness of a run-time mechanism that transparently determines and uses the optimal number of threads on each SMT processor. This technique yields measurable, though limited performance improvement. Despite its assistance, the architectural bottlenecks of SMTs do not allow OpenMP applications to efficiently exploit the additional execution contexts of SMT processors.

The rest of the paper is organized as follows: In section 2 we outline related work. In section 3 we evaluate the execution of OpenMP codes on SMT- and CMP-based multiprocessors and pinpoint architectural bottlenecks using a variety of performance metrics. Section 4 evaluates a simple, yet effective mechanism that automatically determines and exploits the optimal number of execution contexts on SMT-based multiprocessors. In section 5 we outline some implications of the proliferation of hybrid, SMT- and CMP-based multiprocessors for OpenMP. Finally, section 6 concludes the paper.

## 2 Related Work

Earlier research efforts have ported and evaluated OpenMP on specific processor designs, including heterogeneous chip multiprocessors [14], slipstream processors [9] (a form of 2-way chip multiprocessors in which the second core is used for speculative runahead execution) and Cyclops, a fine-grain multithreaded processor architecture introduced by IBM [1]. Our evaluation focuses on commodity processors, with organizations spanning the design space between simultaneous multithreading and chip multiprocessors and a few execution contexts. Although not at the high end of the design space of supercomputing architectures, such processors are becoming commonplace and are natural building blocks for larger multiprocessors. A recent study of OpenMP loop scheduling policies on multiprocessors with Intel's Hyperthreaded processors indicated the need for adaptation of both the degree of concurrency and the loop scheduling algorithms when OpenMP applications are executed on simultaneous multithreading architectures, because of different forms of interferences between threads [18]. Our evaluation corroborates these results and provides deeper insight on the architectural reasons due to which adaptivity is an effective method for improving the performance of OpenMP programs on SMT processors.

## 3 Experimental Evaluation and Analysis

### 3.1 Hardware and Software Environment and Configuration

In order to ascertain the effects of the characteristics of modern processor architectures on the execution of OpenMP applications, we have considered two types of multiprocessors which are becoming more and more popular in today's computing environment, namely multiprocessors based on either SMTs or CMPs. SMTs incorporate minimal additional hardware in order to allow multiple co-executing threads to exploit potentially idle processor resources. The threads usually share a single set of resources such as execution units, caches and the TLB. CMPs on the other hand integrate multiple independent processor cores on a chip. The cores do, however, share one or more outer levels of the cache hierarchy, as well as the interface to external devices.

We used a real, 4-way server based on Hyperthreaded (HT) Intel processors as a representative SMT-based multiprocessor. Intel HT processors are a low-end / low-cost implementation of simultaneous multithreading. Each processor offers 2 execution contexts which share execution units, all levels of the cache, and a common TLB. The experiments targeted at the CMP-based multiprocessors have been carried out on a simulated 4-way system. The simulated CMP processors integrate 2 cores per processor. They are configured almost identically to the real Intel HTs, apart from the L1 cache and TLB which are private, per core on the CMP and shared between execution contexts on the SMT. Note that using private L1 caches and TLBs favors CMPs by providing more effective cache and TLB space to each thread and reducing contention. Therefore, our experimental setup seems to favor CMPs. Note however, that we are evaluating a CMP with in-order issue cores, which are much simpler than the out-of-order execution engines of our real SMT platform. Furthermore, the multicore organization of our simulated CMP enables a chip layout with private L1 caches at a nominal increase in die area [13]. For these reasons, the simulated CMP platform can still be considered as roughly equivalent (in terms of resources) to our real SMT platform. We used the Simics [7] simulation platform to conduct complete system simulations, including system calls and operating system overhead. Table 1 describes the configuration of the two systems in further detail.

**Table 1.** Configuration of the SMT- and CMP-based multiprocessors used throughout the experimental evaluation

|     | Processors | L1 Cache | L2 Cache | L3 Cache | TLB | Main Mem. |
|-----|------------|----------|----------|----------|-----|-----------|
| SMT | 4 x Intel P4 Xeon, 1.4 GHz Hyperthreaded x 2 Execution Contexts per Processor | 8K Data, 12K Trace (Instr.), Shared | 256K Unified, Shared | 512K Unified, Shared | 64 Entries Data, 128 Entries Instr., Shared | 1GB |
| CMP | 4 Processors x 2 P4 Cores per Processor | 2x8K Data, 2x12K Trace (Instr.) Private per Core | 256K Unified, Shared | 512K Unified, Shared | 2x64 Entries Data, 2x64 Entries Instr., Private per Core | 1GB |

We evaluated the relative performance of OpenMP workloads on the two target architectures, using 7 OpenMP applications from the NAS Parallel Benchmarks suite (version 3.1) [11]. We executed the class A problem size of the benchmarks, since it is a large enough size to yield realistic results. At the same time, it is the largest problem class that allows the working sets of all applications to fit entirely in the available main memory of 1GB.

We executed all the benchmarks on the SMT with 1, 2, 4, and 8 threads. The main goal of this experiment set was to evaluate the effects of executing 1 or 2 threads on the 2 execution contexts of each processor. We thus ran our experiments under six different thread placements: i) 1 thread, ii) 2 threads bound on 2 different physical processors, iii) 2 threads bound on the 2 contexts of 1 processor, iv) 4 threads bound on 4 processors, v) 4 threads paired on the execution contexts of 2 processors and vi) 8 threads paired on 4 processors. Each thread is pinned on a specific execution context of a specific processor using the Linux sched_setaffinity system call. The applications were executed using Intel VTune [10] performance analyzer. We recorded both the execution time and a multitude of additional performance metrics attained from the hardware performance counters available in the processor. Such metrics provide insight into the interaction of applications with the hardware, thus they are a valuable tool for understanding the observed application performance.

The same experiments have been repeated on the simulated CMP-based multiprocessor. Full system simulation with Simics introduces an average 7000-fold slowdown in the execution time of applications, compared with the execution on a real machine. We simulated the same application binaries, using the same data sets, however we reduced the number of iterations[1] we ran on the simulator in order to limit the execution time to reasonable levels. More specifically, we executed only 3 of the outermost iterations of each benchmark, discarding the results from the first iteration in order to eliminate transient effects due to cache warmup. The simulator directly provides similar, detailed performance information as Vtune.

All experiments were performed on a dedicated machine in order to rule out data perturbations due to interactions with third-party applications and services. The operating system on both the real and the simulated system was Linux 2.4.25.

### 3.2 Experimental Results

We evaluated the relative performance of the benchmarks on the real SMT-based and the simulated CMP-based multiprocessors when 1 or 2 threads are activated per physical processor, using the different binding schemes described in section 3.1. We monitored a multitude of direct (wall clock time, number of instructions, number of L2 and L3 references and misses, number of stall cycles, number of data TLB misses, number of bus transactions) and derived (CPI, L2 and L3 miss rates) performance metrics. Due to space limitations we only present and discuss the results for L2 miss rates, stall cycles, data TLB misses and execution time.

The results for the L2 miss rate evaluation are depicted in Figure 1. The reported values are for 2 threads per processor and have been normalized with respect to the single-thread per processor execution of each benchmark on the specific architecture and number of processors. This way, the graphs emphasize the effects of using a second

---

[1] All the NAS applications we used are iterative. The computational routines are enclosed in an external, sequential loop.

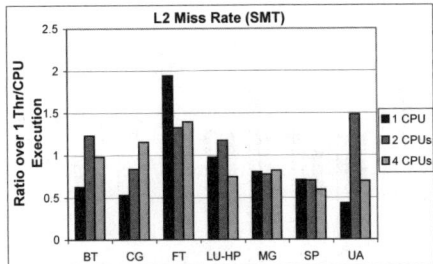 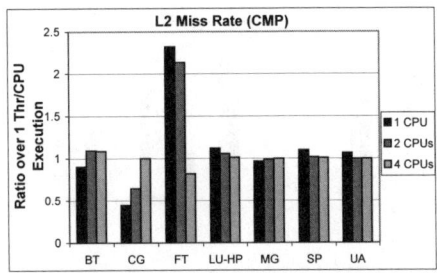

**Fig. 1.** Normalized L2 miss rates of the benchmarks on the SMT and CMP multiprocessor (left and right diagrams respectively). The corresponding 1 thread/processor miss rates on each architecture and number of processors have been used as references for the normalization.

thread per processor. The relative L2 cache performance of applications when 1 and 2 threads are executed on each physical processor depends highly on the specific characteristics of the application. If the working sets of both threads do not fit in the L2 cache, there is an increase in the L2 miss rate, since cross-thread cache-line eviction results in more misses. If, on the other hand, the 2 threads executing on the same processor share data, then each of them will probably benefit from data already fetched to the cache by the other thread.

In most cases, executing 2 threads per processor on the SMT system proved beneficial for L2 cache performance. On average, thread pairing resulted in 1.05 times lower miss rates in comparison with the single-thread per processor execution. An application in which thread cross-eviction appears is FT. The FT threads have large working sets that can not entirely fit into any level of the cache hierarchy. Moreover, the degree of data sharing between threads co-executing on the same processor is low. As a result, miss rates increase significantly if both execution contexts of each processor are activated. Another interesting pattern can be observed in CG. Although the exploitation of the second hyperthread of each processor results in a significant reduction in miss rate in the single processor experiments, as more physical processors are added the trend is reversed. CG has a high degree of data sharing between the threads. If few threads are active, the benefits of the shared cache are evident. However, as more physical processors are added, inter-processor data sharing results in a large number of cache-line invalidations, which eventually outweigh the benefit of intra-processor data sharing.

On the CMP-based multiprocessor the L2 cache miss rate generally appears to be uncorrelated to the exploitation of 1 or 2 execution cores per physical processor. Although the L2 is shared between both cores, the private, per core L1 caches function as a buffer that prevents many memory accesses from reaching the second level of the cache. In fact, the use of a second thread on SMTs results in an increase in the number of L2 cache accesses, due to the inter-thread interference in the L1 cache. More specifically, the number of L2 accesses always increases, by 1.42 times on average, when the second execution context is activated on each physical processor. The private L1 caches in CMPs alleviate this problem. The number of L2 accesses is reduced by an average factor of 1.37 when the second core - and its private L1 cache - are activated on each CPU.

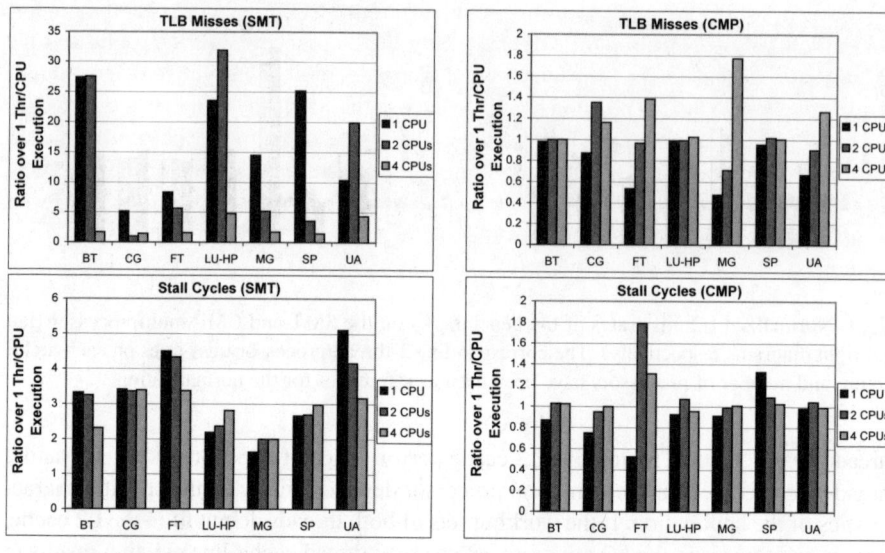

**Fig. 2.** Normalized number of TLB misses (top diagrams) and stall cycles (bottom diagrams) of the benchmarks on the SMT and CMP multiprocessor (left and right diagrams respectively). The corresponding 1 thread/processor stalls and TLB misses on each architecture and number of processors have been used as references for the normalization.

The behavioral patterns observed for CG and FT on the SMT-based multiprocessor are repeated on the CMP as well.

Figure 2 depicts the normalized number of stall cycles and TLB misses. Once again, the reported values have been normalized with respect to the corresponding single-thread per processor execution of each benchmark on the specific architecture and number of processors. The results indicate that using the second execution context of the SMT processors has a significant effect on the number of TLB misses. Their number suffers an up to 27-fold increase when we move from binding schemes that assign 1 thread per processor to those that assign 1 thread per execution context. On average, TLB misses increase by 10.78 times. The 2 threads on each processor often work on different areas of the virtual address space, thus being unable to share TLB entries. Furthermore, the Intel SMT processor has a surprisingly small data TLB (64 entries), which can not achieve a good coverage of the virtual address space of the benchmarks we executed. As a result, the effective per thread size of the shared TLB is reduced drastically when both execution contexts of each processor are activated. The CMP processor provides private TLBs for each core. As a consequence, the number of TLB misses is much more stable than on the SMT system. In fact, the execution of 1 or 2 threads per processor has, on average, no effect on the number of TLB misses.

The behavior in terms of stall cycles also varied significantly between the two architectures. On SMT processors the number of stall cycles represents the cumulative effect of both cycles spent waiting for data to be fetched from any level of the memory hierarchy and cycles during which execution was stalled because of conflicting resource requirements of the threads executing on the different execution contexts of the processor. On CMPs, co-executing threads share only the 2 outer levels of the cache and the interface to external devices, thus the second factor does not contribute to the total number of stall cycles. For all benchmarks executed on the SMT, the number of stall cycles increased – on average by 3.1 times – when the configuration was changed from 1 to 2 threads per processor. The corresponding average overhead on the CMP is a mere 1.03. This is a safe indication that the vast majority of stall cycles on the SMT can be attributed to conflicting requirements of co-executing threads for internal processor resources.

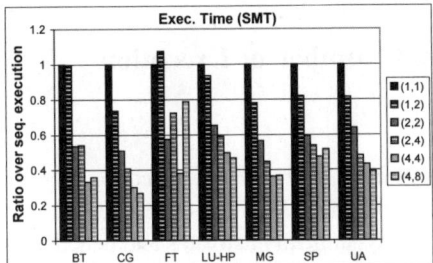 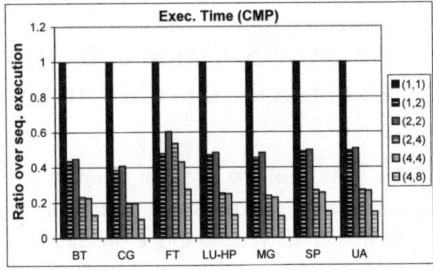

**Fig. 3.** Normalized execution time of the benchmarks on the SMT and CMP multiprocessor (left and right diagram respectively). The single-threaded (sequential) execution time on each architecture is used as a reference for the normalization.

Finally, Figure 3 depicts the results from the execution time of applications on the two target multiprocessor architectures. This time, the reported values have been normalized with respect to the sequential (single-threaded, single-processor) execution time of applications on each architecture. The different binding schemes are labeled as (num_processors, num_threads), where num_processors stands for the number of physical processors onto which the threads are bound and num_threads for the number of threads used for the application execution.

All 7 benchmarks scale well on both the SMT and the CMP as more physical processors are made available to the application. This indicates that potential performance problems under some binding schemes can not be attributed to the scalability characteristics of the benchmarks. In fact for the 2-threaded BT and CG execution on the CMP the speedups are superlinear, due to the availability of cumulatively larger L1 cache and TLB when more than 1 threads are used.

Given a specific number of threads, execution times on the SMT multiprocessor are always lower if the threads are spread across as many physical processors as possible, instead of being placed on both execution contexts of each processor. Moreover, in 7 out of 21 experiments the activation of the second execution context, given a specific number of physical SMT processors, resulted in a reduction of the observed application performance. It should also be pointed out that, even for a given application, it is not always clear whether the exploitation of all execution contexts of each processor is the optimal strategy or not. In the case of SP, for example, exploiting 2 execution contexts per processor is optimal when 1 and 2 processors are available, however it results in performance penalties when all 4 processors are used.

The results are totally different on the CMP-based multiprocessor. In 8 out of 14 cases placing a given number of threads on the cores of as few processors as possible yields higher performance than spreading them across processors. Moreover, the activation of the second core always resulted in performance improvements. The replication of execution units, L1 caches and TLBs on the CMPs allows threads to execute more effectively, without the limitations posed by resource sharing on SMTs. The reduction in resource conflicts due to hardware replication often allows the benefits of inter-processor cache sharing to be reflected in a reduction in execution time.

## 4 Adaptive Selection of the Optimal Number of Execution Contexts for OpenMP on SMTs

The selection of the optimal number of execution contexts for the execution of each OpenMP application is not trivial on SMT-based SMPs. We thus experiment with a performance-driven, adaptive mechanism which dynamically activates and deactivates the additional execution contexts on SMT processors to automatically approximate the execution time of the best static selection of execution contexts per processor. We used a simpler mechanism than the exhaustive search proposed in [18], which avoids modifications to the OpenMP compiler and runtime. Our mechanism identifies whether the use of the second execution context of each processor is beneficial for performance and adapts the number of threads used for the execution of each parallel region. The algorithm introduced in [18] also targets identification of the best loop scheduling policy.

Our method is based on the annotation of the beginning and end of parallel regions with calls to our runtime. The calls can be inserted automatically, by a simple preprocessor. Alternatively, run-time linking techniques such as dynamic interposition can be used to intercept the calls issued to the native OpenMP runtime at the boundaries of parallel regions and apply dynamic adaptation even to unmodified application binaries.

We slightly modify the semantics of the OMP_NUM_THREADS environment variable, using it as a suggestion for the number of processors to be used instead of the number of threads. Moreover, we add a new environment variable (OMP_SMT). If OMP_SMT is defined to be 1 or 2, the application always uses 1 and 2 execution contexts per physical processor respectively. If its value is 0, or the variable is not defined, adaptive execution is activated. In this case, each kernel thread is first bound on a specific execution context upon program startup. On the second and third time each parallel region is encountered, our runtime executes it using 1 and 2 execution contexts per processor and monitors execution time. After the third execution of each region, a decision is made using the timing results from the two test executions. Upon additional invocations of the parallel region, the runtime automatically adjusts the number of threads according to the decision. The first execution of each parallel region is not monitored, in order to avoid any interference in the decision process due to cache warmup effects. The runtime makes decisions independently for each parallel region. The execution of most applications proceeds in phases, with different execution characteristics for each phase. The boundaries of parallel regions often indicate phase changes. Thus, varying the number of threads at the boundaries of parallel regions offers context sensitive adaptation[2].

We evaluated the performance of our adaptive mechanism using the NAS Parallel Benchmarks along with two other OpenMP codes: MM5 [8], a mesoscale weather prediction model, and COBRA [4], a matrix pseudospectrum computation code. We ran each of the benchmarks statically with 1 and 2 threads per processor on 1, 2, 3, and 4 processors. We then executed each benchmark using the adaptive strategy. Even in the experiments using a static number of threads, threads are bound on specific execution contexts in order to avoid unfairly penalizing performance due to suboptimal thread placement decisions of the Linux scheduler. The results are depicted in Figure 4.

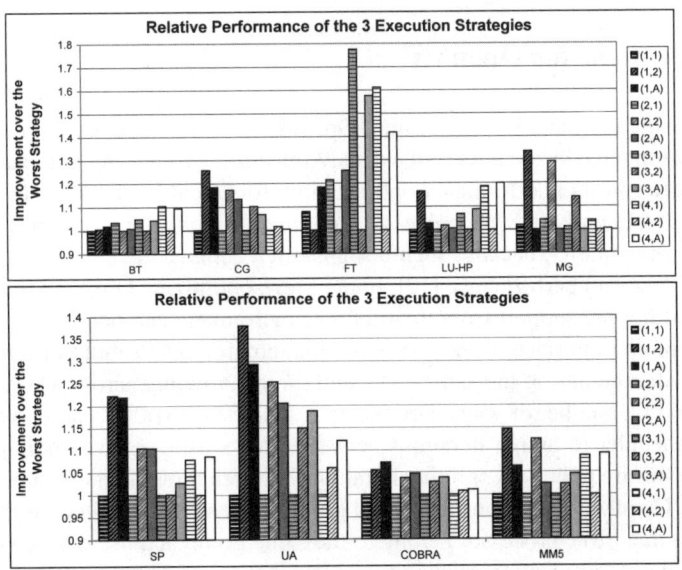

**Fig. 4.** Relative performance of adaptive, 1 and 2 threads per physical processor execution strategies. The execution times have been normalized with respect to the execution time of the worst strategy for each experiment.

Compared with the optimal static number of threads for each case, our approach was only 3.0% slower on average. At the same time, it achieved a 10.7% average speedup over the worse static number of threads for each (benchmark, number of processors) combination. The average overall speedup observed over all static configurations was 3.9%. In 17 out of the total 36 experiments the adaptive mechanism even provided a performance improvement over both static strategies for selecting the number threads. This can be attributed to the flexibility of the adaptive mechanism and its ability to decide the optimal number of threads independently for each parallel region.

---

[2] In fact loop boundaries can offer a better approximation of application phases. OpenMP specifications however, prohibit varying the number of active threads inside a parallel region, thus adaptive mechanisms like ours can not be used to make decisions at a loop-level resolution.

The adaptive technique did not perform well for MG. MG performs only 4 outermost iterations. Given that 3 iterations are needed for the initialization and decision phases, MG executes in adaptive mode for only 1 iteration. However, it does not take many iterations for the adaptive execution to compensate for the overhead of the monitoring phase. CG, for example, performs just 15 iterations and the adaptive strategy is only slightly inferior than the best static strategy.

The performance benefits attained by our simple mechanism are lower than those attained by the combined adaptation of the number of threads and loop schedules in [18]. They indicate that dynamic adaptation can provide some speedup on SMT-based multiprocessors, however the inherent architectural bottlenecks of contemporary SMTs hinder the efficient exploitation of the additional execution contexts.

## 5 Implications for OpenMP

Our study indicates that although scaling OpenMP on CMPs can be effortless, scaling on SMTs is hindered by the effects of extensive resource sharing. We argue that it is still worthwhile to consider performance optimizations for OpenMP on SMTs. In addition to the current Intel family of SMT processors, multicore architectures with SMT cores are also gaining popularity, because such designs often achieve the best balance between energy, die area and performance [12]. In our view, optimizing OpenMP for SMTs entails both additional support from the runtime environment and possible extensions to the programming interface. Clearly, the runtime environment should differentiate between threads running on the same SMT and threads running across SMTs. This can be achieved in a number of ways: For example, a new SCHEDULE clause would allow the loop scheduler to assign iterations between SMTs using a given policy and then use an SMT-aware policy for splitting iterations between threads on the same SMT. Alternatively, OpenMP extensions for thread groups [3] can be exploited, so that threads within the same SMT processor belong to the same group and use their own scheduling and local synchronization mechanisms. Note that using groups in this case does not necessarily imply the use of nested parallelism. SMT-aware programs may utilize just a single level of parallelism but use different policies for executing threads within SMTs. In fact, current SMTs do not allow the exploitation of parallelism with granularity much finer than what can be exploited by conventional multiprocessors [2]. If no extensions to the OpenMP interface are desired, then more intelligence should be embedded in the runtime environment, to dynamically identify threads sharing an SMT and differentiate its internal thread management policies. Although such an expectation is not unreasonable for regular iterative scientific applications, it is difficult to achieve the same level of runtime sophistication for irregular applications.

Regarding portability (of both code and performance), one of the most important problems for implementing an SMT-aware version of OpenMP is thread binding to processors and execution contexts within processors. Clearly, if the programmer wishes to exploit a single level of parallelism in a non-malleable program, the issue of binding is irrelevant. If however the programmer wishes for any reason to utilize SMTs for an alternative multithreaded execution strategy (e.g. for nested parallelism, or for slipstream execution), then it is necessary to specify the placement of threads on processors.

Although the OpenMP community has proposed extensions to handle similar cases (e.g. via an ONTO clause), exposing architecture internals in the programming interface is undesirable in OpenMP. Therefore, new solutions for improving the execution of OpenMP programs on SMTs in an autonomic manner are desirable.

## 6 Conclusions

In this paper we evaluated the performance of OpenMP applications on SMT- and CMP-based multiprocessors. We found that the execution of multiple threads on each processor is more efficient and predictable on CMPs than it is on SMTs due to the higher degree of resource isolation, which results in fewer conflicts between threads co-executing on the same processor. Although adaptive run-time techniques can improve the performance of OpenMP applications on SMTs, inherent architectural bottlenecks hinder the efficient exploitation of these processors.

Our analysis indicated that the interference between co-executing threads in the shared levels of the cache or the shared TLB may prove a determining factor for performance. Driven by this observation, we intend to evaluate run- and compile-time techniques for TLB partitioning on SMTs and cache partitioning on both SMT and CMP architectures. The forthcoming proliferation of processors which combine simultaneous multithreading and chip multiprocessing, such as the IBM Power5, and their use as basic building blocks of multiprocessors will certainly generate a multitude of challenging software optimization problems for system software and application developers.

## Acknowledgements

This work is supported by an NSF ITR grant (ACI-0312980), an NSF CAREER award (CCF-0346867) and the College of William and Mary.

## References

1. Almasi, G., Ayguade, E., Cascaval, C., Castanos, J., Labarta, J., Martinez, F., Martorell, X., Moreira, J.: Evaluation of OpenMP for the Cyclops Multithreaded Architecture. In: Voss, M.J. (ed.) WOMPAT 2003. LNCS, vol. 2716, pp. 147–159. Springer, Heidelberg (2003)
2. Antonopoulos, C.D., Ding, X., Chernikov, A., Blagojevic, F., Nikolopoulos, D.S., Chrisochoides, N.: Multigrain Parallel Delaunay Mesh Generation: Challenges and Opportunities for Multithreaded Architectures. In: Proc. of the 19th ACM International Conference on Supercomputing (ICS 2005), Cambridge, MA, U.S.A (June 2005)
3. Ayguadé, E., Gonzàlez, M., Martorell, X., Oliver, J., Labarta, J., Navarro, N.: NANOSCompiler: A Research Platform for OpenMP Extensions. In: Proc. of the First European Workshop on OpenMP, Lund, Sweden, October 1999, pp. 27–31 (1999)
4. Bekas, C., Gallopoulos, E.: Cobra: Parallel path following for computing the matrix pseudospectrum. Parallel Computing 27(8), 1879–1896 (2001)
5. Dally, W., Hanrahan, P., Erez, M., Knight, T., Laboté, F., Ahn, J., Jayasena, N., Kapasi, U., Das, A., Gummaraju, J., Buck, I.: Merrimac: Supercomputing with Streams. In: Proc. of the IEEE/ACM Supercomputing 2003: High Performance Networking and Computing Conference (SC 2003), Phoenix, AZ (November 2003)

6. Ebcioglu, K., Saraswat, V., Sarkar, V.: The IBM PERCS Project and New Opportunities for Compiler-Driven Performance via a New Programming Model. In: Compiler-Driven Performance Workshop (CASCON 2004) (October 2004)
7. Dahlgren, F., Grahn, H., Karlsson, M., Larsson, F., Lundholm, F., Moestedt, A., Nilsson, J., Stenström, P., Werner, B.: Simics/sun4m: A virtual workstation. In: Proc. of the 1998 USENIX Annual Technical Conference, New Orleans, LA (June 1998)
8. Grell, G.A., Dudhia, J., Stauffer, D.R.: A Description of the Fifth-Generation Penn State/NCAR Mesoscale Model (MM5). NCAR Technical Note NCAR/TN-398 + STR, National Center For Atmospheric Research (NCAR) (June 1995)
9. Ibrahim, K., Byrd, G.: Extending OpenMP to Support Slipstream Execution Mode. In: Proc. of the 17th International Parallel and Distributed Processing Symposium (IPDPS 2003), Nice, France (April 2003)
10. Intel Inc. Intel VTune Performance Analyser (2003), http://www.intel.com/software/products/vtune
11. Jin, H., Frumkin, M., Yan, J.: The OpenMP Implementation of NAS Parallel Benchmarks and its Performance. Technical report nas-99-011, NASA Ames Research Center (October 1999)
12. Kalla, R., Sinharoy, B., Tendler, J.: IBM POWER5 Chip: A Dual-Core Multithreaded Processor. IEEE Micro 24(2), 40–47 (2004)
13. Kumar, R., Jouppi, N., Tullsen, D.: Conjoined-Core Chip Multiprocessing. In: Proc. of the 37th International Symposium on Microarchitecture (MICRO-37), Portland, OR, December 2004, pp. 195–206 (2004)
14. Liu, F., Chaudhary, V.: Extending OpenMP for Heterogeneous Chip Multiprocessors. In: Proc. of the 2003 International Conference on Parallel Processing, Kaohsiung, Taiwan, October 2003, pp. 161–168 (2003)
15. Mitchell, J.: Sun's Vision for Secure Solutions for the Government. National Laboratories Information Technology Summit (June 2004)
16. Olukotun, K., Nayfeh, B., Hammond, L., Wilson, K., Chang, K.: The Case for a Single-Chip Multiprocessor. In: Proc. of the 7th International Conference on Architectural Support for Programming Languages and Operating Systems (ASPLOS 1996) (October 1996)
17. Tullsen, D.M., Eggers, S.J., Levy, H.M.: Simultaneous Multithreading: Maximizing On-Chip Parallelism. In: In Proceedings of the 22nd Intenational Symposium on Computer Architecture, June 1995, pp. 392–403 (1995)
18. Zhang, Y., Burcea, M., Cheng, V., Ho, R., Cheng, V., Voss, M.: An Adaptive OpenMP Loop Scheduler for Hyperthreaded SMPs. In: Proc. of PDCS 2004: International Conference on Parallel and Distributed Computing Systems, San Francisco, CA (September 2004)

# SPEC OpenMP Benchmarks on Four Generations of NEC SX Parallel Vector Systems

Matthias S. Müller

HLRS, University of Stuttgart, D-70550 Stuttgart, Germany
mueller@hlrs.de
http://www.hlrs.de/people/mueller

**Abstract.** We describe the performance characteristics of the SPEC OMP benchmarks on parallel vector supercomputers. Points of interest are vectorization, scalability and the comparison between different generations of the same family of NEC SX vector supercomputers. We relate the different performance development of the 11 different applications to different hardware properties of the machine and also to results of the EPCC microbenchmarks. Of special interest is the fact the the NEC SX parallel architecture is not cache consistent.

## 1 Introduction

The SPEC OMP benchmarks[1] are a set of OpenMP parallel applications that can be used to measure the performance of shared memory systems. Submitted results exist for a variety of vendors. However, since the vendors typically try to submit the best possible result the benchmarks are normally executed with the number of threads that achieve the highest performance on a given system. Therefore only a few scalability results exist. There have been research papers on the performance characteristics on small shared memory machines[3,2], large SMP systems [5,6] and on pseudo vector machines like the Hitachi SR8000 [7]. The goal of this paper was to gain insight into the behavior of this set of applications on parallel vector machines like the NEC SX-series.

## 2 System Description

The NEC SX series is a family of shared memory parallel vector supercomputers. All members of the family have a crossbar that connects the CPUs to the shared memory. An overview of the characteristics can be seen in Tab. 1. In addition, the SX4 has a memory system consisting of SSRAM memory, whereas the SX-5 and SX-6 use DRAM technology. The number of banks in the SX-4 is also significantly higher. To achieve a balanced system with regard to the memory performance the number of CPUs was reduced to 16 and 8 CPUs in the SX-5 and SX-6 respectively. Unlike most SMPs the SX systems are not cache consistent, therefore problems like false-sharing should not exist. The only cache consistency is between the CPU caches and the vector units. The number of memory banks is the same for SX-6 and SX-8. To reach twice the memory bandwidth the bank busy time should be reduced by a factor two. This is not exactly the case, especially for the version with DDR2 RAM that was used here. There is a version with FC-RAM that has better bank busy times. However, there are modifications in the memory system like a bank cache and improved stride 2 memory access that might compensate this effect.

Table 1. Basic performance characteristics of the NEC SX family

|  | SX-4 | SX-5e | SX-6 | SX-6+ | SX-8 |
|---|---|---|---|---|---|
| Peak Performance | 2 GF | 4 GF | 8 GF | 9 GF | 16 GF |
| Memory Bandwidh/CPU | 16 GB/s | 32 GB/s | 32 GB/s | 36 GB/s | 64 GB/s |
| Clock | 125 MHz | 250 MHz | 500 MHz | 563 MHz | 1GHz |
| max Nr. of CPUs | 32 | 16 | 8 | 8 | 8 |
| total Memory Bandwidth | 512 GB/s | 512 GB/s | 256 GB/s | 288 GB/s | 512 GB/s |

## 3 Short Description of the SPEC OMP Benchmark Suite

The SPEC OMP benchmark suite consists of 11 applications. With the exception of gafort they have their roots in the SPECfp 2000 benchmark suite. Just a short description of the application is provided here, for a more detailed description see [1]. Wupwise is a quantum chromodynamics code, it consists of 2400 lines of Fortran code. Swim is a program for shallow water modeling. It is a small program with 435 lines of F77 that is known to be memory intensive. Mgrid is a simple Multigrid solver with 489 lines of F77 computing a three dimensional potential field. It was adopted by SPEC from the NAS Parallel Benchmarks. Applu is a parabolic/elliptic PDE solver consisting of 3980 lines of F77. Five coupled nonlinear PDEs are solved on a 3 dimensional, structured grid. An implicit pseudo-time marching schemed is used, based on two-factor approximate factorization of the sparse Jacobian Matrix. This is functionally equivalent to a nonlinear block SSOR iterative scheme with lexicographic ordering. Galgel performs a fluid dynamics analysis of oscillatory instability. Equake calculates a seismic wave propagation. Apsi is a meteorology code, which calculates pollutant distribution. Gafort is a integer intensive genetic algorithm code. Fma3d is a finite element crash code. Art performs an image recognition using a neural network. It is written in C. Ammp is a molecular dynamics code in the field of computational chemistry, it is also implemented in C.

## 4 Porting to NEC SX Systems

The most difficult part in porting the benchmarks was the compilation of the SPEC tools on the target platform. This step was necessary because no binaries are provided by SPEC. The tools are a collection of perl scripts and GNU utilities like make and diff to control the build and run process. One problem in porting the tools was that the NEC systems only have very limited support of shared libraries and dynamic linking, a feature used by the Perl tools.

The compilation of the 11 application codes was almost straightforward. The only required modification was the replacement of the assignment OMP_LOCK_KIND = selected_int_kind(18) with OMP_LOCK_KIND = 8 in gafort.f90, because the selected integer kind of the first construct was not supported by the compiler.

It should be noted that our goal was not to reach the maximum performance, but to understand the performance characteristics. Many of the benchmarks have been performed on loaded machines, where only gang-scheduling and resource management ensured that the benchmarks were not affected by other applications running on the machine. Nevertheless, resources like the memory crossbar were shared.

All vectorization was done automatically by the compiler. No directives were inserted and no source code modifications were applied to improve the performance.

**Table 2.** Properties of the SPEC codes on the NEC SX

| Name | Lang. | Vratio | Vlen | Memory (MB) |
|---|---|---|---|---|
| 310.wupwise_m | F | 87.34 | 58.74 | 1488 |
| 312.swim_m | F | 99.75 | 253.48 | 1584 |
| 314.mgrid_m | F | 99.14 | 211.04 | 480 |
| 316.applu_m | F | 81.31 | 34.17 | 1520 |
| 318.galgel | F | 92.57 | 45.41 | 272 |
| 320.equake_m | C | 0.06 | 9.6 | 464 |
| 324.apsi_m | F | 76.70 | 23.02 | 1648 |
| 326.gafort_m | F | 40.25 | 59.60 | 1680 |
| 328.fma3d_m | F | 10.29 | 8.95 | 1040 |
| 330.art_m | C | 32.06 | 242.14 | 272 |
| 332.ammp_m | C | 76.67 | 102.79 | 176 |

Tab. 2 shows the reported performance characteristics on the NEC vector-machines. Three codes (swim, mgrid and galgel) show a vectorization ratio of more than 90%. Five codes have a ratio between 40% and 90% (wupwise, applu, apsi, gafort, ammp). The remaining three codes (art, fma3d and equake) show a vectorization below 40%.

The other important issue is the average vector length. Here, swim, mgrid, art and ammp have a vector length above 100. Only equake and fma3d have a very small vector length below 10. The maximum (and best) vector length would be 256.

Looking at the vectorization of the codes it is clear, that only swim, mgrid and maybe galgel will show good performance on the NEC SX. Better results can be achieved with source code modifications, but this was not the target of this work. Instead of the absolute performance, the relative performance between different generations of SX hardware as well as scalability is the main focus. If the balance between different performance properties is constant, the relative performance grow of all applications would be the same. Any difference in performance grow points therefore to different hardware characteristics that dominates the performance for this kind of applications.

## 5 Performance Measurements

For the scalability tests the number of threads were increased from 1 up to the number of CPUs in the machine. The results can be seen in Fig. 1 and Fig. 2, the results of a run with one thread on the SX-4 has been set to 1. Tab. 3 shows the relative performance between the different SX generations for one and eight threads. The efficiency of the parallel execution is shown in Fig. 4 and Fig. 3.

The following list summarizes the observations for the different applications:

**wupwise:** the code scales very well up to 32 threads. Due to the good scalability the total node performance is the same for SX-4, SX-5 and SX-6, because all three generations have the same node peak performance.

**swim:** This code is known to be very memory intensive. The scalability on SX-8 is limited, reaching only a speed-up of 4.5 on 8 threads. This is due to memory degradation on this platform. Tab. 3 shows that the single CPU of SX-8 is 7.02 times faster than a SX-4 CPU, but the 8 thread result is only 4.31 times faster. As explained earlier this is an effect of the relatively long bank busy time of the DDR2 version of SX-8.

**mgrid:** This code is also known to be quite memory intensive, with a slightly more irregular access pattern than swim, due to the multigrid method. The scalability is good on all platforms. There is an excellent performance improvement going from SX-6+ to SX-8. This is probably due to the improved stride 2 memory access.

**applu:** The code has only limited scalability and relatively small performance improvements on new platforms.

**galgel:** For galgel the hardware performance counters show a significant waiting time due to cache misses. This waiting time increases with the number of threads, resulting in poor scalability. For a vector computer cache misses are normally not an issue, but galgel only has a moderate vectorization ratio with a small vector length, thus scalar performance is important. For galgel the efficiency is less for machines with large CPU numbers due to the poor scalability.

**equake:** This code has limited scalability, especially on machines with fast CPUs, indicating high synchronization overhead.

**apsi:** The code shows good scalability and efficiency on all platforms.

**gafort:** This code shows superlinear scaling on SX-6. Again, gafort is one of the codes with small vectorization ration and vector length. Hardware performance counters confirm that this behavior is due to data cache misses.

**fma3d:** The code shows good scalability and efficiency on all platforms. Only on the SX-4 with up to 32 threads, the efficiency is lower.

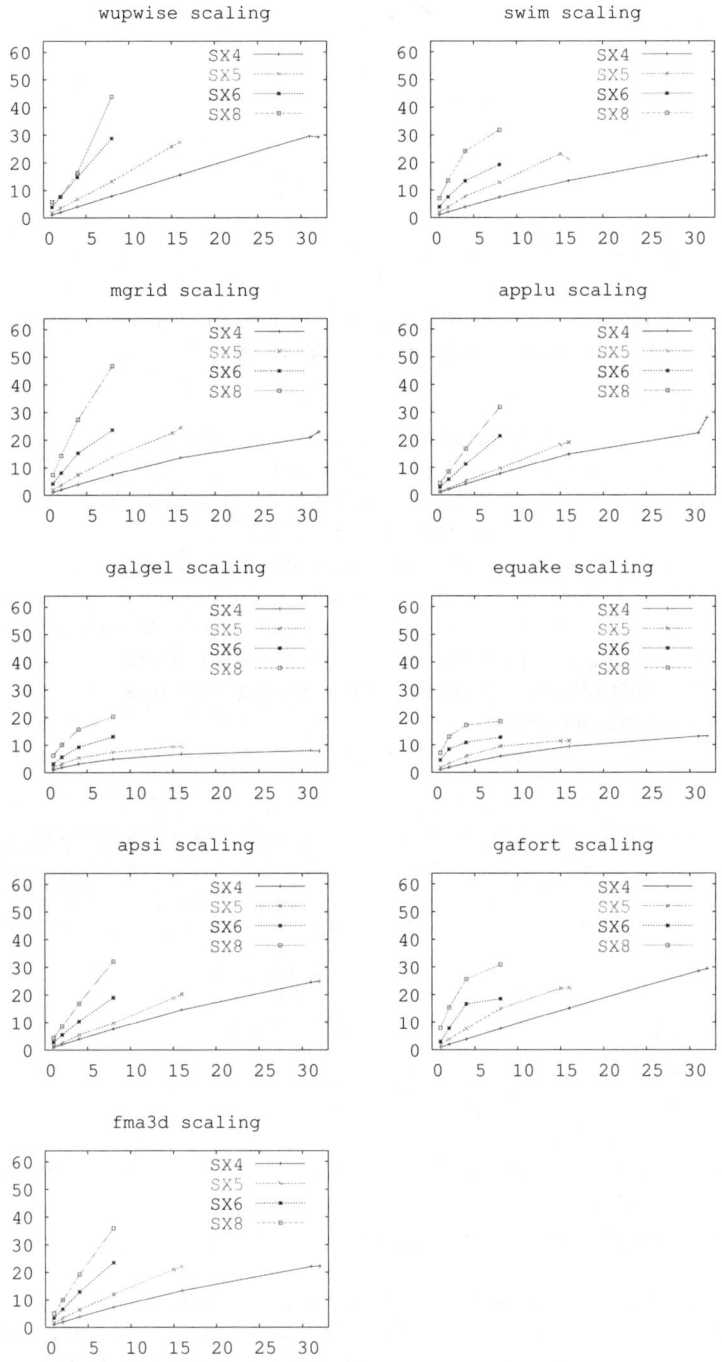

**Fig. 1.** Scalability of SPEC OMPM applications

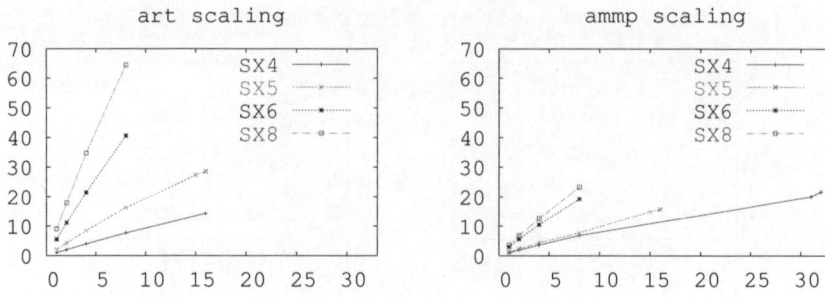

**Fig. 2.** The application art benefits from new architectural features, whereas ammp shows almost the same performance on SX-6 and SX-8

**art:** It was not possible to achieve a successful run with more than 16 threads on the SX-4. Comparing the different generations of SX systems, this code improves better than the peak performance. It clearly benefits from new architectural features introduced by the SX-6.

**ammp:** This is one of the codes that only show limited improvements between different SX generations: only 3.4 out of 8 when moving from SX-4 to SX-8. One reason is that this code applies a lot of OpenMP locks. On the SX6 the lock overhead is 4.3 microseconds, on SX-8 3.5 microseconds (measured by EPCC microbenchmarks [4]). This ratio perfectly relates to the observed ratio of ammp performance.

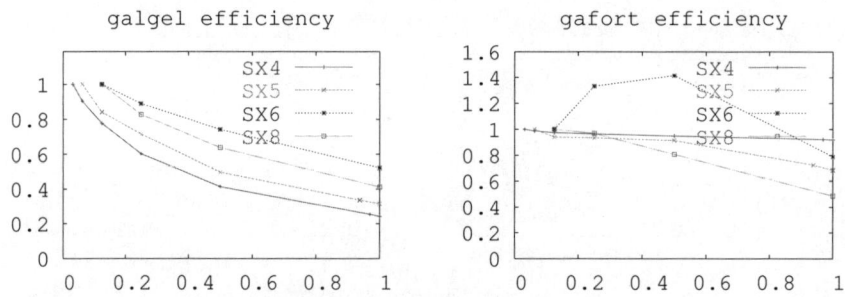

**Fig. 3.** Efficiency of galgel and gafort depending on the fraction of used CPUs

## 6 Summary and Conclusion

With the exceptions of galgel and equake all codes show good scalability. This shows that despite the fact that most current shared memory machines are cache consistent this is not required to achieve good performance with OpenMP. The single node efficiency is about the same for all generations of SX, showing that the platforms are well balanced.

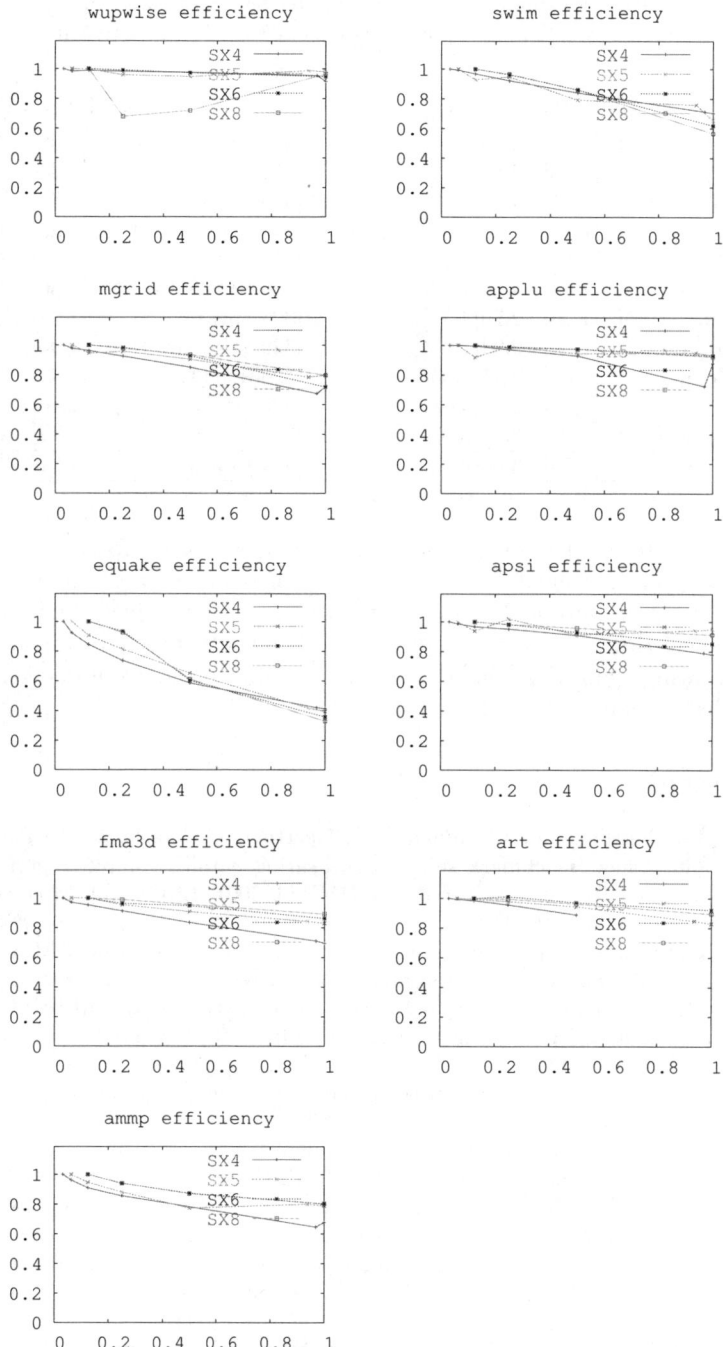

**Fig. 4.** Efficiency of SPEC OMPM applications

**Table 3.** Relative Performance for 1 thread (left) and 8 threads (right) on different SX models. Obviously different applications benefit to different extent from the faster models.

| Name | SX4 | SX5 | SX6+ | SX8 |
|---|---|---|---|---|
| 310.wupwise_m | 1 | 1.75 | 3.80 | 5.66 |
| 312.swim_m | 1 | 2.03 | 3.89 | 7.02 |
| 314.mgrid_m | 1 | 1.91 | 4.11 | 7.32 |
| 316.applu_m | 1 | 1.28 | 2.88 | 4.32 |
| 318.galgel | 1 | 1.85 | 3.12 | 6.16 |
| 320.equake_m | 1 | 1.81 | 4.48 | 7.04 |
| 324.apsi_m | 1 | 1.33 | 2.79 | 4.39 |
| 326.gafort_m | 1 | 2.06 | 2.94 | 7.94 |
| 328.fma3d_m | 1 | 1.66 | 3.41 | 5.03 |
| 330.art_m | 1 | 2.15 | 5.50 | 9.01 |
| 332.ammp_m | 1 | 1.25 | 3.00 | 3.62 |
| arith. mean | 1 | 1.73 | 3.63 | 6.14 |

| Name | SX4 | SX5 | SX6+ | SX8 |
|---|---|---|---|---|
| 310.wupwise_m | 1 | 1.68 | 3.66 | 5.56 |
| 312.swim_m | 1 | 1.74 | 2.61 | 4.31 |
| 314.mgrid_m | 1 | 1.87 | 3.19 | 6.32 |
| 316.applu_m | 1 | 1.25 | 2.76 | 4.11 |
| 318.galgel | 1 | 1.52 | 2.70 | 4.21 |
| 320.equake_m | 1 | 1.60 | 2.17 | 3.16 |
| 324.apsi_m | 1 | 1.27 | 2.49 | 4.21 |
| 326.gafort_m | 1 | 1.95 | 2.40 | 4.01 |
| 328.fma3d_m | 1 | 1.65 | 3.22 | 4.92 |
| 330.art_m | 1 | 2.10 | 5.30 | 8.43 |
| 332.ammp_m | 1 | 1.12 | 2.81 | 3.40 |
| arith. mean | 1 | 1.62 | 3.03 | 4.78 |

The absolute application performance gain was also compared to the peak performance improvement. This shows a lack of sustained performance also for vector computers, although they are known to have relatively high sustained performance. However, there is not much difference between vector and scalar codes, showing again that the architectural improvements from one generation to the next are well balanced.

# References

1. Aslot, V., Domeika, M., Eigenmann, R., Gaertner, G., Jones, W.B., Parady, B.: SPEComp: a new benchmark suite for measuring parallel computer performance. In: Eigenmann, R., Voss, M.J. (eds.) WOMPAT 2001. LNCS, vol. 2104, pp. 1–10. Springer, Heidelberg (2001)
2. Aslot, V., Eigenmann, R.: Quantitative performance analysis of the SPEC OMP2001 benchmarks. In: Scientific Programming (to appear)
3. Aslot, V., Eigenmann, R.: Performance characteristics of the SPEC OMP2001 benchmarks. In: 3rd European Workshop on OpenMP, EWOMP 2001, Barcelona, Spain (September 2001)
4. Bull, J.M.: Measuring synchronization and scheduling overheads in OpenMP. In: First European Workshop on OpenMP (1999)
5. Iwashita, H., Yamanaka, E., Sueyasu, N., van Waveren, M., Miura, K.: The SPEC OMP 2001 benchmark on the Fujitsu PRIMEPOWER system. In: 3rd European Workshop on OpenMP, EWOMP 2001, Barcelona, Spain (September 2001)
6. Saito, H., Gaertner, G., Jones, W., Eigenmann, R., Iwashita, H., Lieberman, R., van Waveren, M., Whitney, B.: Large system performance of SPEC OMP2001 benchmarks. In: Zima, H.P., Joe, K., Sato, M., Seo, Y., Shimasaki, M. (eds.) ISHPC 2002. LNCS, vol. 2327, pp. 370–379. Springer, Heidelberg (2002)
7. Takahashi, D., Sato, M., Boku, T.: Performance evaluation of the hitachi sr8000 using OpenMP benchmarks. In: Zima, H.P., Joe, K., Sato, M., Seo, Y., Shimasaki, M. (eds.) ISHPC 2002. LNCS, vol. 2327, pp. 390–400. Springer, Heidelberg (2002)

# Performance Evaluation of Parallel Sparse Matrix–Vector Products on SGI Altix3700

Hisashi Kotakemori[1], Hidehiko Hasegawa[2], Tamito Kajiyama[1], Akira Nukada[1], Reiji Suda[1], and Akira Nishida[1]

[1] CREST, Japan Science and Technology Agency
Graduate School of Information Science and Technology, University of Tokyo
7-3-1 Hongo, Bunkyo-ku, Tokyo 113-0033, Japan
{kota, kajiyama, nukada, reiji, nishida}@is.s.u-tokyo.ac.jp
[2] CREST, Japan Science and Technology Agency
Graduate School of Library, Information and Media Studies, University of Tsukuba
Tsukuba 305-8550, Japan
hasegawa@slis.tsukuba.ac.jp

**Abstract.** The present paper discusses scalable implementations of sparse matrix-vector products, which are crucial for high performance solutions of large-scale linear equations, on a cc-NUMA machine SGI Altix3700. Three storage formats for sparse matrices are evaluated, and scalability is attained by implementations considering the page allocation mechanism of the NUMA machine. Influences of the cache/memory bus architectures on the optimum choice of the storage format are examined, and scalable converters between storage formats shown to facilitate exploitation of storage formats of higher performance.

## 1 Introduction

Fast solution of linear equations with large sparse coefficient matrices is an essential requirement of advanced computations in science and engineering, and considerable research has been performed on solvers and preconditioners, and high performance implementations have been conducted to that end. We are planning to develop a new library for large-scale sparse matrix solutions that features a wide range of iterative solvers, preconditioners, and storage formats for sequential, shared memory and distributed memory parallel architectures. In the present paper, we discuss the performance of sparse matrix-vector products on a cc-NUMA machine SGI Altix3700.

The matrix-vector product is the most important kernel operation for iterative linear solvers, and its performance has a significant effect on the performance of linear solvers. We will show that satisfactory scalability cannot be attained unless the implementation is aware of the page allocation mechanism of the cc-NUMA machine. In addition, we will show that the storage format of the highest performance may be different for different matrices and on different architectures (CPU and memory), which indicates the importance of the availability of various

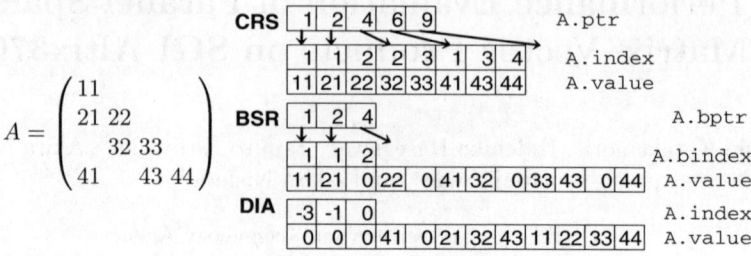

**Fig. 1.** Data structures of CRS, BSR and DIA. Example matrix $A$ is stored in these storage formats. The arrows in the figures of CRS and BSR designate that the elements of (b)ptr are used as indices to (b)index. The BSR of the block size is $r = 2$ and $c = 2$.

storage formats in a library. Since modifying their application programs for different storage formats is a burden on users, subroutines for conversions between storage formats are necessary. We will show that scalable parallel implementations of storage conversion routines enable performance enhancements by use of storage formats of higher performances.

## 2 Storage Formats and Their Implementations

A number of storage formats have been proposed for sparse matrices. They have been proposed for various objectives, such as simplicity, generality, performance, or convenience with respect to a specific algorithm. We implemented seven formats: Compressed Row Storage (CRS), Compressed Column Storage (CCS), Modified Compressed Sparse Row (MSR), Block Sparse Row (BSR), Diagonal (DIA), Ellpack-Itpack generalized diagonal (ELL) and Jagged Diagonal (JDS). In addition to CRS as the baseline format, only BSR and DIA are discussed in the following experiments, because the performance of matrix–vector products in the other formats was lower. In this section, the data structures and the parallel implementations of the matrix–vector products of CRS, BSR and DIA formats are discussed. The structures of the other formats can be found in [1,2,3,4,5]. Figure 1 shows an example matrix $A$ and data structures of CRS, BSR and DIA for $A$. In the following explanation, mathematically $A$ is assumed to be a square $n \times n$ matrix.

### 2.1 Compressed Row Storage (CRS)

The CRS format is shown by three arrays (ptr,index,value). Let $nnz$ be the number of the non-zero elements in matrix $A$. The double-precision array value of length $nnz$ stores the value of non-zero elements of matrix $A$ as they are traversed row-wise. The integer array index of length $nnz$ stores the column indices of the non-zero elements as stored in the array value. The integer array ptr of length $n+1$ stores pointers to the beginning of each row in the arrays value and index.

The following code shows the implementation of the matrix–vector product $y = Ax$ in the CRS format. It is parallelized at the outer loop, and thus (the computations related to) the rows of the matrix are distributed to the threads.

```
#pragma omp parallel for private(i,j,t)
  for(i=0; i<n; i++)  {
    t = 0.0;
    for(j=A.ptr[i];j<A.ptr[i+1];j++)
      t += A.value[j] * x[A.index[j]];
    y[i] = t;
  }
```

## 2.2 Block Sparse Row (BSR)

For BSR, the matrix is split into $r \times c$ submatrices (called blocks), where $r$ and $c$ are fixed integers. BSR stores the non-zero blocks (submatrices with at least one non-zero element) in a manner similar to CRS. Let $nr = n/r$ and $nnzb$ be the number of non-zero blocks in $A$. BSR is shown by three arrays (bptr, bindex, value). The double precision array value of length $nnzb \times r \times c$ stores the elements of the non-zero blocks: the first $r \times c$ elements are of the first non-zero block, and the next $r \times c$ elements are of the second non-zero block, etc. The integer array bindex of length $nnzb$ stores the block column indices of the non-zero blocks. The integer array bptr of length $nr + 1$ stores pointers to the beginning of each block row in the array bindex.

The code of the parallel matrix–vector product for BSR of the $2 \times 2$ block (i.e. $r = 2$ and $c = 2$) is shown below. A larger $r$ reduces the number of load instructions for the elements of the vector $x$, and a larger $c$ works as the unrolling of the inner loop, but this wastes memory and CPU power because of the zero elements in the non-zero blocks.

```
#pragma omp parallel for private(i,j,jj,t0,t1)
  for(i=0; i<nr; i++)  {
    t0 = t1 = 0.0;
    for(j=A.bptr[i];j<A.bptr[i+1];j++)  {
      jj = A.bindex[j];
      t0 += A.value[j*4+0] * x[jj*2+0] + A.value[j*4+2] * x[jj*2+1];
      t1 += A.value[j*4+1] * x[jj*2+0] + A.value[j*4+3] * x[jj*2+1];
    }
    y[2*i+0] = t0; y[2*i+1] = t1;
  }
```

## 2.3 Diagonal (DIA)

DIA is shown by two arrays (index, value). The double precision array value of length $nnd \times n$ stores the non-zero diagonals of the matrix $A$, where $nnd$ is the number of non-zero diagonals. The integer array index of length $nnd$ stores the offsets of each of the diagonals with respect to the main diagonal.

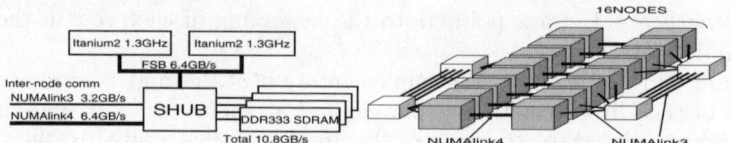

**Fig. 2.** System configuration of the SGI Altix3700. The left-hand side illustrates the inside of each node, and the right-hand side depicts the interconnections among the 16 nodes.

The code of the parallel matrix–vector product for DIA is shown below. In our implementation the storage scheme is modified so that the matrix elements accessed by each thread is stored in a contiguous region of memory. The inner loop is strip-mined with the number of threads, and interchanged with the outer loop.

```
#pragma omp parallel for private(i)
  for(i=0; i<n; i++)
    y[i] = 0.0;
#pragma omp parallel for private(i,j,k,is,ie,n1,n2,jj,ii)
  for(k=0;k<threads;k++)  {
    n1 = n/threads; n2 = n%threads;
    is = k<n2 ? (n1+1)*k : n1*k+n2;
    ie = k<n2 ? is+n1+1  : is+n1;
    for(j=0;j<nnd;j++)  {
      jj = A.index[j]; ii = _max(is,-jj)-_min(ie,n-jj);
      for(i=_max(is,-jj);i<_min(ie,n-jj);i++)
        y[i] += A.value[nn*k*nnd + j*nn + ii++] * x[jj+i];
    }
  }
```

### 2.4 The NUMA Architecture and Data Allocation for NUMA

The experiments reported in Section 3 are carried out on a cc-NUMA machine SGI Altix3700 that consists of 16 nodes, and as illustrated in the left-hand side of Fig. 2, each node has one memory controller called SHUB, to which two Itanium2 Madison 1.3-GHz processors with 16-KB L1 cache, 256-KB L2 cache, and 3-MB L3 cache for each are connected with a 6.4-GB/s shared front-side bus and four modules of 512-MB DDR333 SDRAM are connected with 10.8 GB/s. Two nodes are linked by a 6.4-GB/s NUMAlink4 interconnect, and four nodes are connected to one router through a 3.2-GB/s NUMAlink3, as shown in the right-hand side of Fig. 2.

The data is distributed by the first-touch mechanism, i.e. each page is stored in the memory of the node with a processor that accesses the page first. Because data must be transferred via interconnects when it is accessed by a processor out of the node that owns the data, each page should be assigned to the node with the processor that most often accesses the page in order to attain good performance.

To control page allocation, the arrays for matrix storage are initialized with zeros by the same threads as the matrix-vector products.

## 2.5 Conversion of Storage Format

Routines for the conversions from CRS to the other formats are based on those in the SPARSKIT [3]. Several modifications were necessary to control page allocation. For example, sequential implementation of the conversion from CRS to BSR fills three arrays (bptr, bindex and value) at the same time, but parallel implementation requires bptr to be filled first, because accesses to the arrays bindex and value are distributed referring to bptr as shown in the code in Section 2.2.

# 3 Experimental Results

The experiments are carried out on a cc-NUMA machine SGI Altix3700. An Intel C/C++ Compiler 8.1 is used with option -O3. In the experiments with 16 or fewer threads, the threads are allocated to different nodes using the dplace command, so that the front-side bus is dedicated to a single thread. In the experiments with 32 processors, the bus of each node is shared with the two processors in the node, and the effective memory access performance can be lowered.

Table 1 shows the dimensions, the number of non-zero elements, and the average number of non-zero elements per row for each test matrix used in the experiments. Matrices (a)–(e) are matrices from the Matrix Market[6], and matrix (f) is obtained by finite element discretization of the three–dimensional Poisson equation on a cube.

## 3.1 Parallel Performance

Table 2 shows the execution time in seconds for 1,000 iterations of matrix–vector products in various storage formats and numbers of threads. Other than CRS as the baseline, the storage formats that give the highest performance, designated in bold face digits, are presented.

First note the dependency of the performance on the matrix. For matrices (a), (c), (d), and (e), the best performance is attained by the same BSR_41. However, the relative performance of BSR_41 is less than twice that of CRS for most cases of (c) and (d), but is more than double that of CRS for (a) and (e). For matrix (b) BSR with another block size is the optimum, and for matrix (f) BSR with yet another block size is the best for eight threads or less and DIA is the best for 16 and 32 threads. These results lead to observations that (1) the performance is improved by optimizing the choice of the matrix storage format, and that (2) the best storage format differs for different matrices and machines (here, the number of processors used), and thus the availability of various storage formats in a library package is important.

**Table 1.** Test matrices for the experiments. Matrices (a) to (e) are from Matrix Market, and (f) is obtained by finite element discretization of the three-dimensional Poisson equation on a cube. The average number of the non-zero elements per row is shown in the column "Ave.".

| Name | Application area | Dimension | Nonzeros | Ave. |
|---|---|---|---|---|
| (a) af23560 | flows over airfoils | 23,560 | 484,256 | 20.55 |
| (b) fidapm37 | finite element modeling | 9,152 | 765,944 | 83.69 |
| (c) fidap011 | finite element modeling | 16,614 | 1,091,362 | 65.69 |
| (d) bcsstk30 | structural engineering | 28,924 | 2,043,492 | 70.65 |
| (e) s3dkq4m2 | cylindrical shell | 90,449 | 4,820,891 | 53.30 |
| (f) Poisson | Poisson eq. on a cube | 1,000,000 | 26,463,592 | 26.46 |

**Table 2.** Execution times (in seconds) of 1000 iterations of matrix–vector products (performance relative to CRS in parentheses). The block size for BSR is shown as BSR_rc for $r \times c$ blocks. The fastest implementation for each matrix and parallelism are shown in bold.

| # of threads | | 1 | 2 | 4 | 8 | 16 | 32 |
|---|---|---|---|---|---|---|---|
| Matrix | Format | | | | | | |
| (a) | CRS | 3.79 (1.00) | 1.89 (1.00) | 0.91 (1.00) | 0.46 (1.00) | 0.24 (1.00) | 0.14 (1.00) |
|  | BSR_41 | **1.46 (2.59)** | **0.72 (2.64)** | **0.28 (3.22)** | **0.15 (3.04)** | **0.09 (2.64)** | **0.07 (2.07)** |
| (b) | CRS | 2.53 (1.00) | 1.33 (1.00) | 0.63 (1.00) | 0.32 (1.00) | 0.18 (1.00) | 0.10 (1.00) |
|  | BSR_22 | **2.24 (1.13)** | **1.19 (1.12)** | **0.57 (1.11)** | **0.24 (1.34)** | **0.14 (1.26)** | **0.09 (1.18)** |
| (c) | CRS | 3.87 (1.00) | 1.98 (1.00) | 1.01 (1.00) | 0.48 (1.00) | 0.26 (1.00) | 0.15 (1.00) |
|  | BSR_41 | **2.51 (1.54)** | **1.30 (1.52)** | **0.65 (1.55)** | **0.24 (2.04)** | **0.13 (1.91)** | **0.09 (1.63)** |
| (d) | CRS | 6.81 (1.00) | 3.53 (1.00) | 1.88 (1.00) | 0.97 (1.00) | 0.46 (1.00) | 0.24 (1.00) |
|  | BSR_41 | **4.48 (1.52)** | **2.34 (1.51)** | **1.30 (1.44)** | **0.61 (1.60)** | **0.23 (1.96)** | **0.14 (1.75)** |
| (e) | CRS | 20.87 (1.00) | 10.47 (1.00) | 5.26 (1.00) | 2.71 (1.00) | 1.43 (1.00) | 0.68 (1.00) |
|  | BSR_41 | **9.17 (2.28)** | **4.65 (2.25)** | **2.39 (2.20)** | **1.30 (2.08)** | **0.62 (2.29)** | **0.27 (2.49)** |
| (f) | CRS | 149.50 (1.00) | 74.96 (1.00) | 37.43 (1.00) | 18.76 (1.00) | 9.51 (1.00) | 4.97 (1.00) |
|  | BSR_31 | **85.60 (1.75)** | **43.25 (1.73)** | **21.53 (1.74)** | **10.92 (1.72)** | 5.63 (1.69) | 4.87 (1.02) |
|  | DIA | 178.50 (0.84) | 89.19 (0.84) | 44.34 (0.84) | 16.40 (1.14) | **4.72 (2.02)** | **2.81 (1.77)** |

The speed-up ratios for the parallel matrix–vector products are shown in Table 3. The parallelization speed-ups are nearly ideal in most cases. Super linear speed-ups (speed-up ratios larger than the number of threads) are sometimes observed, possibly due to the improved cache hit rates because of much smaller data size. The speed-ups for 32 threads are much less than twice those for 16 threads, which may be due to sharing the bus with the two processors on a node. The lower speed-up ratios for 32 threads are most obvious for BSR formats, perhaps because BSR requires a greater number of memory accesses.

DIA outperforms BSR only for matrix (f), which has a regular 27-diagonal structure, and thus is stored very efficiently in the DIA format. However, DIA is still slower than BSR for eight threads or less. For 16 threads, the parallelization speed-up ratio of DIA is more than twice the number of threads, which may be ascribed to the lower memory requirement of DIA, which is approximately half

**Table 3.** Speed-up ratios for parallel matrix–vector products

| # of threads | | 1 | 2 | 4 | 8 | 16 | 32 |
|---|---|---|---|---|---|---|---|
| Matrix | Format | | | | | | |
| (a) | CRS | 1.00 | 2.00 | 4.18 | 8.19 | 15.51 | 27.16 |
|  | BSR_41 | 1.00 | 2.04 | 5.19 | 9.59 | 15.77 | 21.69 |
| (b) | CRS | 1.00 | 1.90 | 3.99 | 7.93 | 14.23 | 24.14 |
|  | BSR_22 | 1.00 | 1.88 | 3.91 | 9.42 | 15.90 | 25.18 |
| (c) | CRS | 1.00 | 1.95 | 3.82 | 8.03 | 15.13 | 26.50 |
|  | BSR_41 | 1.00 | 1.93 | 3.83 | 10.60 | 18.75 | 28.03 |
| (d) | CRS | 1.00 | 1.93 | 3.63 | 7.00 | 14.91 | 28.07 |
|  | BSR_41 | 1.00 | 1.91 | 3.45 | 7.34 | 19.22 | 32.21 |
| (e) | CRS | 1.00 | 1.99 | 3.97 | 7.70 | 14.61 | 30.72 |
|  | BSR_41 | 1.00 | 1.97 | 3.83 | 7.04 | 14.73 | 33.63 |
| (f) | CRS | 1.00 | 1.99 | 3.99 | 7.97 | 15.72 | 30.07 |
|  | BSR_31 | 1.00 | 1.98 | 3.97 | 7.84 | 15.20 | 17.58 |
|  | DIA | 1.00 | 2.00 | 4.03 | 10.88 | 37.84 | 63.51 |

**Table 4.** Execution times (in seconds) of 1000 iterations of matrix–vector products. Results of another set of experiments for matrix (f) in BSR_31.

| # of threads | 1 | 2 | 4 | 8 | 16 | 32 |
|---|---|---|---|---|---|---|
| One thread per node | 85.60 | 43.25 | 21.53 | 10.92 | 5.63 | — |
| Two threads per node | — | 73.07 | 36.58 | 18.48 | 9.33 | 4.87 |
| All data on a single node | 85.60 | 80.88 | 149.21 | 224.83 | 264.22 | 267.94 |

**Table 5.** Conversion times $T_{conv}$ (in milliseconds with the threshold numbers of iterations $N_{th}$ in parentheses) for the same sets of experiments as in Table 2

| # of threads | | 1 | 2 | 4 | 8 | 16 | 32 |
|---|---|---|---|---|---|---|---|
| Matrix | Format | | | | | | |
| (a) | BSR_41 | 61.2 (27) | 30.7 (27) | 15.0 (24) | 8.5 (28) | 6.7 (45) | 10.4 (144) |
| (b) | BSR_22 | 96.9 (332) | 50.8 (366) | 24.9 (410) | 12.4 (153) | 7.7 (209) | 8.5 (531) |
| (c) | BSR_41 | 132.8 (98) | 68.1 (100) | 35.4 (99) | 17.8 (73) | 11.1 (92) | 14.1 (251) |
| (d) | BSR_41 | 247.6 (107) | 132.3 (112) | 69.8 (122) | 35.9 (99) | 20.2 (90) | 22.2 (215) |
| (e) | BSR_41 | 575.9 (50) | 292.7 (51) | 148.5 (52) | 78.2 (56) | 47.7 (60) | 53.5 (132) |
| (f) | BSR_31 | 3370.8 (53) | 1720.3 (55) | 1073.5 (68) | 478.6 (61) | 303.8 (79) | 439.2 (4306) |
|  | DIA | 907.4 (-31) | 485.6 (-34) | 270.3 (-39) | 178.0 (76) | 165.7 (35) | 178.8 (83) |

that of BSR. With 32 threads, the parallelization speed-up ratio of BSR drops, which may be ascribed to the higher memory requirements of BSR. These results exemplify the heavy influences of the memory architecture (cache and shared bus) on the optimum choice of the storage format.

Table 4 shows the results of another set of experiments on the matrix (f) in BSR_31. The first line reproduces the results in Table 2. The second line

**Table 6.** Speed-up ratios for parallel conversion from CRS to target storage format

| # of threads | | 1 | 2 | 4 | 8 | 16 | 32 |
|---|---|---|---|---|---|---|---|
| Matrix | Format | | | | | | |
| (a) | BSR_41 | 1.00 | 1.99 | 4.07 | 7.21 | 9.11 | 5.90 |
| (b) | BSR_22 | 1.00 | 1.91 | 3.89 | 7.79 | 12.54 | 11.44 |
| (c) | BSR_41 | 1.00 | 1.95 | 3.75 | 7.47 | 11.92 | 9.38 |
| (d) | BSR_41 | 1.00 | 1.87 | 3.54 | 6.90 | 12.28 | 11.15 |
| (e) | BSR_41 | 1.00 | 1.97 | 3.88 | 7.37 | 12.08 | 10.76 |
| (f) | BSR_31 | 1.00 | 1.96 | 3.14 | 7.04 | 11.10 | 7.68 |
|     | DIA    | 1.00 | 1.87 | 3.36 | 5.10 | 5.48 | 5.07 |

gives the execution times with two threads assigned to each node. Here, the influences of the share of the front-side bus by the processors in a node on the performance are observed, and the absolute performances are much lower than in the previous experiments (expect for 32 threads). The speed-ups relative to the performance with two threads are steady up to 32 threads, which confirms that the shared bus is the reason for the lower speed-up ratio for 32 threads. For the third line of Table 4, the data structure of BSR_31 is constructed by a single thread, and thus all of the data are allocated to a single node. The data for the computations on the other node are accessed via the interconnections, and the resulting performances were poor. This confirms the importance of the data distribution discussed in Section 2.4.

### 3.2 Performance of Storage Format Conversions

Assume that a matrix $A$ is given in the CRS format (e.g. the user prefers the CRS format for some reason) and is to be multiplied to many vectors (e.g. an iterative linear solver is used). If a certain storage format (referred to hereinafter as the target format) is known to attain a higher performance than CRS in matrix–vector products for $A$, then it may be better to convert into the target format before the matrix–vector multiplications.

Let $T_{crs}$ and $T_{tgt}$ be the execution times of the matrix–vector product in the CRS and target formats, respectively, and let $T_{conv}$ be the execution time of the conversion from the CRS format to the target format. Define the threshold number of iterations $N_{th}$ as $N_{th} = \lceil T_{conv}/(T_{crs} - T_{tgt}) \rceil$. If the number of matrix–vector multiplications is at least $N_{th}$, then it is better to use the target format; otherwise, it is better to use CRS format without conversion.

Table 5 tabulates the conversion times $T_{conv}$ (with the threshold numbers of iterations $N_{th}$ in parentheses) for the same set of matrices, storage formats, and numbers of threads as Table 2. Note that the number of threads (if it is 16 or less) has little effect on $N_{th}$ in most cases of BSR. This is the case in which the parallelization speed-up ratios for the matrix–vector products in CRS and in BSR and for the conversion from CRS to BSR are similar. The speed-up ratios of the conversion routines are shown in Table 6.

With 32 threads, however, $T_{conv}$ tends to be longer than that with 16 threads, which can be ascribed to the share of the memory bus by the processors in a node. In this case, a strange phenomenon occurs. For matrix (a), execution by 16 threads takes less time than execution by 32 threads if the number of matrix–vector products is from 134 to 185, because the conversion time is shorter with 16 threads than with 32 threads. Accordingly, the threshold number of iterations $N_{th}$ for 32 threads is much larger than those for 16 threads or less. If the code is added using 32 threads, but following the conversion routine for 16 threads, then processors $2n$ and $2n + 1$ ($n = 0, ..., 15$) become the same node. Therefore, efficient execution is possible, because the data locality changes only slightly for 16 threads or 32 threads.

```
omp_set_num_threads(16);
#pragma omp parallel
cpubind(omp_get_thread_num()*2);
The Storage Format Conversion
omp_set_num_threads(32);
#pragma omp parallel
cpubind(omp_get_thread_num());
The Matrix-Vector Products
```

The conversion times $T_{conv}$ for the DIA format are shorter than those for BSR, perhaps because the amount of data to be stored for DIA is approximately half that for BSR. Negative $N_{th}$ means $T_{dia} > T_{crs}$, thus it is better NOT to convert the matrix into DIA format irrespective of the number of iterations.

## 4 Related Works

There are a variety of portable software packages that are applicable to the iterative solver of sparse linear systems. SPARSKIT [3] is a toolkit for sparse matrix computations written in Fortran. SPARSKIT provides a number of matrix storage formats, each of which has a pair of converters to and from the CRS format. Together with a rich set of matrix computation subroutines, the toolkit contains several sequential iterative solvers implemented based on reverse communication [7]. PETSc [8] is a C library for the numerical solution of partial differential equations and related problems, and can be used in application programs written in C, C++, and Fortran. The library provides an extensible set of matrix storage formats including various specialized formats that can be directly passed to external libraries. PETSc includes parallel implementations of iterative solvers and preconditioners based on MPI. Aztec [9] is another library of parallel iterative solvers and preconditioners and is written in C. Aztec provides two matrix storage formats. The library is fully parallelized using MPI and can be used in applications written in C and Fortran. From the viewpoint of functionality, our library and all three of the libraries mentioned above support different sets of matrix storage formats, iterative solvers, and preconditioners. Moreover, our library is parallelized using OpenMP and takes the cc-NUMA architecture into consideration.

The performance-enhancing techniques of sparse matrix–vector products are reported in [10,11]. A related issue is the selection of the best storage format for a given matrix and machine. In order to address this problem, E. Im [12] and Demmel et al. [13] proposed an automated empirical tuning mechanism for sparse matrix computations that selects an appropriate matrix storage format and solver implementation based on benchmarking data gathered in advance and structural characteristics of non-zero elements in a sparse matrix in hand.

## 5 Conclusions

In the present paper, we have discussed the parallel performance of matrix-vector product routines, which is crucial for high performance implementation of iterative linear solvers, on a cc-NUMA machine SGI Altix3700. Implementations that take into account the page allocation mechanism have attained satisfactory scalabilities. The memory architecture (specifically, the cache and the memory bus) have been observed to greatly affect the performance of matrix-vector products, and, consequently, storage formats that require more memory are influenced more. The baseline format CRS has scaled well up to 32 threads, and the performance of the BSR format that requires the most memory began to decrease at 32 threads, and the DIA format that requires the least memory has become faster for 16 threads or more. In order to maximize the performance of a machine, users must be able to choose an appropriate storage format for each matrix and each machine, and our scalable implementations of matrix-vector products and storage format conversions in a variety of storage formats enable such selection.

The target machine examined herein (SGI Altix3700) is a cc-NUMA machine. We are planning to port and to evaluate our codes to other shared memory parallel machines, including those having UMA (Uniform Memory Access) and SDSM (Software Distributed Shared Memory) architectures. Parallelization for distributed memory parallel machines through MPI and MPI-OpenMP hybrid parallelization is our next goal. We will also work toward high-performance iterative linear solvers using these kernel routines and effective preconditioners for the solvers, with the goal of developing a complete sparse linear solver library for sequential, shared memory and distributed memory parallel architectures.

## Acknowledgements

The authors would like to thank Associate Professor Kengo Nakajima (Tokyo University), Professor Yoshio Oyanagi (Tokyo University) and Dr. Akihiro Fujii (Kogakuin University) who provided valuable advice and discussions regarding the advanced research.

## References

1. Barrett, R., et al.: Templates for the Solution of Linear Systems: Building Blocks for Iterative Methods. SIAM, Philadelphia (1994)
2. Duff, I., Grimes, R., Lewis, J.: Sparse matrix test problems. ACM Trans. Math. Soft. 15, 1–14 (1989)

3. Saad, Y.: SPARSKIT: a basic took kit for sparse matrix computations, version 2, (June 1994), http://www.cs.umn.edu/~saad/software/SPARSKIT/sparskit.html
4. Kincaid, D., Oppe, T., Respess, J., Young, D.: ITPACKV2C User's Guide, Report CNA191. The University of Texas at Austin (1984)
5. Saad, Y.: Krylov subspace methods on supercomputers. SIAM J. Sci. Stat. Comput. 10, 1200–1232 (1989)
6. Matrix Market, http://math.nist.gov/MatrixMarket
7. Dongarra, J., Eijkhout, V., Kalhan, A.: Reverse communication interface for linear algebra templates for iterative methods. Technical Report UT-CS-95-291, University of Tennessee (May 1995)
8. Balay, S., Buschelman, K., Eijkhout, V., Gropp, W., Kaushik, D., Knepley, M., McInnes, L., Smith, B., Zhang, H.: PETSc users manual. Technical Report ANL-95/11, Argonne National Laboratory (August 2004)
9. Tuminaro, R.S., Heroux, M., Hutchinson, S.A., Shadid, J.N.: Official Aztec user's guide, version 2.1. Technical Report SAND99-8801J, Sandia National Laboratories (November 1999)
10. Toledo, S.: Improving the memory-system performance of sparse-matrix vector multiplication. IBM Journal of Research and Development 41(6), 711–725 (1997)
11. Pinar, A., Heath, M.T.: Improving Performance of Sparse Matrix-Vector Multiplication. Supercomputing 99 (1999)
12. Im, E.J.: Optimizing the performance of sparse matrix-vector multiplication. Ph.D. thesis, University of California (May 2000)
13. Demmel, J., Dongarra, J., Eijkhout, V., Fuentes, E., Petitet, A., Vuduc, R., Whaley, R.C., Yelick, K.: Self adapting linear algebra algorithms and software. Proceedings of the IEEE: Special Issue on Program Generation, Optimization, and Adaptation 93(2), 293–312 (2005)

# The OpenMP Language and Its Evaluation

# The OpenMP Memory Model

Jay P. Hoeflinger[1] and Bronis R. de Supinski[2]

[1] Intel, 1906 Fox Drive, Champaign, IL 61820
jay.p.hoeflinger@intel.com
http://www.intel.com

[2] Lawrence Livermore National Laboratory, P.O. Box 808, L-560,
Livermore, California 94551-0808[*]
bronis@llnl.gov
http://www.llnl.gov/casc

**Abstract.** The memory model of OpenMP has been widely misunderstood since the first OpenMP specification was published in 1997 (Fortran 1.0). The proposed OpenMP specification (version 2.5) includes a memory model section to address this issue. This section unifies and clarifies the text about the use of memory in all previous specifications, and relates the model to well-known memory consistency semantics. In this paper, we discuss the memory model and show its implications for future distributed shared memory implementations of OpenMP.

## 1 Introduction

Prior to the OpenMP version 2.5 specification, no separate OpenMP Memory Model section existed in any OpenMP specification. Previous specifications had scattered information about how memory behaves and how it is structured in an OpenMP program in several sections: the parallel directive section, the flush directive section, and the data sharing attributes section, to name a few. This has led to misunderstandings about how memory works in an OpenMP program, and how to use it.

The most problematic directive for users is probably the `flush` directive. New OpenMP users may wonder why it is needed, under what circumstances it must be used, and how to use it correctly. Perhaps worse, the use of explicit flushes often confuses even experienced OpenMP programmers.

Indeed, the SPEC OpenMP benchmark program ammp was recently found to be written with assumptions that violate the OpenMP version 2.0 (Fortran) [1] specification. The programmer apparently assumed that a full-memory flush was implied by acquiring and releasing an OpenMP lock. The OpenMP Fortran 2.0 specification is largely silent on the issue of whether a flush is implied by a lock acquire, probably creating the confusion that led to the error. One must go to the OpenMP C/C++ 2.0 [2] specification to find language that addresses the flush operation in relation to OpenMP locks, and even that language is ambiguous.

---

[*] This work was partially performed under the auspices of the U.S. Department of Energy by University of California LLNL under contract W-7405-Eng-48. UCRL-ABS-210774.

The proposed OpenMP 2.5 specification unifies the OpenMP Fortran and C specifications into a single document with a single set of rules, as much as possible. The OpenMP language committee has tried to provide a coherent framework for the way OpenMP relates to the base languages. One of the basic parts of this framework is the OpenMP memory model.

Up to now, the lack of a clear memory model has not made much difference. In general, compilers have not been very aggressive with code re-ordering optimizations and multiprocessors have been fairly simple in structure. Programs that did not follow the memory model would still usually work. But optimizing compilers are getting more sophisticated and aggressive. OpenMP implementations and machine architectures are getting more complicated all the time. Multi-core processors, NUMA machines and clusters of both are becoming more prevalent, making it all the more important that the nature of the memory behavior of OpenMP programs be clearly established.

In this paper, we describe the OpenMP memory model, how it relates to well-known memory consistency models, and the implications the model has for writing parallel programs with OpenMP. In section 2, we describe the OpenMP memory model, as it exists in the proposed OpenMP 2.5 specification. In section 3, we briefly discuss how the memory usage was addressed in previous OpenMP specifications, and how this has led to programmer confusion. In section 4, we show how the OpenMP memory model relates to existing memory consistency models. Finally, section 5 discusses the implications of using the OpenMP memory model to address distributed shared memory systems for OpenMP.

## 2 The OpenMP Memory Model

OpenMP assumes that there is a place for storing and retrieving data that is available to all threads, called the *memory*. Each thread may have a *temporary view* of memory that it can use instead of memory to store data temporarily when it need not be seen by other threads. Data can move between memory and a thread's temporary view, but can never move between temporary views directly, without going through memory.

Each variable used within a parallel region is either shared or private. The variable names used within a parallel construct relate to the program variables visible at the point of the parallel directive, referred to as their "original variables". Each shared variable reference inside the construct refers to the original variable of the same name. For each private variable, a reference to the variable name inside the construct refers to a variable of the same type and size as the original variable, but private to the thread. That is, it is not accessible by other threads.

There are two aspects of memory system behavior relating to shared memory parallel programs: coherence and consistency [3]. Coherence refers to the behavior of the memory system when a single memory location is accessed by multiple threads. Consistency refers to the ordering of accesses to different memory locations, observable from various threads in the system.

OpenMP doesn't specify any coherence behavior of the memory system. That is left to the underlying base language and computer system. OpenMP does not guarantee anything about the result of memory operations that constitute data races within a program. A data race in this context is defined to be accesses to a single variable by at least two threads, at least one of which is a write, not separated by a synchronization operation. OpenMP *does* guarantee certain consistency behavior, however. That behavior is based on the OpenMP *flush* operation.

The OpenMP flush operation is applied to a set of variables called the *flush set*. Memory operations for variables in the flush set that precede the flush in program execution order must be firmly lodged in memory and available to all threads before the flush completes, and memory operations for variables in the flush set, that follow a flush in program order cannot start until the flush completes. A flush also causes any values of the flush set variables that were captured in the temporary view, to be discarded, so that later reads for those variables will come directly from memory.

A `flush` without a list of variable names flushes all variables visible at that point in the program. A `flush` with a list flushes only the variables in the list.

The OpenMP flush operation is the only way in an OpenMP program, to guarantee that a value will move between two threads. In order to move a value from one thread to a second thread, OpenMP requires these four actions in exactly the following order:

1. the first thread writes the value to the shared variable,
2. the first thread flushes the variable.
3. the second thread flushes the variable and
4. the second thread reads the variable.

```
1:  A = 1

    ...

2:  Flush(A)
```

**Fig. 1.** A write to shared variable A may complete as soon as point 1, and as late as point 2

The flush operation and the temporary view allow OpenMP implementations to optimize reads and writes of shared variables. For example, consider the program fragment in Figure 1. The write to variable A may complete as soon as point 1 in the figure. However, the OpenMP implementation is allowed to execute the computation denoted as "..." in the figure, before the write to A completes. The write need not complete until point 2, when it must be firmly lodged in memory and available to all other threads. If an OpenMP implementation uses a temporary view, then a read of A during the "..." computation in Figure 1 can be satisfied from the temporary view, instead of going all the way to memory for the value. So, flush and the temporary view together allow an implementation to hide both write and read latency.

A flush of all visible variables is implied 1) in a barrier region, 2) at entry and exit from parallel, critical and ordered regions, 3) at entry and exit from combined parallel work-sharing regions, and 4) during lock API routines. The flushes associated with the lock routines were specifically added in the OpenMP 2.5 specification, a distinct change to both 2.0 specifications, as discussed in the following section. A flush with a list is implied at entry to and exit from atomic regions, where the list contains the object being updated.

The C and C++ languages include the `volatile` qualifier, which provides a consistency mechanism for C and C++ that is related to the OpenMP consistency mechanism. When a variable is qualified with `volatile`, an OpenMP program must behave as if a flush operation with that variable as the flush set were inserted in the program. When a read is done for the variable, the program must behave as if a flush were inserted in the program at the sequence point prior to the read. When a write is done for the variable, the program must behave as if a flush were inserted in the program at the sequence point after the write.

Another aspect of the memory model is the accessibility of various memory locations. OpenMP has three types of accessibility: shared, private and threadprivate. Shared variables are accessible by all threads of a thread team and any of their descendant threads in nested parallel regions.

Access to private variables is restricted. If a private variable X is created for one thread upon entry to a parallel region, the sibling threads in the same team, and their descendant threads, must not access it. However, if the thread for which X was created encounters a new parallel directive (becoming the master thread for the inner team), it is permissible for the descendant threads in the inner team to access X, either directly as a shared variable, or through a pointer. The difference between access by sibling threads and access by the descendant threads is that the variable X is guaranteed to be still available to descendant threads, while it might be popped off the stack before siblings can access it. For a threadprivate variable, only the thread to which it is private may access it, regardless of nested parallelism.

| #pragma omp parallel private(x) shared(p0,p1) | | | |
|---|---|---|---|
| **Thread 0** | | **Thread 1** | |
| x = ...; | | x = ...; | |
| p0 = &x; | | p1 = &x; | |
| /* references in the following line are not allowed: */ | | | |
| ...*p1 ... | | ...*p0 ... | |
| #pragma omp parallel shared(x) | | | |
| **Thread 0** | **Thread 1** | **Thread 0** | **Thread 1** |
| ... x ... | ... x ... | ... x ... | ... x ... |
| ...*p0 ... | ...*p0 ... | ...*p1 ... | ...*p1 ... |
| /* the following are not allowed: */ | | | |
| ...*p1 ... | ...*p1 ... | ...*p0 ... | ...*p0 ... |

**Fig. 2.** Access to a private variable by name or through a pointer is allowed only on the thread to which the variable is private, and its descendant threads

## 3 Memory Usage Descriptions in Previous Specifications

OpenMP specifications prior to OpenMP 2.5 barely addressed the OpenMP memory model. In the 2.0 C/C++ spec, the memory model was discussed in a paragraph in the execution model section, and in some text in the description of the flush directive. The 2.0 Fortran spec includes similar text in the description of the flush directive. It has no equivalent paragraph in the execution model section, although a paragraph in the section on the shared clause serves this purpose. The "data sharing attribute clauses" section in the C/C++ 2.0 spec and the "data scope attribute clauses" section of Fortran 2.0 describe the affects of the shared and private clauses.

The scattered text of the 2.0 specifications collectively gives an impression of memory behavior without being comprehensive. Nowhere in the 2.0 or earlier specs was there a mention of a temporary view of memory, but processor registers were mentioned. The proposed 2.5 specification has made this temporary view more general, which allows other forms of temporary memory.

Another issue related to the memory model is whether flushes are implied by the OpenMP lock API routines. The Fortran 2.0 spec is silent on the issue. However, those routines are not mentioned in the list of places where a flush is implied, so it is clear that the intention was that the lock routines do not imply flushes. The C/C++ 2.0 spec is likewise silent, but says "There may be a need for flush directives to make the values of other variables consistent."

The lack of a clear statement in previous specs with respect to flushes for the lock API routines has caused significant confusion. A very common mistake made by programmers is to forget to insert appropriate flushes when locks are being used.

| Thread 0 | Thread 1 |
|---|---|
| `omp_set_lock(lockvar);`<br>`count++;`<br>`omp_unset_lock(lockvar);` | |
| | `omp_set_lock(lockvar);`<br>`count++;`<br>`omp_unset_lock(lockvar);` |

Fig. 3. Threads cooperating through locks to increment a shared variable `count`

Consider the example in Figure 3. Most programs are written in this fashion, but without an implied flush in the `omp_set_lock` or `omp_unset_lock` routines, this program may not work as expected. This is because OpenMP semantics do not require a read of `count` from memory before the increment operation, or a flush of `count` to memory after it. Both threads are allowed to operate only on their temporary view of `count`. Even worse, the compiler might very well in-line the calls and reorder the update of `count` such that it is no longer in the locked region since there is no dependence between the calls and the variable `count`.

```
Thread 0                          Thread 1
omp_set_lock(lockvar);
#pragma omp flush(count)
count++;
#pragma omp flush(count)
omp_unset_lock(lockvar);
                                  omp_set_lock(lockvar);
                                  #pragma omp flush(count)
                                  count++;
                                  #pragma omp flush(count)
                                  omp_unset_lock(lockvar);
```

**Fig. 4.** A failed attempt to correctly use variables inside a locked region

```
Thread 0                          Thread 1
omp_set_lock(lockvar);
#pragma omp flush(count,lockvar)
count++;
#pragma omp flush(count,lockvar)
omp_unset_lock(lockvar);
                                  omp_set_lock(lockvar);
                                  #pragma omp flush(count,lockvar)
                                  count++;
                                  #pragma omp flush(count,lockvar)
                                  omp_unset_lock(lockvar);
```

**Fig. 5.** A correct way to write a locked update according to OpenMP 2.0

Including flushes of count inside the locked region, as in Figure 4, ensures that the most recent value for count is read, and that memory is updated with the write. However, it still does not address the compiler reordering problem. Essentially, these flushes on count do not ensure any ordering with operations on lockvar. The compiler is still free to reorder the call to omp_set_lock with respect to the flushes and the increment of count because they don't refer to the same variables.

The programmer would need to write the code as in Figure 5 to both prevent reordering with respect to the lock calls, and to keep the global value of count up to date. That is, the programmer must ensure ordering between the two variables by including both in the flush list.

A no-list flush is implicit for the lock API routines in the proposed 2.5 spec. Thus, code written in the natural manner of Figure 3 will work as most programmers expect. As mentioned above, the SPEC OpenMP code ammp was written in this manner (see Figure 6).

```
#ifdef _OPENMP
        omp_set_lock(&(a1->lock));
#endif
        a1fx = a1->fx;
        a1fy = a1->fy;
        a1fz = a1->fz;
        a1->fx = 0;
        a1->fy = 0;
        a1->fz = 0;
        xt = a1->dx*lambda +a1->x - a1->px;
        yt = a1->dy*lambda +a1->y - a1->py;
        zt = a1->dz*lambda +a1->z - a1->pz;
#ifdef _OPENMP
        omp_unset_lock(&(a1->lock));
#endif
```

**Fig. 6.** SPEC OpenMP benchmark ammp source code that demonstrates failure to use flush directives with OpenMP locks (incorrect prior to specification version 2.5)

## 4 The OpenMP Memory Consistency Model

OpenMP provides a relaxed consistency model that is similar to *weak ordering* [7][8]. Strong consistency models enforce program order, an ordering constraint that requires memory operations to appear to execute in the sequential order specified by the program. For example, a memory model is sequentially consistent if "the result of any execution is the same as if the operations of all the processors were executed in some sequential order, and the operations of each individual processor appear in this sequence in the order specified by its program" [3]. The OpenMP memory model specifically allows the reordering of accesses within a thread to different variables unless they are separated by a flush that includes the variables. Intuitively, the temporary view is not always required to be consistent with memory. In fact, the temporary views of the threads can diverge during the execution of a parallel region and flushes (both implicit and explicit) enforce consistency between temporary views.

Memory consistency models for parallel machines are based on the ordering enforced for memory accesses to different locations by a single processor. We denote memory access ordering constraints by using the "$\rightarrow$" (ordering) notation applied to reads (R), writes (W), and synchronizations (S). For instance, for reads preceding writes in program execution order, constraining them to maintain that order would be denoted $R \rightarrow W$. Sequential consistency requires all memory accesses to complete in the same order as they occur in program execution, meaning the orderings $R \rightarrow R$, $R \rightarrow W$, $W \rightarrow R$, and $W \rightarrow W$. It also requires the effect of the accesses by all threads to be equivalent to performing them in some total (i.e., sequential) order.

Sequential consistency is often considered difficult to maintain in modern multiprocessors. The program order restriction prevents many important compiler optimizations that reorder program statements [4]. Frequently, sequentially consistent multiprocessors do not complete a write until its effect is available to all other processors.

Relaxed consistency models remove the ordering guarantees for certain reads and writes, but typically retain them around synchronizations [4][5][6]. There are many types of relaxed consistency. The OpenMP memory model is most closely related to weak ordering. Weak ordering prohibits overlapping a synchronization operation with any other shared memory operations of the same thread, while synchronization operations must be sequentially consistent with other synchronization operations. Thus, the set of orderings guaranteed by weak ordering is the following: {S→W, S→R, R→S, W→S, S→S}.

Relaxed consistency models have been successful because most memory operations in real parallel programs can proceed correctly without waiting for previous operations to complete. Successful parallel programs arrange for a huge percentage of the work to be done independently by the processors, with only a tiny fraction of the memory accesses being due to synchronization. Thus, relaxed consistency semantics allows the overlapping of computation with memory access time, effectively hiding memory latency during program execution.

OpenMP makes no guarantees about the ordering of operations during a parallel region, except around flush operations. Flush operations are implied by all synchronization operations in OpenMP. So, an optimizing compiler can reorder operations inside a parallel region, but cannot move them into or out of a parallel region, or around synchronization operations. The flush operations implied by the synchronization operations form memory fences. Thus, the OpenMP memory model relaxes the order of memory accesses except around synchronization operations, which is essentially the definition of weak ordering.

The programmer can use explicit flushes to insert memory fences in the code that are not associated with synchronization operations. Thus, the OpenMP memory consistency model is a variant of weak ordering.

The OpenMP memory model further alters weak ordering by allowing flushes to apply only to a subset of a program's memory locations. The atomic construct includes an implied flush with a flush set consisting of only the object being updated. An optimizing compiler can reorder accesses to items not in the flush set with respect to the flush. Further, no ordering restrictions between flushes with empty flush set intersections are implied. In general, using a flush set implies that memory access ordering is only required for that set. The correct use of flush sets can be very complicated and we urge OpenMP users to avoid them in general.

The ordering constraint of OpenMP flushes is modeled on sequential consistency, similar to the restrictions on synchronization operations in weak ordering and lazy release consistency [7][8][9]. Specifically, the OpenMP memory model guarantees:

1. If the intersection of the flush-sets of two flushes performed by two different threads is non-empty, then the two flushes must be completed as if in some sequential order, seen by all threads;
2. If the intersection of the flush-sets of two flushes performed by one thread is non-empty, then the two flushes must appear to be completed in that thread's program order;
3. If the intersection of the flush-sets of two flushes is empty, then the threads can observe these flushes in any order.

If an OpenMP program uses synchronization constructs and flushes to avoid data races, then it will execute as if the memory was sequentially consistent.

## 5  Future Implications of the OpenMP Memory Model

Messaging latency to remote nodes in a modern computational cluster is hundreds or thousands of times higher than latency to memory for modern processors (see Figure 7). This latency makes the traditional method of enforcing sequential consistency (requiring a thread that issues a write of a shared variable to wait for the value to be visible to all threads) prohibitively expensive for OpenMP clusters. Fortunately, the OpenMP memory consistency model allows latency hiding of memory operations. This freedom is useful for a hardware shared memory (HSM) OpenMP implementation, but it is essential for a distributed shared memory (DSM) version of OpenMP, which simply has more memory latency to hide.

> latency to L1: 1-2 cycles
> latency to L2: 5 - 7 cycles
> latency to L3: 12 - 21 cycles
> latency to memory: 180 – 225 cycles
> Gigabit Ethernet - latency to remote node: ~28000 cycles
> Infiniband - latency to remote node: ~23000 cycles

**Fig. 7.** Itanium® latency to caches compared with latency to remote nodes

So, we claim that the relaxed memory model of OpenMP, with its ability to do cheap reads and hide the latency of writes, enables DSM OpenMP implementations. Without the ability to hide a cluster's enormous memory latency, DSM OpenMP implementations might only be useful for embarrassingly-parallel applications.

Even taking advantage of latency hiding, a DSM OpenMP implementation may be useful for only certain types of applications. In an Intel® prototype DSM OpenMP system, called Cluster OMP, we have found that codes in which certain characteristics dominate are very difficult to make perform well, while codes with other dominant characteristics can have good performance.

Codes that use flushes frequently tend to perform poorly. This means that codes that are dominated by fine-grained locking, or codes using a large number of parallel regions with small amounts of computation inside, typically have poor performance. Frequent flushes emphasize the huge latency between nodes on a cluster, since they reduce the frequency of operations that can be overlapped.

Codes dominated by poor data locality are also unlikely to perform well with DSM OpenMP implementations. Poor data locality for the Cluster OMP system means that memory pages are being touched by multiple threads. This implies that more data will be moving between threads, over the cluster interconnect. This data movement taxes the cluster interconnection network more than for a code with good data locality. The more data being moved, the more messaging overheads will hurt performance.

On the other hand, in experiments with the Cluster OMP system, we have observed that certain applications achieved speedups that approach the speedups obtained with an HSM system (see Figure 8). We have seen that computation with good data locality and little synchronization dominates the highest performing codes.

The applications we tested were gathered from Intel® customers who were participating in a technology preview of the prototype system. We can't reveal details of the codes, but the application types were:

1. a particle simulation code
2. a magneto-hydro-dynamics code
3. a computational fluid dynamics code
4. a structural simulation code
5. a graph processing code
6. a linear solver code
7. an x-ray crystallography code

Figure 8 shows the performance results we obtained for these codes. The speedup is shown for both the OpenMP and Cluster OMP versions of each code. In addition, the ratio of those speedups is shown, in the form of the Cluster OMP speedup as a

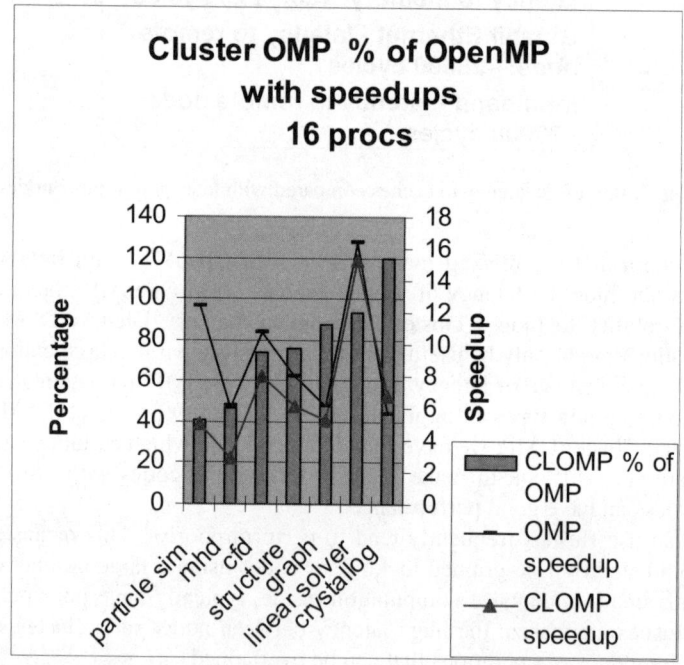

**Fig. 8.** Raw speedup of Cluster OMP on a cluster and OpenMP on a hardware shared memory machine, plus speedup percentage of Cluster OMP versus OpenMP for a set of codes

percentage of the OpenMP speedup. For these seven codes, five of them achieved a Cluster OMP speedup that was more than 70% of the performance of the OpenMP performance.

We have found that the applications for which Cluster OMP works well are usually those that have a large amount of read-only shared data and a small amount of read-write shared data. If, in addition, the data access patterns are irregular, then the applications would be more difficult to write using a direct messaging API, such as MPI.

These characteristics are typical of an emerging class of applications, known as RMS workloads (Recognition, Mining, and Synthesis) [10]. These workloads involve applications that typically use massive amounts of input data, such as pattern recognition, parallel search, data mining, visualization, and synthesis. We speculate that a DSM implementation of OpenMP may be useful for applications of this type.

## 6 Conclusion

The proposed OpenMP 2.5 spec unifies and clarifies the OpenMP memory model. The refined memory model description alleviates some of the confusion over the use of the `flush` directive and simplifies the correct use of OpenMP. Further, the proposed OpenMP 2.5 spec has added an implicit no-list flush to the lock API routines, making their use more intuitive.

OpenMP enforces a variant of weak ordering, as clearly demonstrated in the memory model description. This has performance implications for programs run on HSM systems, because it allows the compiler to apply optimizations and reorder code in a program. It also has important implications for reasonable performance of OpenMP on future DSM systems.

## References

1. OpenMP Architecture Review Board: OpenMP Fortran Application Program Interface, Version 2.0. OpenMP Architecture Review Board (2000)
2. OpenMP Architecture Review Board: OpenMP C and C++ Application Program Interface, Version 2.0. OpenMP Architecture Review Board (2002)
3. Lamport, L.: How to Make a Multiprocessor Computer that Correctly Executes Multiprocessor Programs. IEEE Transactions on Computers 28(9), 690–691 (1979)
4. Adve, S.V., Gharachorloo, K.: Shared Memory Consistency Models: A Tutorial. IEEE Computer 29(12), 66–76 (1996) (Also: WRL Research Report 95/7, Digital Western Research Laboratory, Palo Alto, California, 1995)
5. Hennessy, J.L., Patterson, D.A.,, C.A.: Computer Architecture A Quantitative Approach, 2nd edn. Morgan Kaufman Publishers, Inc., San Francisco (1996)
6. Gharachorloo, K.: Memory Consistency Models for Shared-Memory Multiprocessors. PhD thesis, Stanford University (1995)
7. Dubois, M., Scheurich, C., Briggs, F.: Memory Access Buffering in Multiprocessors.
8. Adve, S.V., Hill, M.D.: Weak Ordering – A New Definition. In: Proceedings of the 17th Annual International Symposium on Computer Architecture, May 1990, pp. 2–14 (1990)
9. Keleher, P.: Lazy Release Consistency for Distributed Shared Memory. PhD thesis, Department of Computer Science, Rice University (January 1995)
10. http://www.intel.com/technology/computing/archinnov/teraera/

# Evaluating OpenMP on Chip MultiThreading Platforms

Chunhua Liao, Zhenying Liu, Lei Huang, and Barbara Chapman

Computer Science Department, University of Houston, Texas
{liaoch, zliu, leihuang, chapman}@cs.uh.edu

**Abstract.** Recent computer architectures provide new kinds of on-chip parallelism, including support for multithreading. This trend toward hardware support for multithreading is expected to continue for PC, workstation and high-end architectures. Given the need to find sequences of independent instructions, and the difficulty of achieving this via compiler technology alone, OpenMP could become an excellent means for application developers to describe the parallelism inherent in applications for such architectures. In this paper, we report on several experiments designed to increase our understanding of the behavior of current OpenMP on such architectures. We have tested two different systems: a Sun Fire V490 with Chip Multiprocessor technology and a Dell Precision 450 workstation with Simultaneous MultiThreading technology. OpenMP performance is studied using the EPCC Microbenchmark suite, subsets of the benchmarks in SPEC OMPM2001 and the NAS parallel benchmark 3.0 suites.

## 1 Introduction

OpenMP has been successfully deployed on small-to-medium shared memory systems and large-scale DSMs, and is evolving over time. The OpenMP specification version 2.5 public draft [18] was released by the Architecture Review Board (ARB) in November 2004. It merged C/C++ and FORTRAN and clarified some concepts, especially with regard to the memory model. OpenMP 3.0 is expected to follow, and to consider a variety of new features. Among the many open issues are some tough challenges including determining how best to extend OpenMP to SMP clusters, how best to support new architectures, making hierarchical parallelism more powerful, and making it easier to write scalable code. In this paper, we explore some aspects of current OpenMP performance on two recent platforms: a Sun Fire V490 [24] with Chip Multiprocessor capability and a Dell Precision 450 workstation with Simultaneous Multithreading technology.

As computer components decrease in size, architects have begun to consider different strategies for exploiting the space on a chip. A recent trend is to implement Chip MultiThreading (CMT) in the hardware. This term refers to the simultaneous execution of two or more threads within one chip. It may be implemented through several physical processor cores in one chip (a Chip Multiprocessor, CMP) [17], a single core processor with replication of features to maintain the state of multiple threads simultaneously (Simultaneous multithreading, SMT) [26] or the combination of CMP and SMT [10]. OpenMP support for these new microarchitectures needs to be evaluated and possibly enhanced.

In this paper, we report on the behavior of some OpenMP benchmarks on each of the two systems mentioned above: the Sun Fire V490 exploiting CMP technology and the Dell Precision 450 workstation with SMT technology. The remainder of the paper is organized as follows. In Section 2, we discuss the architectures that are used in this experiment, and comment on their implications for OpenMP. In Section 3, we then describe the methodology of how to run the benchmarks, followed by our results and a discussion of them in Section 4 and 5. Finally, the paper presents related work and reaches some conclusions in Section 6 and 7.

## 2 Chip MultiThreading and Its Implications for OpenMP

CMT is emerging as the dominant trend in general-purpose processor design [23]. In a CMT processor, some chip-level resources are shared and further software approaches need to be explored to maximize overall performance. CMT may be implemented through Chip Multiprocessor (CMP) [14], Simultaneous Multithreading (SMT) [11] or the combination of CMP and SMT [10]. In this section, we give a brief overview of CMP and SMT first, and discuss their implications for OpenMP.

*Chip Multiprocessing* enables multiple threads to be executed on several physical cores within one chip. Each processor core has its own resources as well as shared ones. The extent of sharing varies from one implementation to another. For example, the UltraSPARC IV [27]'s two cores are almost completely independent except for the shared off-chip data paths while the Power4 [14] processor has two cores with shared L2 cache to facilitate fast inter-chip communication between threads.

*Simultaneous MultiThreading* combines hardware multithreading with superscalar processor technology to allow several independent threads to issue instructions to a superscalar's multiple function units each cycle [26]. SMT permits all thread contexts to simultaneously compete for and share processor resources; it uses multiple threads to compensate for low single-thread instruction-level parallelism.

CMP and SMT are two closely related technologies. They can be simply seen as two different extents of sharing of on-chip resources among threads. However, they are also significantly different because the various types of resource sharing have different implications for application performance, especially when the shared pipelines on SMT are compared with the private pipelines on CMP. Moreover, new multi-threaded chips such as Power5 [10], tend to integrate both CMP and SMT into one processor. This kind of integration brings even deeper memory hierarchy and more complex relationship between threads.

The CMP and SMT technology introduces new opportunities and challenges for OpenMP. The current flat view of OpenMP threads is not able to reflect these new features and thus may need to be revisited to ensure continuing applicability in the future. Previous research on SMT [21, 26, 28, 29] has developed some strategies for efficient sharing of key resources, especially caches. In OpenMP, we may need to identify sibling[1] threads to perform the work cooperatively with proper scheduling and load balancing mechanisms. We need to perform research to explore optimiza-

---

[1] We use the term "sibling" to refer to cores in the same CMP, and to logical processors in the same physical processor for SMT.

tions to avoid inter-thread competition for shared resources, and to select the best number of cores from a group of multiple cores. A CMP system with only one multi-core processor is really a slim implementation of SMP on a chip. While chip level integration has the benefits of fast synchronization and lower latency communication among threads, the shared resources may lead to conflicts between threads and unsatisfactory performance. Secondly, for SMPs composed of several multicore processor systems, the relationship between processing units is no longer strictly symmetric. For example, cores within one processor chip may have faster data exchange speed than cores crossing processor boundaries. Multithreading based on those cores has to take this asymmetry into account in order to achieve optimal performance. Altogether, the integration, resource sharing and hierarchical layout in SMP systems with CMT bring additional complexity to the tasks of data set partition, thread scheduling, work load distribution, and cache/memory locality maintenance.

## 3 Methodology

We have chosen a Sun Fire V490 with UltraSPARC IV [27] processors and a Dell Precision 450 workstation with Xeon [11] processors as test beds to explore the impact of CMT technology. In this section, we describe these two machines, benchmarks that we ran and how to execute the benchmarks. We attempt to understand the scalability of OpenMP applications on the new platforms, and the performance difference between SMPs with CMT and traditional SMPs via the designed experiments.

### 3.1 Sun Fire V490 with UltraSPARC IV and Dell Precision 450 with Xeon

Sun UltraSPARC IV was derived from earlier uniprocessor designs (UltraSPARC III) and the two cores do not share any resources, except for the off-chip data paths. The UltraSPARC IV processor is able to execute dual threads based on two 14-stage, 4-way superscalar UltraSPARC III [8] pipelines of two individual cores. Each core has its own private L1 cache and exclusive access to half of an off-chip 16MB L2 cache. The L1 cache has 64KB for data and 32K for instructions. L2 cache tags and a memory controller are integrated into the chip for faster access.

The Sun Fire V490 server for our experiments has four 1.05 GHz UltraSPARC IV processors and 32 GB main memory. The basic building block is a dual CPU/Memory module with two UltraSPARC IV cores, an external L2 cache and a 16 GB interleaved main memory. The Sun Fireplane$^{TM}$ Interconnect is used to connect processors to Memory and I/O devices. It is based on a 4-port crossbar switch with a 288-bit (256-bit data, 32-bit Error-Correcting Code) bus and clock rate of 150 MHz. The maximum transfer rate is thus 4.8 GB/sec. The Sun Fire V490 is loaded with the Solaris 10 operating system [22] and Sun Studio 10 [25] integrated development environment which support OpenMP 2.0 APIs for C/C++ and FORTRAN 95 programs. Solaris 10 allows superusers to enable or disable individual processor cores via its processor administration tool. An environment variable SUNW_OMP_PROBIND is available that lets users bind OpenMP threads to processors. More CMP-specific features are described in [15].

Xeon processors have Simultaneous MultiThreading (SMT) technology, which is called HyperThreading [11]. HyperThreading makes a single physical processor appear as two logical processors; most physical execution resources including all 3 levels of caches, execution units, branch predictors, control logic and buses are shared between the two logical processors, whereas state resources such as general-purpose registers are duplicated to permit concurrent execution of two threads of control. Since the vast majority of micro-architecture resources are shared, the additional hardware consumes less than 5% of the die area.

The Dell Precision 450 workstation, on which we carry out the experiments, has dual Xeon 2.4 GHZ CPUs with 512K L2 cache, 1.0GB memory and HyperThreading technology. The system runs Linux with kernel 2.6.3 SMP. The Omni compiler [19] is installed to support OpenMP applications and GCC 3.3.4 acts as a backend compiler. Linux Kernel 2.6.3 SMP [30] in our 2-way Dell Precision workstation has a scheduler which is aware of HyperThreading. This scheduler can recognize that two logical processors belong to the same physical processor, thus maintaining the load balance per physical CPU, not per logical CPU. We confirmed this via simple experiments that showed that two threads were mostly allocated to two different physical processors unless there was a third thread or process involved.

### 3.2 Experiments

Since it was not clear whether the Solaris 10 on our Sun Fire V490 with 4 dual-core processors is aware of the asymmetry among underlying logical processors, we used the Sun Performance *Analyzer* [25] to profile a simple OpenMP Jacobi code running with 2, 3, and 4 threads. We set the result timelines to display data per CPU instead of per thread in *Analyzer*, and found that Solaris 10 is indeed aware of the differences in processors of a multicore platform and tries to avoid scheduling threads to sibling cores. Therefore we can roughly assume the machine acts like a traditional SMP for OpenMP applications with only 4 or less active threads. For applications using 5 or more threads, there must be sibling cores working at the same time. As a result, any irregular performance change from 4 to 5 threads might be related to the deployment of sibling cores.

The EPCC Microbenchmark [4] Suite, SPEC OMPM2001 [2], and NAS parallel benchmark (NPB) 3.0 [9] were chosen to discover the impact of CMT technology on OpenMP applications. The EPCC microbenchmark is a popular program to test the overhead of OpenMP directives on a specific machine while the SPEC OMPM2001 and NAS OpenMP benchmarks are used as representative codes to help us understand the likely performance of real OpenMP applications on SMP systems with CMT.

For experiments running on the Sun Fire V490 machine, we compiled all the selected benchmarks using the Sun Studio 10 compiler suite with the generic compilation option "-fast -xopenmp" and ran them from 1 to 8 threads in multi-user mode. This round of experiments gave us a first sense of the OpenMP behavior on the CMP platform and exposed problematic benchmarks. After that, we ran the problematic benchmarks using 2 threads when only 2 cores were enabled: the two cores were either from the same CMP or from different processors (an approximate traditional SMP). For both cases, Sun performance tools were used to collect basic performance metrics as well as related hardware counter information in order to find the reasons for the performance difference between a CMP SMP and a traditional SMP. The experiments on the Dell Precision workstation were designed in the same fashion whenever possible. For example, we measured the performance of the EPCC microbenchmarks using 2 logical processors of the same physical processor and on 2 physical processors with HyperThreading disabled. This way, we can understand the influence of HyperThreading better.

## 4 Results and Analysis

This section illustrates the results of the experiments on the two machines using the selected benchmarks, and our analysis. Some major performance problems are explained with the help of performance tools. We are especially interested in the effects of the sibling cores and sibling processors. In other words, particular attention is paid to performance gaps when the number of threads changes from 4 to 5 on the Sun Fire V490 and from 2 to 3 on the Dell Precision 450.

### 4.1 The EPCC Microbenchmark Suite

#### 4.1.1 Sun Fire V490

Fig. 1 and 2[2] shows the results of the EPCC microbenchmark on the Sun Fire V490. Most OpenMP directives have higher overhead than on the Sun HPC 3500 [4], a traditional SMP machine, since the default behavior of idle threads has changed from *SPIN* to *SLEEP*. PARALLEL, PARALLEL FOR and REDUCTION have similar overhead on the Sun Fire V490. For mutual exclusion synchronization results shown in Fig. 2, LOCK and CRITICAL show similar overhead and scale well. The ORDERED directive has a high cost when the full eight threads are used, although it scales as well as LOCK and CRITICAL when less than eight threads are involved. The exception of ORDERED may come from hardware, OpenMP compiler implementation, or scheduler in Solaris 10. The

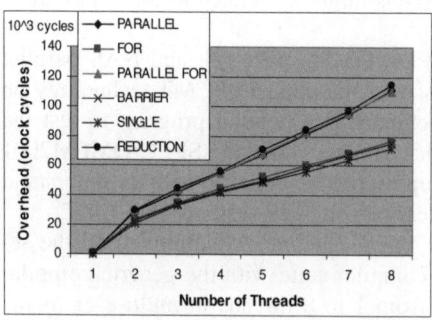
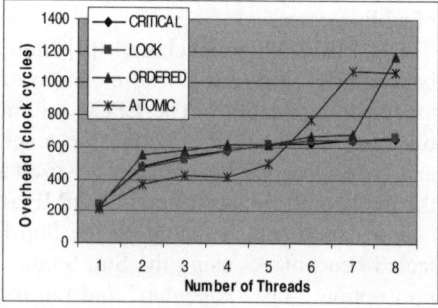

**Fig. 1.** Synchronization overhead on Sun Fire V490

**Fig. 2.** Mutual exclusion + ORDERED overhead on Fire 490

---

[2] We display the overhead of ORDERED and mutual exclusive directives in Fig. 2 as they have the same order of magnitude overhead.

ATOMIC directive is noticeably cheaper than LOCK and CRITICAL when the number of threads is up to five, however, it is more expensive than the other two if we use more than five threads. We checked the corresponding assembly code of an ATOMIC construct (see Table 1), and found that the Sun Studio 10 compiler uses runtime library calls **__mt_b_atomic_** and **__mt_e_atomic_** to start and finish an ATOMIC operation, rather than a single hardware primitive. The *Analyzer* indicated that the execution time of those two calls is sensitive to the physical layout of processor cores: ATOMIC does not have good performance if both sibling cores are involved.

**Table 1.** An OpenMP source and the corresponding assembly code segment for an ATOMIC construct

| An OpenMP code segment | The corresponding assembly code for the ATOMIC construct |
|---|---|
| ... <br> float aaaa=0; <br> #pragma omp parallel private(j) <br> { <br>    for (j=0; j<innerreps/nthreads; j++){ <br>    #pragma omp atomic <br>      aaaa += 1; <br>    } <br> } .... | call     __mt_b_atomic_    ! params = ! Result = <br> nop <br> ld    [%sp+92],%f2 <br> add    %l4,1,%l4 <br> ld    [%l5],%f0 <br> fadds    %f2,%f0,%f0 <br> call     __mt_e_atomic_    ! params = ! Result = <br> ... |

We show OpenMP loop scheduling results on the Sun Fire V490 in Fig. 3. Block cyclic scheduling (STATIC, n) and block scheduling (STATIC) have similar performance when the chunk size is not too small. The cost of GUIDED decreases noticeably when the chunk size is incremented up to 1024.

Altogether, the synchronization overhead for OpenMP directives does not show sensitivity to the change from 4 to 5 threads. We further designed some experiments to closely examine the effect of different combination of cores. We compare the

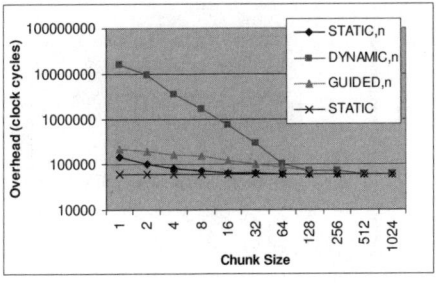 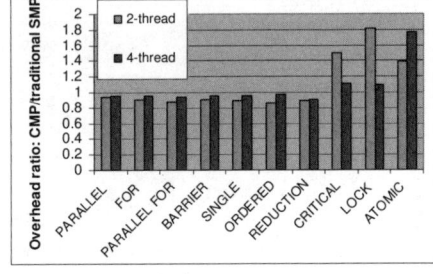

**Fig. 3.** Scheduling overhead on Sun Fire V490

**Fig. 4.** Synchronization overhead ratio: CMP/ traditional SMP

benchmark performance on two sibling cores and non-sibling cores (in fact, the former reflects the effects of CMP and the latter the effects of the traditional SMP), and on four cores from two processors and from four different processors, respectively. In order to ensure the desired core/processor layout, we take advantage of the *prsadm* utility from Sun Solaris 10 to turn off the cores that we will not use. Fig 4 shows the overhead ratio for OpenMP directives on cores belonging to the same processor(s) and different processor(s). OpenMP directives on CMP take slightly less time than on a traditional SMP except for three mutual exclusion directives: CRITICAL, LOCK and ATOMIC. The overall overhead difference tends to be smaller (close to ratio of 1) when more threads are used, except for the ATOMIC directive.

Therefore, we may conclude that sibling cores do not bring significantly faster synchronization or fewer overhead for OpenMP constructs on the Sun Fire V490 machine. This is because the sibling cores in the UltraSPARC IV do not share L1 and/or L2 cache to facilitate faster communications.

### 4.1.2 Dell Precision 450

We depict the results obtained by using the EPCC microbenchmarks to measure OpenMP synchronization overhead on the Dell Precision 450 in Fig. 5 and Fig. 6, and

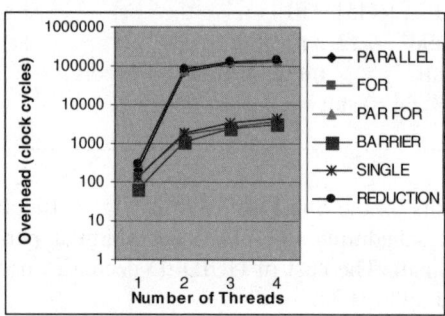

**Fig. 5.** Synchronization overhead on Dell Precision 450

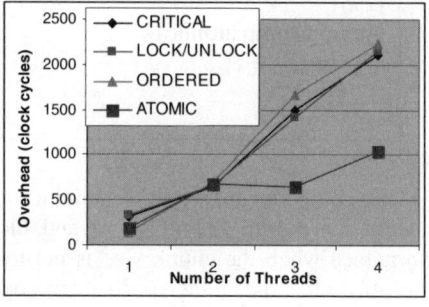

**Fig. 6.** Mutual exclusion + ORDERED overhead on Dell Precision 450

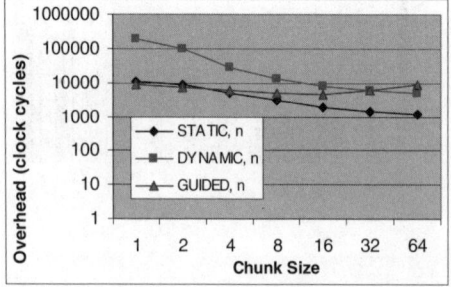

**Fig. 7.** Scheduling overhead on Dell Precision 450

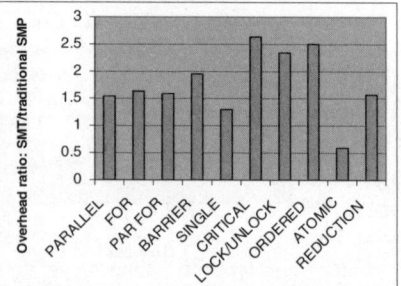

**Fig. 8.** Synchronization overhead ratio: SMT vs. traditional SMP

the results for OpenMP scheduling overhead in Fig. 7. Overall, the results are similar to those on a traditional SMP system. In Fig. 8, we display the overhead ratio between the two sibling processors and two physical processors. Only the overhead of ATOMIC is smaller in the case of sibling processors. The OpenMP synchronization directives are mostly implemented using *spin-wait*, which leads to more competition for shared resources on Xeon systems.

### 4.2 SPEC OMPM2001 and NAS NPB 3.0

#### 4.2.1 Sun Fire V490

We show the speedup of a subset from SPEC OMPM2001 and NPB 3.0 with class B dataset on the Fire V490 machine in Fig. 9 and Fig. 10 separately. In Fig. 9, WUPWISE, EQUAKE, and APSI show good scalability on Sun Fire V490, with only slight changes after 4 threads. AMMP scales as poorly as described on traditional SMPs [20]. The APPLU demonstrates super-linear speedup for no less than 6 threads, which start to provide enough L2 caches to accommodate its critical dataset as mentioned in [20]. As can be seen in Fig. 10, most NAS OpenMP benchmarks show a more consistent behavior during the change from 4 to 5 threads than those from SPEC. Only EP achieves linear speedup because its dataset is much smaller than the size of L2 cache.

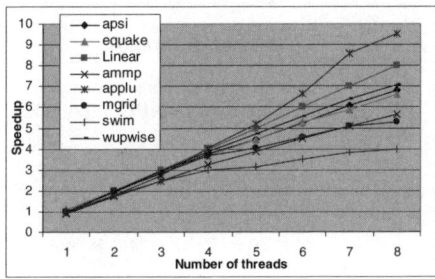
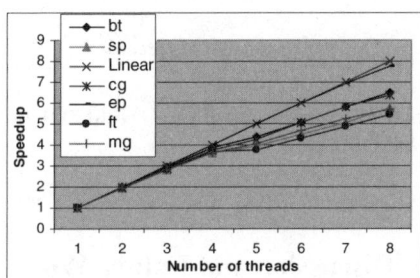

**Fig. 9.** Speedup of SPEC OMPM2001 on Sun Fire V490

**Fig. 10.** Speed up of NAS NPB 3.0 on Sun Fire V490

Several benchmarks do not scale very well on this platform, particularly SWIM from SPEC and FT from NPB 3.0. We profiled SWIM and FT using 2 threads on sibling and non-sibling cores, and the profiling results confirmed the negative impact of multicores, as sibling cores did cause longer L2 stalls in the major loops. Both SWIM and FT are memory-intensive so that they cannot benefit from the multicore architecture. For the medium dataset, the major loops of SWIM perform computations on fourteen arrays, each of which contains 3802 * 3802 double precision floating point numbers; hence the total memory requirement is 38022*8*14=1.51G bytes. Similarly, the FT benchmark with class B dataset also requires over 1 GB memory during its execution. Moreover, non-contiguous memory loads and stores of FT's major function cause high cache miss rates and competition for the shared data path of the sibling cores. Therefore, performance degradation occurs when the number of involved threads is increased from 4 to 5.

### 4.2.2 Dell Precision 450

We tested a subset of the SPEC OMPM2001 and NAS NPB 3.0 benchmarks on our dual-Xeon Linux workstation. Results are given in Fig. 11 and Fig. 12 respectively. EQUAKE from SPEC is a memory-intensive application in which more SMT threads lead to performance degradation due to the memory competition. The Xeon's halved memory bandwidth does not fulfill the demands of more threads [6]. We also observed that MGRID did not have a speedup with 2 threads. The reason is that the OS schedules the two threads onto a single physical processor and resource conflicts occur. From our experiments, it seems the HyperThreading-aware scheduler in Linux kernel 2.6.3 is not well implemented. We did not get good speedup for the NAS benchmarks except EP, which requires much less memory than others.

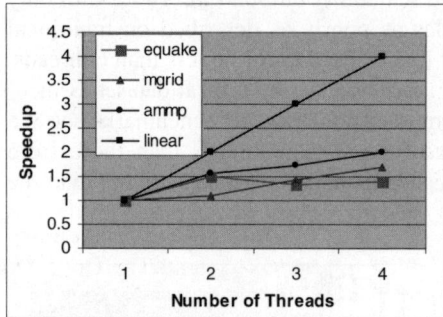

**Fig. 11.** Speedup of SPEC OMPM2001 on Dell Precision 450

**Fig. 12.** Speed up of NAS NPB 3.0 on Dell Precision 450

## 5 Discussion and Future Work

The Sun Fire V490 is a successful platform for OpenMP as one of the first generation CMP (multicore) machines. We find its overall scalability to be comparable to a traditional SMP machine since each core has the similar capability as a regular uniprocessor. Most OpenMP applications from SPEC OMPM2001 and NPB 3.0 scale very well. For memory-intensive applications using 5 and more threads, scalability may be compromised to an acceptable degree due to the competition for the shared data path between sibling cores. There should be a threshold for the applications' memory demand to be intensive enough to cause performance degradation on a specific machine. Unfortunately, the EPCC microbenchmark's results did not show profitable faster synchronization among sibling cores within one processor in this machine. We believe it is mostly due to the fact that the L2 cache is not really shared between sibling cores. Meanwhile, the Solaris operating system schedules OpenMP threads very well taking the asymmetry between cores into consideration. The Sun Studio Performance Analyzer is a very useful and handy tool to help us understand the underlying reasons for performance problems. However, the caller-callee tab leaves some room to be improved and we used our own tool [7] for better understanding the benchmarks.

Compiler optimizations are also key factors in OpenMP performance on the Sun platforms. For example, we observed a performance degradation from 4 to 5 threads for Jacobi code compiled with "–fast –xopenmp" even for a small data set (3 500*500 arrays and 1000 iterations). The *Analyzer* showed that the Sun Studio 10 compiler performed loop unrolling on the major loops and inserted PREFETCH instructions into them. Data cache stall information collected from hardware counters hints that some PREFETCH operations initiated from two sibling cores stressed the shared data path connected to L2 cache and memory, thus incurred longer stall cycles than usual. But it is not always the case for all PREFETCH operations. Further work is needed to understand the exact conditions and effects of these traditional optimizations with regard to the OpenMP performance on machines with CMP capabilities.

We observed that a straightforward OpenMP implementation for traditional SMP architecture may not achieve good scalability on the Xeon system. The main reasons are memory bandwidth bottleneck [6] and competition for the shared computing resources, which degrade the overall SMT performance. More OpenMP scheduling policies [29] and precomputation and a prefetch approach via a helper thread [28] which utilize new SMT features are able to lead to a better performance. Our experiments also showed that a HyperThreading-aware OS is important for maintaining load balance and efficiently utilizing the resources of an SMT system. The EPCC microbenchmarks results also indicated that the overhead of OpenMP synchronization implementation in a SMT system is higher than that in an SMP system. The OpenMP implementation needs to take the SMT features into account.

Our experiments for CMP and SMT were carried out separately to understand the implications of CMT. We plan to analyze those OpenMP benchmarks and consider the interaction between CMP and SMT when we have access to a machine with both technologies. The ultimate objective is to obtain a quantitative model, which is capable of predicting and explaining the OpenMP performance on CMT machines, considering machine parameters, application features and compiler optimizations.

# 6 Related Work

CMP technology exists in IBM Power4, Sun UltraSPARC IV, AMD Opteron [1], and some embedded systems [13]. The research on Power4 [5] showed that the improved data locality of the L2 cache made the SP program from NAS NPB scale from 16 to 32; otherwise, the program runs on 32 processors are slower. [12] gives an overview of the Sun Fire E25K with UltraSPARC IV and its OpenMP support; in particular, the base and peak performance of SPEC OMPL benchmarks using almost maximum threads was demonstrated and compared with a traditional SMP. A system-on-chip (SOC) design from Cradle Technologies, Inc. integrates processors into one chip and OpenMP was selected in [13] to deal with the heterogeneity of CMP: OpenMP is extended to support Cradle's Digital Signal Engine (DSE) and optimized via data prefetching and privatization. [3] explores the performance impact of asymmetric multicore architectures using a wide range of applications on a hardware prototype.

Simultaneous multithreading techniques can be dated back to 10 years ago as a means to improve the utilization of superscalar processors [26]. Work related to compiler support for SMT has focused on the problems of synchronization, memory allocation and program optimizations for shared caches. Researchers explored the performance of symbiosis [21], a group of two sequential programs, or a parallel program other than OpenMP running on SMT. In particular, hand-written program transformations, namely dynamic tiling, copying and blocking, were presented in [16] to partition the shared caches on SMT processors and the performance gain of a parallel program other than OpenMP or multiple programs is 16-29%. For OpenMP programs, speculative precomputation and thread-level parallelism are used together [28] to achieve more efficient execution on real SMT processors using a runtime approach, however, it is hard to control the parameters (runahead distance and the span of the precomputation chunk) of speculative precomputation, and to transform OpenMP programs to include speculative precomputation. [29] presented an adaptive OpenMP loop scheduler on top of the Omni OpenMP compiler. Its runtime system adds affinity and trapezoidal scheduling, and enables the dynamic selection of the number of threads on each processor for each parallel region.

# 7 Conclusions

SMP technology is increasingly widely used. SMT and/or CMP technology are alternative strategies for increasing performance on a chip and exploiting space on the die to get better throughput. Applications need to be able to profit from the additional power by using multiple threads of execution and by adapting to best utilize the memory hierarchy. However, it will be hard for a compiler to automatically derive suitably large regions of code to obtain good performance. OpenMP is a relatively straightforward way to specify multithreading in a code and appears to be ideal for this purpose. But OpenMP was not designed with this kind of system and memory hierarchy specifically in mind, and there are to date few reports on the performance of OpenMP codes on such platforms. More experience, an analysis of the architectures and reconsideration of compilation strategies, are needed to determine the amount of speedup that can be expected, to derive suitable approaches to parallelization for these architectures, and to decide whether there are any modifications to OpenMP itself that would facilitate the execution of OpenMP applications on this kind of hardware.

# Acknowledgements

We thank Sun Microsystems Inc. for loaning the Sun Fire V490 machine to the Computer Science Department at University of Houston (UH). Chandler Wilkerson from UH coordinated the loan and provided timely installation and technical support with the help of Tony Curtis in the High Performance Computing Center of UH. Nawal Copty from Sun Microsystem Inc. helped us to run some benchmarks and to understand some results.

# References

1. AMD Multi-Core: Introducing x86 Multi-Core Technology & Dual-Core Processors from AMD (2005), http://multicore.amd.com/
2. Aslot, V., Domeika, M., Eigenmann, R., Gaertner, G., Jones, W.B., Parady, B.: SPEComp: A New Benchmark Suite for Measuring Parallel Computer Performance. In: Eigenmann, R., Voss, M.J. (eds.) WOMPAT 2001. LNCS, vol. 2104, pp. 1–10. Springer, Heidelberg (2001)
3. Balakrishnan, S., Rajwar, R., Upton, M., Lai, K.: The Impact of Performance Asymmetry in Emerging Multicore Architectures. In: 32nd Annual International Symposium on Computer Architecture (ISCA) (June 2005)
4. Bull, J.M., O'Neill, D.: A Microbenchmark Suite for OpenMP 2.0. In: Proceedings of the Third European Workshop on OpenMP (EWOMP 2001), Barcelona, Spain (September 2001)
5. Frumkin, M.: Efficiency and Scalability of an Explicit Operator on an IBM POWER4 System. Technical Report NAS-02-008, NASA Ames Research Center (August 2002)
6. Guiang, C., Purkayastha, A., Milfeld, K., Boisseau, J.: Memory performance of dual-processor nodes: comparison of Intel Xeon and AMD Opteron memory subsystem architectures. In: Proceedings for ClusterWorld Conference & Expo 2003, San Jose, CA (June 2003)
7. Hernandez, O., Liao, C., Chapman, B.M.: Dragon: A Static and Dynamic Tool for OpenMP. In: Chapman, B.M. (ed.) WOMPAT 2004. LNCS, vol. 3349, pp. 53–66. Springer, Heidelberg (2005)
8. Horel, T., Lauterbach, G.: UltraSPARC-III: designing third-generation 64-bit performance. IEEE Micro 19(3), 73–85 (1999)
9. Jin, H., Frumkin, M., Yan, J.: The OpenMP Implementation of NAS Parallel Bench-marks and its Performance. Technical Report NAS-99-011. NASA Ames Research Center (1999)
10. Kalla, R., Sinharoy, B., Tendler, J.: IBM POWER5 chip: a dualcore multithreaded processor. IEEE Micro 24(2), 40–47 (2004)
11. Koufaty, D., Marr, D.T.: Hyperthreading Technology in the NetBurst Microarchitec-ture. IEEEMicro (2003)
12. Lee, M., Whitney, B., Copty, N.: Performance and Scalability of OpenMP Programs on the Sun FireTM E25K Throughput Computing Server. In: Chapman, B.M. (ed.) WOMPAT 2004. LNCS, vol. 3349, pp. 19–28. Springer, Heidelberg (2005)
13. Liu, F., Chaudhary, V.: Extending OpenMP for heterogeneous chip multiprocessors Parallel Processing. In: Proceedings of International Conference on Parallel Processing, October 2003, pp. 161–168 (2003)
14. Moore, C.: POWER4 System Microarchitecture. Microprocessor Forum (2000)
15. Nagarajayya, N.: Improving Application Efficiency Through Chip Multi-Threading, (March 10, 2005), http://developers.sun.com/solaris/articles/chip_multi_thread.html
16. Nikolopoulos, D.S.: Code and Data Transformation for Improving Shared Cache Performance on SMT Processors. In: Veidenbaum, A., Joe, K., Amano, H., Aiso, H. (eds.) ISHPC 2003. LNCS, vol. 2858, pp. 20–22. Springer, Heidelberg (2003)
17. Olukotun, K., et al.: The Case for a Single-Chip Multiprocessor. In: Intl. Conf. on Architec-tural Support for Programming Languages and Operating Systems, pp. 2–11 (1996)
18. OpenMP Application Program Interface, Version 2.5, public draft (November 2004)

19. Sato, M., Satoh, S., Kusano, K., Tanaka, Y.: Design of OpenMP compiler for an SMP cluster. In: Proc. of the 1st European Workshop on OpenMP, September 1999, pp. 32–39 (1999)
20. Saito, H., Gaertner, G., Jones, W., Eigenmann, R., Iwashita, H., Lieberman, R., Wa-veren, M.V., Whitney, B.: Large System Performance of SPEC OMP2001 Benchmarks. In: Zima, H.P., Joe, K., Sato, M., Seo, Y., Shimasaki, M. (eds.) ISHPC 2002. LNCS, vol. 2327, pp. 370–379. Springer, Heidelberg (2002)
21. Snavely, A., Mitchell, N., Carter, L., Ferrante, J., Tullsen, D.: Explorations in Symbiosis on Two Multithreaded Architectures. In: Workshop on Multi-Threaded Execution, Architecture, and Compilers (M-TEAC99) (January 1999)
22. Solaris 10, http://www.sun.com/software/solaris/
23. Spracklen, L., Abraham, S.G.: Chip Multithreading: Opportunities and Challenges. In: 11th International Symposium on High-Performance Computer Architecture (HPCA-11), pp. 248–252 (2005)
24. Sun Fire™ V490 and V890 Servers Architecture, http://www.sun.com
25. Sun Studio 10, http://www.sun.com/software/products/studio/index.xml
26. Tullsen, D., Eggers, S., Levy, H.: Simultaneous Multithreading: Maximizing On-Chip Parallelism. In: Intl. Symp. on Computer Architecture, pp. 392–403 (1995)
27. UltraSPARC®IV Processor Architecture Overview, http://www.sun.com
28. Wang, T., Blagojevic, F., Nikolopoulos, D.S.: Runtime Support for Integrating Precomputation and Thread-Level Parallelism on Simultaneous Multithreaded Processors. In: The Seventh Workshop on Languages, Compilers, and Run-time Support for Scalable Systems (LCR 2004), Houston, TX (October 2004)
29. Zhang, Y., Burcea, M., Cheng, V., Ho, R., Voss, M.: An Adaptive OpenMP Loop Scheduler fro Hyperthreded SMPs. In: Proc. of International Conference on Parallel and Distributed Systems (PDCS-2004), San Francisco, CA (September 2004)
30. Pranevich, J.: The Wonderful World of Linux 2.6, http://www.kniggit.net/wwol26.html

# Experiences Parallelizing a Web Server with OpenMP

Jairo Balart, Alejandro Duran, Marc Gonzàlez,
Xavier Martorell, Eduard Ayguadé, and Jesús Labarta

CEPBA-IBM Research Institute
Departament d'Arquitectura de Computadors
Universitat Politècnica de Catalunya
Jordi Girona, 1-3, Barcelona, Spain
{jbalart,aduran,marc,xavim,eduard,jesus}@ac.upc.edu

**Abstract.** Multi–threaded web servers are typically parallelized by hand using the pthreads library. OpenMP has rarely been used to parallelize such kind of applications, although we foresee that it can be a great tool for network servers developers. In this paper we compare how easy is to parallelize the Boa web server using OpenMP, compared to a pthreads parallelization, and the performance achieved. We present the results of a parallelization based on OpenMP 2.0, the dynamic sections model and pthreads.

## 1 Introduction and Motivation

OpenMP [1] has successfully been used to parallelize a great number of applications in the scientific domain. Extensive work has been done to fulfill the needs of scientific applications in shared memory environments. The parallelism that appears in numerical applications has significantly influenced the definition of the OpenMP API. Most work distribution schemes are specifically designed to support the main source of parallelism of scientific applications: parallel loops. For synchronization mechanisms programmers can use barrier synchronizations, mutual exclusion and atomic synchronizations.

But scientific applications are not the only niche where parallelism can be used to increase the performance of applications. In fact, with new generations of architectures containing multiple cores, parallelism will be exploited in applications with characteristics and needs dramatically different from those of the scientific world. We believe the OpenMP community should start studying these characteristics. This study will either determine if the current OpenMP API suites these new applications or whether it needs changes to support them efficiently.

We have studied the feasibility of using OpenMP to parallelize a web server to start exploring the characteristics that will be found in these new applications. Web servers are inherently parallel applications as the different requests are unrelated and free of dependences between them. Thus, the requests can be handled in parallel. Web servers have been traditionally parallelized with *threading* techniques (mainly *pthreads*). Our objective is to specify useful extensions if the OpenMP API does not easily support the parallelism found in web servers.

We have selected the Boa [2] web server as the platform to be parallelized. Different parallel strategies have been developed. An OpenMP version allowed us to evaluate the programming effort and the resulting performance using the current OpenMP standard. Also, we developed a version that uses the proposed *dynamic sections* [3] constructions to check their usefulness. Finally, a manual *pthread* based version was developed to do a comprehensive comparison. All three versions were evaluated against the original version which is single threaded.

The structure of the paper follows: section 2 describes the contributions of this paper to the state of the art. Section 3 overviews the structure of the Boa web server. In section 4 we describe our parallelized versions of Boa. Section 5 describes the experiments performed and the results obtained. Section 6 discusses our experiences with the different parallel versions. And finally, in section 7 we present the conclusions of this work and we describe some lines for future study.

## 2 Related Work

Several authors have compared OpenMP versus pthreads. These comparisons have been in scientific applications or comparisons between basic language constructions. Kuhn et al. compared the primitives provided by both models concluding that OpenMP was easier to use but they also pointed out some problems with irregular applications [4]. Lee and Downar compared both languages for a nuclear reactor transient code [5]. They obtained similar performance with both OpenMP and pthreads but OpenMP was easier to use. Breshears and Luong compared both models in the context of a Coastal Ocean Circulation Model [6]. Their conclusion was that OpenMP was easier to use yielding the same performance than pthreads. Dedu et al. compared both models with some algorithms from the artificial intelligence field [7]. They found that although for regular applications OpenMP was easier to use and obtained the same performance than pthreads for irregular applications OpenMP was difficult to use.

Some other works have tried to extend OpenMP to be able to cope with irregular applications. Asenjo et al. explored some techniques to deal with pointers and traversal of structures [8]. Shah et al. introduced the *workqueueing model* [9]. This proposal extends the OpenMP programming model with an alternative work distribution scheme based on the definition of *queues of work* from where the executing threads extract work. The extension targets algorithms traversing memory and linked data structures. The proposal has been successful but introduces dramatic changes in the OpenMP execution model. We previously proposed to use *dynamic sections* [3] to minimize the changes of the *workqueueing model*, in which *dynamic sections* are based. We further extend the semantics of the *dynamic sections* in this work.

Different works have evaluated the usefulness of threaded web servers. For example, Roper et al. compared different web servers concluding that multi–threaded servers can outperform traditional event driven servers [10] which are not threaded. Jeong et al. showed in their work [11] that the use of multiple CPUs with dynamic content increases the throughput obtained and reduces the response time. Using multiple CPUs with static content does not increase throughput but response time still decreases. The benefits of using multiple CPUs in SSL enabled web servers were shown by Guitart et al.[12]. Other successful multi–threaded servers include Apache [13], SEDA [14] and Flash [15]. All these works used a system thread packages, mainly pthreads. This work, instead, explores the suitability of OpenMP as a language for the development of a parallel web server.

## 3 The Boa Web Server

The Boa web server architecture is a single threaded event–driver HTTP server architecture. This kind of architecture, unlike traditional web servers, does not fork for each incoming connection. Instead, Boa comes with an integrated task scheduler that handles multiple requests concurrently but not in parallel. Boa multiplexes all ongoing requests, trying to maximize the throughput and minimize the response time. The scheduler uses two request queues: the *ready* queue keeps those requests available for further processing. The *blocked* queue keeps those requests waiting for any data dependence to be satisfied. Iteratively, the server traverses the *ready* queue and further processes each request. The server uses a round–robin technique to avoid large requests starving other *ready* requests. Boa logically divides each request in smaller chunks of work and each time a request is processed a single chunk is consumed. Figure 1 shows a simplified code that processes the requests from the *ready* queue. After a request is processed

1. It is kept in the *ready* queue because it should be further processed ( i.e. it has more chunks and all data are available ).
2. It is moved to the *blocked* queue because the server detected an unsatisfied dependence ( e.g. it needs to read data from a socket ).
3. It is freed because the last chunk of the request was consumed.

```
1  for each request in the ready queue
2  {
3      update_time;
4      result = process_step(request);
5      accept new requests (if any);
6      if ( result == BLOCK )        block(request);
7      else if ( result == FINISHED ) free(request);
8      else keep it in the queue
9  }
```

**Fig. 1.** Request processing loop pseudo–code

The server traverses the *blocked* queue and for each request it checks if the request dependences are satisfied using the information it collects with the *select* system call. When a request has no further dependences it is moved again to the *ready* queue.

Figure 2 shows the structure of the main loop of the server. This loop is infinite and in each iteration the server

1. processes any pending signal.
2. traverses the *blocked* queue to check the dependences of blocked requests.
3. establishes pending new connections using the *accept* system call.
4. traverses the *ready* queue to process more chunks of the unblocked requests.
5. calls *select* to obtain information about new connections and the status of incoming and outgoing data.

```
1   while (1)
2   {
3       process signals (if any)
4       move requests from blocked to ready using select result
5       accept new connections (if any)
6       process requests in the ready queue
7       select system call
8   }
```

**Fig. 2.** Main loop pseudo–code

Boa tries to reduce the number of issued system calls by *mmaping* local files into the server memory. A cache of *open files* (i.e active maps), which it is checked each time a new file is requested, avoids *mmaping* twice the same file.

The server performs all input/output with non-blocking system calls to ensure that no single request blocks the processing of others that are ready.

## 4 Parallelizing Boa

The main source of parallelism in the Boa web server is the possibility of overlapping the computations related to different requests. As explained in section 3 the requests are placed in the *ready* and *blocked* queues and the server iteratively traverses these queues. Parallel processing is possible by processing each element of the queues in parallel. The server can also do different tasks in parallel (e.g. accepting new requests and processing new requests ). All the versions described exploit these sources of parallelism.

Different points in the server require serialization. First, modification of global variables (e.g. the number of active connections ). Second, manipulations of the *ready* and *blocked* queues. Third, acceptance of new connections. Fourth, access to the cache of open files. And last, write access to the server log files to avoid mixing the output of different threads.

The original version uses a lot of *static* variables insides functions. These variables were converted to extra parameters of the function they were in. Another possible option was to use per thread variables ( e.g. *threadprivate* in OpenMP ).

We have developed three parallel versions: a *pthreads* version, a pure OpenMP version and a version using *dynamic sections* which are non–standard.

## 4.1 Pthreads Parallel Version

The *pthread* parallel version exploits the possibility of processing different requests in parallel, as they are unrelated. The parallelization uses a producer–consumer approach. One thread executes all the tasks in the main Boa loop, described in Figure 2, except requests processing. This thread is the producer of new *ready* requests. The remaining threads consume the *ready* requests and process them as explained in section 3. Figure 3 shows the code the consumer threads execute. The *pthread* version uses the same round–robin mechanism of the serial version. So, each time a thread extracts a request a single chunk is consumed and the request may be queued again to the *ready* queue.

This version uses different mutex locks to protect accesses to the *ready* queue, accesses to the *blocked* queue, accesses to global variables, and writing to the server log files. The cache of open files is also protected by a mutex lock per entry, to maximize concurrency, and a global mutex lock for global cache variables.

```
1   for ( ; ; ) {
2     while( not pending requests );
3     pthread_mutex_lock(&ready_lock);
4     if ( pending requests ) {
5        req = dequeue(request_ready);
6     }
7     pthread_mutex_unlock(&ready_lock);
8     if ( req )
9        process_request(req);
10  }
```

**Fig. 3.** Code for thread consumers in the *pthread* Boa version

## 4.2 OpenMP Standard Parallelization

For our OpenMP parallelization we targeted the request processing loop Figure 1. We wanted to distribute all the requests in the *ready* queue among the available threads by using a workshare. The number of requests in the queue can vary during its traversal (e.g. if a request is free ). Because in OpenMP all threads must see the same iterations we splitted the loop in two new loops: one does not remove requests the queue while the other does. In the first new loop, the server processes each request in the *ready* queue. The result of each processing is stored in a new field of the request structure. This loop was parallelized using a *parallel* construct and we used a *single* workshare to distribute the different iterations. Using the result from the first loop the server modifies the queues in the second loop which must be done single-threaded. Figure 4 shows the OpenMP parallelization of the new loops. When new requests are accepted they are added to the beginning of the queue. Before the server accepts any request all threads grab the head of the queue. This guarantees they will traverse the same elements.

Access to the cache of open files was protected with a critical section for the global cache variables and an OpenMP lock per cache entry. We used another critical construction to guarantee correct access to the log files. Several critical sections protect access to Boa global variables.

```
1   #pragma omp parallel
2   {
3
4           get head of ready queue
5           #pragma omp barrier
6           for each request in the ready queue
7           {
8               #pragma omp master
9                   update time;
10              #pragma omp single nowait
11                  request.result = process step(request);
12              #pragma omp master
13                  accept new requests (if any);
14          }
15  }
16
17  for each request in the ready queue
18  {
19          if ( request.result == BLOCK )                  block(request);
20          else if ( request.result == FINISHED )          free(request);
21          else keep it in the queue
22  }
```

**Fig. 4.** OpenMP request processing loop pseudo–code

### 4.3 Dynamic Sections Parallelization

The previous presented parallelizations were in previous section were mainly possible because the serial version had already embedded a complex code that dealt with request blocking and their scheduling ( i.e. the ready queue ). Our question now is: could OpenMP make this work easy to the programmer?

The available parallelism can be seen as a collection of tasks. We have used the *dynamic sections* proposed extension to express this parallelism easily. Under this model a single thread is in charge of performing the serial work ( accepting requests, extracting them from the blocked queue, ... ) while the remaining threads execute the parallel tasks that are created (i.e. a dynamic section).

In this version, we have removed part of the integrated schedule: the ready queue has been removed. Instead of queueing requests in the *ready* queue now the threads create new dynamic sections. A new dynamic section is created

– When a new request is accepted, a new section is created for the first chunk of the request.
– When a request has completed a chunk, if it was not the last, a new section is created for the next chunk. This dynamic section is created inside another. We have extended the original model to allow nesting of SECTION constructs.
– When a request is unblocked, because their dependences are fulfilled, a new section is created for the next chunk.

Figure 5 shows our parallelization of the code of the main loop using the *dynamic sections*.

As in the previous version, several critical constructions protect accesses to the open files cache, access to the global variables and access to the server log files.

```
1   #pragma omp parallel
2   #pragma omp sections dynamic
3   while (1)
4   {
5           process signals (if any)
6           foreach request from the blocked queue {
7                   if ( request dependences are met ) {
8                           extract from the blocked queue
9   #pragma omp section captureprivate(request)
10                          serve_request(request)
11                  }
12          }
13          if ( new connection ) {
14                  accept it
15  #pragma omp section captureprivate(new connection)
16                  server_request (new connection)
17          }
18          select system call
19  }
```

**Fig. 5.** Main loop pseudo–code with *dynamic sections*

## 5 Evaluation

### 5.1 Environment

For our experiments we used a 4-way Intel Xeon at 1.4GHz with 2GB of RAM to run the web server and a 2-way Intel Xeon at 2.4 GHz with 2GB of RAM to run the benchmark client. All the machines were running a 2.6 Linux kernel. The network that connected the machines was a switched Gigabit network.

### 5.2 Workload Generator

We used Httperf[16] to generate the different workloads for the experiments. This tool allows the creation of a continuous flow of HTTP requests to the server machine. The tool accepts as one of its parameters the number of clients per second. For each client, it opens a session with the server through a persistent HTTP connection. Then a series of requests are issued by the client, some of them pipelined, some spaced by a *think time*. Another parameter of Httperf is the sessions database from where clients get the requests they ask for and the think times to wait. We have used a database extracted from the Surge[17] workload generator.

The scenario produced by Surge is a static content workload characterized by short session lengths and low computational costs for each request serviced. The Surge distribution is based on a model developed from the observation of real web server logs.

### 5.3 Experiments

We evaluated all different versions of the Boa web server using the Surge workload with different loads of clients. These load configurations ranged from a low load of clients (10 per second) to heavy load of clients (800 clients per second). In the following plots, we labeled the different Boa versions as follows:

- *original boa* refers to the unmodified single–thread Boa server.
- *boa-pthreads* refers to the parallel version that uses *pthreads*.
- *boa-omp* refers to the parallel version that uses standard OpenMP constructions.
- *boa-dsections* refers to the parallel version that uses *dynamic sections*.

All parallel versions were run with 2 and 4 threads.

Figure 6 shows for each load of clients the throughput obtained by each Boa version. All versions, except the *boa-omp* version, obtained a similar throughput up to a workload of 700 clients per second. The *boa-omp* version was outperformed because the server did not run as much time in parallel as the other versions do. Doubling the number of threads in the parallel versions did not result in a noticeable increase of throughput because Surge is not CPU–intensive workload as it works with static content. With 700 clients per second the limit of the Gigabit network was reached and all versions throughput deteriorated as they were saturated. Saturation happens earlier using more threads because contention in shared resources have a greater impact. The Boa server uses a mechanism that minimizes the effect of saturation limiting the number of active connections. We disabled this mechanism on purpose so we could find the point at which each parallel version saturates.

Figure 7 shows the average response time for each load of clients and each Boa version. All parallel versions achieved a lower response time than the original Boa version. With a load of 800 clients per second response time was reduced as much as by three. Doubling the number of threads in *boa-dsections* and *boa-pthreads* was useful reducing the response time. With four threads *boa-dsections* and *boa-pthreads* behaved so closely that their lines are overlapped. With two

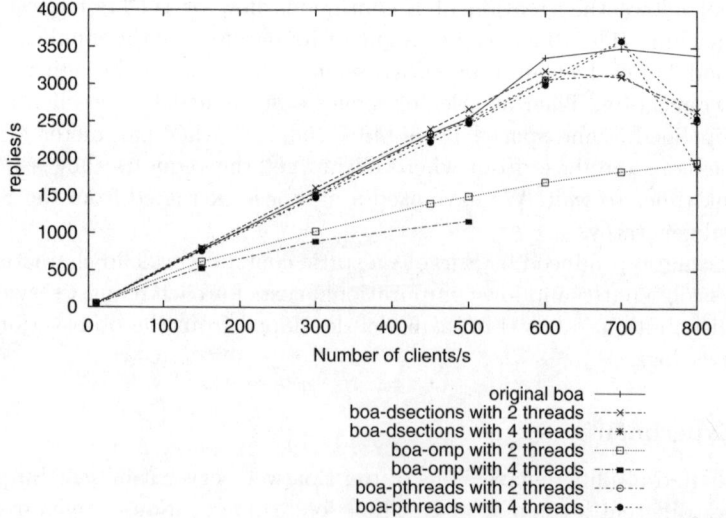

**Fig. 6.** Throughput(replies per second )

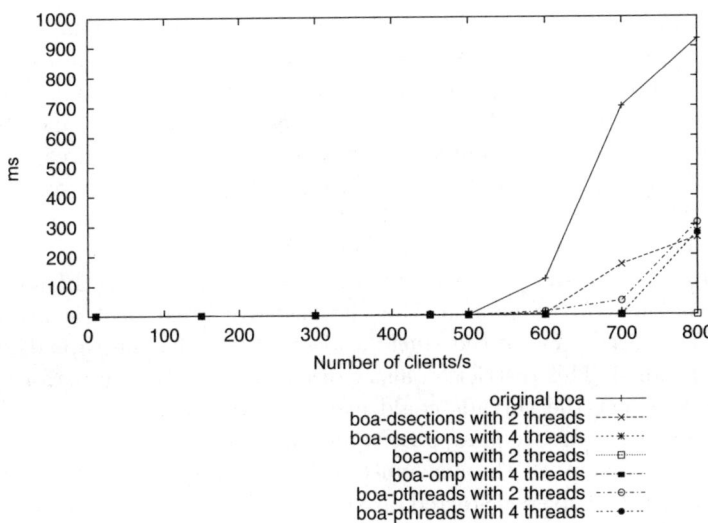

**Fig. 7.** Response time ( ms. )

threads they behaved similarly except with a load of 700 clients per second where *boa-pthreads* response time was lower. The average response time for the *boa-omp* version was very low, near to zero. This result is misleading because the server was rejecting more than 75% of the requests.

## 6 Comparison

In this section we compare the programming effort required by the three parallel versions: *pthreads*, OpenMP and *dynamic sections*.

One of the most time consuming tasks in parallelizing all the versions, was removing the static variables in local functions. Due to the large amount of *static* variables that may appear in serial C codes,s compilers could provide a command line option that transformed any variable with *static* storage to a variable with *thread private* storage. This option would reduce the time spent in parallelizing large C codes as a first step in the parallelization.

Another effort, common in all versions, was protecting shared data with critical sections and locks. This work in was quicker done with OpenMP than with *pthreads* as you only need to add the appropriate directive instead of having to declare a mutex variable and using *lock* and *unlock* calls. Nevertheless, when you need a critical section for each element of a structure (e.g. the open files cache) for improving performance, it is as difficult in OpenMP as it is in *pthreads* because you need to use omp_locks. We think a possible way to solve this problem would be allowing dynamically named critical sections (e.g. cache_lock[i]). With these kind of critical sections the same code protecting a single entry of a complex structure would work for all the entries while each one will still have its own lock.

The *pthread* version required several modifications to the original source code. Code for the consumer threads was developed. And, although it was not very difficult because there was only one type of task to consume (i.e. requests) it required some expertise to handle access to the queue correctly and efficiently. Some other changes were required to the code of the producer thread including: creating the consumers, initializing the locks, and removing the request processing loop. Also, mutex locks were added to protect the access to the *ready* and *blocked* queues. The overall effort was moderate.

The OpenMP version did not require as many changes in the source code compare as the *pthread* version. But, it needed a great deal of attention to maintain the correctness of the *single* workshare (i.e. that all threads executed all the iterations). This restriction also caused the reduction in performance. In the other versions, the threads could execute a task as soon as it was ready, or even run in parallel request processing and accepting new connections. In the OpenMP version all threads must wait until all the requests, that were available when the traversal started, are processed even if there are new requests to process. This suggests that current workshares are not well suited for handling irregular parallelism.

But both the *pthread* version and the OpenMP version were easier to develop because the original version had integrated a complex code that enabled request concurrency. Otherwise, the programming effort would have been greater. On the other hand, the *dynamic sections* version was simpler as the programmer did not need to code the management of the *ready* tasks. Even from a simpler version of the serial version the programming effort with *dynamic sections* would have been minor.

*Dynamic sections* also allow to easily mix different kinds of parallel tasks. While the *pthreads* code would became more and more complex if it had to deal with different kinds of parallel tasks the *dynamic section* complexity would remain constant.

# 7 Conclusions and Future Work

In this paper, we have explored the use of OpenMP to parallelize a web server. We have shown how, adding a few directives, the request processing loop of the web server can be parallelized. But this simple version did not perform efficiently. We used the proposed *dynamic sections* to implement a simpler parallel web server (i.e. without application level task management ). Evaluation showed that this version had a performance, in throughput and average response time, as good as the performance obtained by the server developed with *pthreads*. But in the *pthread* version the programmer needed to develop a specific task management for the application whereas the *dynamic sections* version simplified the programming.

In the future, we will apply OpenMP to other web scenarios where studies have pointed out that there can improvements in throughput by using multiple processors: SSL enabled applications[12] and dynamic content applications [11].

## Acknowledgements

Authors will like to thank Vincenç Beltran, David Carrera and the eDragon team [18] for their valuable help and for allowing us to use their resources. This research has been supported by the Ministry of Science and Technology of Spain under contract TIN2004-07739-C02-01.

## References

1. OpenMP Organization. Openmp fortran application interface, v. 2.0 (June 2000), http://www.openmp.org
2. Doolittle, L., Nelson, J.: Boa webserver site, http://www.boa.org
3. Balart, J., Duran, A., González, M., Martorell, X., Ayguadè, E., Labarta, J.: Nanos mercurium: a research compiler for openmp. In: Proceedings of the European Workshop on OpenMP 2004 (October 2004)
4. Kuhn, B., Petersen, P., O'Toole, E.: Openmp versus threading in c/c++. Concurrency - Practice and Experience 12, 1165–1176 (2000)
5. Lee, D.J., Downar, T.J.: The application of posix threads and openmp to the u.s. nrc neutron kinetics code parcs. In: Eigenmann, R., Voss, M.J. (eds.) WOMPAT 2001. LNCS, vol. 2104, pp. 69–83. Springer, Heidelberg (2001)
6. Breshears, C., Luong, P.: Comparison of openmp and pthreads within a coastal ocean circulation model code. In: Workshop on OpenMP Applications and Tools (July 2000)
7. Dedu, E., Vialle, S., Timsit, C.: Comparison of openmp and classical multithreading parallelization for regular and irregular algorithms. In: Software Engineering Applied to Networking and Parallel/Distributed Computing (SNPD), pp. 53–60 (2000)
8. Asenjo, R., Corbera, F., Gutiérrez, E., Navarro, M.A., Plata, O., Zapata, E.L.: Optimization techniques for irregular and pointer-based programs. In: Proceedings of the 12th EuroMicro Conference on Parallel, Distributed and Network-Based Processing (PDP 2004), February 2004, pp. 11–13 (2004)
9. Shah, S., Haab, G., Petersen, P., Throop, J.: Flexible control structures for parallellism in openmp. In: 1st European Workshop on OpenMP (September 1999)
10. Ishihara, T., Keen, A.W., Maris, J.T., Wohlstadter, E., Olsson, R.A.: Cow: A cooperative multithreading web server. In: The 2002 International Conference on Parallel and Distributed Processing Techniques and Applications (PDPTA 2002), June 2002, pp. 991–996 (2002)
11. Jeong, J., Park, S., Nang, J.: Performance analysis of a multithreaded web server on multiprocessor. In: Proceedings of the International Conference on Parallel and Distributed Processing Techniques and Applications (PDPTA 2000), June 2000, pp. 1885–1889 (2000)
12. Guitart, J., Beltran, V., Carrera, D., Torres, J., Ayguadé, E.: Characterizing secure dynamic web applications scalability. In: 19th International Parallel and Distributed Symposium (IPDPS 2005) (April 2005)
13. Apache web server, http://www.apache.org/
14. Welsh, M., Culler, D., Brewer, E.: Seda: An architecture for well-conditioned, scalable internet services. In: Symposium on Operating Systems Principles (SOSP 2001), October 2001, pp. 230–243 (2001)

15. Pai, V.S., Druschel, P., Zwaenepoel, W.: Flash: An efficient and portable web server (1999)
16. Mosberger, D., Jin, T.: httperf – a tool for measuring web server performance. In: First Workshop on Internet Server Performance, June 1998, pp. 59–67 (1998)
17. Barford, P., Crovella, M.: Generating representative web workloads for network and server performance evaluation. In: Measurament and Modeling of Computer Systems, pp. 151–160 (1998)
18. eDragon Research Group site, http://www.ciri.upc.edu/edragon

Second International
Workshop on OpenMP
IWOMP 2006, June 12–15,
Reims, France

# Advanced Performance Tuning

# Automatic Granularity Selection and OpenMP Directive Generation Via Extended Machine Descriptors in the PROMIS Parallelizing Compiler[*]

Walden Ko and Constantine D. Polychronopoulos

Center for Supercomputing Research and Development, University of Illinois at
Urbana-Champaign, 1308 W. Main St., Urbana, Illinois 61801
{w-ko, cdp}@csrd.uiuc.edu

**Abstract.** This paper describes Extended Machine Descriptors (EMD) in the PROMIS multigrain compiler. EMDs extend the concept of Machine Descriptors to encompass coarser granularities of parallelism, thereby enabling high-level transformations to be quickly retargeted. Code generation for parallel runtime libraries like OpenMP benefit from the granularity control and automatic directive generation provided by EMDs. An overview of EMDs, their implementation in the PROMIS compiler, and a demonstration of their utility for portability and retargetability are described in this paper.

## 1 Introduction

The recent emergence of dual and multi-core processors in mainstream computers marks the beginnings of the widespread availability of fine as well as coarse-grain parallelism in entry level system, and hence common applications. Dual core processors are only the start of a trend in processor design leading to multiple CPU cores on a chip such as the Sun Microsystems Niagara processor. As these resources provide greater functionality, chip multiprocessors with many heterogeneous, specialized cores are foreseeable with some already in use in networking and embedded applications. These systems provide the flexibility to utilize parallelism ranging from fine-grain instructions up to coarse-grain loop-level and function level threads.

As is common when new technologies move into the mainstream, there is a disconnect between the functionality of the hardware and the capabilities of the software tools available for them. As a result, applications may not fully benefit from the power of multicore systems. OpenMP is one solution for improving the programmability of and enabling parallel execution on these systems. By allowing programmers to identify and exploit the parallelism in their programs, OpenMP provides the means to enable parallelism and maximize the use of these resources. However, even portable programming models like OpenMP can benefit from enhancements that enable them to be *adaptive* to different resources. Within the same family of architectures, there can be variations in the number of CPU cores and the functionality of resources.

---

[*] This research was supported in part by grants from the National Science Foundation under grant No. NSF CCR 00-85917 ITR and a research gift from Intel Corp.

Differences in multicore versus multichip arrangements or the availability of SIMD parallelism within the processors of parallel computer systems are some of the important considerations in parallelization. This is where a compiler solution for controlling and adapting the type and granularity of parallelization can improve support for these new architectural features.

PROMIS [1] is a multilingual, retargetable compiler designed from its inception for multigrain parallelism. Because retargetable compilation is one of the primary objectives, Extended Machine Descriptors (EMD) are an important part of the framework for selecting and controlling the granularity of parallelism. Granularity selection allows parallelization to be tailored to the needs of each architecture, going beyond instruction set retargeting to achieve resource-aware parallelization at coarser granularities. The EMD also provides the means to describe the directives of programming models like OMP to allow automatic generation of code that exploits the resources of parallel programming libraries.

This paper describes Extended Machine Descriptors in the PROMIS compiler and shows how EMDs are used to describe the resources in these heterogeneous systems. The paper also demonstrates the utility of an external description framework for rapidly retargeting a compiler to different systems and architectures. First, a brief overview of the PROMIS compiler is given in Section 2, followed by an introduction of the EMD and its use and implementation in the PROMIS compiler in Section 3. Section 4 describes the use of EMDs for pass control and resource aware parallelization. EMDs are the means by which the capabilities of the compiler are specialized for the systems targeted and how resource-aware compilation is achieved.

## 2 Internal Representation and Organization

The PROMIS internal representation (IR) is based on the Hierarchical Task Graphs (HTG) [2]. HTGs are designed to expose both structured (e.g. loop) and unstructured parallelism at multiple levels of granularity. The explicit hierarchy in the representation clearly delineates loops and other program structures, making it useful for exposing and transforming parallelism.

An example of an input program and its corresponding HTG is shown in Figure 1. The Hierarchical Task Graph (HTG) is a directed acyclic graph $G = (V, E)$, where $V$ is a set of nodes and $E$ is a set of edges that represent control flow through the nodes. The set $V$ contains five types of nodes:

1. Start node: A node without incoming edges that dominates all other nodes in $V$.
2. Stop nodes: A node without outgoing edges post-dominates all other nodes in $V$.
3. Simple nodes: Tasks without subtasks. These are equivalent to program statements or instructions.
4. Loop nodes: Represent loops whose bodies are themselves HTGs
5. Compound nodes: Represent sub-HTG's other than loops. A compound node X represents the HTG, $G(X) = (V(X), E(X))$. These nodes are used for single-entry, single-exit constructs other than loops (*e.g.*, basic blocks and subroutines).

Parallelism is represented independently at each level of the HTG. For example, LoopNode structures have a flag to indicate parallel loops. Vector types and operations are represented directly in the IR by extending the core IR types beyond basic integers and words for wide vector words and operations on multiword operands. Finally, instruction-level parallelism is represented by multiple operations bundled into a single compound operation node.

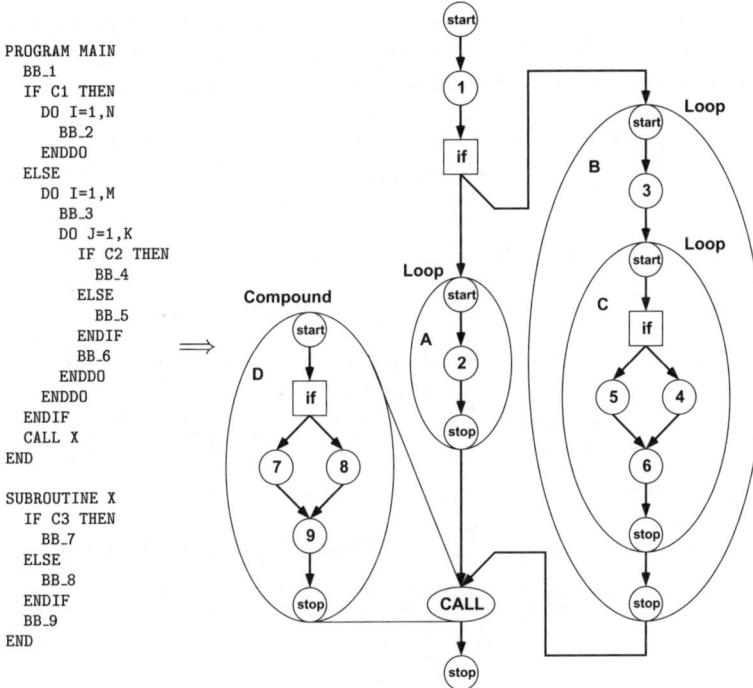

**Fig. 1.** HTG graph

Many of the classic loop transformations are available in PROMIS including unrolling, peeling, distribution, blocking, fusion, interchange, and normalization. To identify loop parallelism, a resource-independent DO-to-DOALL pass identifies loops that are free of cross-iteration dependences. Private and shared variables in the loop are marked during the detection phase for later use by the code generator. Finally, an OpenMP pass filters loops and automatically generates DOALL loops with OpenMP *parallel for* directives (Figure 2).

An example of a loop conversion is illustrated in Figure 3. This is the second nontrivial nested loop of the BESTAB function from *su2cor* in the SPEC95 Fortran benchmark suite. Both loops in the nests are found to be DOALL and converted in the C code output. As a result, *omp parallel for* directives are automatically generated in the output code before each loop. In addition, the inner iteration variable is explicitly declared private in the outer loop as is necessary for OpenMP C scoping rules.

Both C and Fortran input are converted to a single unified representation in the IR. When enabled, OpenMP directive generation occurs automatically for parallel regions. These regions are tagged with the corresponding data for each directive. The information is then propagated through the IR and later emitted by the code generator. This process operates under the umbrella of the Machine Descriptions, which provide granularity information for selecting the parallelism to expose and syntax for OpenMP directives.

In addition to loop parallelization, PROMIS supports autoscheduling [2], a method for utilizing the compiler's static analysis capabilities to expose parallelism at all levels of granularity. Parallelism at any level (loop, task, instruction, *etc.*) is identified and partitioned into functions in preparation for runtime scheduling using task queues.

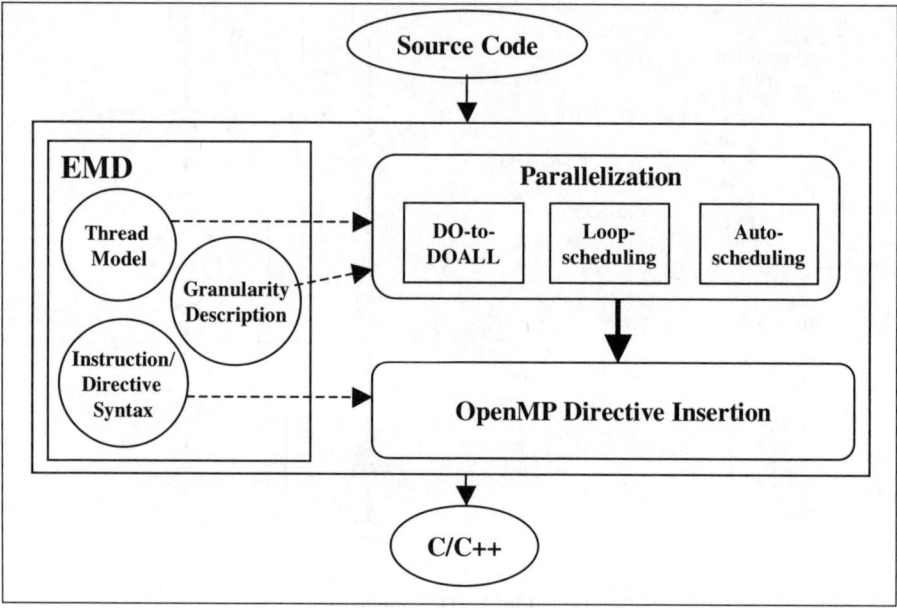

**Fig. 2.** Transformation Ordering in PROMIS

One of the features of autoscheduling is its ability to utilize unstructured or functional parallelism in a program in addition to loop and instruction-level parallelism. The compiler is responsible for identifying the appropriate parallelism to expose (based on granularity requirements specified in the Machine Descriptor) and calculating necessary control and data dependence conditions necessary for each task to begin execution (called *execution conditions*).

Individual function blocks are created for each task along with variables representing execution conditions for that task. The Intel workgroup extensions to OpenMP provide the task queues needed to schedule and execute the threads in the resulting code [3]. The `intel omp task` directive is used to specify individual tasks to be inserted into task queues created with the `intel omp taskq` directives. The execution conditions are maintained with code that is generated as part of the static task creation phase during compilation. At runtime, counters are also initialized to track the total number of execution conditions for each task. Once this counter is null, a task is enqueued. Upon completion, execution conditions for any dependent tasks are cleared, the counter is decremented, and additional tasks may be ready for execution. Further details on the implementation are provided in [4].

```
     DO 200 IBSQ=0,200
        DO 200 ND=2,20
        ...
200  CONTINUE
```

```
#pragma omp parallel for private(su2cor_f__SL__NDX16)
for /* DOALL*/ (su2cor_f__SL__IBSQX16 = 0; su2cor_f__SL__IBSQX16 <= 200;
           su2cor_f__SL__IBSQX16 += 1) {
    #pragma omp parallel for
    for /* DOALL */ (su2cor_f__SL__NDX16 = 2; su2cor_f__SL__NDX16 <= 20;
               su2cor_f__SL__NDX16 += 1) {
        ...
    }
}
```

**Fig. 3.** Parallelized loop from su2cor. Fortran block above; tagged C output below

## 3 Extended Machine Descriptions

Extended Machine Descriptors (EMD) in PROMIS are motivated by the need to qualify the type and amount of parallelism that can be exploited on a given target architecture. High level information about the machine architecture can both drive source level parallelization or qualify what type of parallelism is to be exploited and how, given an already parallelized source or binary and a target architecture. The EMD describes relevant characteristics of the architecture, providing direction in adapting transformations to the targeted system. The information goes beyond information on opcodes and functional unites found in typical machine descriptors [5] [6].

The EMD contains an Instruction Description describing the mapping of opcodes and Hardware Description specifying resource parameters. However, the focus of this paper is the component relevant to parallelization, a description of resources called Node Mappings. Node Mappings are textual descriptions of HTG nodes that describe mappings of coarse-grain parallelism to resources. Each mapping specifies a particular HTG node structure that, if found in the IR and identified as parallel, can be mapped to the resource identified in the description. This is similar to the purpose of opcode mappings, but applied at a coarser granularity. These mappings control how transformations are activated and applied to the program being analyzed.

The essential part of a Node Mapping is the textual specification of HTG nodes. Each type of HTG node has an equivalent text description with specific parameters and possible subnodes. Any configuration of nodes and subnodes that is possible in the IR is allowed in a node specification, allowing a resource's capabilities to be described in the form of a particular HTG node. The mapping informs the compiler where a parallel node of that type can be mapped.

The general form of a node description is shown in Figure 4. This description consists of a name for the node type, a list of parameters, and a list of subnodes of that node. For example, a typical parameter for a loop node is a trip count (*FixedIterationLoop* node). There are also keywords for instructions (OperationBlock), loops with indeterminate trip count (General Loop), basic blocks (HtgBBlock), and entire function bodies (PromisFunction) just as there are such nodes in the HTG. Subnode within blocks may be used to specify operations. An example is a list of operations within a loop.

```
node_type {
  [parameterlist],
  include:
    [subnodelist]
}
```

**Fig. 4.** Form of a Node Description

A Node Mapping consists of a node description enclosed in a block with additional parameters like the type of parallelism. The name for a mapping is used later to identify physical resources. In Figure 5, a FixedIterLoop is used to specify a resource for vector add operations. The loop must contain at least 4 iterations for the node to be vectorizable.

```
Resource VectorUnit {
  Type: SIMD,
  Node:
    FixedIterLoop {
      length: 4,
      include:
        OperationBlock {
          include:
            op_add
        }
    }
}
```

**Fig. 5.** Node Mapping for a SIMD unit

In addition to exact node matches, mappings may contain a modifier, *minimum*, to specify a minimum granularity requirement. A match is satisfied for this type of mapping if a node contains at least the specified operations. Nodes with additional operations will also match. For example, the Mapping in Figure 6 specifies that a LoopNode containing at least two *add* and two *multiply* operations is the minimum granularity needed for a loop node to be parallelized for OpenMP for this particular mapping.

A Node Mapping is a declaration of a type. The identifiers used for each description tie a resource to the mapping. As with opcode mappings, a node mapping name is like an identifier for a functional unit, which may support multiple types of operations. However, nodes—not individual operations—are being mapped in this case. Each of the mappings describes a node structure that, if found to be parallel, can be mapped to that resource. The mapping is then bound to a physical resource through a subsequent instantiation that describes how many of each unit exists and may participate in work sharing of parallelism.

```
Resource Processor {
   Type: OMP,
   MINIMUM,
   Node:
     LoopNode{
        include:
          OperationBlock {
            include:
              op_add, op_add, op_mul, op_mul
          }
     }
}
Processor P1, P2, P3, P4;
```

**Fig. 6.** Node Mapping specifying a loop with a minimum granularity

The declaration in Figure 6 binds the resources, P1-P4, to the Processor mapping. These same identifiers are then used in the Hardware Description to link these mappings to physical characteristics like cache sizes and memory bus latency and throughput values. The Hardware Description provides additional hardware parameters (*i.e.*, cache size, latencies, etc.) for each identified resource.

Mappings are stored in a node mapping file in text form. The mappings are parsed at compile time and are accessible through methods calls to the MachineDescription module. Because the mappings correspond to resources and IR nodes, systems with similar resources may share similar same mappings. Their description files might differ only in the number of each type of resource and the instruction set description. This allows the mappings to be easily reused across systems.

## 4 Pass Selection and Control

Information provided by Node Mappings is used in selecting the transformations that are applied to nodes. Mappings allow the transformation search space to be pruned and limited only to those that are relevant to valid mappings. Pseudocode for this process is shown in Figure 7. First, all mappings are grouped by Node Type (*e.g.*, FixedIterLoops, HtgNodes, *etc.*) in *GenerateMappingGroups()*. Matching nodes are then found for each type of group in the list. This is followed by mapping of the type of parallelism associated with a node. Finally, because autoscheduling is orthogonal to parallelization of individual nodes, it is performed as the last step after parallelization of nodes is complete.

Transformation selection and granularity control occur during the matching within the function *GetMatchingNodes()*. Ideally, nodes satisfy a mapping precisely or require only minor modification to meet the minimum granularity specified by a mapping. These nodes can be directly mapped to system resources. However, mappings more often serve as templates describing the options that are available in the system. Matching is then the process of determining whether it is possible to transform and map a node.

```
Function parallelize() {
   mappings := GenerateMappingGroups();

   forall (m in mappings) {
      nodes := GetMatchingNodes(m);

      forall (n in nodes) {
         switch (m.type) {
         case SIMD:
            DoSIMD(m,n);
            break;
         case loop_par:
            DoLoopPar(m,n);
            break;
         ...
         }
      }
   }
   Autoschedule();
}
```

**Fig. 7.** Parallelization algorithm

One step in matching is the use of various patterns to improve the likelihood of successful matches. The loop from Figure 3 is shown again in Figure 8. The inner loop assigns values to 4 different members of a structure. If this is an insufficient number of operations to be parallelized, the match fails. However, the cumulative number of iterations in the inner loop does allow a match of the outer loop. In such a case, only the outer loop is parallelized (Figure 8). As the number of node variations and idioms for alternate patterns like these increases, the likelihood of a match improves.

The mappings also provide parameters needed by transformations after granularity selection. For example, the mapping from Figure 4 describes *add* operations on vectors of size 4. If both the vectorization and parallelization mappings exist and a full set of vector operation mappings is available, mixed parallelism is feasible. To vectorize the inner DOALL loop of Figure 8 with this information, the operations are distributed from other statements in the loop. The length of the vector is then used as a parameter to strip mine the resulting single-operation loop and create vectors of the correct size (Figure 9).

This is also true of other transformations. Fusion merges loops and creates a single loop with a larger granularity. Interchange allows a DOALL to be shifted to outer loops of a nest for parallelization or the inner loop for vectorization. Heuristics for selecting these transformations are controlled with information provided in the EMD.

```
#pragma omp parallel for private( j )
for /* DOALL*/ (i = 0; i <= 200; i += 1) {
    #pragma omp parallel for
    for /* DOALL */ (j = 2; j <= 20; j += 1) {
        BESS1.BESS1_BESPOL.CD0[j-2][i] = CDX16[j-2][i+1];
        BESS1.BESS1_BESPOL.CD1[j-2][i] = -CDX16[j-2][i]/3 -.5*CDX16[j-2][i+1] +
                                          CDX16[j-2][i+2] –CDX16[j-2][i+3]/6
        BESS1.BESS1_BESPOL.CD2[j-2][i] =  0.5*CDX16[j-2][I] + CDX16[j-2][i+2] –
                                          CDX16[j-2][i+1];
        BESS1.BESS1_BESPOL.CD3[j-2][i] = CDX16[j-2][i+3] – CDX16[j-2][i]/6 +
                                          0.5*CDX16[j-2][i+1] – CDX16[j-2][i+2];
    }
}
```

**Fig. 8.** Single loop parallelization

```
#pragma omp parallel for private( j )
for /* DOALL*/ (i = 0; i <= 200; i += 1) {
    for /* DOALL */ (j = 0; j <= 15; j += 4) {
        T1[j:j+3] = -CDX16[j:j+3][i] / 3;
        T2[j:j+3] = .5*CDX16[j:j+3][i+1];
        T3[j:j+3] = CDX16[j-2][i+3] / 6;

        T5[j:j+3] = T1[j:j+3] – T2[j:j+3];
        T6[j:j+3] = CDX16[j:j+3][i+2] – T3[j:j+3];
        BESS1.BESS1_BESPOL.CD1[j:j+3][i] =  T5[j:j+3] + T6[j:j+3];

    }
    for /* DOALL */ (j = 2; j <= 20; j += 1) {
        ...
    }
}
```

**Fig. 9.** Loop distribution and vectorization of the second statement from Figure 8

In most cases, additional transformations are required, for example, to satisfy alignment constraints and split complex operations. However, this is the basic process for matching IR nodes to parallelization heuristics. The benefit of this external configurability is most apparent for systems with a myriad of resources and multiple granularities of parallelism [9] [10] [11]. A change in the EMD is all that is required to adapt the compiler to multiple heterogeneous processors, cores, and SIMD units. For systems with the same microarchitecture that differ only in the multiplicity of cores or processors, modification of the Instruction and Hardware descriptions are not even necessary. A modification in the number Node Mapping types that are instantiated is all that is needed to adapt the mapping heuristics to different resources.

## 5 Summary

Extended Machine Descriptors in the PROMIS compiler are designed to improve both the effectiveness of parallelization and the efficiency of parallel execution by enabling automatic multigrain program parallelization to be adaptive subject to the capabilities of the target architecture. Node Mappings are the means of describing the IR structures targeted. They specify mappings of program parallelism to the resources in the system. The ability to control pass selection from an external description file allows a single compiler binary to be adapted and fine-tuned to the resources of each system, thereby improving its effectiveness. In this paper we showed how we use EMDs to guide pass selection to automatically parallelize code and generate OpenMP directives.

## References

1. Saito, H., Stavrakos, N., Polychronopoulos, C., Nicolau, A.: The Design of the PROMIS Compiler—Towards Multilevel Parallelization. International Jounal of Parallel Programming 28(2) (2000)
2. Girkar, M.: Functional Parallelism: Theoretical Foundations and Implementation, PhD Thesis, University of Illinois at Urbana-Champaign (1992)
3. Su, E., Tian, X.-M., Girkar, M., Haab, G., Shah, S., Petersen, P.: Compiler Support of the Workqueueing Execution Model for Intel SMP Architectures. In: Proc. Of European Workshop on OpenMP (EWOMP) (September 2002)
4. Koopmans, F.: On the Implementation and Effectiveness of Hierarchical Autoscheduling in the PROMIS Compiler, Master's Thesis, University of Illinois at Urbana-Champaign (2002)
5. Stallman, R.: Using and Porting GNU CC. Free Software Foundation (1992)
6. Gyllenhaal, J.C.: HMDES version 2.0 specification. Technical Report IMPACT-96-3, University of Illinois (1996)
7. Curtis-Maury, M., Ding, X., Antonopoulos, C.D., Nikolopoulos, D.: An Evaluation of OpenMP on Current and Emerging Multithreaded/Multicore Processors. In: Proc. of the First International Workshop on OpenMP (IWOMP 2005), Eugene, OR (June 2005)
8. Zhang, Y., Burcea, M., Cheng, V., Voss, M.: An Adaptive OpenMP Loop Scheduler for Hyperthreaded SMP's. In: Proc. of PDCS-2004: International Conference on Parallel and Distributed Computing Systems, San Francisco, Ca (September 2004)
9. Kongetira, P., Aingaran, K., Olukotun, K.: Niagara: A 32-way Multithreaded Sparc Processor. IEEE-Micro 25(2) (March/April, 2005)
10. Borkar, S., Dubey, P., Kahn, K., Kuck, D., Mulder, H., Pawlowski, S., Rattner, J.: Platform 2015: Intel Processor and Platform Evolution for the Next Decade. White Paper (2005), http://www.intel.com/go/platform2015
11. Kahle, J.A., Day, M.N., Hofstee, H.P., Johns, C.R., Maeurer, T.R., Shippy, D.: Introduction to the Cell Multiprocessor. IBM Journal of Research and Development 49(4/5) (July/September, 2005)

# Nested Parallelization of the Flow Solver TFS Using the ParaWise Parallelization Environment

Steve Johnson[1], Peter Leggett[1,2], Constantinos Ierotheou[2], Alexander Spiegel[3], Dieter an Mey[3], and Ingolf Hörschler[4]

[1] Parallel Software Products, London, UK
[2] Parallel Processing Research group, University of Greenwich, London SE10 9LS, UK
[3] Center for Computing and Communication, RWTH Aachen University
[4] Institute for Aerodynamics, RWTH Aachen University

**Abstract.** The Navier-Stokes Solver TFS developed by the Institute of Aerodynamics of the RWTH Aachen University is currently used in a multidisciplinary project to simulate the air flow through the human nose. In order to reduce the runtime of the expensive computations the ParaWise/CAPO automatic parallelization environment has been used to assist in the nested OpenMP parallelization of the TFS multi-block code targeted at Sun Microsystems shared memory parallel systems. Further manual tuning improved the scalability of the OpenMP approach.

## 1 Introduction

The ParaWise/CAPO automatic parallelization environment [1,2] has been used to assist in the OpenMP parallelization of the TFS multi-block code targeted at Sun Microsystems shared memory parallel systems [6]. A series of parallel versions have been created based on differing sources of loop parallelism to evaluate the performance. The aim was to assess the quality of the parallelization from the environment, determine what interaction with the environment is needed and also what manual code changes are necessary. A parallel version of TFS that can scale to large numbers of processors is the ultimate goal of this work. The parallel code versions produced and assessed were:-

1 The initial version produced by ParaWise/CAPO with no user interaction.
2 An improved version where the ParaWise and CAPO browsers were used to investigate, add to and alter information to improve the parallelization.
3 An addition to the previous version where the outer loop of the ijk loop nests within a block (typically the k loop) is chosen to exploit parallelism.
4 A version where parallelism is exploited at the loops that iterate over the blocks in the multi-block code.
5 A nested version exploiting both inter-block and intra-block parallelism, using nested OpenMP directives.

## 1.1 The TFS Flow Solver

The Navier-Stokes Solver TFS is a program library which has been developed by the Institute of Aerodynamics of the RWTH Aachen University during the last 10-15 years. It is well prepared for vectorization as it uses one-dimensional arrays to store 3-dimensional geometries. The numerical method is second-order accurate on a multi-block structured grid with general curvilinear coordinates. The package is currently used in a multidisciplinary project to simulate the air flow through the human nose [3,4]. The goal is to get a better understanding of the functioning of the human nose and then to provide a work flow for computer assisted surgery allowing the physician to first perform a virtual surgery in a virtual reality environment. This would then give the opportunity to verify the success of such an operation by another computer simulation before the actual surgery on the patient. The nested version contains some 16,000 lines of Fortran code including 534 OpenMP directives in 79 parallel regions.

## 1.2 The Machinery

The main target for the parallelization effort was the Sun Fire E25K SMP machine (SFE25K). For further performance comparisons and studies a few other machines were used as well as listed in the following table. Whereas the Sun Fire

Table 1. List of the computer systems used for the performance studies

| Machine model (abbreviated) | Processors | Operating system | Compiler | Remark |
|---|---|---|---|---|
| Sun Fire E 25K (SF25K) | 72 UltraSPARC IV 1.05 GHz dual core | Solaris 10 | Sun Studio 11 | ccNUMA, each processor board has 8 cores+local memory |
| Sun Fire E6900 (SFE6900) | 24 UltraSPARC IV 1.2 GHz dual core | Solaris 9 | Sun Studio 11 | Flat memory |
| Sun Fire E2900 (SFE2900) | 12 UltraSPARC IV 1.2 GHz dual core | Solaris 10 | Sun Studio 11 | Flat memory |
| Sun Fire V40z (SFV40z) | 4 Opteron 875 2.2 GHz dual core | Solaris 10 | Sun Studio 11 | ccNUMA, each dual core-processor has a local memory |
| NEC SX-8 (NECSX8) | 8 NEC SX-8 2.0 GHz vector unit | SX-OS | NEC | SMP vector system with flat memory |

E6900 (SFE6900), the Sun Fire E2900 (SFE2900) and the NEC SX-8 (NECSX8) have a rather flat memory system, the Sun Fire 25K (SFE25K) and even more the Opteron-based Sun Fire V40z (SFV40z) have a ccNUMA architecture. On the SFE25K the two stage cache coherence protocol and the limited bandwidth of the backplane lead to a reduction of the global memory bandwidth and to an increased latency when data is not local to the accessing process. On the 4-socket Opteron-based SFV40z system data and cache coherence traffic is transferred using the HyperTransport links, which connect the four processors in a ring.

Whereas the local memory access is very fast, multiple simultaneous remote accesses can easily lead to grave congestions of the HyperTransport links. Since version 9 update 1, the Solaris operating system provides the Memory Placement Optimization Facility (MPO) which allows use of first-touch or random placement strategies, and also provides a low-level API to explicitly migrate data to where it is used next (next-touch strategy). As the TFS code is ideally suited for vectorization we are also interested in combining vectorization and OpenMP parallelization on the NEC SX-8 shared-memory parallel vector system. Unless otherwise mentioned, all the results quoted in this paper were obtained on the SFE25K. During the parallelization work the Sun Studio 9 Fortran compiler was used and then for the performance improvement efforts the most recent Sun Studio 11 Fortran compiler was employed.

## 2 Production and Performance of the Initial Parallel Version

The initial version was produced without any user interaction, just using the ParaWise interprocedural, value based dependency analysis and the ParaWise OpenMP code generator [7,2]. The aim of ParaWise is to exploit as much application code computation that can execute in parallel, so if all outer loops in a loop nest have dependencies that inhibit parallelism, any parallelism from inner loops in the loop nest is exploited. Parallelism exploited at an inner loop level can have a detrimental effect on performance as the frequency of OpenMP runtime overhead can become significant. Most automatic compilers employ machine dependent metrics to determine at runtime if parallel execution is desirable, forcing serial execution when the OpenMP overhead is deemed to exceed the benefit of parallel computation within the loop. This is not at present used in ParaWise so slowdown can be exhibited by some loops, but subsequent user interaction to uncover parallelism in outer loops and other features in ParaWise can be used to avoid such cases. The parallelism exploited in this parallelization (version 1) was mainly from within a block from one of the $i,j$ or $k$ loops that operate over the dimensions of a block (although not always the outermost of these loops) and also from grid level loops. The code was produced automatically by analysing the application code and generating the parallel OpenMP code all within the ParaWise environment. This included addition of privatization and reduction clauses, nowait clauses to avoid loop end synchronization when legal, generation of parallel regions containing as many parallel loops as possible in an interprocedural context. An example of the automatically generated OpenMP code with no user interaction is shown in figure 1. It shows a chosen parallel loop in routine *AUSM* and the lists of PRIVATE and SHARED variables determined by ParaWise. It also shows the associated parallel region which is in routine *UPWIND* surrounding the call to *AUSM*. All loops in *AUSM* and another loop in *UPWIND* are contained within this parallel region to promote efficient execution. This first version did exhibit some limited speed improvement on some dedicated parallel systems, however, on the heavily loaded SFE25K system at Aachen, parallel executions were slower than the serial version for the medium size input data.

## 2.1 Improving the Initial Parallel Version with User Interaction

ParaWise was specifically designed with the understanding that user interaction is desirable and, often, essential for the production of effective parallel code. In the TFS code, one of example is the nature of the arrays used to index a large work array ($DA$) that is dynamically allocated in the beginning of the program and then frequently passed to subroutines through parameter lists and used throughout the program in the implementation of the multi-block mesh and the mesh dimension strides within a block that are calculated for each block as it is processed. It is obviously not possible for ParaWise to automatically determine the nature of the pointer arrays, so the user must provide this information. For the mesh dimension strides within a block, the way this is implemented in TFS is beyond the current capabilities of ParaWise. The user can exploit their knowledge of the nature of these variables to address the parallelization inhibitors determined by ParaWise and displayed in its browsers. The user can examine serial loops using ParaWise. These include *Totally Serial* loops (where a loop is serial and is not nested within a parallel loop and also does not contain any parallel loops) and *Covered Serial* loops (where a surrounding loop is parallel or some contained loops are parallel). Parallelism may be prevented by several inhibitors, these include (i) Loop carried True dependencies (where a value is assigned in an iteration of the loop and used in a subsequent iteration); (ii) Loop carried Anti dependencies (usage of a value in an iteration where the used memory location is overwritten in a later iteration) but only if loop in/out dependencies also exist; (iii) Loop carried Output dependencies (where the same memory location is assigned in several iterations) but only if loop in/out dependencies also exist; (iv) Loop in/out dependencies (where values assigned before the loop are used by computations within the loop or where values assigned within the loop are used after the loop) but only if loop carried Anti or Output dependencies also exist and FIRSTPRIVATE or LASTPRIVATE clauses cannot satisfy the code requirements. For the True dependence inhibitors, the user may be able to determine that no values are passed between iterations of the loop from either understanding of the implemented variable references or from knowledge of the algorithm being followed in that loop. For Anti and Output dependencies, it may be that those dependencies from one iteration of the loop to another do not exist (allowing the associated variable to be shared) or it may be that the uses after the loop actually receive their data from assignments after the loop and not from inside the loop (allowing the associated variable to be privatized), although it can also be the case that parallel execution is not legal as the inhibitors actually exist. For this improvement of the initial parallel version of the TFS code, no information about the multi-block nature of the application was exploited so only intra-block parallelism was investigated. Most of the interaction related to the linearised one-dimensional loops within a block, particularly due to the lack of information about the dimension stride variables in linearized one-dimensional array references of the three-dimensional data. Typically, Anti and/or Output dependencies between iterations of the loop were set along with usages of the associated variable after the loop where it was the dependencies between iterations of the loop that should not exist. These inhibitors can be simply removed in the ParaWise Why Directive browser which details all inhibitors to parallelism for the selected loop. Details about the inhibitors can be examined using the full range of ParaWise browsers where, in particular, the Dependence Graph browser can be used to easily study in detail the interprocedural dependencies where assignments and usages exist down deep call trees from inside the loop. In Figure 2 the $k$ loop is currently serial as a loop carried output dependence for array $DUMM1$ (a temporary or workspace array) is set along with uses of $DUMM1$ after the loop has completed. An investigation of the indices of $DUMM1$ reveals that $NI$ is computed in a call to $SETNMX$ using the function $IPROD1$ which prevented the ParaWise dependence analysis proving the non-existence of the output dependence. With knowledge of the meaning of the $NI$ array and/or an understanding of what the loop is doing as its part of the TFS algorithm, the user can remove the output dependence, allowing the loop to execute in parallel. The ParaWise generated OpenMP code (version 2) did exhibit some speedup on small numbers of processors, but very little scalability.

## 2.2 Exploiting Parallelism at the Outer Loop Relating to a Mesh Dimension of a Block

It is often desirable to exploit parallelism in a mesh dimension as these can involve a reasonably large number of iterations and therefore can be a profitable source of parallelism. To implement this, the user can use the CAPO Why Directive browser *New_Type* option to force non-$ijk$ loops that are currently in the Chosen Parallel category to execute in serial, allowing contained parallel $ijk$ loops to be chosen (using the setting shown in Figure 3). This version exhibited significant speedup on dedicated parallel systems (albeit on small numbers of processors) indicating the potential to scale well on larger numbers of processors. When run on the SFE25K system at Aachen, however, although speedup on 4, 8 and 16 processors was obtained, the performance was disappointing compared to that seen on a dedicated parallel system. The timings were also very erratic for repeated executions, indicating that the performance was very dependent on the overall load on the system. Using the Sun Performance Analyzer to determine the cause of this poor performance, the reasonably frequent interprocessor synchronization was, we believe, being affected by the machine load. An investigation revealed than the overhead was related to the block-block interactions, where many block pairs have no connections but all were involved end of loop and end of parallel region synchronizations. The code generated was manually altered to move the parallel regions out of the loop over blocks with only a few synchronizations required for those interacting block-block pairs. Although some speedup is exhibited (4.19 on 8 threads), the scalability is severely

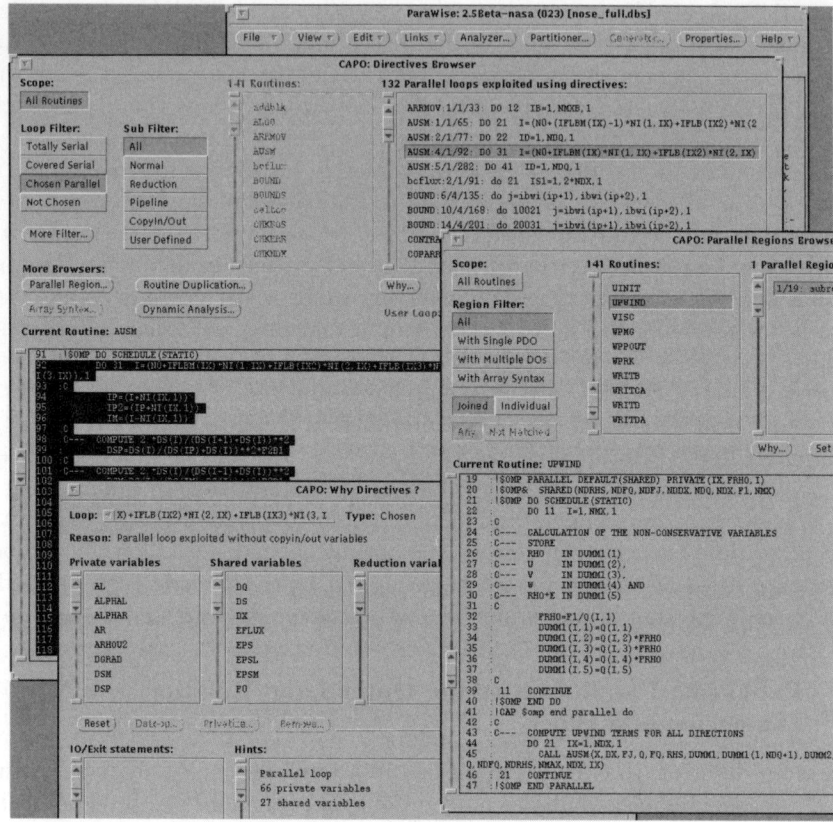

**Fig. 1.** CAPO showing a parallel region in routine UPWIND and a parallel loop in routine AUSM with the automatically determined PRIVATE and SHARED variables

restricted (4.86 on 16 threads). The most significant factor is still the block-block interactions where the amount of computation involved is still relatively small even when compared to the greatly reduced OpenMP runtime overheads.

### 2.3 Exploiting Parallelism from the Block Loops

Parallelism can also be exploited from the loops that operate over the blocks. To create a version that exploits inter-block parallelism, the previous *ijk*-loop parallel version was used as a starting point. The multi-block nature of the code is implemented by integer arrays read into the application code at runtime, preventing dependence analysis from determining independence between the iterations of the block loops. As a result, the CAPO browsers are again used to produce most of the parallel version with manual code changes used to complete the parallel version. Only block loops that did not contain I/O were considered. All the relevant loops involved the *DA* array with dependencies as inhibitors to

Nested Parallelization of the Flow Solver TFS    223

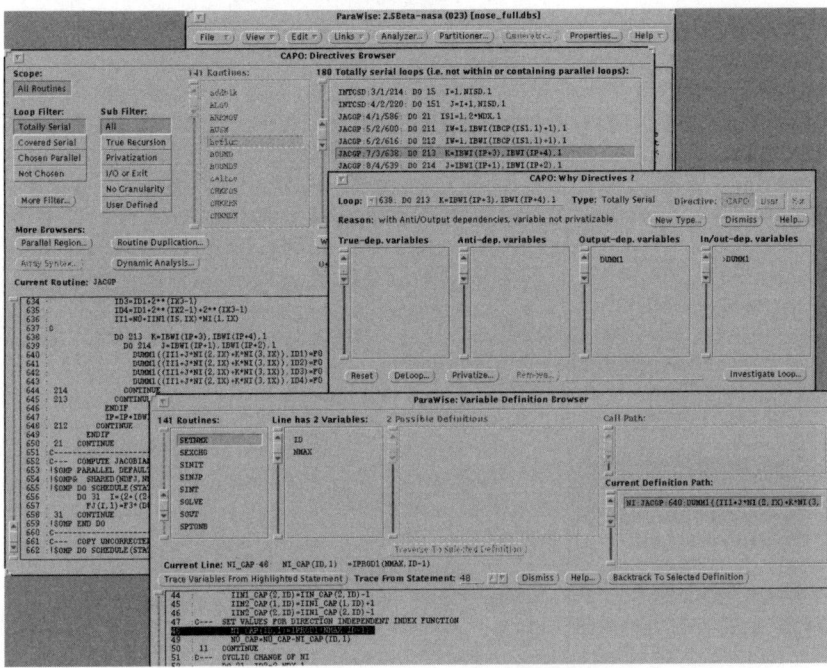

**Fig. 2.** Using ParaWise to investigate serial loops. Here the loop in JACGP is serial due to the variable DUMM1 having both output and loop out dependencies.

parallelism and some loops also involved other variables as parallelism inhibitors. For these other variables, many inhibitors were removed as either dependencies between iterations are known not to exist or loop in/out dependencies are known not to exist. A few variables were involved in reduction style operations and these can be set in the CAPO Why Directive browser *New_Type* option (Figure 3).

Some of the routines called in the block loops contained code controlled by a condition to only execute for the first block. These mainly related to initializing variables that are produced from summations of contributions from every block. To facilitate parallel execution, these initializations were manually moved to before the block loop. For the *DA* array inhibitors, although the dependencies between iterations of the block loop for call arguments relating to field data of the TFS multi-block mesh should not exist, some arguments using *DA* related to workspace in the called routine which was re-used by every block. Those sections of *DA* used for mesh data must be SHARED as they are used throughout the application code. However, the sections of *DA* used for workspace need to be PRIVATE due to the re-use for every block and where they are not used after the loop completes. The ParaWise Dependence Graph and Arguments browser were used to determine which call arguments relate to workspace to assist the process, but the code alterations required to handle the shared and private sections of *DA* were implemented manually after generation of code from CAPO. Inside the

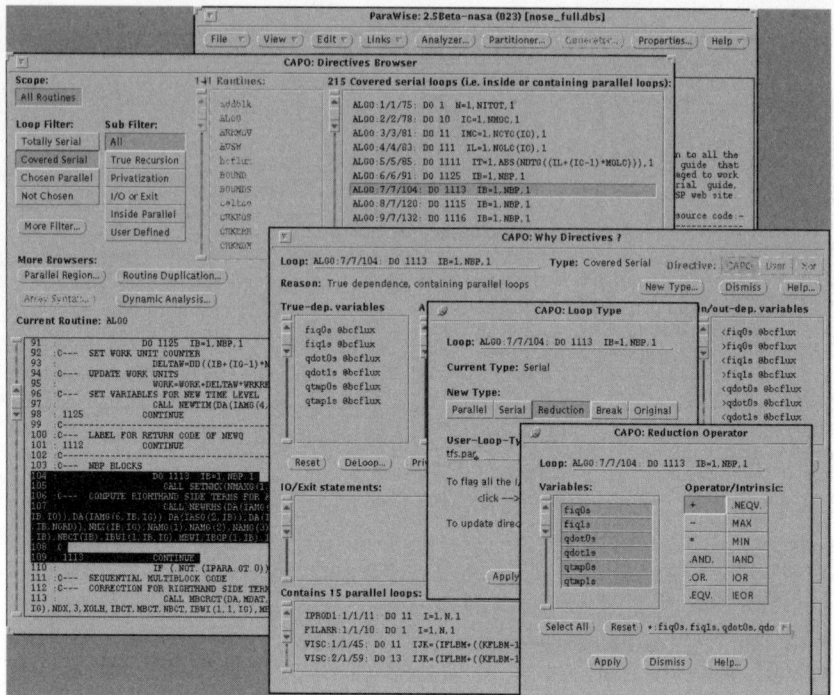

**Fig. 3.** Using ParaWise to investigate serial loops. Here the loop in JACGP is serial due to the variable DUMM1 having both output and loop out dependencies.

subroutines called within the block loops, the renaming of various sections of *DA* allow simple expression of SHARED or PRIVATE variables. Outside of this, in the loop containing the block loop, no renaming is available. One solution is to introduce new arrays for the workspace and alter the call arguments so that the *DA* array is not used. As the target for the parallel TFS code is for it to execute efficiently on large numbers of processors, the extra memory with this approach only relates to the master thread as all slave threads will allocate the workspace for their private copies anyway. For block loops that contain I/O that were not considered for parallelism, the *ijk*-loops within a block are considered instead. Additionally, the block-block interaction loop in routine *EXCHNG* has dependencies that force serial execution. Version 4 of the application does display some speedup on the SFE25K system at Aachen with 2.88 on 4 threads, but only 3.1 on 8 threads. Obviously, as the size of blocks and the corresponding workload for each block varies significantly (a 20+ times difference between quickest and slowest blocks was measured), a static schedule for the block loops was inefficient as compared to using a dynamic schedule. In this case, as the test meshes provided contain only 32 blocks, scalability is restricted to 32 processors, but for efficiency with the variations in block workload, around 8 processors (or a proportion of the number of blocks allowing a few blocks per processor) are needed to allow reasonable load balance.

## 2.4 Nested Inter and Intra Block Parallel Version

Version 5 was created by merging the two previous inter-block and intra-block parallel versions and using nested OpenMP parallelism to allow a more efficient parallel execution, each level using a small number of processors. In this new version the use of THREADPRIVATE directives for common blocks was not possible. Instead, the affected variables were passed into the necessary routines as additional arguments and then defined as PRIVATE for the outer (inter block) parallel region, but shared by the subsequent inner (intra block) parallel region. Another requirement was to turn nested OpenMP parallelism on and off as required by the parallelism in the code. For most of the application code, where the outer block loop is parallel and inner mesh dimension loops are also parallel, OpenMP nested parallelism is enabled and the number of threads set to the square root of the total number of threads requested by the user (as both levels are given the same number of threads). For the code sections containing only a single level of parallelism, the number of threads is set to be the number requested by the user. As expected, the results for the nested version were superior to either of the previous parallel versions, with a speedup of 7.39 on 16 processors and 8.71 on 25 processors. The overheads on larger numbers of processors relate to OpenMP runtime overheads and load imbalance between different blocks. Contributions to this include the poorly performing block-interaction loop that must execute in serial (so that all threads are employed for the contained ijk loops), with the additional impact of one or two small code sections that were left as serial because the OpenMP overheads led to slowdown when parallelism was exploited. The code sections which exploit inter-block level parallelism, but no intra-block level parallelism only represented a tiny proportion of runtime in serial, but on larger numbers of processors their impact on speedup is more significant (Amdahl's law).

## 3 Improving Scalability of the Nested Parallel Program

Figures 4-6 depict the best effort speedup of the intra-block, the inter-block and the nested parallel versions respectively on the machines listed in table 1. It can clearly be seen that the nested approach leads to an increased speedup for all of the larger SMP machines. The maximum is close to 20 when using 64 threads on the SFE25K machine, whereas the speedup of both single-level approaches is less than 10 in all cases. On the NECSX8 vector machine, vectorization replaces the loop level OpenMP parallelization and the block-level parallelization delivers a speedup of 2.7 when using 8 threads. Nevertheless, the efficiency of the nested version is not yet satisfying on the SFE25K. A speedup of 17.5 already seems to be a reasonable result on the SFE6900 (48 threads) and the SFE25K (32 threads), but we wanted to further improve the absolute speed of TFS for

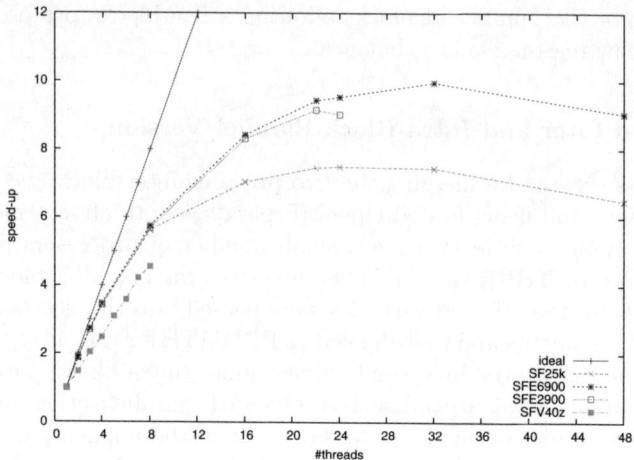

**Fig. 4.** Speedup of the intra-block version of TFS

up to 144 threads on the SFE25K. We also looked at the SFV40z because of the interest in ccNUMA effects at a smaller scale.

**Sorting blocks by size.** Because of the complex geometry of the human nose, the blocks of the computational grid are considerably varying in size: The largest block has about 15 times more grid points than the smallest block and accounts for about 10% out of the 2,200,000 grid points. This fact limits the attainable speedup on the block level to 10. Splitting the larger blocks was not considered at this point. The first approach of selecting a dynamic schedule for all of the block-level loops in order to handle the resulting load imbalance already works reasonably well. But if a relatively large block is scheduled to one thread at the end of the loop, the other threads might finally be idle. Sorting the blocks in decreasing order, such that the smallest block is scheduled last, leads to a first slight improvement in runtime of 5 to 12 percent for 9 or more threads on the SFE25K and 13.5 percent for 8 threads on the SFV40z.

**Thread balancing.** The idea of the dynamic thread balancing scheme has previously been used to solve load imbalance of hybrid (MPI + OpenMP) programs [5], so the number of threads of the inner teams were adjusted to the size of the corresponding blocks. This leads to an improvement of more than 10% on the SFE25K when using 121 threads, as the scalability of the loop level parallelization is limited and cannot overcome the difference in size of the blocks.

**Grouping blocks.** As the block sizes remain constant during the whole runtime of the program, the blocks can be explicitly grouped and accordingly distributed to a given number of threads on the outer parallel level in order to reduce the overhead of the dynamic schedule and to avoid idle threads. Surprisingly this did not lead to a measurable performance improvement on the SFE25K. Further investigations using hardware counters to measure the number of L2 cache misses revealed that threads working on smaller blocks profit more from the large size

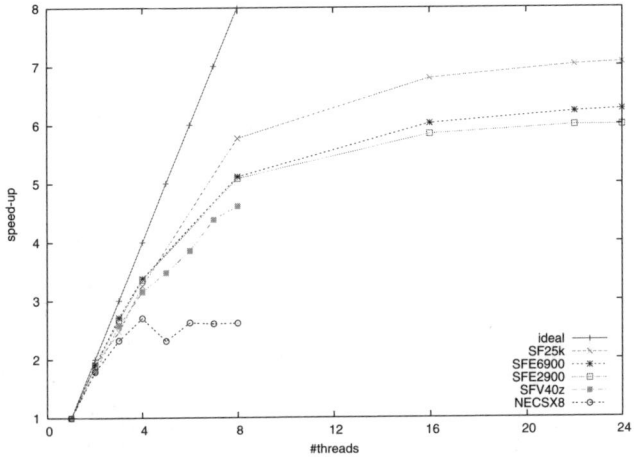

**Fig. 5.** Speedup of the inter-block version of TFS

(8 MB) of the external L2 of the UltraSPARC IV-based machines than larger blocks and therefore ran at a much higher speed. When employing a single thread on the loop level, the thread working on the smallest block ran at 351 MFlop/s and the one working on the largest block ran at 225 MFlop/s. This of course aggravates the load imbalance. Grouping and distributing the blocks was profitable on the SFV40z as the varying block size did not impact the MFlop/s rate, because of the smaller L2 cache (1 MB) of the Opteron processor. The performance improved by 6.7% when using 8 threads.

**Memory locality.** Further hardware counter measurements indicated that L2 cache misses lead to a high percentage of remote misses and that the global memory bandwidth consumption of the code on the SFE25K was close to the maximum value, which we observed when stressing the memory system with the Stream benchmark [8] in earlier experiments with random memory placement used by the Solaris 8 operating system. The first touch memory placement policy provided by more recent versions of the Solaris operating system an improvement by a factor of 5 can be achieved for this benchmark. We concluded that an improvement in the memory locality would also have a positive impact on the performance of TFS on the SFE25K. In order to improve memory locality we bound threads to processors and also used the madvise Solaris system call after a warm-up phase of the program in order to explicitly migrate data to where it is used (next-touch mechanism). Surprisingly, this was only profitable when applied to a single level of parallelism. It lead to an improvement of roughly 10% on the SFE25K for higher thread counts, but was particularly beneficial on the SFV40z - the improvement was up to a factor of 1.9 for loop level parallelization. Unfortunately, applying these techniques to the nested version was not profitable at all, because the current Sun implementation of nested parallelization with OpenMP employs a pool of threads. These threads are dynamically assigned whenever an inner team is forked. Therefore the threads of the inner

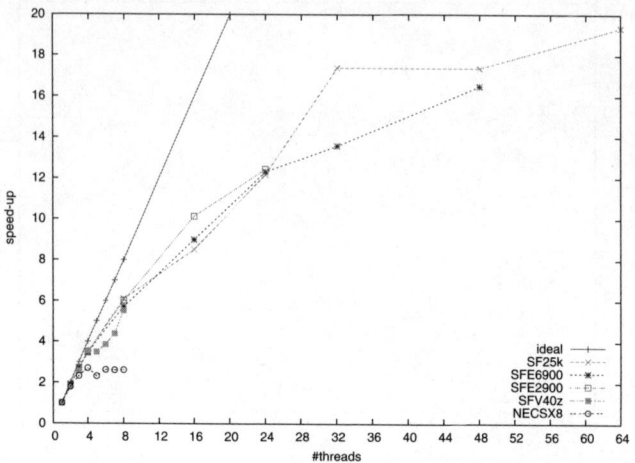

**Fig. 6.** Speedup of the nested parallel version of TFS

**Table 2.** Run times in seconds for 3 iterations, best efforts

| Machine | Serial runtime | #threads block level | #threads loop level | Parallel runtime | Remark |
|---|---|---|---|---|---|
| SFE25K | 341.51 | 8 | 4 | 19.76 | Sorted blocks, first touch placement |
|  |  | 8 | 8 | 17.76 | Sorted blocks, random placement |
| SFE6900 | 312.17 | 8 | 6 | 18.92 | Sorted blocks, first touch placement |
| SFV40z | 147.61 | 8 | 1 | 26.19 | Balanced blocks, binding, next touch |
| NECSX8 | 15.67 | 8 | Vector | 5.785 | Dynamic schedule |

teams frequently loose their data affinity. Table 2 contains the runtime of the most successful combinations of thread numbers on both parallelization levels and the strategy used.

## 4 Summary of Experience of Using the ParaWise Environment for the Parallelization of the TFS Code

The ParaWise environment was used to perform most of the parallelization of the TFS code, and greatly assisted in the subsequent manual tuning. Most of the loops and routines in the code were automatically parallelized by ParaWise, never requiring user attention. Most of the work in improving the parallelism detected and in selecting the most profitable loops for parallel execution was performed within the environment, allowing this to be achieved in a short time period (a few hours) without any need for debugging the generated parallel code. The Sun Performance Analyzer was used to determine which loops were performing poorly focusing the user's effort in the CAPO browsers to the crucial code sections. For the mesh dimension intra-block parallel version, manual intervention was only required in the final tuning phase where the amount of computation in the parallel loops was small when compared to the related OpenMP overheads. For the inter-block parallel version, most of the parallelization was performed within the ParaWise environment. Additionally, the requirement of OpenMP to define variables as PRIVATE or SHARED forces alteration to the calls using sections of the large work array $(DA)$ as part of the mesh and workspace. Currently, this needs to be performed manually, although an algorithmic approach that could be automated in CAPO may be possible in future versions. By carefully choosing the number of threads on each parallelization level and assigning blocks to the threads of the outer team the speed up of the nested version of OpenMP could be improved. The inter-block parallelization also works efficiently in combination with a vectorization on the loop level on the NEC SX8 parallel vector machine, which is particularly well suited for the TFS program. Furthermore, thread binding and data migration, which is supposed to permit higher performance on ccNUMA machines, was only profitable for a single level of parallelization because of the way nested OpenMP parallelism is implemented currently by the Sun Studio compilers. Thus we were not able to obtain a higher speedup than 20 by using more than 64 threads on the Sun Fire E25K.

## References

1. ParaWise automatic parallelisation environment, PSP Inc., http://www.parallelsp.com
2. Jin, H., Frumkin, M., Yan, J.: Automatic generation of OpenMP directives and it application to computational fluid dynamics codes. In: Valero, M., Joe, K., Kitsuregawa, M., Tanaka, H. (eds.) ISHPC 2000. LNCS, vol. 1940, pp. 440–456. Springer, Heidelberg (2000)
3. Hörschler, I., Meinke, M., Schröder, W.: Numerical simulation of the flow field in a model of the nasal cavity. Computers & Fluids 32, 39–45 (2003)
4. Hörschler, I., Brücker, C., Schröder, W., Meinke, M.: Investigation of the impact of the geometry on the nose flow. European Journal of Mechanics–B/Fluids (in press), http://dx.doi.org/10.1016/j.euromechflu.2005.11.006
5. Spiegel, A., an Mey, D., Bischof, C.: Hybrid Parallelization of CFD Applications with Dynamic Thread Balancing. In: Dongarra, J., Madsen, K., Waśniewski, J. (eds.) PARA 2004. LNCS, vol. 3732, Springer, Heidelberg (2006)
6. Johnson, S., Ierotheou, C.: Parallelization of the TFS multi-block code from RWTH Aachen using the ParaWise/CAPO tools, PSP Inc, TR-2005-09-02 (2005) http://www.parallelsp.com/downloads/TechnicalReports/TR-2005-09-02.pdf
7. Johnson, S.P., Cross, M., Everett, M.: Exploitation of Symbolic Information In Interprocedural Dependence Analysis. Parallel Computing 22, 197–226 (1996)
8. McCalpin, J.D.: STREAM: Sustainable Memory Bandwidth in High Performance Computers., http://www.cs.virginia.edu/stream/

# Performance Characteristics of OpenMP Language Constructs on a Many-core-on-a-chip Architecture

Weirong Zhu, Juan del Cuvillo, and Guang R. Gao

Department of Electrical and Computer Engineering
University of Delaware, Newark, Delaware 19716, U.S.A
{weirong,jcuvillo,ggao}@capsl.udel.edu

**Abstract.** Recent emerging many-core-on-a-chip architectures present massive on-chip parallelism through hardware support for multithreading. In order to achieve fast development of parallel applications that exploit this massive intra-chip parallelism to achieve highly sustainable performance, suitable programming models are needed. OpenMP, the industry de facto standard for writing parallel programs on shared memory systems, could become a reasonable candidate. To increase our understanding of the behavior and performance characteristics of OpenMP programs on many-core-on-a-chip architectures, this paper presents a performance study of basic OpenMP language constructs on the IBM Cyclops-64 architecture, which consists of 160 hardware thread units in a single chip. Compared with previous work on conventional SMP systems [1], the overhead of OpenMP language constructs on C64 many-core architecture is at least one order of magnitude lower.

## 1 Introduction

Although advances in IC processing technology have led to hundreds of millions (now reaching 1 billion) of transistors to be fabricated on a single silicon die, the delivered performance versus number of transistors integrated in a chip for conventional single-thread wide-issue superscalar architectures keep declining over time. In order to utilize the transistor budget and mitigate the effects of high interconnect delay, multi-core or many-core-on-a-chip architectures are emerging. Instead of devoting the entire die to a single and complex processor, this new generation of architectural technology proposes to integrate a large number of tightly-coupled simple processor cores on a single chip. The many-core-on-a-chip architecture naturally exploits the thread-level and process-level parallelism, which are expected to be widespread in future applications and multiprocessor-aware operating system and environments [2].

Cyclops-64 (C64) [3,4] is a petaflop supercomputer project under development at IBM T.J. Watson Research Laboratory. The C64 chip architecture employs the many-core-on-a-chip approach by integrating 160 processing cores on a single chip. To the best of our knowledge, the C64 project is one of the most ambitious projects currently under development. Unlike other academia projects, a Cyclops-64 system is planned to be delivered in 2007.

Given the intra-chip parallelism presented by a many-core-on-a-chip architecture, such as C64, it is important and challenging to provide high level parallel programming models for application developers to efficiently map the inherent parallelism in applications to a large number of on-chip processing cores. As a de facto industry standard for writing parallel programs on shared memory systems, OpenMP [5] is considered as one of the possible candidates. Parallel application developers express parallelism, work sharing, and synchronization through the OpenMP language constructs. For the purpose of understanding the behavior and performance characteristics of OpenMP-based parallel programs on many-core architectures, it is important to evaluate the performance of OpenMP basic language constructs, whose overhead accounts for up to 12% of the total execution time in some instances [1].

To conduct a prototype study on high level parallel programming models, we ported the Omni-1.6 OpenMP compiler [6] to C64, and optimized the Omni OpenMP runtime system to adapt to the C64 hardware features [7]. In this paper, based on the number reported by the EPCC microbenchmarks [8], we measure and evaluate the performance characteristics of major OpenMP language constructs on a C64 many-core-on-a-chip architecture with up to 160 cores. In addition, we compare our results to previous work on conventional SMP systems and find remarkable differences. In some instances, the overhead on C64 is one order of magnitude lower.

With our study we provide insight regarding the following aspects of software development on many-core architectures: (1) we provide application developers a better understanding of the behavior of OpenMP programs on a many-core architecture; (2) we give library and compiler developers hints regarding possible optimizations and/or language extensions specific to many-core architectures, specifically, to efficiently exploit multi-level memory hierarchies and fast intra-chip synchronization mechanisms; (3) using the OpenMP runtime library optimization as an example to understand the pros and cons of the C64 architecture, we provide software developers hints on how to write and optimize programs for this type of architecture. To the best of our knowledge, this paper is the first attempt that measures and evaluates the performance characteristics of OpenMP language constructs on many-core-on-a-chip architecture with up to 160 cores.

## 2 Cyclops-64 Architecture

The Cyclops-64 (C64) [3,4] is designed to serve as a dedicated petaflop compute engine for running high performance applications. A C64 system is built out of tens of thousands of C64 chips connected through a 3D-mesh network. The C64 chip employs the many-core-on-a-chip technology by integrating 160 hardware thread units, half as many floating point units, the same number of embedded SRAM memory banks, and the communication hardware in the same piece of silicon (see Figure 1).

A thread unit, the C64 computation cell, is a simple 64-bit, single issue, in-order RISC processor operating at a moderate clock rate (500MHz). Efficient support for thread level execution, such as thread sleep/wakeup, is incorporated in the thread unit. Resource virtualization mechanisms are not provided by the hardware. For instance, thread execution is non-preemptive, and there is no virtual memory manager.

The three-level (SP, on-chip SRAM, off-chip DRAM) memory hierarchy of the C64 chip is visible to the programmer. C64 does not employ data cache. Instead, a portion

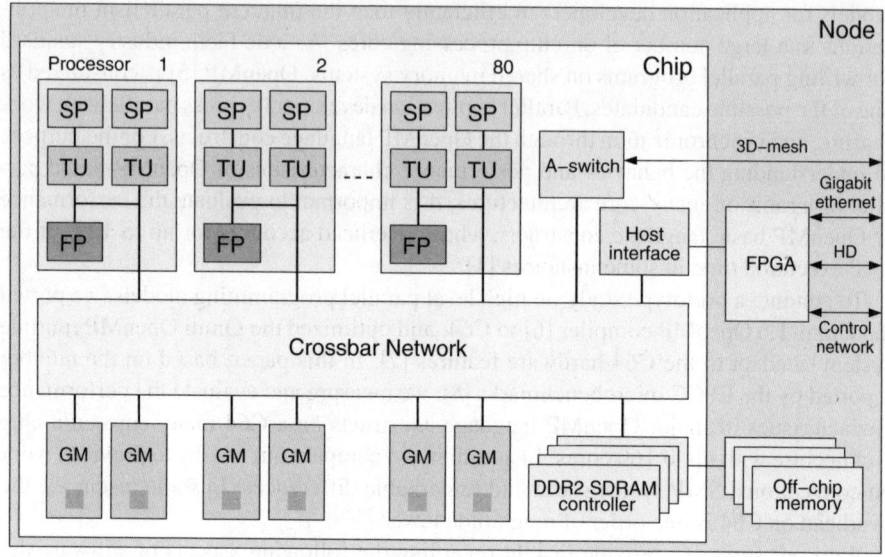

**Fig. 1.** Cyclops-64 node

of each SRAM bank can be configured as scratch-pad memory, which provides a fast temporary storage to exploit locality under software control. The integration of thread units and memory banks on a single chip is further leveraged with a rich set of hardware supported in-memory atomic instructions. Atomic instructions in C64 only block the memory bank where they operate upon, while the remaining banks proceed servicing other requests. This functionality facilitates the scalability of multithreading programs with intensive synchronization operations.

C64 also employs the Network-on-Chip (NoC) concept, all on-chip resources are connected to an on-chip crossbar network, which sustains a 4GB/s bandwidth per port per direction, 384 GB/s per direction in total. Besides the crossbar network, all the thread units within a chip connect to a 16-bit signal bus, which provides a means to efficiently implement barriers.

## 3 Experimental Infrastructure

As shown in Figure 2, the C64 system software toolchain [9] is the infrastructure for software and application prototype development on the incoming C64 system. The toolchain provides binary utilities (assembler, linker, etc.), GNU CC compilers (3.2.3 and 4.0.2), standard C and math libraries that are derived from those in newlib-1.10.0. A microkernel and the TiNy Threads[TM](TNT) runtime system are customized for the unique features of the C64 architecture [10]. The TNT library provides user and library developers an efficient Pthread-like API for thread level parallel programming purpose. The OpenMP compiler and runtime environment is ported from Omni-1.6 [6]. We investigated and optimized the Omni OpenMP runtime library by exploring C64 hardware

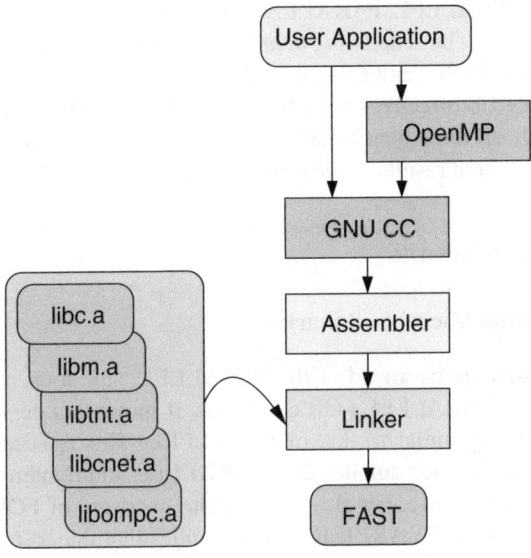

**Fig. 2.** Cyclops-64 software toolchain

features, such as the explicitly visible and programmable memory hierarchy, the efficient in-memory atomic instructions, the thread level execution support, and the fast barrier synchronization through the on-chip signal bus [7].

All the experiments are conducted on a functionally accurate simulator (FAST) [11]. FAST is an execution-driven, binary-compatible simulator of a multi-chip C64 system. It accurately reproduces the functional behavior and count of hardware components of a C64 system. In addition, it generates timing information that accounts for the main sources of pipeline delays and stalls such as contention in memory, the crossbar, and/or other functional units. Although not cycle accurate, this information has proven to be useful for performance estimation, characterization and application tuning as well [11]. FAST has been extensively used by the C64 architecture design team at IBM for the purpose of chip design verification, and dozens of system software developer and application scientists for early application development.

## 4  EPCC Microbenchmarks

In order to understand the performance behavior of an OpenMP application, we use EPCC microbenchmarks [8] to measure the overheads of OpenMP language constructs. The basic methodology employed by EPCC is as follows. First, a reference time is obtained by executing a loop (or loop nests) sequentially without using any OpenMP directive. Then, the overhead is calculated by comparing this reference time with the execution time of the same code extended with OpenMP constructs.

There are three components of the EPCC microbenchmark. The *synchronization benchmark* measures the overhead of OpenMP work-sharing and mutual exclusion directives, such as PARALLEL, PARALLEL FOR, BARRIER, CRITICAL, ATOMIC, and REDUCTION etc.. The *scheduling benchmark* compares different scheduling policies – STATIC, DYNAMIC, and GUIDED. The *array benchmark* measures the overhead of the PARALLEL directives with the PRIVATE, FIRSTPRIVATE, and COPYIN clauses. We execute all three benchmarks on a single C64 chip with up to 128 threads and report the experiment results in the next section.

## 5 Experimental Results

### 5.1 Synchronization Microbenchmark

Figure 3(a) compares the overhead of the PARALLEL , the loop, and the combined parallel work-sharing PARALLEL FOR constructs. It shows that the PARALLEL FOR construct has overhead similar to that of PARALLEL. This is because the overhead of the FOR construct is much smaller than PARALLEL and remains almost constant. From Figure 3(a) and (b), we can also see that the overhead of FOR is only slightly higher than the overhead of BARRIER, which implies that the cost of FOR is mainly due to the implicit BARRIER at the end of the loop.

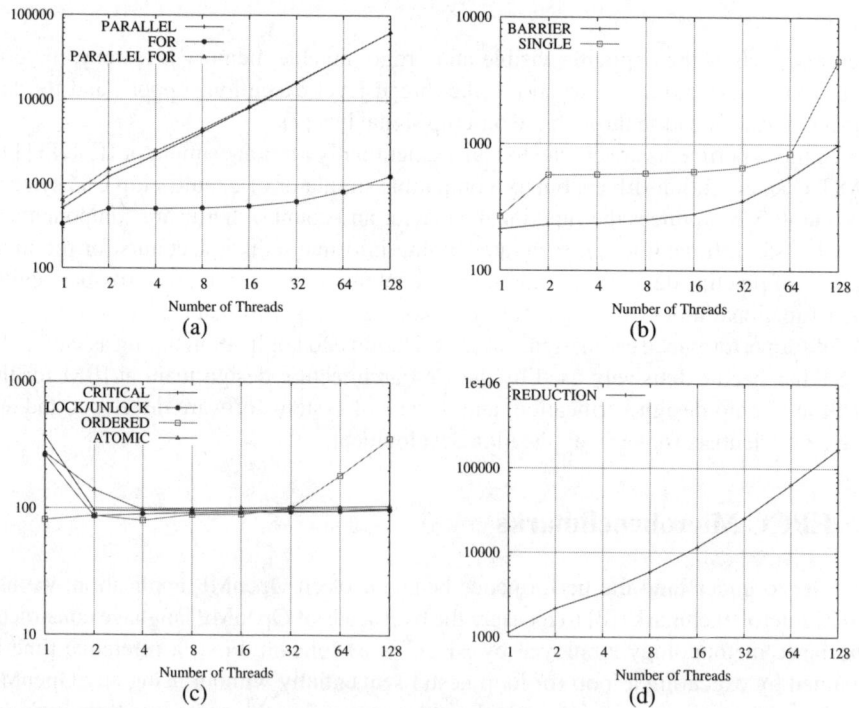

**Fig. 3.** Overhead (cycles) of Synchronization Directives (a) PARALLEL, FOR, PARALLEL FOR (b) BARRIER, SINGLE (c) Mutual Exclusion (d) Reduction

Note the high overhead of the SINGLE directive, especially when the number of threads increases to 128. This is because the implementation of SINGLE is very expensive in order to guarantee the semantics of SINGLE. The memory contention incurred to complete the SINGLE operation rises dramatically when the number of threads increases. SINGLE also suggests an implicit barrier.

Because the OpenMP runtime library is carefully designed and tuned to map to the C64 hardware features, and the hardware components of C64 are tightly coupled in a single chip, the PARALLEL and BARRIER constructs incur much lower overhead than on conventional SMP systems. For example, a previous study [1] shows that the overhead of the PARALLEL construct reaches 120 microseconds (108,000 cycles) when running with 70 threads on a 72-node Sun Fire 15K system. Even while running with 128 threads, the same construct only presents a 63,020 cycles overhead. This observation implies that the thread management on a C64 like many-core architecture is much more efficient than common SMP environments.

We customized the well-known linked-list-based MCS spin-lock algorithm [12] to implement the low level lock acquisition and release primitives in the OpenMP runtime library [7]. Unlike common SMP systems where the overhead of lock increases with the number of threads, Figure 3(c) shows that the overhead of mutual exclusion constructs in OpenMP remain within the same range without increasing dramatically. Even for 128 threads, the CRITICAL directive costs only 154 cycles.

The of overhead of the REDUCTION construct increases exponentially, as shown in Figure 3(d). As future work, the reduction operation can be optimized in the runtime library by taking advantage of the C64's rich set of in-memory atomic instructions, which can perform certain operations, such as addition, subtraction, and various logical operations, atomically in memory. From our previous experiences with other benchmarks, such as Table Toy [11], we expect to improve the performance of REDUCTION dramatically.

## 5.2 Scheduling Microbenchmark

In OpenMP, there are three means for scheduling loop iterations among threads: STATIC, DYNAMIC, and GUIDED [5]. Please note that EPCC only reports the overhead of the GUIDED($n$) scheduling policy for small values of $n$. Figure 4 compares different loop scheduling policies when running on 1 to 128 threads. It is apparent that STATIC and STATIC(128) always incur the lowest overhead in all cases. For the STATIC($n$) policy, STATIC(1) causes the largest overhead, and the overhead decreases to the overhead of STATIC with increasing chunk size. Actually, the overhead of STATIC and STATIC(n) increases slowly for runs from 2 threads to 64 threads. When 128 threads are executed concurrently, the overhead is much larger than running with 64 threads because of the high memory contention.

DYNAMIC(1), which is the most fine-grained scheduling policy, generates huge overheads (3,621 microseconds) when running on 128 threads. This is because the small chunk size causes frequent dynamic scheduling function calls, whose execution time is counted as the overhead. As a result, the overhead of static scheduling is multiple orders of magnitude smaller than dynamic scheduling.

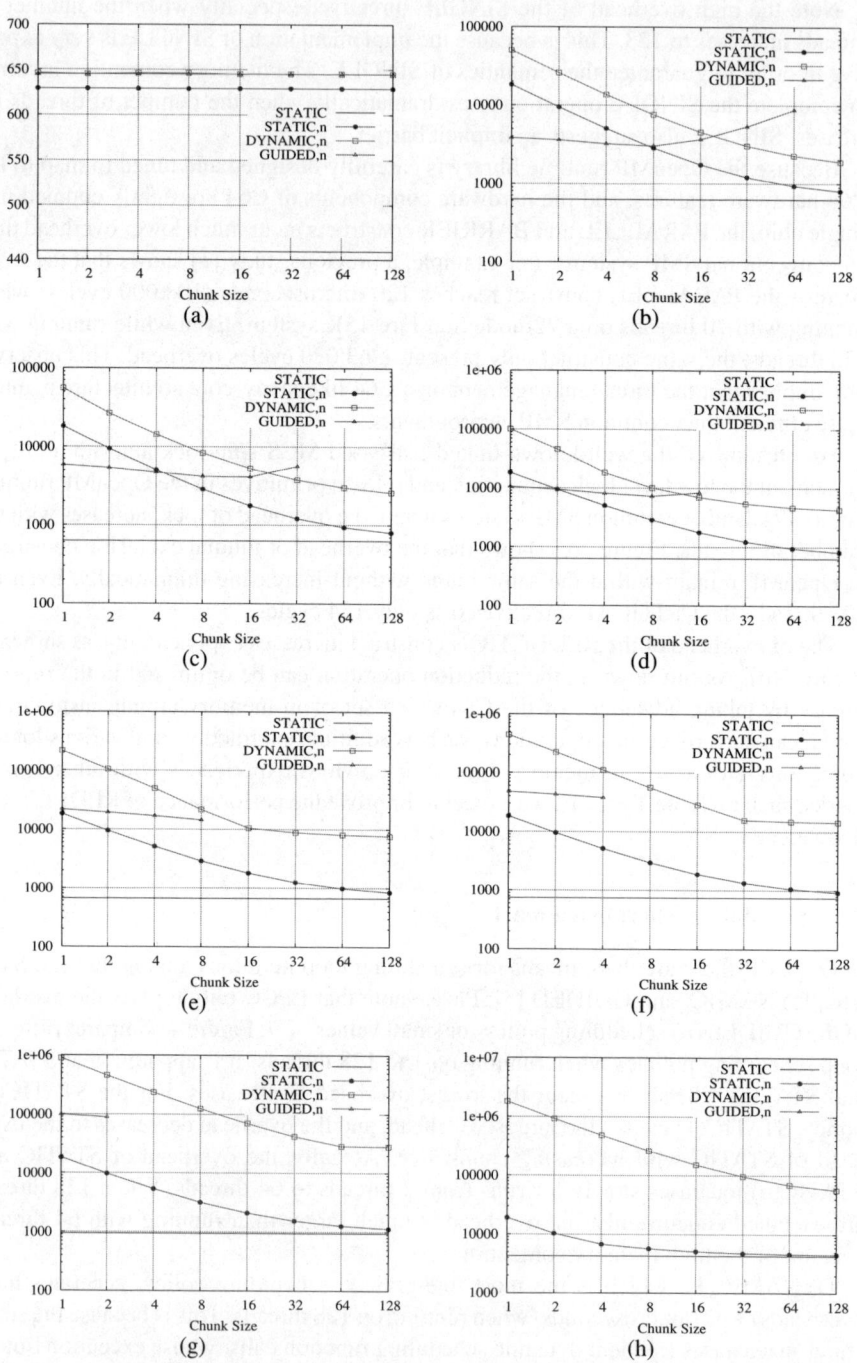

**Fig. 4.** Overhead (cycles) of Scheduling Policies with (a) 1 Thread (b) 2 Threads (c) 4 Threads (d) 8 Threads (e) 16 Threads (f) 32 Threads (g) 64 Threads (h) 128 Threads

The overhead of the GUIDED($n$) scheduling is always better than the DYNAMIC($n$). The GUIDED($n$) policy starts with a large chunk size, then gradually decreases it to $n$. Figure 4 also demonstrates that the STATIC policy always incurs lower overhead than the GUIDED policy. The overheads measurement suggests that on C64 OpenMP programmer should consider the STATIC scheduling policy as the first option for loop scheduling, given the tasks can be statically balanced. Only if the benefit of dynamic load balancing surpasses the scheduling overhead, the dynamic and guided scheduling policy are worth being chosen.

In the OpenMP runtime library, the dynamic and guided scheduling functions are implemented to frequently access the thread descriptor, and sometimes access the master thread's descriptor by acquiring a lock first. By taking advantage of the explicit programmable multi-level memory hierarchy of C64, we place the thread descriptor of each work thread into its own scratchpad memory, which guarantees very fast accesses, i.e., 1 cycle for a store, 2 cycles for a load. The master thread's descriptor is placed in on-chip global memory, whose access latency is longer than scratchpad but smaller than off-chip memory. By leveraging the C64's in-memory atomic instruction and thread level execution support, the lock/unlock primitives used to guarantee the mutual exclusion for accessing the master thread's descriptor are efficiently implemented as demonstrated in Figure 3(c) [7]. Therefore, compared with common SMP systems, the overhead of loop scheduling is at least an order of magnitude lower on a C64-like many-core-on-a-chip architecture. For example, as reported in [1], when running on a 72-node Sun Fire 15K, the DYNAMIC(1) incurs an overhead of around 27M cycles (30,000 microseconds) with 24 threads, while on C64 it costs 0.44M cycles with 32 threads, and 1.8M cycles with 128 threads. The overhead of STATIC scheduling is 9,000 cycles with 24 threads on a Sun Fire 15K [1], but only 743 cycles with 32 threads, and 4,298 cycles with 128 threads on C64.

### 5.3 Array Microbenchmark

The *array microbenchmark* measures the overhead of the PARALLEL directive with the PRIVATE, FIRSTPRIVATE, and COPYIN clauses. In the current design of C64 system software, the stack of a thread is placed in its own scratchpad memory and the size of the stack is limited. As a result, in our experiments, we can only run the benchmark with an array size smaller than or equal to 729. As a work in progress, the C64 toolchain will provide support for automatic stack extension, a feature that allows applications that require more stack than available to continue running at the expense of performance. When the stack area is exhausted, the runtime system automatically relocates the stack into off-chip memory. Notice the relocation is performed very quickly, as it requires setting a few registers and copying a few locations from the stack (but not all). If at a later point, the stack shrinks, the runtime system undoes the changes and sets the program stack back to scratchpad memory. However, in order to achieve good performance, it is not recommended to declare large arrays on the stack (as automatic variables), or make deep recursive function invocations in the program.

As shown in Figure 5, the PRIVATE and FIRSTPRIVATE clauses have similar overheads (the overhead of FIRSTPRIVATE is slightly higher). Compared with the PARALLEL constructs without any data-sharing attribute and data copying clauses, it is

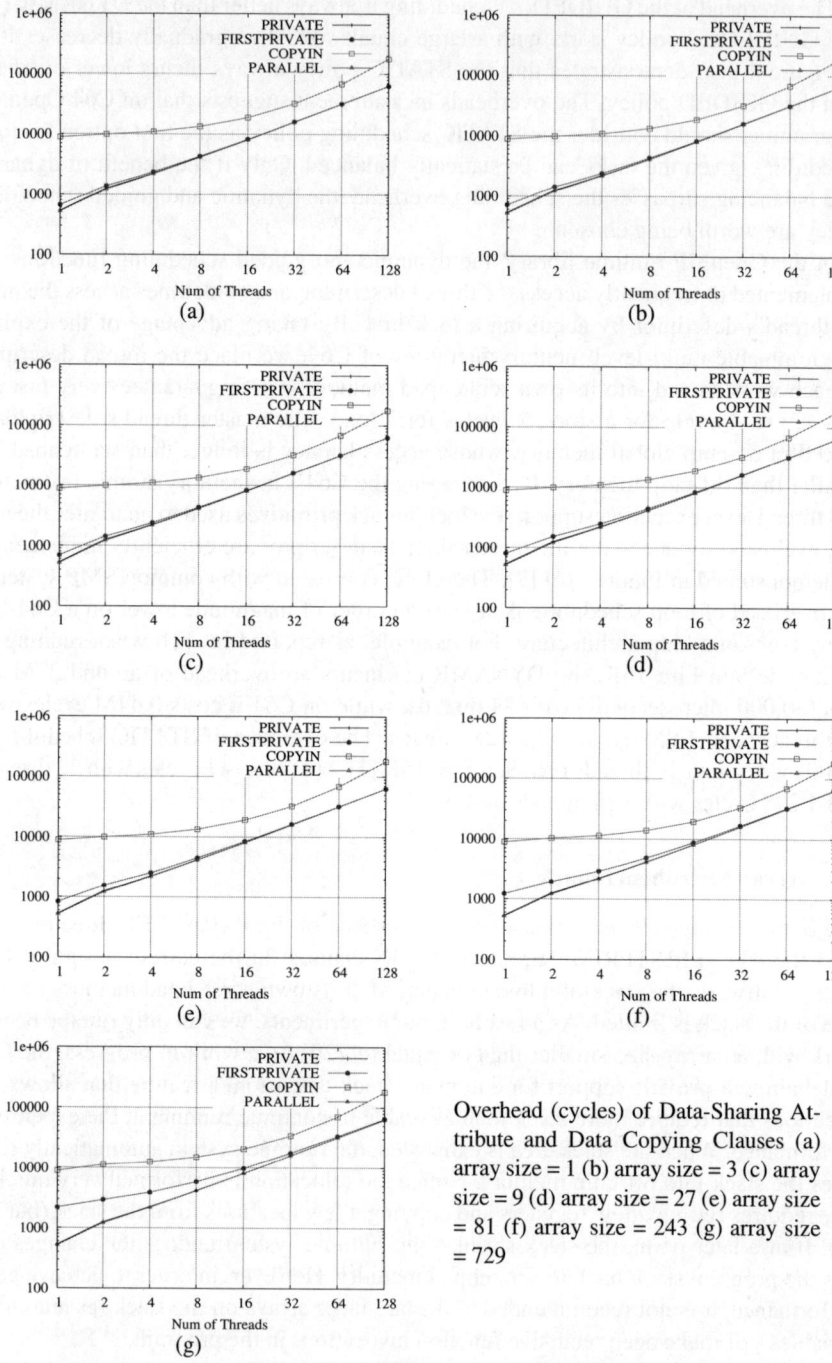

Overhead (cycles) of Data-Sharing Attribute and Data Copying Clauses (a) array size = 1 (b) array size = 3 (c) array size = 9 (d) array size = 27 (e) array size = 81 (f) array size = 243 (g) array size = 729

Fig. 5.

also clear that the curves of PRIVATE and FIRSTPRIVATE almost match the curve of PARALLEL constructs. This means attaching the PRIVATE or FIRSTPRIVATE clause to the PARALLEL construct incurs negligible costs. In both cases, the compiler directly allocates the private array in the stack of each thread, which incurs no overhead at runtime.

For FIRSTPRIVATE, the C library function *bcopy* is used to initialize the private array by copying the contents of a global array. In the standard C library of C64, routines like *memcpy*, and *bcopy*, are optimized and fine tuned. They are aware of the explicit memory hierarchy. The C64 load and store multiple instructions are used to exploit the memory bandwidth and save cycles from not issuing multiple instructions. In addition, the instruction sequences are manually scheduled to hide memory accessing latencies. Since the array size used in our experiments is small, the copying is performed very efficiently. Therefore, no significant overhead is observed for FIRSTPRIVATE.

From Figure 5, the COPYIN clause generates one order of magnitude larger overhead than the other two clauses. By attaching the COPYIN clause to the PARALLEL directive, the Omni OpenMP compiler generates codes that dynamically allocate the storage for thread private data. The heap manager allocates the thread private data in the on-chip global memory. There are also overheads from lock/unlock operations for using the memory allocator. Moreover, since the data is allocated in the global memory at runtime, the latency of memory accesses in the loop body is much higher than accessing scratchpad memory. This is the reason why COPYIN has much larger overhead than PRIVATE and FIRSTPRIVATE. This suggests a scope for possible optimizations either in the compiler or the runtime system.

## 6 Related Work

Previous work [7] demonstrated a set of optimizations on the Omni OpenMP runtime library by exploiting C64 hardware features. We introduced the optimization techniques and demonstrated the effectiveness by showing the performance improvement of OpenMP synchronization constructs compared to the unoptimized OpenMP runtime library. This paper presents the measurement and evaluation of all major OpenMP language constructs, including synchronization directives, scheduling policies, and array clauses, with the optimized runtime library on C64. We also compare our results to those previously reported on conventional SMPs. The purpose of this work is to provide the application programmers, compiler and library developers a better understanding of the behavior of OpenMP programs on a many-core architecture.

Most of the previous work on performance characterization of OpenMP were conducted on the general purpose commercial shared memory SMP systems [8,6,13,14,15,1]. Liao et. al. [16] evaluated OpenMP on chip multithreading platforms. However, the chip multiprocessor (UltraSPARC III) evaluated in the paper only has two cores. To the best of our knowledge, this paper is the first attempt to measure and evaluate the performance characteristics of OpenMP language constructs on a C64-like (160 cores) like many-core-on-a-chip architecture.

In [17,18], the authors presented the experiment results of OpenMP NAS benchmarks on an experimental Cyclops architecture. It is worth noting that this experimental architecture was a preliminary design of the Cyclops architecture and it is never to be built, while the first C64 system is planned to be delivered in 2007. Also, this experimental Cyclops architecture included data caches in the design, and the C64 system employs scratchpad memory technology instead of data cache. Neither [17] nor [18] conducted performance characterization of the OpenMP language constructs, since that was not the purpose of those two papers.

## 7 Conclusion and Future Work

Multi-core or many-core-on-a-chip architecture tends to be widely accepted in the near future. Given the massive intra-chip parallelism, a high level parallel programming model is needed for fast and efficient application development. OpenMP is considered as one reasonable candidate. In order to help the application developer and system software designer to increase the understanding of the performance behavior of OpenMP programs on many-core-on-a-chip architecture, this paper reports the performance characteristics of OpenMP language constructs on the Cyclops-64 chip architecture, which integrates 160 cores in a single chip. As for the future work, we would like to evaluate the performance of OpenMP on C64 with application kernels and benchmarks, such as NAS parallel benchmarks, and the SPEC OMP Benchmark suite.

## Acknowledgments

We would like to acknowledge the support from IBM, in particular, Monty Denneau, who is the architect of the IBM Cyclops-64 architecture, ETI, the Department of Defense, the Department of Energy (DE-FC02-01ER25503), the National Science Foundation (CNS-0509332), and other government sponsors. We would also like to acknowledge other members of the CAPSL group at University of Delaware, in particular Ziang Hu, Yuan Zhang, Geoff Gerfin, and Brice Dobry.

## References

1. Fredrickson, N.R., Afsahi, A., Qian, Y.: Performance characteristics of OpenMP constructs, and application benchmarks on a large symmetric multiprocessor. In: Proceedings of the 17th annual international conference on Supercomputing (ICS 2003), pp. 140–149. ACM Press, New York (2003)
2. Hammond, L., Nayfeh, B.A., Olukotun, K.: A single-chip multiprocessor. Computer 30(9), 79–85 (1997)
3. Denneau, M., Warren Jr., H.S.: 64-bit Cyclops principles of operation part I. Technical report, IBM Watson Research Center, Yorktown Heights, NY (2005)
4. Denneau, M., Warren Jr., H.S.: 64-bit Cyclops principles of operation part II: Memory organization, the A-switch, and SPRs. Technical report, IBM Watson Research Center, Yorktown Heights, NY (2005)
5. OpenMP Architecture Review Board: OpenMP C and C++ application program interface. Technical Report 2.5, OpenMP Architecture Review Board (2005), http://www.openmp.org/specs

6. Kusano, K., Satoh, S., Sato, M.: Performance evaluation of the Omni OpenMP compiler. In: Valero, M., Joe, K., Kitsuregawa, M., Tanaka, H. (eds.) ISHPC 2000. LNCS, vol. 1940, pp. 403–414. Springer, Heidelberg (2000)
7. del Cuvillo, J., Zhu, W., Gao, G.R.: Landing OpenMP on Cyclops-64: An efficient mapping of OpenMP to a many-core system-on-a-chip. In: Proceedings of the 3rd ACM International Conference on Computing Frontiers, Ischia, Italy (2006)
8. Bull, J.M.: Measuring synchronization and scheduling overheads in OpenMP. In: Proceedings of First European Workshop on OpenMP, Lund, Sweden (1999)
9. del Cuvillo, J., Zhu, W., Hu, Z., Gao, G.R.: Toward a software infrastructure for the Cyclops-64 cellular architecture. In: Proceedings of 20th International Symposium on High Performance Computing Systems and Applications, St. John's, Newfoundland and Labrador, Canada (2006)
10. del Cuvillo, J., Zhu, W., Hu, Z., Gao, G.R.: TiNy Threads: A thread virtual machine for the Cyclops64 cellular architecture. In: Fifth Workshop on Massively Parallel Processing, in conjuction with 19th International Parallel and Distributed Processing Symposium (IPDPS 2005), Denver, Colorado, USA, p. 265 (2005)
11. del Cuvillo, J., Zhu, W., Hu, Z., Gao, G.R.: FAST: A functionally accurate simulation toolset for the Cyclops64 cellular architecture. In: Workshop on Modeling, Benchmarking, and Simulation (MoBS2005), in conjuction with the 32nd Annual International Symposium on Computer Architecture (ISCA2005), Madison, Wisconsin (2005)
12. Mellor-Crummey, J.M., Scott, M.L.: Algorithms for scalable synchronization on shared-memory multiprocessors. ACM Transactions on Computer Systems 9(1), 21–65 (1991)
13. Bull, J.M., O'Neill, D.: A microbenchmark suite for OpenMP 2.0. SIGARCH Comput. Archit. News 29(5), 41–48 (2001)
14. Berrendorf, R., Nieken, G.: Performance characteristics for OpenMP constructs on different parallel computer architectures. Concurrency - Practice and Experience 12(12), 1261–1273 (2000)
15. Prabhakar, A., Getov, V., Chapman, B.: Performance comparisons of basic OpenMP constructs. In: Proceedings of the 4th International Symposium on High Performance Computing, Kansai Science City, Japan, pp. 413–424 (2002)
16. Liao, C., Liu, Z., Huang, L., Chapman, B.: Evaluating OpenMP on chip multithreading platform. In: First International Workshop on OpenMP, Eugene, Oregon, USA (2005)
17. Almasi, G., Ayguadé, E., Cascaval, C., José Castanos, J.L., Martínez, F., Martorell, X., Moreira, J.: Evaluation of OpenMP for the Cyclops multithreaded architecture. In: Voss, M.J. (ed.) WOMPAT 2003. LNCS, vol. 2716, pp. 69–83. Springer, Heidelberg (2003)
18. Ródenas, D., Martorell, X., Ayguadé, E., Labarta, J., Almási, G., Caşcaval, C., Castaños, J., Moreira, J.: Optimizing NANOS openMP for the IBM Cyclops multithreaded architecture. In: 19th IEEE International Parallel & Distributed Processing Symposium (IPDPS 2005), Denver, Colorado, USA (2005)

# Improving Performance of OpenMP for SMP Clusters Through Overlapped Page Migrations

Woo-Chul Jeun[1], Yang-Suk Kee[2], and Soonhoi Ha[1]

[1] Electrical Engineering and Computer Science, Seoul National University,
Seoul, Korea
{wcjeun, sha}@iris.snu.ac.kr
http://peace.snu.ac.kr/research/parade/
[2] Computer Science & Engineering, University of California,
San Diego, USA
yskee@csag.ucsd.edu

**Abstract.** Costly page migration is a major obstacle to integrating OpenMP and page-based software distributed shared memory (SDSM) to realize the easy-to-use programming paradigm for SMP clusters. To reduce the impact of the page migration overhead on the execution time of an application, the previous researches have mainly focused on reducing the number of page migrations and hiding the page migration overhead by overlapping computation and communication. We propose the 'collective-prefetch' technique, which overlaps page migrations themselves even when the prior approach cannot be effectively applied. Experiments with a communication-intensive application show that our technique reduces the page migration overhead significantly, and the overall execution time was reduced to 57%~79%.

## 1 Introduction

Since the OpenMP specification [1] was proposed as a standard shared-memory programming model in 1998, there have been a variety of efforts to adopt OpenMP as an easy-to-use programming paradigm for parallel processing platforms, which range from small chip-level systems [2] to Grids. As commodity off-the-shelf symmetric multi-processors (SMPs) and high-speed network devices are widely deployed, SMP clusters have become an attractive platform for high performance computing. Accordingly, there have been many studies [3][4][5][6][7][8][9][10] on applying the OpenMP programming model to cluster systems. An intuitive approach to realizing OpenMP for SMP clusters is to use an SDSM (Software Distributed Shared Memory) system, which transparently provides a single shared address space across distributed memory. Specifically, most of them utilize page-based SDSM systems, which keep memory consistency with a user-level page fault signal handler.

The execution time of an application on a page-based SDSM system can be decomposed into computation time, page migration overhead, synchronization overhead, and signal handler overhead. We define a page migration as a procedure that sends a page request to the home node and receives the page reply for the request, and define the page migration overhead as the net time to complete all page migrations. It is known that SDSM systems suffer from poor performance due to high synchronization overhead and excessive accesses to the remote pages [11]. Therefore, the common challenge that the prior studies on OpenMP for SMP clusters confronted was how to overcome these intrinsic performance bottlenecks of the conventional SDSM systems.

Their specific efforts were how to reduce page migration overhead and synchronization overhead. Specifically, their solution techniques can be categorized into five: implementing synchronization directives efficiently [5], reducing the shared address space [6][8], lessening the number of page migrations [4][5][6][8][9][10], reducing the page migration delay with fast communication HW and page update protocol [7], and hiding the page migration overhead by overlapping computation time and page migration overhead [6][10]. Some of these techniques are complementary, and consequently can be applied independently to improve performance.

In this paper, we assume that the number of page migrations is already minimized by other techniques. To lessen the page migration overhead for a given number of page migrations, previous studies have mainly focused on hiding the page migration overhead by overlapping the computation time and the page migration overhead. However, they are effective only if computation time is large enough to hide the page migration overhead.

Specifically, this paper proposes a 'collective-prefetch' technique to lessen this page migration overhead. This technique analyzes the page access patterns and prefetch remote pages by overlapping page migrations themselves. Note that our technique is different from the prior techniques in that it can be used even when computation cannot overlap communication. When this technique is applied to our target SDSM system, experiments with a communication-intensive application show that the page migration overhead was reduced to 30%~72% and overall execution time was reduced to 57%~79%.

This paper is organized as follows. In Section 2, motivation of our research is presented with a communication-intensive application. Then, we explain the proposed technique in Section 3. We detail our implementation and present the experimental results using an OpenMP programming environment in Section 4. Finally, we draw a conclusion with some idea of future research direction.

## 2 Motivational Example

We used ParADE [5] as our target system. ParADE is an OpenMP-based parallel programming environment that consists of OpenMP translator, thread-safe SDSM based on HLRC (home-based lazy release consistency) protocol [11] with migratory home and multiple-writer protocol, and MPI library such as MPI/Pro. By executing the program in a hybrid model of software DSM and MPI, ParADE shows better performance than pure SDSM-based environment such as Omni/SCASH.

Our motivational application is FT [12] that contains a computational kernel of a 3-D Fast Fourier Transform (FFT)-based spectral method. Table 1 shows the number of page migrations and execution time breakdown of FT on ParADE for 2 ~ 8 nodes. To isolate the page migration overhead clearly, we use one computation thread for each node to avoid overlapping computation and communication. The page migration overhead is over about 65% of the total execution time, much larger than the computation time. In consequence, overlapping computation time with page migration overhead has little effect with this application. The possible improvement is no more than the computation time, about 13%~26%. Note that this poor performance is not due to any inefficiency of ParADE implementation. When we ran the same FT program on Omni/SCASH, the execution time was much longer than ParADE.

**Table 1.** The number of page migrations and execution time breakdown of FT class A on ParADE

| Nodes | Synchro-nization | Handler | Computation | Migration | Execution Time(sec) | Page migrations of each node (times) |
|---|---|---|---|---|---|---|
| 2 | 1.54% | 7.49% | 26.18% | 64.79% | 116.29 | 115,968~116,399 |
| 4 | 2.50% | 6.52% | 17.62% | 73.36% | 87.27 | 87,872 ~ 88,397 |
| 8 | 5.62% | 5.91% | 12.87% | 75.60% | 59.16 | 51,648 ~ 52,187 |

Like most page-based SDSM systems, ParADE uses a segmentation fault signal handler for page migrations [13]. Figure 1 demonstrates the memory consistency mechanism of a page-based SDSM. Any unprivileged access of a computation thread (at Node 0) to a page invokes the segmentation fault signal handler and then the handler fetches the valid page negotiating with the owner (Node 1) of the page. The problem is that the handler is blocked for a long time, waiting for the reply. This makes the application experience long memory access latency during runtime.

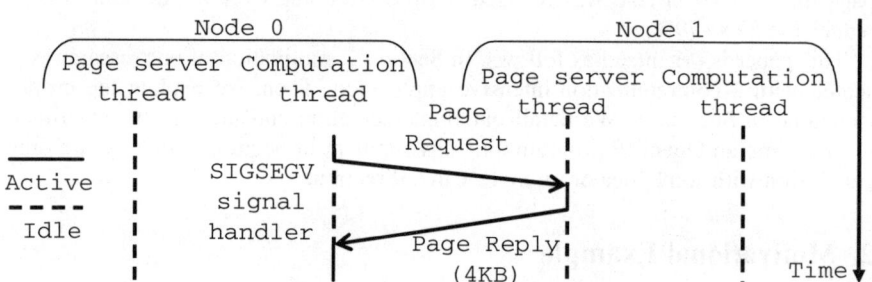

**Fig. 1.** Page migration between nodes in the ParADE. Time runs down the page.

The commonly used approaches to overcome such long latency problem are pipelining and prefetching. Pipelining and prefetching have been proposed in different contexts such as Web [14], HPF [15], and SDSM without OpenMP [16]. These techniques enable computation and communication to be overlapped. In contrast, we overlap page migrations themselves. Our technique does not reduce the latency of each page migration, but reduce the page migration overhead. Furthermore, it can be effective even when computation workload is small.

## 3 Overlapping Page Migrations

In this section, we present the proposed technique with the inspector-executor model [17]. Our technique consists of two steps: generating the inspector-executor code and translating OpenMP code. Without any modification of an OpenMP source code, our OpenMP translator automatically generates the inspector-executor code from the source code. The inspector checks all references to the shared memory and collects page migrations. The executor overlaps page migrations and executes the actual parallel loop.

The OpenMP translator replaces the original parallel loop with the inspector-executor code at the first step. Then, the OpenMP translator translates the inspector-executor code at the second step.

### 3.1 Inspector-Executor Code

The OpenMP translator does a pre-processing step to generate the inspector-executor code before the OpenMP translates a parallel loop. For example, Figure 2 shows a simple OpenMP code segment where x and y are shared.

```
#pragma omp parallel for default(shared)
for(i=0;i<10000;i++) {
  x[i] = y[i];
}
```

**Fig. 2.** A target parallel loop in the original OpenMP code

The OpenMP translator analyzes the parallel loop and generates the inspector-executor code before the OpenMP translation. Figure 3 shows the inspector code for the parallel loop in Figure 2. The inspector collects page migration requests before the executor overlaps page migrations. The codes have the same parallel loop structure as the original loop except that they just identify the pages to be accessed without executing the entire code block. A simple macro, CHECK_SHARED_ADDRESS, is used to inform the OpenMP runtime system of the pages that need to be fetched remotely. The inspector collects the information about the pages to be migrated before the real page migrations occur.

```
#pragma omp parallel for default(shared)
for(i=0;i<10000;i++) {
  CHECK_SHARED_ADDRESS(&x[i]);
  CHECK_SHARED_ADDRESS(&y[i]);
}
```

**Fig. 3.** Inspector code generated by OpenMP translator

Figure 4 shows the executor code for the parallel loop in Figure 2. The executor code consists of a prefetch function call and the actual parallel loop. do_prefetch() function initiates the page migrations using the proposed page migration techniques discussed below.

```
        do_prefetch();
        #pragma omp parallel for default(shared)
        for(i=0;i<10000;i++) {
            x[i] = y[i];
        }
```

**Fig. 4.** Executor code generated by OpenMP translator

### 3.2 Multiple Prefetch Technique

Figure 5(a) shows a simplified original page migration scenario between two nodes. The master thread creates a computation thread, which asks page requests one by one. The computation thread sequentially performs computation and page migrations, and joins the master thread after the parallel region ends.

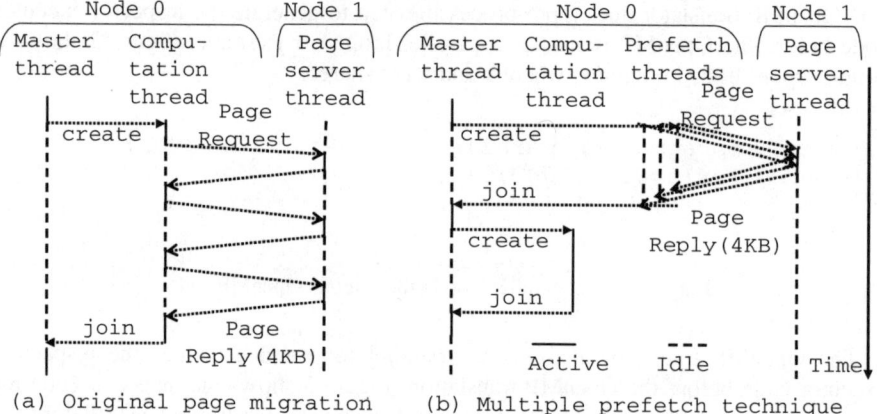

**Fig. 5.** Original page migration and multiple prefetch technique

The proposed technique reduces the page migration overhead by overlapping page migrations. An intuitive approach is to invoke multiple prefetch threads concurrently. We refer to this approach as multiple prefetch technique. The master thread creates prefetch threads in do_prefetch() function before creating the computation thread. Figure 5(b) shows that three prefetch threads overlap page migrations for a home node. While each prefetch thread completes page migrations one by one, a prefetch thread can send a page request to a home node even when other prefetch threads are waiting for page replies from the home node. Also, prefetch threads can concurrently send page requests to multiple nodes. Ideally prefetch threads may overlap page migrations as many as the number of prefetch threads, which is not true in reality. All prefetch threads join the master thread after all page migrations end. Then, the master thread creates a computation thread, which does computation without page migrations and joins the master thread later.

The drawback of this basic approach is difficult to decide the proper number of prefetch threads for the best performance. Though too many prefetch threads cause large context-switching overhead, very small number of threads cannot fully utilize the overlapping effect. Moreover, the waiting time grows longer if multiple page requests arrive at one node simultaneously. Moreover, prefetch threads compete with the page server thread in each node so lengthen the delay of page server thread.

### 3.3 Collective Prefetch Technique

We propose the collective prefetch technique to overcome the problems of multiple prefetch technique. This technique is to send a list of page request once to a home node instead of sending page requests one by one. The master thread creates a collective prefetch thread that sends a page request list to the home node and receives pages from the home node continuously, as shown in Figure 6. The page server thread creates a collective page server thread. As all page requests initially arrive at a home node, the collective page server thread can send page replies continuously without idle time between page replies. This reduces waiting time for multiple page migrations to waiting time for one page migration. Each thread joins its parent thread after all page migrations are completed. Then, the master thread creates a computation thread, which does computation without page migrations and joins the master thread later.

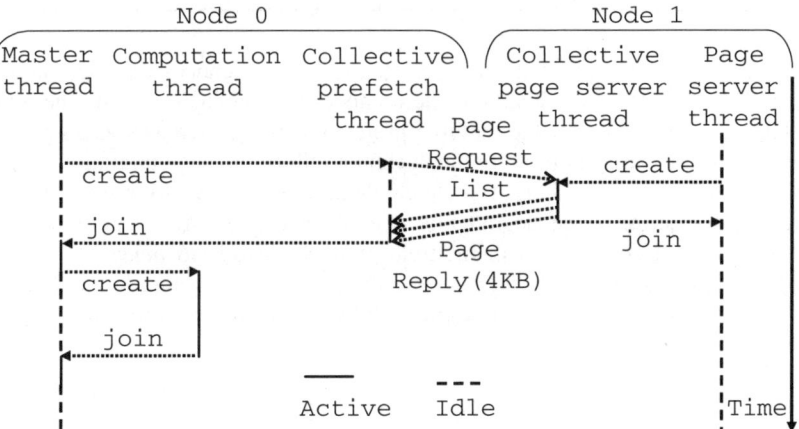

**Fig. 6.** Collective prefetch technique

## 4 Experiments

### 4.1 Experiment Environment

The experiment platform is a Linux cluster consisting of 8 nodes interconnected by a Gigabit Ethernet switch. Each node has dual 2.4GHz Intel Pentium 4 Xeon processors and 1GB memory. We used Red Hat 8.0 Linux 2.4.18-14 SMP kernel for ParADE and Red Hat 7.3 Linux 2.4.18-3 SMP kernel for Omni/SCASH because the SCore 5.6.1 package [18] installs Red Hat 7.3 and Omni/SCASH together.

We evaluated our technique with the FT kernel in the 3.0 NPB suite. We ported the original FORTRAN program to the C version for ParADE. We used the gcc compiler with the optimization level 3 (O3).

### 4.2 Results

We implemented the proposed technique into ParADE and analyzed the execution time of FT. Table 2 shows the page migration overhead of FT on ParADE with one computation thread per node, varying the number of prefetch threads for the multiple prefetch technique. The performance of the original ParADE without the proposed technique is used for the baseline performance.

**Table 2.** Page migration overhead of FT on ParADE with one computation thread (seconds)

| Nodes | Original | Multiple-prefetch (# of prefetch threads) | | | | | | Collective-prefetch |
|---|---|---|---|---|---|---|---|---|
| | | 1 | 2 | 3 | 4 | 5 | 6 | |
| 2 | 75.34 | 68.84 | 43.56 | 33.59 | 29.49 | 24.93 | 23.18 | 21.94 |
| 4 | 64.02 | 62.30 | 49.92 | 46.69 | 45.44 | 45.85 | 45.64 | 31.58 |
| 8 | 44.72 | 58.79 | 51.23 | 47.44 | 46.57 | 46.84 | 47.22 | 24.45 |

The table shows that the collective-prefetch technique reduces the page migration overhead to 30%~55% consistently as the number of nodes increases. As the number of prefetch threads increases, the multiple-prefetch technique reduces the page migration overhead more. But multiple-prefetch cannot reduce the page migration overhead additionally when the number of prefetch threads is over 6. Moreover, multiple-prefetch shows performance degradation over 8 nodes. The main cause is that the page server thread competes with multiple prefetch threads to delay its responses. Though the multiple-prefetch technique reduces the page migration overhead to 30%~70% on 2~4 nodes, the multiple-prefetch technique increases the page migration

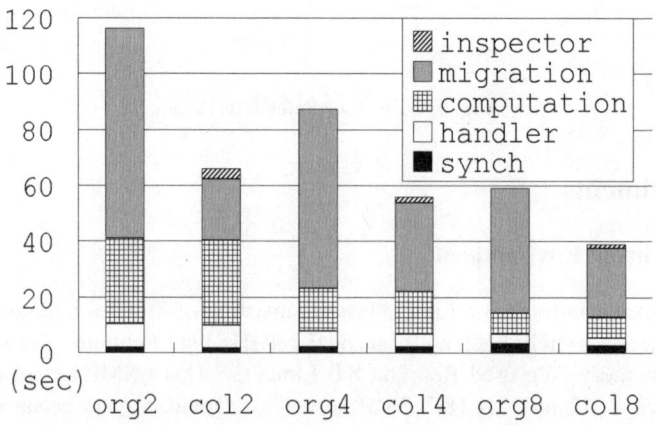

**Fig. 7.** Execution time breakdown of FT on ParADE with one computation thread (seconds)

overhead to 105% on 8 nodes. Note that the collective prefetch technique always outperforms the multiple-prefetch technique.

The collective-prefetch technique reduces the execution time to 57%~66% for 2~8 nodes. Figure 7 shows the execution time breakdown of FT on ParADE for 2~8 nodes. Original ParADE and collective-prefetch are represented as 'org' and 'col'. 'inspector' indicates the overhead of the inspector to check page requests and to make a list of page request for our technique. The overhead is less than 6% in all cases. Since the synchronization overhead, computation time, and the handler overhead are constant over the time, the change of the page migration overhead determines the change of the total execution time.

Table 3 shows the page migration overhead of FT on ParADE with two computation threads per node. The performance of the original ParADE with two computation threads is used for the baseline performance. If we use more than one computation thread, computation and page migration are somewhat overlapped. The collective-prefetch reduces the page migration overhead to 52%~72% consistently even when the number of nodes increases. However, multiple-prefetch shows the same characteristics as when the number of computation threads is one. The difference between them is that the multiple-prefetch shows performance degradation over 4 nodes, earlier than the case of one computation thread. Though the multiple-prefetch technique reduces the page migration overhead to 54% on 2 nodes, the multiple-prefetch technique increases the page migration overhead to 101% and 120% on 4~8 nodes.

**Table 3.** Page migration overhead of FT on ParADE with two computation threads (seconds)

| Nodes | Original | Multiple-prefetch (# of prefetch threads) | | | | | | Collective-prefetch |
|---|---|---|---|---|---|---|---|---|
| | | 1 | 2 | 3 | 4 | 5 | 6 | |
| 2 | 44.81 | 70.18 | 45.20 | 34.60 | 29.63 | 26.98 | 24.25 | 23.22 |
| 4 | 44.56 | 63.11 | 49.70 | 46.23 | 45.27 | 45.51 | 45.19 | 32.11 |
| 8 | 39.60 | 59.95 | 50.39 | 47.40 | 47.85 | 45.86 | 47.79 | 24.41 |

Though the number of computation threads varies from one to two, we expect that the page migration overhead is almost constant regardless of the number of computation threads because the number of page migrations is mainly dependent on the number of nodes of FT. The page migration overheads in table 2 and table 3 meet our expectation. However, the original ParADE shows reduced page migration overhead because two computation threads can hide page migrations. To evaluate the effect of overlapping computation and page migration, we increased the number of computation threads from 1 to 5 on a SMP node, as shown in Table 4. Though the execution time decreases as computation threads are created up to 4, the execution time starts increasing with more computation threads. In addition, FT shows non-deterministic behavior as more computation threads are used. FT often takes long execution time over 100 seconds. Such non-deterministic behavior is believed due to the fact that many computation threads affect thread scheduling and the page server thread is interrupted by them.

**Table 4.** Execution time of FT on ParADE with varying computation threads (seconds)

| Nodes | # of computation threads | | | | |
|---|---|---|---|---|---|
| | 1 | 2 | 3 | 4 | 5 |
| 2 | 116.29 | 69.76 | 61.08 | 56.05 | 53.03 |
| 4 | 87.27 | 58.28 | 53.49 | 54.26 | 55.08 |
| 8 | 59.16 | 51.50 | 43.33 | 42.44 | 44.70 |

The collective-prefetch reduces the execution time to 65%~79%. Figure 8 shows the execution time breakdown of FT on ParADE for 2~8 nodes. The change of the page migration overhead determines the change of the execution time. The difference of the execution times in Figure 7 and Figure 8 is the computation time reduced by two computation threads.

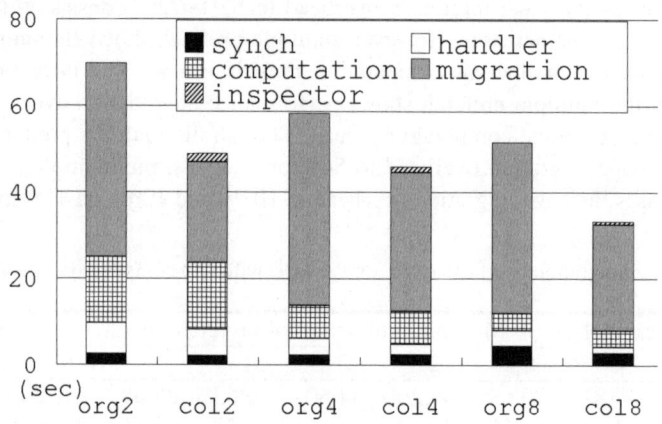

**Fig. 8.** Execution time breakdown of FT on ParADE with two computation thread (seconds)

In summary, the collective-prefetch reduces the page migration time of FT from 45~75 seconds to 22 ~ 32 seconds consistently even when the number of nodes increases. In percentage notation, the collective-prefetch technique reduces page migration overhead to 30%~72%, and reduces the execution time of FT to 57%~79%.

## 5 Related Work

In our experiments, we invoke multiple computation threads to overlap computation and page migration. There are other approaches to overlap overlapping computation and page migration overhead. Seung-Jai et al. [6] used dynamic scheduling to balance the execution time of nodes considering the page migration overhead. More computation is performed on a node experiencing less page migration.

J.J. Costa et al. [10] presented 'presend' technique similar to our technique. 'presend' analyzes access patterns and prefetches necessary pages for the next parallel loop as soon as the current parallel loop ends. In other words, 'presend' learns the sequence of loops and registers the memory regions that are accessed by the parallel loop. After a thread makes the page access list once, it sends the pages specified in the list whenever it finishes a loop. 'presend' overlaps the page migration overhead with the execution of "serial section" before the next parallel loop. So if the "serial section" is not large enough, the page migration overhead is not effectively hidden. And they only focused on overlapping the computation and the page migrations. Though they did not consider the effect of overlapping page migrations themselves, our experiments shows that overlapping page migrations themselves can improve the performance of an application. And 'presend' makes the page list once for a parallel loop. It can make the page prediction difficult if a parallel loop accesses the shared array given as a parameter of a function. For example, in FT, as the argument of the function changes, a parallel loop accesses different memory region. In even that case, our approach can compute the correct page list because we newly make the page list every time.

T. Mowry et al. [16] proposed another technique for an SDSM system, where prefetch requests are invoked before the computation. Unless a requested page arrived earlier than the computation thread needs, it still experiences the page migration overhead. In summary, the prior works to overlap computation and page migration are effective only when there is sufficient computation workload to hide the overhead.

## 6 Conclusion

Long latency of page migrations has been the major performance bottleneck of OpenMP on page-based SDSM for clusters. This paper shows that the proposed collective-prefetch technique could reduce the page migration overhead effectively even when the computation can not hide the page migration overhead by the previous works. Experiments with a communication-intensive application show that our technique reduces the page migration overhead significantly, and the overall execution time was reduced to 57%~79%.

The technique presented in this paper has been implemented manually. We are currently working on automating the translation in our OpenMP translator. Another research issue is to develop a hybrid technique between the proposed one and the prior techniques considering computation-to-communication ratio of an application.

## Acknowledgments

This work was supported by National Research Laboratory Program (No.M1-104-00-0015), Brain Korea 21 Project and IT leading R&D Project funded by Korean MIC. The ICT at Seoul National University provides research facilities for this study.

## References

1. OpenMP C and C++ API, Version 1.0 (1998), http://www.openmp.org
2. Liu, F., Chaudhary, V.: A Practical OpenMP Compiler for System on Chips. In: Voss, M.J. (ed.) WOMPAT 2003. LNCS, vol. 2716, pp. 54–68. Springer, Heidelberg (2003)

3. Hu, Y., Lu, H., Cox, A., Zwaenepoel, W.: OpenMP for Networks of SMPs. J. Parallel and Distributed Computing 60(12), 1512–1530 (2000)
4. Sato, M., Harada, H., Ishikawa, Y.: OpenMP compiler for a Software Distributed Shared Memory System SCASH. In: Proc. of WOMPAT 2000 (2000)
5. Kee, Y., Kim, J., Ha, S.: ParADE: An OpenMP Compiler for a Software Distributed Shared Memory Systems. In: Proc. of IEEE/ACM Supercomputing (2003)
6. Min, S., Basumallik, A., Eigenmann, R.: Optimizing OpenMP Programs on Software Distributed Shared Memory Systems. Int. J. Parallel Programming. 31(3), 225–249 (2003)
7. Tao, J., Karl, W., Trinitis, C.: Implementing an OpenMP Execution Environment on Infiniband Clusters. In: Proc. of IWOMP 2005 (2005)
8. Chun, H., Xuejun, Y.: Performance Analysis and Improvement of OpenMP on Software Distributed Shared Memory Systems. In: Proc. of EWOMP 2003 (2003)
9. Matsuba, H., Ishikawa, Y.: OpenMP on the FDSM software distributed shared memory. In: Proc. of EWOMP 2003, pp. 71–78 (2003)
10. Costa, J.J., Cortes, T., Martorell, X., Ayguade, E., Labarta, J.: Running OpenMP applications efficiently on an everything-shared SDSM. In: Proc. of IPDPS 2004 (2004)
11. Li, K., Hudak, P.: Memory coherence in shared virtual memory systems. ACM Transactions on Computer Systems 6(4), 321–359 (1989)
12. Baily, D., Saphir, W., van der Wijngaart, R., Woo, A.: The NAS Parallel Benchmarks. Technical Report, NAS-95-020 (1995)
13. Kee, Y., Kim, J., Ha, S.: Memory Management for Multi-Threaded Software DSM Systems. Parallel Computing 30, 121–138 (2004)
14. Padmanabhan, V., Mogul, J.: Using Predictive Prefetching to Improve World Wide Web Latency. SIGCOMM Computer Communication Review (1996)
15. Muller, M.: Compiler-Generated Vector-based Prefetching on Architectures with Distributed Memory. In: Jer, W., Krause, E. (eds.) High Performance Computing in Science and Engineering 2001, Springer, Heidelberg (2001)
16. Mowry, T., Chan, C., Lo, A.: Comparative Evaluation of Latency Tolerance Techniques for Software Distributed Shared Memory. In: Proc. of HPCA-4 (1998)
17. Koelbel, C., Mehrotra, P.: Compiling Global Name-Space Parallel Loops for Distributed Execution. IEEE Transaction on Parallel and Distributed Systems 2(4), 440–451 (1991)
18. The PC Cluster Consortium: The SCore cluster system, http://www.pccluster.org

# Aspects of Code Development

# Adding New Dimensions to Performance Analysis Through User-Defined Objects

Gabriele Jost[1], Oleg Mazurov[2], and Dieter an Mey[3]

[1] Sun Microsystems, 3295 NW 211th Terrace, Hillsboro, OR, 97124, USA
Gabriele.Jost@sun.com
[2] Sun Microsystems, 16 Network Circle, Menlo Park, CA, 94025, USA
Oleg.Mazurov@sun.com
[3] RWTH Aachen University, Aachen, Germany
anmey@rz.rwth-aachen.de

**Abstract.** Performance analysis of parallel applications requires a high degree of flexibility for the user to navigate through huge amounts of performance trace data. Many different statistics need to be calculated and compared in order to derive meaningful conclusions. In this paper we propose user-defined objects as a means to customize the analysis for different hardware platforms, programming paradigms, or application areas. We describe how the concept is realized in a development version of the Sun™ Studio Performance Analyzer and demonstrate its usefulness for the analysis of a nested OpenMP application.

## 1 Introduction

Performance analysis of large-scale scientific applications poses the challenge of meaningful interpretation of a large amount of performance data. A plethora of factors influence the performance of a parallel application, such as the hardware platform, the system software, and the programming model. Poor performance will usually be due to an intricate interaction of many components. This requires that many different metrics are calculated, attributed to different components and compared to each other. The type of metrics and components will depend not only on the compute system, but also on the programming paradigm and even the type of application. This requires a high degree of flexibility within a performance analysis system to collect performance data, calculate metrics, and allow for mapping of these metrics onto specific entities, such as subroutine calls or program counters. In an experimental version of the Sun™ Studio Performance Analyzer we have introduced the concept of user-defined objects, which provide means for a user to introduce new entities to map performance metrics on. This way the analysis process can be tailored to specific needs arising from hardware, programming paradigm, and system and application software. The purpose of this paper is to describe how user-defined objects are supported by our analysis system and to demonstrate the usefulness of this feature when analyzing the performance of a nested OpenMP application on two different types of SMP (Symmetric Multiprocessor) systems.

The rest of the paper is structured as follows: Section 2 gives an overview of the Sun™ Performance Analyzer. Section 3 describes the nested OpenMP application used in our study. Timings and performance analysis are presented in Section 4. Related work is discussed in Section 5, where we also draw our conclusions.

## 2 The Sun™ Performance Analyzer

The Sun™ Performance Analyzer [5] is a general-purpose application level performance analysis tool. Performance data is collected during a program run and stored in an experiment data file. A GUI and a command line interface are available to navigate through the collected data. Clock and hardware counter based profiling are provided on multiple platforms: SPARC, x86, x86-64, Solaris, Linux.

The Sun™ Performance Analyzer provides support for many parallel programming paradigms, such as OpenMP, MPI and hybrid MPI/OpenMP. Special efforts have been made to allow for the analysis of OpenMP programs. The concept of an OpenMP specific *state of a thread* was introduced. Examples for such states are *Work, Reduction, Wait for work,* or *explicit/implicit barrier,* depending on the OpenMP construct the thread is currently executing. At runtime, the performance data collection module communicates with the OpenMP runtime library in order to obtain specific information about the state of all OpenMP threads.

Two special metrics are computed for OpenMP programs: Time spent in OMP-work and time spent in OMP-wait. The sum of the two is always equal to the total time spent by a program across all parallel threads. The metrics are based on OpenMP thread states as reported by the OpenMP runtime library. For example:

| OpenMP thread state | Time attributed to |
|---|---|
| Work | OMP-work |
| Reduction | OMP-work |
| Wait for work | OMP-wait |
| Implicit barrier | OMP-wait |
| Explicit barrier | OMP-wait |

Other types of data recorded are hardware counter overflows, thread identifiers, CPU identifiers, a high-resolution timestamp, and a call stack. By default, the recorded data is attributed to functions, source lines, and disassembly instructions. The code generated by the Sun™ Studio compilers for OpenMP constructs has the body of a parallel loop abstracted into a separate function, which may be called from either the master or the slave threads. A naming convention allows the tools to associate those functions with the original function from which the construct was extracted. There are two modes to examine the recorded trace data. In *user mode* the call stack is reconstructed to reflect the users view by hiding internals of the OpenMP runtime library. All parallel instances of a function or a parallel loop have the same origin and can be traced back to the main program. In *machine mode*, the call stack is presented as recorded, including details about the OpenMP runtime library.

A typical task that the user conducts during performance analysis is the filtering of collected performance data. The analyzer provides support for the user to define filters. The user achieves this by specifying a filter expression, which is evaluated for each data record to determine whether or not it is to be included. An example is a filter to calculate metrics only if the thread is in the OpenMP Work state mentioned above.

In addition to filtering, the user will want to attribute the calculated metrics to certain entities. Most commonly, performance analysis systems provide means to map metrics onto function calls, source and assembly lines, and program counters.

In an experimental version of the Sun$^{TM}$ analyzer we have added a feature to allow for user-defined program objects. The objects are specified via a command language. A customized object can be constructed from basic data records such as thread identifier (THRID), CPU identifier (CPUID), time stamp (TSTAMP), or virtual and physical memory addresses (VADDR, PADDR). The analyzer can compute any available metric for such objects and the object can be used to define data filters.

For example, each thread within one process has a unique identifier THRID, which is recorded in the performance trace file. This identifier can be used to define a new object as follows:

```
Thread: THRID
```

An object Seconds is useful to see how performance metrics evolve over time. Such an object can be defined as follows:

```
Seconds: (TSTAMP/1000000000),
```

where TSTAMP is the time stamp recorded in the trace file.

It is possible to define objects involving 2 or more basic record types. For example, time stamps and thread identifiers can be combined to provide for a linearized 2-dimensional mapping of a metric for all threads at a certain time, such as:

```
Thread_Second: THRID + 1000 * Seconds.
```

The syntax also allows for the definition of hardware specific objects such as processor boards, memory boards, or cache lines. An object representing the processor board can be defined in terms of a CPUID, for example:

```
Processor_Board: (CPUID&0x1fc)>>2
```

An object representing a certain area of memory can be defined in terms of physical memory addresses, such as:

```
Mem_Seg:(PADDR>=0x1C000000000)&&(PADDR<0x1C800000000).
```

Last but not least, the program address space, which consists of the *virtual memory* allocations created by the operating system on behalf of the application, can be expressed in terms of virtual addresses, as in:

```
Program_Seg:(VADDR>=0x7B000000)&& (VADDR<0x80000000)
```

On the Solaris operating system the program address space can be displayed at runtime using the *pmap* command.

## 3 The Flow Solver TFS

The Navier-Stokes Solver TFS is a program library, which has been developed in the Institute of Aerodynamics of the RWTH Aachen University during the last 10-15 years. It is well prepared for vectorization as it uses one-dimensional arrays to store 3-dimensional geometries. It uses block-structured grids with general curvilinear coordinates and has also been parallelized on the block level using message passing such as supported by MPI [8]

The package is currently used in a multidisciplinary project to simulate the airflow through the human nose [2], [3]. The goal is first to get a better understanding of the functioning of the human nose and then to provide a work flow for computer assisted surgery allowing the physician to first perform a virtual surgery in a virtual reality environment. This provides the opportunity to verify the success of such an operation by another computer simulation before the actual surgery on the patient.

As the geometry of the human nose is quite complex, the simulation uses a mathematical grid consisting of 32 blocks, which vary in size considerably. So far the block level has not been parallelized using MPI for this problem case. The program can also be parallelized on a loop level using OpenMP. Nested parallelization, employing OpenMP on loop and block levels has been accomplished using the ParaWise automatic parallelization environment [7].

Selecting a dynamic schedule for the loop was the first approach to handle the resulting load imbalance. It yielded a better workload distribution than static scheduling. There were still situations, however, where a relatively large block was scheduled to one thread at the end of the loop, forcing all the other threads to be idle during this time. As the block sizes remain constant during the whole runtime of the program, the blocks can easily be sorted and optimally distributed to a given number of threads on the outer parallel level in order to reduce the overhead of the dynamic schedule and to avoid idle threads. This strategy was employed for nested OpenMP parallelism. More details are described in [6].

## 4 Performance Analysis Case Study

For our study we used two UltraSPARC IV (US IV) based Sun™ (SF) Fire systems located at the Sun Microsystems Benchmarking Center in Hillsboro, OR. The following provides a brief description of the systems. One system is a SF E6900 node with 24 US IV processors and 100GB of shared memory. The other system is a SF E25K node with 72 US IV processors and 144GB of shared memory. The US IV consists of two superscalar 64-bit processor cores with 2 levels of separate cache each. The level 1 cache resides on chip and has 64 KB for data, the level 2 cache is off chip and has 8 MB for data and instructions. The processors have a 1350 MHz clock rate. The SF Fire E6900 has an almost flat memory system, in other words about 235ns (local access) and 274ns (remote access). The SF 25K nodes provide a cc-NUMA memory system where data locality is important. The latency for memory access within a board is about 248ns and remote access has a latency of approximately 500ns.

Two important performance issues, which arise in multilevel parallel programs, are

- workload balancing and
- remote memory access on cc-NUMA architectures.

We will show how to address these issues during performance analysis using customized objects.

### 4.1 Detecting Load Imbalance

The timings for the TFS benchmark when employing single level OpenMP parallelization using 16 threads are shown in Fig 1. The timings were obtained with nested parallelism enabled, but employing all 16 threads on the outer parallel region. The figure shows that the timings are approximately the same on both systems under consideration. The scalability is poor with a speed-up of about 5 for 16 threads. Table 1 shows the performance metrics attributed to functions calls.

The fact that about 584 seconds out of 1255 seconds are spent in barrier time hints at a load-balancing problem. In order to determine whether workload imbalance is a

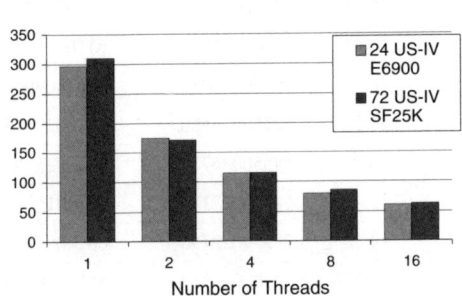

**Fig. 1.** Timings in seconds for TFS employing single level parallelism

**Table 1.** Performance metrics attributed to subroutine calls

| User CPU secs | OMP-work | OMP-wait | Name |
|---|---|---|---|
| 1255.81 | 489.47 | 815.54 | <Total> |
| 584.08 | 0 | 588.14 | <OMP-implicit_barrier> |
| 207.35 | 0 | 209.41 | <OMP-idle> |
| 124.03 | 130.52 | 0 | $d1S1010.ausm_ |
| 60.8 | 61 | 0 | $d1dK6930.mbcrct_ |
| 30.19 | 33.96 | 0 | $d1dR7030.addblk_ |
| 23.48 | 24.73 | 0 | $d1hH14108.visc_ |

problem, we would like to see performance metrics attributed per thread. We use the `Thread` object discussed in Section 2 and choose as metric OMP-Work time and the number of executed instructions per thread. Further more, we employ the filtering capability to calculate the metric only when the thread performs work. The results are shown in *Table 2*

**Table 2.** Performance metrics attributed to threads

| User CPU secs | OMP-Work secs | #Instructions | # Instructions in OMP-work | |
|---|---|---|---|---|
| 1255.81 | 489.47 | 1078610415814 | 211490226130 | <Total> |
| 71.81 | 80.12 | 35590034473 | 34010034291 | Thread_1 |
| 79.86 | 36.57 | 62210026312 | 13530014562 | Thread_2 |
| 79.76 | 24.82 | 69520022695 | 10070010710 | Thread_3 |
| 80.09 | 12.1 | 76530018343 | 5470006118 | Thread_4 |
| 79.87 | 12.94 | 79480021875 | 5950006704 | Thread_5 |
| 79.68 | 18.5 | 73190020725 | 8180008633 | Thread_6 |
| 78.24 | 46.11 | 57760030882 | 18150019242 | Thread_7 |
| 79.64 | 12.65 | 76500018686 | 6180006560 | Thread_8 |
| 79.26 | 16.37 | 78120022989 | 6860007064 | Thread_9 |
| 77.55 | 47.55 | 60160036650 | 21810025045 | Thread_10 |
| 77.42 | 56.01 | 54850036724 | 24470025183 | Thread_11 |
| 78.2 | 35.42 | 64470027186 | 15100015265 | Thread_12 |
| 79.26 | 14.28 | 78950022738 | 6620007468 | Thread_13 |
| 79.31 | 12.77 | 76280018887 | 5810006739 | Thread_14 |
| 76.86 | 47.71 | 59790036156 | 21780024275 | Thread_15 |
| 79.11 | 15.55 | 75210020493 | 7500008271 | Thread_16 |

There is no imbalance in CPU time, but there is an imbalance in OMP-work time. The question arises whether this is due to the actual computational workload of the threads or whether it is due to increased memory access time by certain threads. The instructions per thread (column 3 in *Table 2*) do not indicate a noticeable imbalance. Threads do execute instructions while waiting for work. What we are really interested in are the instructions that are executed during *OpenMP Work* time. This metric can be calculated by specifying a filter as discussed in Section 2. As can be seen from column 4 in Table 2, the number of instructions during *OpenMP Work* time indicates an imbalance in the computational workload. Threads 1,2,3,7,10,11,12 and 15 execute many more instructions than threads 4,5,6,8,9,13, 14 and 16. The example shows that it is important to distinguish metrics between different thread states.

In order to determine where the imbalance occurs we use the *Thread_Second* view described restricted to OMP-wok, as described in Section 2. Fig 2 shows a graphical display of the view, with three charts displaying metrics *User CPU Time, # Instructions*, and *Data Cache Stall Time* from top to bottom. The view zooms in on a series of objects showing a representative pattern. For each metric, 16 consecutive bars show the metric for different threads accumulated during one second. The image shows areas where all threads execute few instructions and areas with high instruction counts. For the areas with high instruction counts, there are notable differences in the number of instructions between the threads. By selecting such an area (indicated by the vertical line in the image), the user can navigate to the corresponding call-stack information that was collected during the run and find that it occurred within a nested parallel region. OpenMP offers two approaches to improve the workload balance: dynamic work scheduling via the SCHEDULE clause or the use of nested OpenMP parallelism. We tried both techniques. The timings for the different approaches are shown in Fig. 3. These timings were obtained employing 4 threads on the outer parallel region and using the load balancing algorithm of the application as described in Section 3 to set the number of threads on the inner level. The figure shows that nested OpenMP parallelism outperforms dynamic work scheduling on the E6900. The instructions during OpenMP Work

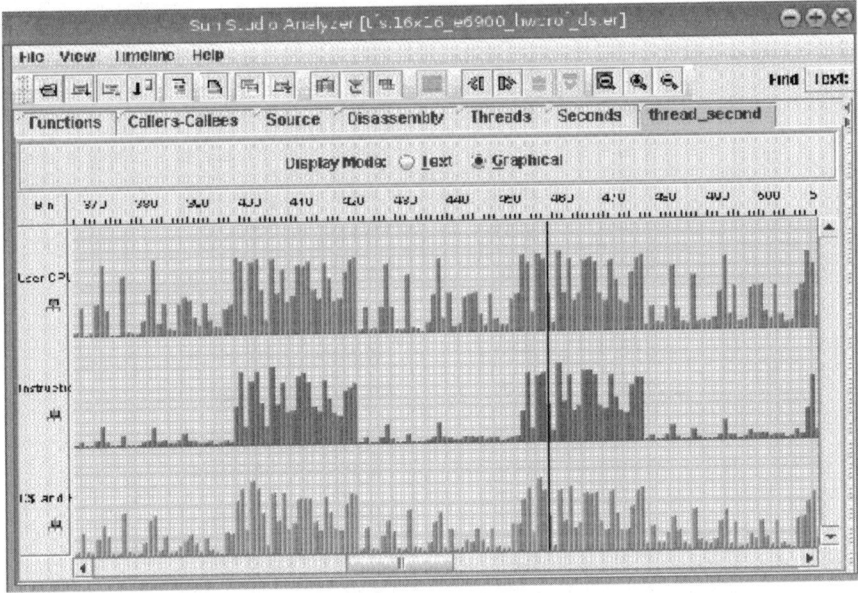

**Fig. 2.** The image shows performance metrics displayed per thread and second. The upper chart displays User CPU Time, the middle chart displays the number of Instructions, and the lower chart displays Data Cache Stall Time. In each chart, 16 subsequent vertical bars show the metric for each of the 16 threads for a particular second of execution time. The image shows that there are time intervals with high and low numbers of instructions.

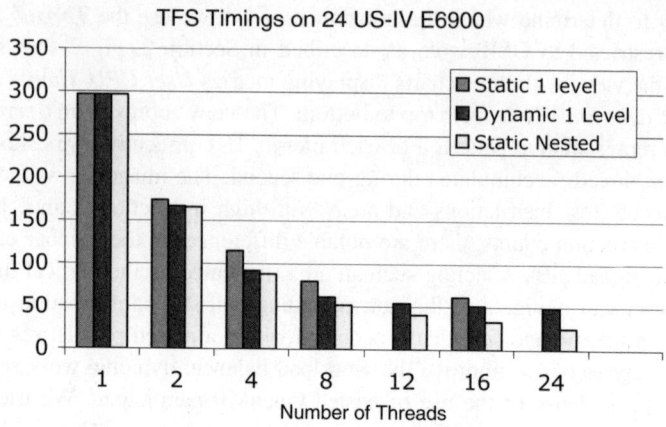

**Fig. 3.** Timings for TFS employing different work scheduling strategies. The timings for nested parallelization are reported for the best nesting combination.

**Table 3.** Performance metrics after improving the workload balance using nested OpenMP

| User Time sec | OMP Work sec | # Instructions in OMP Work | Name |
|---|---|---|---|
| 40.45 | 54.06 | 19590017616 | Thread_1 |
| 28.96 | 30.44 | 12490013246 | Thread_2 |
| 26.16 | 27.46 | 11480012732 | Thread_3 |
| 27.26 | 28.74 | 13700015667 | Thread_4 |
| 27.09 | 29.23 | 12670013394 | Thread_5 |
| 26.77 | 29.01 | 12600013109 | Thread_6 |
| 26.11 | 28.34 | 13110014213 | Thread_7 |
| 26.06 | 28.51 | 12910014123 | Thread_8 |
| 27.15 | 29.71 | 12830013552 | Thread_9 |
| 26.2 | 28.96 | 12820013572 | Thread_10 |
| 25.01 | 27.6 | 12850013872 | Thread_11 |
| 23.91 | 26.39 | 12380014310 | Thread_12 |
| 26.69 | 29.69 | 13180013639 | Thread_13 |
| 24.98 | 27.62 | 13040014575 | Thread_14 |
| 27.88 | 28.98 | 12750013506 | Thread_15 |
| 24.05 | 27.01 | 12910013664 | Thread_16 |

time show a much better distribution when employing nested OpenMP parallelism as can be seen in *Table 3*. The table also indicates a significant decrease in CPU time per thread.

## 5.3 Examination of Memory Access

On cc-NUMA architectures such as the SF E25K the placement of memory and threads onto processor and memory boards has a great impact on the performance. While nested OpenMP yielded a performance increase on the SF E25K based system, it lacked the scalability observed on the E6900.

Objects `Processor_Board` and `Memory_Board` can be defined using the CPUID and PADDR entries in the data records employing techniques as described in Section 2. Mapping data cache stall time onto these objects yields the information displayed in *Table 4*. The user can define a filter to count metrics only when memory is accessed, which is not local to a processor board by a logical expression of the form:

`Processor_Board (ID_list) && !Memory_Board (ID_list)`

Mapping *Data Cache Stall Cycles* and *Remote Data Cache Stall Cycles* onto threads yields the results shown in *Table 5*. The table indicates that threads 1 to 8 show stall time due to remote memory access. The same methods used to determine areas of workload imbalance can be employed to find out where the memory stall time occurs.

One problem we encountered when trying to improve memory locality in our application was, that the thread team composition for the execution of the inner parallel regions changes during the course of the execution. This is due to the implementation of nested OpenMP parallelism in the current Sun$^{TM}$ compiler, which is based on requesting threads from an available pool at the entry to inner parallel regions. We are working on enhancements to thread runtime library to allow for the definition of fixed thread teams, working together on inner parallel regions during the program's execution. At this time we do not yet have results available which demonstrate the impact on the performance.

**Table 4.** Data cache stall cycles mapped onto memory and processor boards

| DC Stall Cycles | Memory Board |
|---|---|
| 97.7 | Memory_Board Memory Object 0 |
| 87.5 | Memory_Board Memory Object 1 |
| 0.4 | Memory_Board Memory Object 6 |
| 0 | Memory_Board Memory Object 10 |
| **DC Stall Cycles** | **Processor Board** |
| 48.1 | Processor_Board Memory Object 9 |
| 47.1 | Processor_Board Memory Object 8 |
| 46.2 | Processor_Board Memory Object 1 |
| 44.3 | Processor_Board Memory Object 0 |

**Table 5.** Data cache stall cycles mapped onto threads

| Total DC Stall Cyles | Remote DC Stall | |
|---|---|---|
| 13.21 | 3.02 | Thread_1 |
| 11.48 | 4.89 | Thread_2 |
| 11.11 | 5.58 | Thread_3 |
| 12.71 | 9.1 | Thread_4 |
| 12.91 | 7.09 | Thread_5 |
| 12.44 | 5.85 | Thread_6 |
| 12.28 | 5.34 | Thread_7 |
| 11.93 | 6.28 | Thread_8 |
| 12.31 | 0 | Thread_9 |
| 12.55 | 0 | Thread_10 |
| 12.11 | 0 | Thread_11 |
| 12.59 | 0 | Thread_12 |
| 12.81 | 0 | Thread_13 |
| 12.37 | 0 | Thread_14 |
| 12.62 | 0 | Thread_15 |
| 12.79 | 0 | Thread_16 |

## 6 Related Work

There are a number of commercial and research performance analysis tools that have been developed over the years. We can only name a few of them and we will focus on research projects. An example for a commercial product is the Intel® Trace Analyzer and Collector [4] which offers runtime event tracing and graphical analysis of OpenMP, MPI, and hybrid MPI/OpenMP applications. It displays metrics for an arbitrary time interval, subroutine execution metrics and call-tree statistics. An example of a research project is TAU (Tuning and Analysis Utilities) [12] which was developed at the University of Oregon. It is a freely available set of tools for analyzing the performance of the C, C++, Fortran and Java programs. The current study distinguishes itself from previous work on TAU in that it describes how user-defined objects can be helpful to analyze real-world applications based on OpenMP. There are older systems such as SIMPLE [1] developed at the University of Erlangen, Germany. It consists of a performance analysis environment for parallel and distributed systems and is based on monitoring concurrent interdependent activities. SIMPLE provides a command language for filtering, statistics calculation, and visualization generation. It is, however, a research project and not integrated into a commercial software product. The Paraver visualization and analysis tool [11] was developed at CEPBA-UPC (European Center of Parallelism of Barcelona-Technical University of Catalonia). It has its own tracing module, OMPItrace [9], and provides very extensive analysis capabilities, including great flexibility in defining and mapping of performance metrics. While Paraver provides means for user-defined metrics, it does not, to our knowledge, allow user-defined objects. Another difference to the techniques described in this paper is, that Paraver requires the specification of filters and mappings by designing configuration files via a graphical user interface. We feel that using a command language as described in this paper provides a more elegant and flexible approach.

## 7 Conclusions

In this paper we have described the concept of user-defined objects in performance analysis, which adds another degree of flexibility to analyze parallel programs. We have implemented the feature in an experimental version of the Sun$^{TM}$ Studio Performance Analyzer, by allowing the specification of objects and filters via a command language. The usefulness was demonstrated for the analysis of a nested OpenMP application. We found that by defining objects to be used for filtering and mapping of performance metrics, performance analysis can be customized to particular programming paradigms, hardware platforms, and application areas. In addition, saving the definition of customized objects and filters provides means to transfer knowledge from an expert to the novice user, thereby bridging the gap between tool developer and casual user.

## Acknowledgements

We would like to thank Nicolai Kosche of Sun Microsystems for his excellent suggestions on how to define customized objects and for many enlightening discussions on performance analysis. We also thank Nawal Copty of Sun Microsystems who carefully reviewed our paper and provided feedback to improve its quality.

## References

1. Dauphin, P., Hofmann, R., Klar, R., Mohr, B., Quick, A., Siegle, M., Sötz, F.: ZM4/SIMPLE: a General Approach to Performance-Measurement and -Evaluation of Distributed Systems. In: Casavant, T.L., Singhal, M. (eds.) Readings in Distributed Computing Systems, pp. 286–309. IEEE Computer Society Press, Los Alamitos (1994)
2. Hörschler, I., Brücker, C., Schröder, W., Meinke, M.: Investigation of the Impact of the geometry on the nose flow. ScienceDirect - European Journal of Mechanics – B/Fluids (2005)
3. Hörschler, I., Meinke, M., Schröder, W.: Numerical simulation of the flow field in a model of the nasal cavity. Computers & Fluids 32, 39–45 (2003)
4. Intel® Trace Analyzer and Collector, http://www.intel.com/cd/software/products/asmo-na/eng/cluster/tanalyzer/index.htm
5. Itzkowitz, M.: The Sun$^{TM}$ Studio Performance Tools. Sun Microsystems, http://developers.sun.com/prodtech/cc/articles/perftools.html
6. Johnson, S., Ierotheou, C., Spiegel, A., an Mey, D., Hoerschler, I.: Nested Parallelization of the Flow Solver TFS using ParaWise/CAPO tools. In: IWOMP 2006 (submitted, 2006)

7. Johnson, S., Ierotheou, C.: Parallelization of the TFS multi-block code from RWTH Aachen using the ParaWise/CAPO tools. Parallel Software Products Inc., TR-2005-09-02 (2005)
8. MPI Forum: MPI: A Message Passing Interface. Int. Journal of Supercomputing Applications 8(3/4) (1994)
9. OMPItrace User's Guide, https://www.cepba.upc.es/paraver/manual_i.htm
10. OpenMP Fortran/C Application Program Interface, http://www.openmp.org/
11. Paraver, http://www.cepba.upc.es/paraver/
12. TAU: Tuning and Analysis Utilities, http://www.cs.uoregon.edu/research/paracomp/tau

# Performance Instrumentation and Compiler Optimizations for MPI/OpenMP Applications*

Oscar Hernandez[1], Fengguang Song[2], Barbara Chapman[1], Jack Dongarra[2], Bernd Mohr[3], Shirley Moore[2], and Felix Wolf[3]

[1] University of Houston, Computer Science Department
Houston, Texas 77204, USA
{oscar, chapman}@cs.uh.edu

[2] University of Tennessee, Computer Science Department
Knoxville, Tennessee 37996, USA
{song, dongarra, shirley}@cs.utk.edu

[3] Forschungszentrum Jülich, ZAM
52425 Jülich, Germany
{b.mohr, f.wolf}@fz-juelich.de

**Abstract.** This article describes how the integration of the OpenUH OpenMP compiler with the KOJAK performance analysis tool can assist developers of OpenMP and hybrid codes in optimizing their applications with as little user intervention as possible. In particular, we (i) describe how the compiler's ability to automatically instrument user code down to the flow-graph level can improve the location of performance problems and (ii) outline how the performance feedback provided by KOJAK will direct the compiler's optimization decisions in the future. To demonstrate our methodology, we present experimental results showing how reasons for the performance slow down of the ASPCG benchmark could be identified.

## 1 Introduction

Many tools have been created to help find performance bottlenecks and tune parallel codes written using the hybrid MPI/OpenMP programming model. Yet tool use remains labor-intensive and fragmentary. Our goal is to improve the application development and tuning process by creating an integrated environment for parallel program optimization that reduces the manual labor and guesswork of existing approaches. We are developing strategies that enable the application developer, compiler and performance tools to collaborate and that generate code based upon a variety of sources of feedback. To demonstrate our ideas, in this paper we describe the integration of existing, open source software - the OpenUH compiler [10] and the automatic trace analysis tool KOJAK [19] - into a single, coherent environment, called COPPER, for collaborative application

---
* This material is based upon work supported by the National Science Foundation under grant No. 0444363 and 0444468.

tuning. The integrated application optimization environment permits a variety of user and tool interactions to solve performance problems, such as compiler assistance in the instrumentation of an application, and provision of high-level feedback information by performance tools to direct the compiler in further optimizations.

In Section 2 we present related work of tools that support different aspects of performance feedback analysis and optimizations. In Section 3 we give an overview of our system and describe its components. Section 4 describes how OpenUH and KOJAK interact and APIs used to accomplish this interaction. Section 5 describes the feedback optimization facility in OpenUH and how it can be extended to support OpenMP optimizations. Section 6 describes the initial evaluation of our system and experimental results using the ASPCG kernel. In Section 7, we present our conclusions and plans for future work.

## 2 Related Work

Profile-guided optimization (PGO), where a program is compiled and executed to collect execution profiling information that is later used to perform optimization in a subsequent compilation step, has been exploited by modern static compilers to achieve significant speedups [4]. On the other hand, dynamic feedback optimizations have been mainly explored for runtime optimizations [2] [6] [5], dynamic compilation [3] and the Java environment [1], where profiling and sampling information is collected to direct run time optimizations. Unfortunately most of these systems do not support an integrated environment for MPI/OpenMP code optimizations and do not take advantage of automatic performance analysis that searches large amounts of performance data to locate inefficiencies on a higher level of abstraction.

Existing state-of-the-art performance tools addressing the combined use of MPI and OpenMP in a single application, such as TAU [11], VGV [7], and VAMPIR [14] deliver valuable performance feedback, albeit on a relatively low level of abstraction. Higher-level information obtained from an automatic analysis of trace data is provided by KappaPi [8], but only for pure message passing applications. Aksum [16], Paradyn [12], and Periscope [15] offer automatic performance analysis features for both MPI and OpenMP applications, but make certain assumptions about the deployment infrastructure that make them less suitable candidates for integration with the OpenUH compiler.

## 3 Overview

Figure 1 depicts the overall architecture of the envisioned COPPER environment and how it relates to the application optimization process. The process starts with the instrumentation of OpenMP constructs on the source-code level using a preprocessor called OPARI [13]. In the next step, the application is compiled by OpenUH and, at the same time, the compiler inserts instrumentation into the user code to generate traces for KOJAK. After application termination,

KOJAK analyzes the resulting trace file and provides higher-level feedback that is returned to OpenUH's feedback optimization module. The initial version of the COPPER environment, as described in this article, so far only integrates OpenUH and KOJAK by using compiler-based instrumentation. The KOJAK feedback is currently returned to the end user instead of the compiler. We hope to close the automatic feedback loop in the near future. Also, in a later step we plan to include the PerfSuite profiling tool [9] into the feedback loop for initial performance assessment and analysis of the OpenMP runtime system.

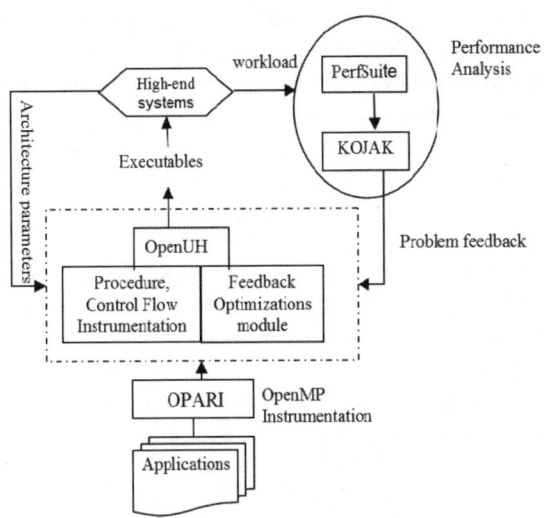

**Fig. 1.** COPPER architecture

The following sections describe each of the components more in detail.

### 3.1 OpenUH

OpenUH [10] is a compiler suite that supports C/C++ and Fortran 90/95 with OpenMP and/or MPI on the IA-64 running Linux. OpenUH is based on Open64, originally developed by SGI, subsequently maintained by Intel and now supported commercially by Pathscale for Opteron architectures. OpenUH is available as an open source compiler. The major functional parts of the compiler are the front ends, the inter-language interprocedural analyzer (IPA), and the middle-end/back end, which is further subdivided into the loop nest optimizer, auto-parallelizer (with an OpenMP optimization module), global optimizer (or whole program optimizer), and code generator. OpenUH currently supports OpenMP 2.0. OpenMP is lowered during different compilation phases. The output code makes calls to a portable run-time library based on Pthreads. The description of how OpenMP is translated in OpenUH can be found in [10].

## 3.2 KOJAK

KOJAK is an automatic performance analysis system for MPI, OpenMP, and hybrid applications written in C/C++ or Fortran. It is based on automatic pattern search in event traces and uses different interoperable components to support the analysis cycle from trace generation to visualization of analysis results.

In the COPPER environment, user regions are automatically instrumented by the OpenUH compiler. The OpenUH compiler supports the instrumentation of `function`, `loop`, `conditional branch`, and `compare and goto` program units. OPARI [13] performs automatic instrumentation of OpenMP constructs according to the POMP profiling interface for OpenMP. Section 4.1 describes OPARI and POMP in more detail. Instrumentation of MPI functions is fully automated by interposing an MPI wrapper library based on the PMPI profiling interface.

At runtime, the instrumented executable generates a single trace file that can be searched off-line for inefficiency patterns using the EXPERT analyzer [19]. The patterns concentrate on wait states resulting from suboptimal parallel interaction. Those can appear in MPI point-to-point or collective communication when processes have to wait for data sent by other processes or in OpenMP when threads reach a barrier at different points in time or when threads compete for the ownerships of locks. Low CPU and memory performance can also be analyzed by adding hardware counter information to event records.

The analysis process transforms the traces into a compact XML representation that maps higher-level performance problems onto the call tree and the hierarchy of system resources, such as nodes, processes, and threads. The XML file can be viewed in the CUBE performance browser (see Figure 4) or, alternatively, automatically processed by third-party tools using the CUBE API.

## 4 Tool Interactions

This section describes how KOJAK and the OpenUH compiler instrument OpenMP constructs and user regions, respectively. The profiling interface and advantages of using KOJAK high-level feedback are also presented.

### 4.1 OpenMP Instrumentation

Similar to the MPI profiling interface PMPI, Mohr et al. [13] defined a portable API that exposes OpenMP parallel execution to performance tools. The performance interface is called "POMP". The POMP API consists of callback functions that are called before and after all OpenMP constructs and runtime routines. The callbacks can be inserted into the program during OpenMP compilation, through a source or binary instrumentation tool, or activated by an instrumented OpenMP runtime system.

OPARI [13] is a source-to-source translation tool that can add POMP instrumentation to C, C++, and Fortran programs. When reading a parallel program containing OpenMP directives, KOJAK automatically invokes OPARI to insert

POMP performance calls where appropriate. As a final step of the instrumentation, KOJAK links the application with a library implementing the POMP API to generate appropriate events and write them to the trace buffer. Thus, with the help of the PMPI library and the OpenUH user-region instrumentation, our approach provides a fully automatic solution to the instrumentation of OpenMP and mixed-mode MPI/OpenMP applications.

### 4.2 OpenUH Instrumentation and Profiling Interface

Compile-time instrumentation has several advantages over source-level and object-level instrumentation. We can use the compiler analysis to detect regions of interest where we can measure and instrument certain events to support different performance metrics. Also, the instrumentation can be performed at different compilation phases, allowing certain optimizations to take place before the instrumentation. These capabilities play a significant role in the reduction of instrumentation points, improve our ability to deal with program optimizations, and reduce the instrumentation overhead and size of performance trace files.

The instrumentation module can be invoked at six different phases during compilation, which are before and after three major stages in the translation: interprocedural analysis, loop nest optimizations, and SSA/DataFlow optimizations. For example, if the user decides to instrument the source code after the interprocedural analysis phase, program transformations such as procedure inlining will reduce the instrumentation points for callsites and the compiler will instrument the body of the procedure being inlined.

The OpenUH compiler provides an interface to enable the instrumentation of three types of user regions: functions, conditional branches, and loops. The instrumentation of other types of regions such as MPI operations and OpenMP constructs are avoided so that they can be handled by the profiling libraries of PMPI and POMP. Procedure and control flow instrumentation is essential to relate the PMPI and POMP results to the execution path of the application. We plan to integrate the POMP instrumentation in later versions of OpenUH. For now we are using OPARI to do it. Additionally, the user has the option to instrument his code after OpenMP gets translated to threading code (by setting a special compiler flag). In this case the OpenMP runtime system specific calls are additionally instrumented by the compiler.

Each user-region type is further divided into several sub-categories whenever possible. For instance, a loop type may be `do loop`, `while do loop` or `do while loop`. Conditional branches may be of type `if then`, `if then else`, `true branch`, `false branch`, or `cselect`. The name of the sub-category is communicated back to KOJAK through the profiling interface and later displayed in the call-tree view of the KOJAK GUI. This detailed presentation provides users with a fine-grained control flow graph in which taken branches, untaken branches, and specific loops are all displayed.

The compiler instrumentation is done by first traversing an intermediate representation of a program to locate different program constructs. The compiler locates starting and exit points of constructs such as procedures, branches and loops to insert specific profiling calls at these points. The compiler profiling interface API is defined as follows. Argument `begn_ln` represents the beginning line of the region, and `end_ln` represents the end line of the region. `type` indicates the specific subtype of the region. `pu_name` is the name of a function or subroutine.

- Functions to initialize and finalize the profiling library:

  `void profile_init(void)`
  `void profile_finish(void)`

- Subroutine entry and exit functions:

  `void profile_invoke(char *pu_name, INT32 begn_ln,`
  `                    INT32 end_ln, char *file_name)`

  `void profile_invoke_exit(void)`

- Conditional branch entry and exit functions:

  `void profile_branch(BranchSubType type, INT32 begn_ln,`
  `                    INT32 end_ln, char *file_name)`

  `void profile_branch_exit(void)`

- Loop entry and exit functions:

  `void profile_loop(LoopSubType type, INT32 begn_ln,`
  `                  INT32 end_ln, char *file_name)`

  `void profile_loop_exit(void)`

The interface has been implemented in KOJAK as part of the trace library to generate appropriate events upon region entry and exit.

### 4.3 KOJAK High-Level Feedback

To optimize a parallel application based on the MPI/OpenMP programming model, developers usually want to see whether the application is load imbalanced, where the synchronization overhead lies, and where there are opportunities to overlap computation and communication. For each parallel region, KOJAK is able to display the execution time broken down by call path and thread or process. The identification of wait states in combination with the distinction of different OpenMP constructs and user-region types simplifies the comparison of loop scheduling strategies and the validation of the program design. OpenMP *parallel for* and *parallel sections* constructs are two common cases where KOJAK can be used.

When the hardware counter feature is enabled, performance data supplied by the PAPI library is also recorded in the trace file. The hardware counter information will help us to correct CPU and memory anomalies such as unevenly distributed L1/L2 cache miss rate, significantly high TLB misses, and long stalls of the pipeline across a group of threads.

In a later stage of the integration project, the OpenUH compiler will read the analysis results from the XML file and automatically perform optimizations. The feedback mechanism to be used for this purpose is described below.

## 5 Compiler Optimizations for OpenMP

OpenUH has a basic facility for performing feedback-directed optimizations, which are accomplished via the automatic insertion of instrumentation and compiler annotation of the profiling results in the intermediate representation. It is able to use this facility to improve procedure inlining and branch prediction and to improve the cost modeling analysis for loop nests. The current feedback infrastructure supports profiling results at the procedure and control flow level of the procedure. Our main focus has been to extend this infrastructure to support hybrid OpenMP/MPI code optimizations. Our initial focus explores loop transformations to alleviate the problem of parallel loop load imbalances. Feedback information from KOJAK gives the compiler cost modeling information related to the parallel overhead of a given transformation. For example, if KOJAK reports a synchronization problem (e.g. long waits at a barrier) because of load imbalances, the compiler will assign a high cost for parallel synchronization overhead. Based on dependence analysis and interprocedural array region (array data-flow analysis) information, the compiler will try out a set of loop permutations, including loop interchange, loop tiling and outer loop unrolling, to alleviate the problem. The compiler also applies different strategies, including attempting to determine which loop in the nest is most profitable to parallelize and determining an appropriate scheduling strategy, and chunk size.

## 6 Experimental Results

To illustrate our approach, we performed an experimental analysis of the different process/thread allocation strategies for the ASPCG sparse linear algebra kernel [18]. Large sparse linear systems that are used in simulation of turbulent flow calculation in complex geometries typically have several million unknowns. The ASPCG kernel from Virginia Tech solves such systems with a preconditioned conjugate gradient(PCG) method. The iterative CG method uses a two-level additive Schwarz preconditioner which adopts a domain decomposition approach to compute solutions on subdomains. The code is available in serial, MPI, OpenMP, and hybrid versions. For our experiments, we used the hybrid version and compiled ASPCG with the OpenUH compiler.

The experiments were performed on NCSA's SGI Altix machine which consists of two SMP systems running the Linux operating system. Each system has 512 Intel Itanium2 1.6 GHz processors. Figures 2, 3, and 4 show performance analysis results for the PCG method preconditioned by the additive Schwarz method for the problem size of $8194 \times 8194$. The GUI windows displayed include three trees. The left tree represents performance properties, the middle tree represents call paths of the source code, and the right tree represents system resources. The numbers represent percentages of the overall execution time.

We conducted two experiments on 32 processors with 8 processes 4 threads each and 4 processes 8 threads each respectively (i.e., $8 \times 4$ and $4 \times 8$ configurations). The walltime is about 35 minutes to execute the $8 \times 4$ experiment

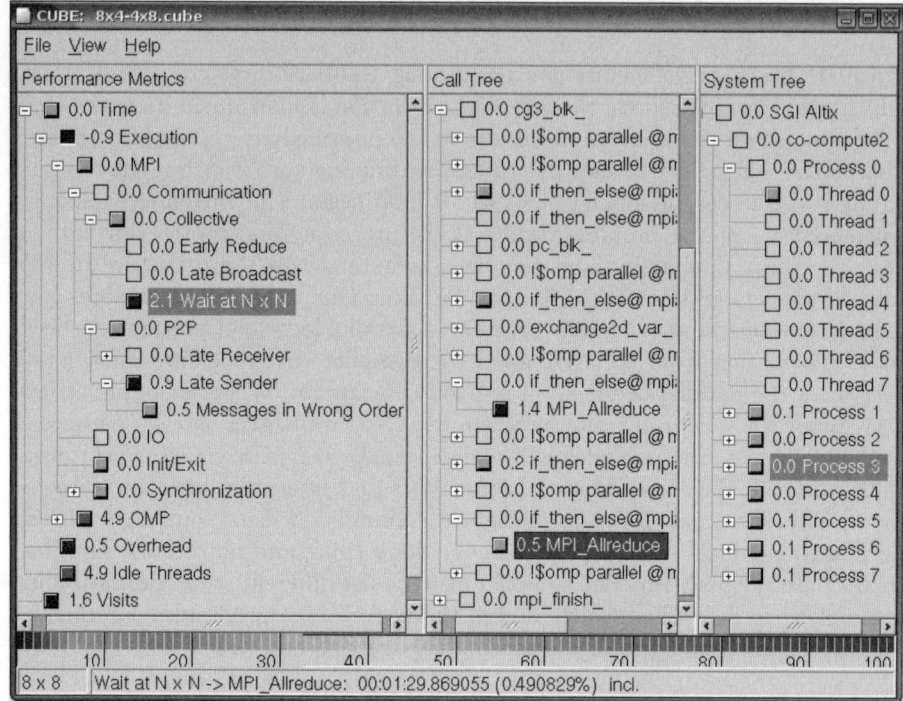

**Fig. 2.** Difference experiment of ASPCG between 8×4 run and 4×8 run. Unlike Figure 3, this figure focuses on the difference in the MPI communication.

and about 25 minutes for the $4 \times 8$ run. A virtual difference experiment was computed by subtracting the $4 \times 8$ experiment from the $8 \times 4$ experiment using our performance algebra tool [17]. Positive values represent performance losses of the $8 \times 4$ experiment, and negative values represent gains of $8 \times 4$. By comparing the two experiments, we find the $8 \times 4$ run is slower than the $4 \times 8$ run by 12.8% (4.9% lies in OpenMP constructs and 3.5% in MPI communication). In order to show the difference experiment in more detail, we use two figures, Figure 2 and Figure 3, to display the difference located in MPI communication and OpenMP constructs, respectively. As shown in Figure 3, the $8 \times 4$ experiment spent 3.5% more time in MPI communication than the $4 \times 8$ experiment did. After expanding the tree node of "MPI", we can see detailed information in Figure 2. The reason why the $8 \times 4$ experiment is slower is because the greater number of processes induced bigger overhead in "Wait at $N \times N$" and longer blocking time in "Late Sender" and "Messages in wrong order". In addition, the increased loss in the metric "Idle Threads" (4.9%) is a consequence of the higher MPI time when the MPI calls were made during sequential phases of execution. Prolonged sequential executions cause slave threads to sit idle, which is a typical problem of hybrid programs.

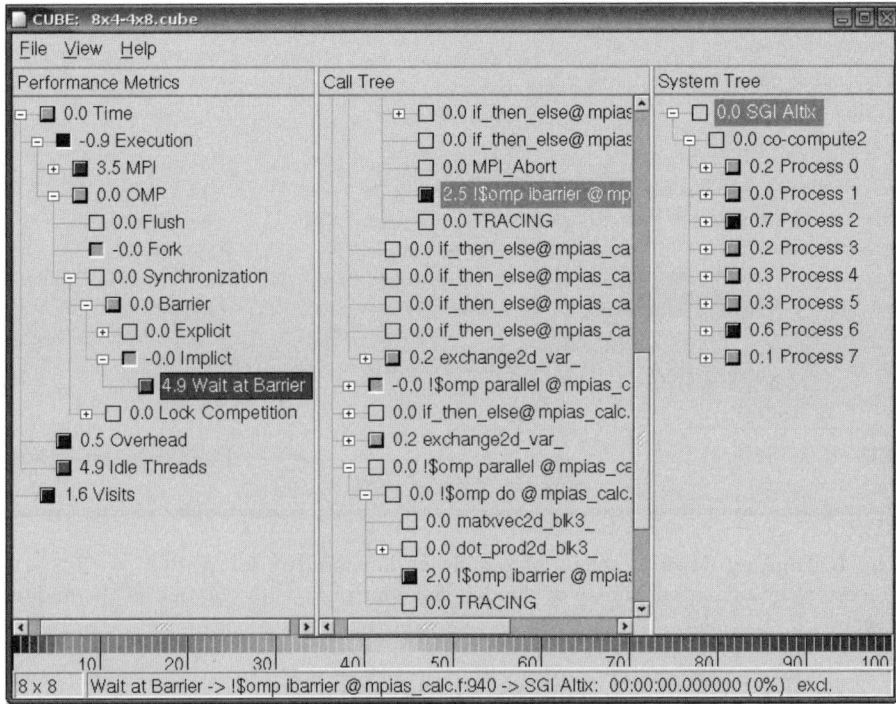

**Fig. 3.** Difference experiment of ASPCG between $8 \times 4$ run and $4 \times 8$ run. Unlike Figure 2, this figure focuses on the difference in the OpenMP constructs.

To continue to investigate the reason for the greater cost in OpenMP (shown in Figure 3), the selected property in the left tree reveals a performance problem: 4.9% of the total execution time was spent waiting in front of OpenMP barriers. The waiting time is significant and it typically indicates the problem of load-imbalance. The scheduling strategy we used was `static`. The $4 \times 8$ experiment is more load-balanced than the $8 \times 4$ one since the workload of the former is more evenly distributed among 8 threads than that of the latter which uses only 4 threads to compute the same amount of work.

Figure 4 displays the result for the $16 \times 2$ ASPCG experiment and demonstrates that more fine-grained control flow information can help users distinguish different instances of function calls within a region. As an example, the "Late Sender" property shown in Figure 4 takes 5.8% of the total execution time. In examining the "Late Sender" problem, we hope to identify which `MPI_Wait` in P2P communication spent most of the time blocking. With the help of conditional branch regions, we are able to identify the locations of the most expensive calls. The four expanded nodes with blue boxes in the middle tree account for 86% of the "Late Sender" inefficiency. By clicking the right-button of the mouse, one can look at the corresponding source code and modify it to eliminate the problem.

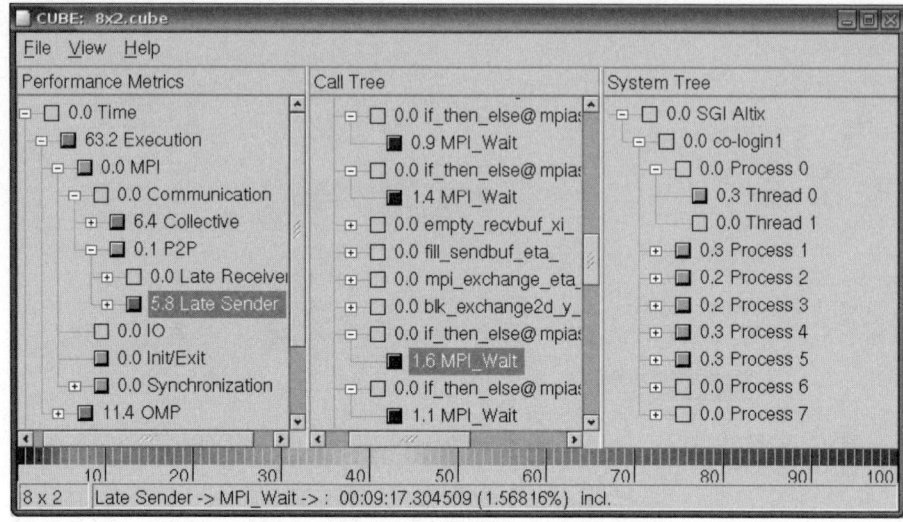

**Fig. 4.** Tracking down locations of significant late senders for ASPCG with 8 × 2 processors by taking advantage of the detailed control flow information in the middle tree

## 7 Conclusions and Future Work

Large-scale parallel applications on advanced architectures with deep memory hierarchies rarely achieve a moderate fraction of the theoretical peak performance. Our work provides a framework to automatically analyze and optimize performance of hybrid MPI/OpenMP applications through integration of an optimizing compiler (OpenUH) with a performance analysis tool (KOJAK). While KOJAK automates the process of instrumenting MPI functions and OpenMP constructs, OpenUH automatically instruments user regions down to the flowgraph level at different compilation stages. The compile-time instrumentation allows a variety of optimizations before the instrumentation and using the results of static analysis will help reduce the instrumentation overhead and the amount of trace data. KOJAK gives performance feedback at a significantly higher level than traditional tools. Such high-level information will enable the OpenUH compiler to adjust the performance model parameters and determine the most effective optimizing strategies to optimize parallel loops.

So far we are able to instrument regions of conditional branches and loops. The support for switch regions is under development. The instrumentation of loops can produce a huge amount of event trace data when confronted with a deep loop nest. An approach to turn the instrumentation of loops on and off or use sampling mechanisms is being considered. The source-to-source translation approach of OPARI has some limitations. We hope to integrate the POMP instrumentation in later versions of the OpenUH compiler. Our future research will mainly focus on how to provide useful compiler-oriented feedback and how the compiler can adapt optimization strategies accordingly. In this context, we plan to extend the current set of KOJAK performance properties to support advanced loop optimizations and scheduling, which might also include the monitoring of OpenMP runtime events beyond those specified in POMP.

## Acknowledgments

We would like to thank Rick Kufrin at NCSA for giving us access to the SGI test platform and the team of Danesh Tafti at Virginia Tech for providing the ASPCG benchmark.

## References

1. Adl-Tabatabai, A.-R.: The StarJIT Compiler: A Dynamic Compiler for Managed Runtime Environments. Intel Technology Journal 7, 19–31 (2003)
2. Ayguadé, E., Blainey, B., Alejandro.: Is the Schedule Clause Really Necessary in OpenMP? In: Voss, M.J. (ed.) WOMPAT 2003. LNCS, vol. 2716, pp. 147–160. Springer, Heidelberg (2003)
3. Burcea, M., Voss, M.J.: A Runtime Optimization System for OpenMP. In: Voss, M.J. (ed.) WOMPAT 2003. LNCS, vol. 2716, pp. 42–53. Springer, Heidelberg (2003)
4. Chen, W., Bringmann, R., Mahlke, S., et al.: Using Profile Information to Assist Advanced Compiler Optimization and Scheduling. In: Banerjee, U., Gelernter, D., Nicolau, A., Padua, D.A. (eds.) LCPC 1992. LNCS, vol. 757, Springer, Heidelberg (1993)
5. Dang, F.H., Rauchwerger, L.: Speculative Parallelization of Partially Parallel Loops. In: Languages, Compilers, and Run-Time Systems for Scalable Computers, pp. 285–299 (2000)
6. Hancock, D.J., Mark Bull, J., et al.: An Investigation of Feedback Guided Dynamic Scheduling of Nested Loops. In: ICPP Workshop (2000)
7. Nagel, W., Hoeflinger, J., Kuhn, B.: An Integrated Performance Visualzer for MPI/OpenMP Programs. In: Eigenmann, R., Voss, M.J. (eds.) WOMPAT 2001. LNCS, vol. 2104, Springer, Heidelberg (2001)
8. Jorba, J., Margalef, T., Luque, E.: Automatic Performance Analysis of Message Passing Applications Using the KappaPI 2 Tool. In: Di Martino, B., Kranzlmüller, D., Dongarra, J. (eds.) EuroPVM/MPI 2005. LNCS, vol. 3666, pp. 293–300. Springer, Heidelberg (2005)
9. Kufrin, R.: Perfsuite: An Accessible, Open Source Performance Analysis Environment for Linux. In: Proc. of the Linux Cluster Conference, Chapel Hill, North Carolina (April 2005)
10. Liao, C., Hernandez, O., Chapman, B., Chen, W., Zheng, W.: OpenUH: An Optimizing, Portable OpenMP Compiler. In: 12th Workshop on Compilers for Parallel Computers (January 2006)
11. Malony, A.D., Shende, S.: Performance Technology for Complex Parallel and Distributed Systems. In: Kacsuk, P., Kotsis, G. (eds.) Quality of Parallel and Distributed Programs and Systems, pp. 25–41. Nova Science Publishers, Inc., New York (2003)

12. Miller, B., Callaghan, M., Cargille, J., et al.: The Paradyn Parallel Performance Measurement Tool. IEEE Computer 28(11), 37–46 (1995)
13. Mohr, B., Malony, A., Shende, S., Wolf, F.: Design and Prototype of a Performance Tool Interface for OpenMP. The Journal of Supercomputing 23, 105–128 (2002)
14. Nagel, W., Weber, M., Hoppe, H.-C., Solchenbach, K.: VAMPIR: Visualization and Analysis of MPI Resources. Supercomputer 63, XII(1), 69–80 (1996)
15. PERISCOPE, http://wwwbode.cs.tum.edu/~gerndt/home/research/periscope/periscope.htm
16. Seragiotto Júnior, C., Geissler, M., Madsen, G., Moritsch, H.: On Using Aksum for Semi-Automatically Searching of Performance Problems in Parallel and Distributed Programs. In: Proc. of PDP 2003, Genua, Italy (February 2003)
17. Song, F., Wolf, F., Bhatia, N., Dongarra, J., Moore, S.: An Algebra for Cross-Experiment Performance Analysis. In: Proc. of the International Conference on Parallel Processing (ICPP), Montreal, Canada (August 2004)
18. Wang, G., Tafti, D.K.: Performance Enhancement on Microprocessors with Hierarchical Memory Systems for Solving Large Sparse Linear Systems. Int. J. of Supercomputing Applications and High Performance Computing 13(1), 63–79 (1999)
19. Wolf, F., Mohr, B.: Automatic Performance Analysis of Hybrid MPI/OpenMP Applications. Journal of Systems Architecture 49(10-11), 421–439 (2003)

# Supporting Nested OpenMP Parallelism in the TAU Performance System

Alan Morris, Allen D. Malony, and Sameer S. Shende

Performance Research Laboratory,
Department of Computer and Information Science
University of Oregon, Eugene, OR, USA,
{amorris,malony,sameer}@cs.uoregon.edu

**Abstract.** Nested OpenMP parallelism allows an application to spawn teams of nested threads. This hierarchical nature of thread creation and usage poses problems for performance measurement tools that must determine thread context to properly maintain per-thread performance data. In this paper we describe the problem and a novel solution for identifying threads uniquely. Our approach has been implemented in the TAU performance system and has been successfully used in profiling and tracing OpenMP applications with nested parallelism. We also describe how extensions to the OpenMP standard can help tool developers uniquely identify threads.

**Keywords:** OpenMP, nested parallelism, TAU.

## 1 Introduction

OpenMP research systems have supported nested parallelism since its introduction in the OpenMP standard (e.g., [12,13]), and most commercial compilers now support nested parallelism in their products. Although some commercial packages provide tools for debugging and performance analysis in the presence of nested parallelism (e.g., Sun Studio [18] and Intel [20]), the recent OpenMP 2.5 specification [21] does not provide sufficient support for developing portable performance measurement and analysis tools with nested parallelism awareness. This deficiency is being discussed in the OpenMP tools community [16] and hopefully will be addressed in future OpenMP specifications.

In the meantime, there is interest in studying how performance measurement systems can determine nesting context during execution in order to capture performance data for threads and interpret the data vis à vis nesting level. In this paper we present the current problem in Section §2 and discuss two possible solutions in Section §3. Based on this approach, we developed an improved, novel method for the TAU performance system [1]. This is described in Section §4. The TFS application [14,15] from RWTH Aachen is used as a case study for the TAU solution. Section §5 provides a detailed performance analysis of a nested parallel execution of TFS. The TAU parallel profile displays clearly show the thread nesting relationships.

The present issues for portable performance measurement of OpenMP nested parallel execution is, as remarked above, hopefully temporary. In Section §6 we outline discussions underway in the OpenMP tools community and what might be expected in the future to address the problem. Conclusions are given in Section §7.

## 2 Issues with Nested Parallelism in OpenMP

OpenMP allows for nested parallel regions during execution. Nested parallelism can be enabled and disabled through the use of the OMP_NESTED environment variable or by calling the *omp_set_nested()* routine. A simple example is given below:[1]

```
#include <omp.h>
#include <stdio.h>

void report_num_threads(int level) {
    printf("Level %d: omp_get_num_threads()=%d",
        level, omp_get_num_threads());
    printf(", omp_get_thread_num()=%d\n",
        omp_get_thread_num());
}

int main(int argc, char **argv) {
    #pragma omp parallel num_threads(2)
    {
        report_num_threads(1);
        #pragma omp parallel num_threads(2)
        {
            report_num_threads(2);
        }
    }
    return(0);
}
```

```
% OMP_NESTED=0 ./a.out
Level 1: omp_get_num_threads()=2, omp_get_thread_num()=0
Level 2: omp_get_num_threads()=1, omp_get_thread_num()=0
Level 1: omp_get_num_threads()=2, omp_get_thread_num()=1
Level 2: omp_get_num_threads()=1, omp_get_thread_num()=0

% OMP_NESTED=1 ./a.out
Level 1: omp_get_num_threads()=2, omp_get_thread_num()=0
Level 1: omp_get_num_threads()=2, omp_get_thread_num()=1
Level 2: omp_get_num_threads()=2, omp_get_thread_num()=0
Level 2: omp_get_num_threads()=2, omp_get_thread_num()=1
Level 2: omp_get_num_threads()=2, omp_get_thread_num()=0
Level 2: omp_get_num_threads()=2, omp_get_thread_num()=1
```

**Fig. 1.** An example illustrating nested OpenMP parallelism, the output (right) is obtained by executing the program on the left

Figure 1 illustrates the effects of nested OpenMP parallelism. When nested parallelism is enabled, both the inner and outer regions will have 2 threads in each team, whereas without nested parallelism, only the outer region have 2 threads. Here, we also see that the *omp_get_thread_num()* runtime call cannot be used for unique thread identification.

Nested OpenMP parallelism poses a challenge to traditional performance analysis tools. The above example is useful in pointing out the issues. Typically, a performance tool will attempt to measure the performance for each thread individually. To do so, there must be an agreement between the application and performance tool for proper thread identification to occur and measured events

---

[1] Adapted from http://docs.sun.com/source/819-0501/2_nested.html.

appropriately assigned. Tools often require that the user configure them with the application's thread package, be it pthreads, sproc, or OpenMP. When the tool and the application use the same underlying thread package, proper thread identification can be done.

However, when nested parallelism is used in OpenMP, the nesting context is not available to the performance interface. It may appear that nested parallelism can be statically analyzed and a tool such as Opari [4,8] could insert additional instrumentation providing nesting information. However, static analysis is insufficient to track threads as the varied interactions of execution paths at runtime can create arbitrary nesting depths. A runtime solution for thread identification is necessary.

Native thread libraries provide thread local storage (TLS) that performance tools use to track thread identities. The OpenMP runtime library provides the *omp_get_thread_num()* API call that returns the thread number or identifier within the active team of threads. Unfortunately, the value returned from this call lies between 0 and *omp_get_num_threads()-1*, which is not the total number of threads in the program, but the number of threads in the current active team. TAU and other tools have traditionally used this call for thread identification. When an instrumented section of code is entered, the profiling library identifies the calling thread and performs measurements associated with that thread. We say unfortunately because this approach does not allow the performance measurement system to uniquely identify threads when nested parallelism is active. When using nested OpenMP parallelism, multiple teams may be active at any one time and more than one thread will return the *same* index from *omp_get_thread_num()*.

Nested parallelism in OpenMP offers the additional challenge of mapping the per-thread performance data back to the nested parallelism abstractions in the source code. To do so, it is necessary to distinguished the performance measurements for each thread with the nesting context.

## 3 Solutions

The problem of thread identification in nested OpenMP programs is widespread in the community, for purposes other than performance evaluation. As such, several solutions have been proposed.

### 3.1 Extending the OpenMP API

A promising solution to this problem is to extend the OpenMP specification itself to allow for more in depth query and knowledge of nested parallelism. Dieter an Mey, RWTH [9] proposed an extension to the OpenMP specification in the form of a runtime library calls that return the current nesting level, the number of threads within that level, and the current thread identifier at that level. This provides enough information to uniquely identify threads for performance measurement purposes as well as information necessary for the proper mapping of the runtime execution to the application developer's abstractions for nested parallelism in the application.

Ultimately, we hope that the OpenMP specification will be extended in this manner, and we will update TAU to use the new runtime calls when these become widely available.

### 3.2 Native Thread Library Hooks

Another method of tracking threads in the face of nested OpenMP parallelism involves bypassing the OpenMP runtime system entirely and instead tracking the threads based on the underlying thread implementation. For example, if the OpenMP thread package is implemented using the native pthread library, the tool could use the pthreads API for thread identification, invoking functions such as *pthread_self()*. Regular thread local storage would be available as well.

A major drawback to this approach is the lack of portability. The underlying thread library must be known and accessible to the performance tool. On some systems, the underlying thread substrate may be inaccessible and such an approach cannot be guaranteed to work universally. We favor approaches that follow the higher-level abstract OpenMP API.

### 3.3 Additional OpenMP Instrumentation

Alexander Spiegel, RWTH [10] proposed another solution to this problem, in which the master and worker threads of each parallel team exchange information through a global shared space which is locked by the master. At the start of parallel regions, code needs to be inserted such that the master stores the data, locks the shared space, then after a barrier, the entire team of threads reads the shared data, and after another barrier, the master unlocks it. In this way, each new parallel region inherits data from the parent region, and a proper mapping can take place.

A tool such as Opari can be extended to support the additional instrumentation required for this type of thread synchronization at the application-level. This approach has the advantage that it tracks the full nesting information, so at any given time, the performance tool can know which thread identifier from each team and each level of nesting is executing. This allows for a better mapping of thread level performance information back to the source code.

The drawback of this approach is of course the additional synchronization and locking at parallel region boundaries. We have not performed studies to measure the overhead involved, but we estimate that it might be significant in some programs.

## 4 TAU's Solution

The TAU performance system supports performance evaluation of OpenMP programs at the region, construct, and routine level. The Opari tool is used to inserting instrumentation based on the POMP [4] interface at OpenMP regions and for OpenMP constructs. PDT [3] is used to instrument OpenMP source at the routine level. During measurement, TAU uses the OpenMP thread synchronization calls for updating the shared performance data structures. Construct-based measurement uses globally accessible timers to aggregate construct-specific performance costs over all OpenMP regions. For region-based measurement, the region descriptor is used to select specific performance data for that context.

In our earlier work [4,5], TAU relied upon the *omp_get_thread_num()* OpenMP API call to distinguish threads for indexing into the runtime performance data structures. Unfortunately, this method is inadequate for nested parallelism due to the issues discussed above. Instead, we need a mechanism to uniquely identify the current thread.

Our approach is to use *#pragma threadprivate()*. Though the values of a threadprivate variable are not guaranteed to persist between parallel regions, we are at least guaranteed that no two currently active threads will point to the same address space for a given threadprivate variable. Using this scheme, TAU can then uniquely identify threads even in the presence of nested parallelism.

The approach requires a single threadprivate variable that is initialized inside the TAU library (when TAU is built using OpenMP threading). This variable and/or its address can be used to distinguish it from other threads executing concurrently. When the TAU runtime system encounters a thread that it has not seen before, it registers the thread and assigns it an identifier on a first come, first serve basis.

In contrast to the other proposed approaches, this method has the advantage of faster speed, as no runtime API calls are made in identifying a thread. The thread registration in the TAU runtime system is done only when a given thread is seen for the first time, so there is no additional overhead at parallel region boundaries. The main drawback with this method is that we are unable to identify the nesting depth or specify a team identifier for a given thread. A thread is assigned a unique identifier, but not necessarily the same identifier between subsequence invocations, and this typically does not map back to any explicit parallelism in the source code. Nevertheless, as we see below, the method produces performance data that can be mapped back to the source code itself and does expose nested parallelism where it occurs.

## 5 A Case Study

To demonstrate TAU's support for nested parallelism, we conducted a case study with TFS [14,15], a computational fluid dynamics code developed at Aerodynamics Institute at RWTH Aachen. TFS was initially parallelized using ParaWise [19] to generate an intra-block parallel version where the parallel loops operated over a single dimension within a block. Then, a second version was developed that used parallel loops to iterate over blocks. Finally, a hybrid, multi-level version was developed which combined the intra and inter block version using nested OpenMP parallelism. In recent performance testing, the developers reported a speedup of 21 for TFS using nested OpenMP parallelism on a SunFire 25K system [17].

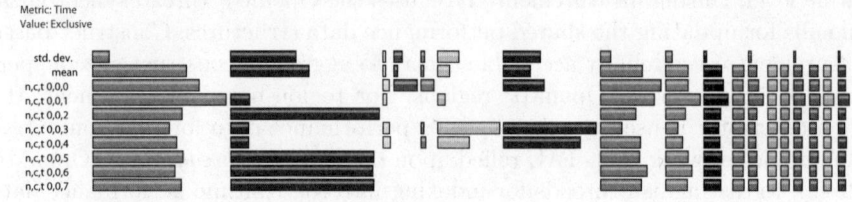

**Fig. 2.** Flat Profile for TFS

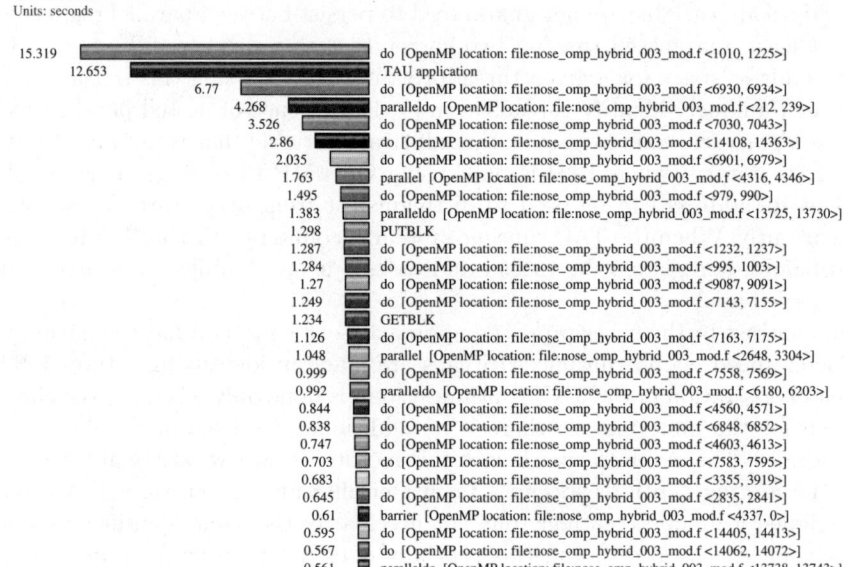

**Fig. 3.** Mean Profile for TFS

We integrated TAU in TFS and instrumented its source code using the TAU's compilation scripts [1]. This process required only a single modification to the TFS build system where the name of the compiler used in the makefile was changed from FC=f90 to FC=tau_f90.sh. This script act as compiler wrapper, allowing for automatic instrumentation of Fortran programs. In the case of OpenMP code, it will automatically invoke the Opari tool to instrument OpenMP constructs and regions by rewriting OpenMP directives using the POMP interface. The code is then parsed by PDT to create files that contain source-level information about the routine names, and their respective entry and exit locations. The script then instruments the entry and exit points using TAU's *tau_instrumentor* utility. Finally, the instrumented code is linked with the TAU library.

We ran TFS on the Sun Fire machines at RWTH Aachen using 8 threads. With TAU's support for nested OpenMP parallelism, we can run the instrumented

Supporting Nested OpenMP Parallelism in the TAU Performance System 285

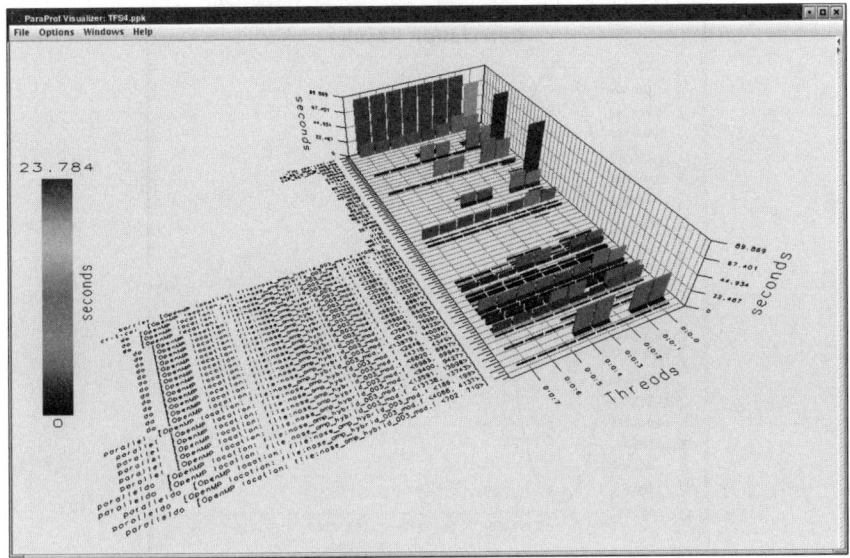

**Fig. 4.** 3D display of TFS performance data (inclusive time)

version of TFS without the thread identifier clash that occurred previously. The flat profile for TFS is shown in Figure 2. Each thread is represented by a row in the graph, and each timer/region is a column. The second column timer (red) is .TAU application, which in this case represents the global time that a thread spent idle (not doing useful work).

Figure 3 shows the mean data for all threads. Note that the timer names for parallel regions and constructs contain the source filename and line number. This data is provided by the Opari tool through the POMP interface.

There is a clear pattern in the data wherein threads 0, 1, 3, and 4 do similar processing, and threads 2, 5, 6 and 7 are also very similar. This pattern is also visible in the three dimensional display of ParaProf shown in Figure 4. The three axes are the threads, the timers (functions), and the exclusive time spent in the given timer.

TAU's PerfExplorer [6] tool is able to automatically discover patterns such as this. PerfExplorer is a performance data mining package that operates on a relational database using statistical packages such as Weka and R. Shown in Figure 5, PerfExplorer performs a correlation analysis and splits the threads into clusters. These are the same clusters that we observed from the flat profile bar graphs.

For a total runtime of 90 seconds, the second cluster which includes threads 2, 5, 6, and 7 are idle for about 22 seconds each, whereas the other slave threads, numbers 1, 3, and 4 spend only about 3 seconds idle. Each cluster executes different functions that may be seen in ParaProf's callgraph displays (not shown). With this knowledge, the application developer can note the regions that each thread executes and map the execution back to the source code.

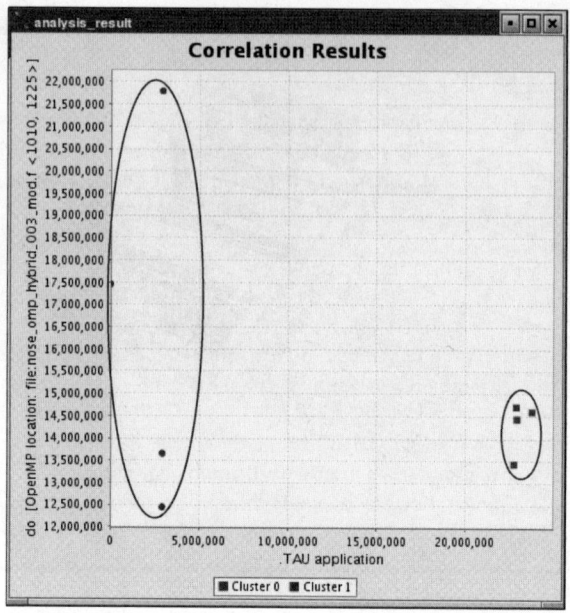

**Fig. 5.** Clustering of threads in TFS

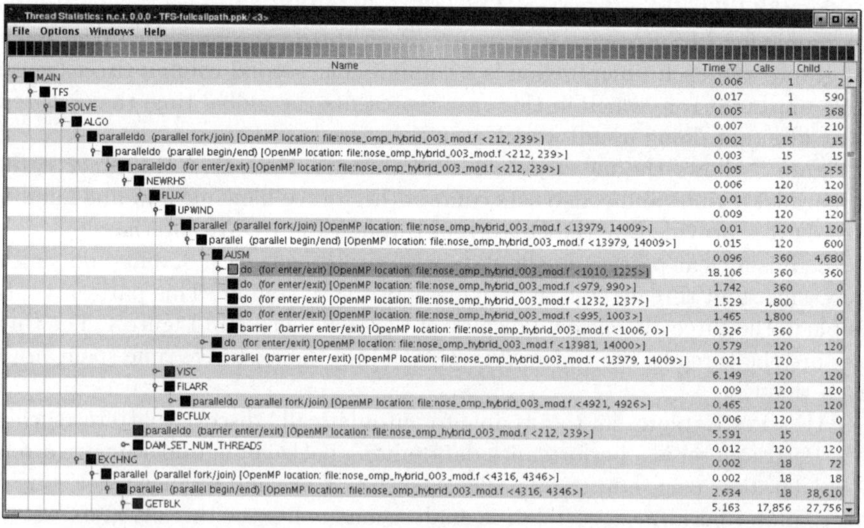

**Fig. 6.** Nested Parallelism in TFS

Using TAU's callpath profiling capability, the nested parallelism present in TFS is easily decomposed. Figure 6 shows where time was spent at each level of nesting. The function ALGO started a parallel region, and deeper in the callpath, the function AUSM also started a parallel region.

## 6 Future Work

As noted earlier, there is active discussion underway in the OpenMP tools forum as to what the appropriate interface should be for execution time tools such as for performance measurement. We hope that runtime system functions will be made available for thread identification and nesting context to be queried.

For TAU's application, we would like to support a higher level mapping of thread identifiers back to the application developer's model of nested parallelism. In the case of non-nested parallelism, TAU provides a clear picture of the performance of each thread in each team. This picture is currently not as clear in the nested case because we have only a single number to identify a thread. We anticipate adding support for thread naming in TAU wherein a thread is identified in the nested OpenMP case by its nesting depth and identifier, or by the identifier in each team where it originated (such as "thread $0 \rightarrow 3 \rightarrow 2$"). The runtime, hopefully provided in a future OpenMP specification, will provide the necessary information.

## 7 Conclusion

Performance tools that measure per-thread performance data must be able to uniquely identify threads of execution at runtime. This is complicated in the presence of nested parallelism when thread identities queried from the runtime system are not unique and do not provide information about nesting context. In this paper, we describe the nested parallelism problem currently faced by tools for portable OpenMP performance analysis. Several possible solutions are discussed. The approach we implemented in the TAU performance system uses thread private storage to create a unique thread identifier. The approach is portable and has been validated with both Sun and Intel's OpenMP compilers. We demonstrated its capabilities with the TFS application.

## Acknowledgments

Research at the University of Oregon is sponsored by contracts (DE-FG02-05ER25663, DE-FG02-05ER25680) from the MICS program of the U.S. Dept. of Energy, Office of Science.

## References

1. Shende, S., Malony, A.: The TAU Parallel Performance System. In: International Journal of High Performance Computing Applications, Summer 2006. ACTS Collection Special Issue (2006)
2. Malony, A., Shende, S.: Performance Technology for Complex Parallel and Distributed Systems. In: Kotsis, G., Kacsuk, P. (eds.) Distributed and Parallel Systems, From Instruction Parallelism to Cluster Computing, Third Workshop on Distributed and Parallel Systems (DAPSYS 2000), pp. 37–46. Kluwer, Dordrecht (2000)

3. Lindlan, K., Cuny, J., Malony, A., Shende, S., Mohr, B., Rivenburgh, R., Rasmussen, C.: A Tool Framework for Static and Dynamic Analysis of Object-Oriented Software with Templates. In: SC 2000 conference (2000)
4. Mohr, B., Malony, A.D., Shende, S., Wolf, F.: Towards a Performance Tool Interface for OpenMP: An Approach Based on Directive Rewriting. In: Proceedings of Third European Workshop on OpenMP (EWOMP 2001) (September 2001)
5. Mohr, B., Malony, A.D., Shende, S., Wolf, F.: Design and Prototype of a Performance Tool Interface for OpenMP. The Journal of Supercomputing 23, 105–128 (2002)
6. Huck, K.A., Malony, A.D.: PerfExplorer: A Performance Data Mining Framework for Large-Scale Parallel Computing. In: Proceedings of SC 2005 conference, ACM, New York (2005)
7. Browne, S., Dongarra, J., Garner, N., Ho, G., Mucci, P.: A Portable Programming Interface for Performance Evaluation on Modern Processors. International Journal of High Performance Computing Applications 14(3), 189–204 (2000)
8. Mohr, B., Wolf, F.: KOJAK - A Tool Set for Automatic Performance Analysis of Parallel Applications. In: Kosch, H., Böszörményi, L., Hellwagner, H. (eds.) EuroPar 2003. LNCS, vol. 2790, pp. 1301–1304. Springer, Heidelberg (2003)
9. an Mey, D.: Proposed Light Weight Extensions to the OpenMP Specification, Private Communication (December 2005)
10. Spiegel, A.: Proposed Solution to Identifying Threads Uniquely in Nested OpenMP threads, Private Communication (October 2005)
11. OpenMP, http://www.openmp.org/drupal/
12. Tanaka, Y., Taura, K., Sato, M., Yonezawa, A.: Performance Evaluation of OpenMP Applications with Nested Parallelism. In: Languages, Compilers, and Run-Time Systems for Scalable Computers, pp. 100–112 (2000)
13. Gonzalez, M., Ayguade, E., Martorell, X., Labarta, J., Navarro, N., Oliver, J.: NanosCompiler: Supporting Flexible Multilevel Parallelism Exploitation in OpenMP. Concurrency - Practice and Experience 12(12), 1205–1218 (2000)
14. Fares, E., Meinke, M., Schröder, W.: Numerical Simulation of the Interaction of Flap Side-Edge Vortices and Engine Jets. In: Proceedings of the 22nd International Congress of Aeronautical Sciences, ICAS 0212 (September 2000)
15. Fares, E., Meinke, M., Schröder, W.: Numerical Simulation of the Interaction of Wingtip Vortices and Engine Jets in the Near Field. In: Proceedings of the 38th Aerospace Sciences Meeting and Exhibit, AIAA Paper 20002222 (January 2000)
16. OpenMP tools mailing list, Omp-tools@openmp.org, http://openmp.org/mailman/listinfo/omp-tools
17. Hörschler, I., Johnson, S.P., an Mey, D.: 100 (Processor) Years Simulation of Flow through the Human Nose using OpenMP (2006), http://www.rz.rwth-aachen.de/computing/events/2005/sunhpc_colloquium/07_Hoerschler.pdf
18. Sun Studio compilers (2006), http://developers.sun.com/prodtech/cc
19. ParaWise (2006), http://www.parallelsp.com/parawise.htm
20. Intel compilers (2006), http://www.intel.com/cd/software/products/asmo-na/eng/compilers
21. OpenMP API Specification 2.5 (May 2005), http://www.openmp.org/drupal/mp-documents/spec25.pdf

# Parallelization of a Hierarchical Data Clustering Algorithm Using OpenMP

Panagiotis E. Hadjidoukas[1,*] and Laurent Amsaleg[2]

[1] Department of Computer Science, University of Ioannina, Ioannina, Greece
phadjido@cs.uoi.gr
[2] IRISA/INRIA, Campus de Beaulieu, 35042 Rennes cedex, France
lamsaleg@irisa.fr

**Abstract.** This paper presents a parallel implementation of CURE, an efficient hierarchical data clustering algorithm, using the OpenMP programming model. OpenMP provides a means of transparent management of the asymmetry and non–determinism in CURE, while our OpenMP runtime support enables the effective exploitation of the irregular nested loop–level parallelism. Experimental results for various problem parameters demonstrate the scalability of our implementation and the effective utilization of parallel hardware, which enable the use of CURE for large data sets.

## 1 Introduction

Data clustering is one of the fundamental techniques in scientific data analysis and data mining. The problem of clustering is to partition the data set into segments (called clusters) so that intra–cluster data are similar and inter–cluster data are dissimilar. Clustering algorithms are very computation demanding and, thus, require high–performance machines to get results in a reasonable amount of time. In this paper, we present a parallel implementation of CURE (Clustering Using REpresentatives) [4], a well–known hierarchical data clustering algorithm, using OpenMP [2]. CURE is a very efficient clustering algorithm with respect to the quality of clusters: it can identify arbitrary–shaped clusters and handle high–dimensional data. However, its worst–case time complexity is O ($n^2$logn), where n is the number of points to be clustered. Although sampling and partitioning can allow CURE to handle larger data sets, the algorithm is not applicable to today's huge data bases because of its quadratic time complexity.

Our general goal is the development of an efficient parallel data clustering algorithm that targets shared memory multiprocessors, clusters of computers and computational grids. This paper focuses only on the shared memory architecture. Although CURE provides high quality clustering, a parallel version was not available due to the asymmetric and non–deterministic parallelism of the clustering algorithm. OpenMP, however, manages to resolve successfully these

---

[*] This work was done while the first author was a postdoctoral researcher at IRISA/INRIA, Rennes.

issues due to the dynamic assignment of parallel tasks to processors. In addition, our previous research work has already resulted in a portable OpenMP environment that supports multiple levels of parallelism very efficiently. Thus, we are able to satisfy the need for nested parallelism exploitation in order to achieve load balancing. Our experimental results demonstrate significant performance gains in PCURE, the parallel version of CURE, and effective utilization of the shared memory architecture both on small–scale SMPs and high performance multiprocessors.

A first survey of parallel algorithms for hierarchical clustering using distance based metrics is given in [9]. Most parallel data clustering approaches target distributed memory multiprocessors and their implementation is based on message passing [1,7,8,11,10]. None of them has been applied to a pure hierarchical data clustering algorithm. As we will show experimentally, static parallelization approaches are not applicable to CURE due to the non–deterministic behavior of its algorithm. In addition, message passing would require significant programming effort to handle the highly irregular and unpredictable data access patterns in CURE.

The rest of this paper is organized as follows: Section 2 presents our modifications to the main clustering algorithm of CURE and its parallelization using OpenMP directives. Experimental results are reported in Section 3. Finally, Section 4 presents some conclusions and our ongoing work.

## 2 CURE Data Clustering Algorithm

### 2.1 Introduction

Cure is a bottom–up hierarchical clustering algorithm, but instead of using a centroid–based approach it employs a method that is based on choosing a well–formed group of points to identify the distance between clusters. For each intermediately computed cluster, CURE chooses a constant number, r of well scattered points. These points are used to identify the shape and size of the cluster. The next step of the algorithm shrinks the selected points towards the centroid of the cluster by a pre–determined fraction a. Varying this fraction between 0 and 1 can help CURE to identify different types of clusters. Using the shrunken position of these points to identify the cluster, the algorithms then finds the clusters with the closest pairs of identifying points. These clusters are chosen to be merged as part of the hierarchical algorithm. Merging continues until the desired by the user number of clusters, k, remain. A k–d tree is used to store information about the clusters and the points that belong to them.

Figure 1 outlines the main clustering algorithm: since CURE is an hierarchical agglomerative algorithm, initially every data point is considered as a separate cluster with one representative, the point itself. The algorithm computes initially the closest cluster for each cluster. Next, it starts the agglomerative clustering, merging the closest pair of clusters until only k clusters remain. According to the merge procedure, the centroid of the new cluster is the weighted mean of the two merged clusters. Moreover, to reduce the time complexity of the algorithm, the

1. Initialization: Compute distances and find nearest neighbors pairs for all clusters

2. Clustering: Perform hierarchical clustering until the predefined number of clusters k has been computed

   While (number of remaining clusters > k) {
      a. Find the pair of clusters with the minimum distance
      b. Merge them:
         i. new size = size1 + size2
         ii. new centroid = a1*centroid1 + a2*centroid2,
            where a1 = size1/new size and a2 = size2/new size
         iii. find c new representative points
      c. Update nearest neighbors pairs for the clusters
      d. Reduce the number of remaining clusters
      e. If conditions are satisfied, apply pruning of clusters
   }

3. Output the representative points of each cluster

**Fig. 1.** Outline of CURE

authors propose an improved merge procedure where the new c representative points are chosen between the 2c points of the two clusters merged.

The worst–case time complexity of CURE is O $(n^2 \log n)$, where n is the number of points to be clustered. In order to allow CURE to handle very large data sets, CURE uses a random sample of the database. Sampling improves the performance of the algorithm since the sample can be designed to fit in main memory, eliminating thus significant I/O costs, and also contributes in the filtering of outliers. To speed up the clustering process when the sample size increases, CURE partitions and partially clusters the data points in the partitions of the random sample. Instead of using a centroid to label the clusters, multiple representative points are used, and each data point is assigned to the cluster with the closest representative point. The use of multiple points enables the algorithm to identify arbitrarily shaped clusters. Empirical work with CURE discovered that the algorithm is insensitive to outliers and can identify clusters with interesting shape. Moreover, sampling and partitioning speed up the clustering process without sacrificing cluster quality.

## 2.2 Implementation

Our parallel implementation of CURE was inspired by the source code of Dr. Han and has been enhanced to handle large data sets. The algorithm uses a linear array of records that keeps information about the size, the centroid and the representative points of each cluster. Taking into consideration the improved procedure for merging clusters and that the labeling of data is a separate process, we do not construct a k–d tree. Instead, when two clusters (entries in the array) are merged, we store the information for the resulted cluster in the entry of the first cluster

```
1.  init_nnbs () {
2.      for (i=0; i<npat; i++) find_nnb(i, &nnb[i].index, &nnb[i].dist);
3.  }
4.
5.  update_nnbs(int pair_low, int pair_high) {
6.      for (i=pair_low+1; i<npat; i++) {
7.          if (entry i has been invalidated) continue;
8.          if (entry i had neighbor pair_low or pair_high)
9.              find_nnb (i, &nnb[i].index, &nnb[i].dist);
10.         else if (pair_high < i)
11.             if ((dist = compute_distance(pair_high, i))) < nnb_dist[i])
12.                 { nnb[i].index = pair_high; nnb[i].dist = dist; }
13.     }
14. }
15.
16. find_nnb(int i, int *index, double *distance) {
17.     min_dist = +inf, min_index = -1;
18.     for (j=0; j<i; j++) {
19.         if (entry j has been invalidated) continue;
20.         if ((dist = compute_distance(i, j)) < min_dist)
21.             { min_dist = dist; min_index = j };
22.     }
23.     *index = min_index; *distance = min_dist;
24. }
```

Fig. 2. Pseudocode of the most computation demanding routines in Cure

and simply invalidate the second one. This design decision not only speeds up the sequential algorithm but also results in significantly less memory consumption.

The algorithm also maintains per–cluster information about the index of the closest cluster and the minimum distance to it. To avoid duplication of this information, the algorithm searches for the closest neighbor of a given cluster only in entries with a smaller index. Therefore, the maintenance of this information for clusters with a larger index requires more computational time. In order to optimize cache memory accesses, we store this information in a separate array. The rationale of this decision is that many entries of this array are likely to be updated in each step of the algorithm (2c), in contrast to the array of clusters, for which only two entries are updated (2b).

Figure 2 presents the pseudocode of the most computation demanding routines in the clustering algorithm, which correspond to the initialization phase (init_nnbs, lines 1–3) and the update of the nearest neighbors (update_nnbs, lines 5–14). Special features of this algorithm are its asymmetry and non–determinism, due to the way it finds closest clusters and the gradual decrease of the number of valid entries (clusters). This asymmetry is reflected in the find_nnb routine, utilized by both the initialization and the update procedure. In addition, non–determinism appears only in the update procedure, since during the initialization phase of the algorithm all clusters are valid.

## 2.3 Parallelization Using OpenMP

The several loops of its clustering algorithm constitute CURE a strong candidate for parallelization using OpenMP, which can enable its use on large data sets. Moreover, this parallelization is possible because there are not data dependencies between the iterations of each loop. Moreover, a k-d tree is not maintained, the construction of which would increase the sequential portion of the algorithm and its traversal would complicate the parallelization.

To avoid false sharing between processors, data structures of the algorithm are aligned on cache line boundaries. Thus, the efficiency of the parallel implementation mainly depends on the even distribution of computations to the processors. This, however, is a challenging task because the computational cost of loop iterations cannot be pre-estimated due to the high irregularity of the algorithm. To tackle the above problems, we parallelized the loops at lines 2 and 6 using the guided and dynamic scheduling policy respectively.

In addition, we parallelized the loop in the find_nnb routine (line 18), using again the guided schedule clause. This loop, which computes for a given cluster the index of its closest neighbor and the distance to it, actually corresponds to a reduction operation not supported directly by OpenMP. To optimize the parallel implementation of this procedure, we fist spawn a team of threads and then execute the loop in parallel, allowing each worker thread to keep the partial results, i.e. minimum distance and index, in its local memory (stack). When the iterations of the loop have been exhausted, the reduction operation takes place, with each thread checking and updating the global result within a critical section.

The total execution time for searching the minimum distance pair is negligible compared to the update phase but can become significant for large data sets. However, instead of applying direct parallelization to the search procedure, we integrated it into the update phase, following the same reduction-like approach as for the search of the closest neighbor. This decision maintains data locality, which would be destroyed by sequential execution, and avoids the overheads of an additional spawning of parallelism for this non-scalable and memory-bound section of code. Finally, pruning has many similarities with the update procedure and has been parallelized accordingly.

## 3 Experimental Results

One of CURE's strong advantages is that provides good quality of clustering. Our parallel implementation of CURE has not altered the original algorithm, which means that in all cases, it produces identical results. In this section, we focus on the performance improvements in the data clustering algorithm with respect to its execution time. Our goal is to exploit parallel processing in order to tackle the quadratic time complexity of agglomerative data clustering (CURE) rather than to alter the algorithm itself.

Table 1 outlines the three data sets that have been used for our experiments. Data Set 1 (DS1) is the data set used in [4] and consists of two dimensional records. Data Sets 2 and 3 (DS2, DS3) come from image recognition and have

**Table 1.** Experimental data sets

| Features: | # Records | Dimensionality | # Clusters |
|---|---|---|---|
| Data Set 1 (DS1) | 100K | 2 | 100 |
| Data Set 2 (DS2) | 100K | 24 | Unknown |
| Data Set 3 (DS3) | 100K | 4875 | Unknown |

been extracted from large image databases, while their dimensionality is 24 and 4875 respectively.

We conducted our experiments on a dual–processor Hyper–Threading (HT) enabled Intel Xeon system running at 2.4GHz with 1GB memory and 512K secondary (L2) cache. The operating system is Debian Linux (2.4.12) and for the OpenMP parallelization, we have used our custom OpenMP implementation that consists of the portable OMPi compiler [3] and the NANOS runtime library, while the native compiler is GNU GCC (3.3.4). As we have shown in [5], this configuration reduces significantly the runtime overheads and results in an OpenMP execution environment with efficient support for nested parallelism.

### 3.1 Scale–Up Experiments

We studied the scalability of PCURE for three different input parameters: data set size, number of representative points and dimensionality. We provide measurements on the HT–enabled Xeon system running PCURE using 1 and 4 OpenMP threads.

- **First Scale–Up Experiment**: Figure 3 illustrates the execution time of the clustering algorithm for various sizes of the three data sets, using the default parameters of CURE (R = 10, a = 0.3). The clustering algorithm stops when the number of clusters has been reduced to 1% of the initial number of records. In all figures, we observe the quadratic complexity of the algorithm and the high speedup that is achieved on the two hyper-threaded processors due to the OpenMP parallelization.
- **Second Scale–Up Experiment**: Figure 4 depicts the execution times of the clustering algorithm for various numbers of representative points (from 1 to 32). The shrinking factor has been set to its default value (a = 0.3), while for DS1 and DS2 we have used 50K records and for DS3 5K records. We observe an almost linear increase of the execution time with respect to the number of representatives and, as before, the effective exploitation of the underlying parallel hardware.
- **Third Scale–Up Experiment**: Figure 5 illustrates the linear increase of execution time with respect to data dimensionality. We have used 5K records from DS3, varying appropriately the number of features, from 2 to 1024. We also observe that performance speedup is improved drastically as

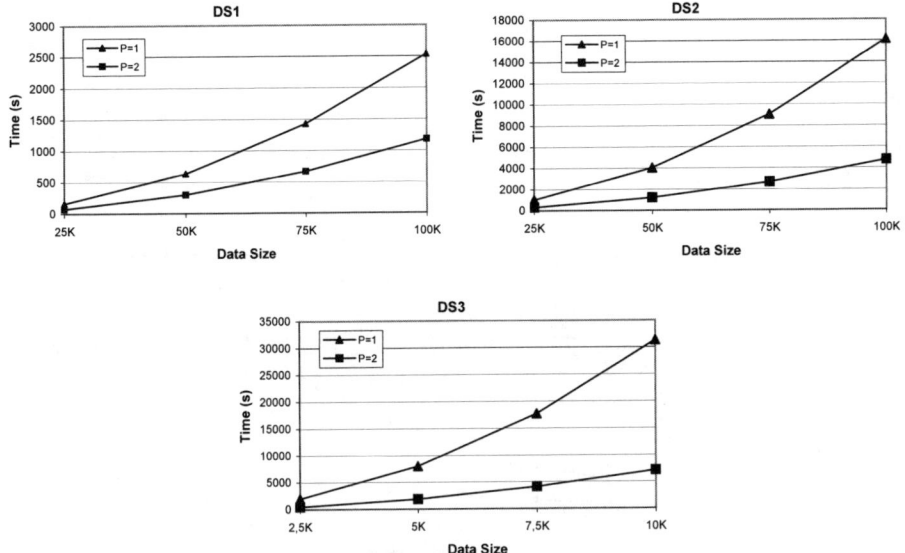

**Fig. 3.** Execution Time / Data set size

dimensionality increases. This is due to to the negligible parallelization overheads, opposed to the total execution time, and the more effective utilization of cache memory for higher dimensions.

Although PCURE is highly parallel, the obtained speedup varies significantly according to the problem parameters. An important factor that affects performance is the parallelization overheads, especially for spawning and joining parallelism in every clustering step. These overheads, however, become negligible for large data sets or high dimensionality. Another factor is the bandwidth of the memory subsystem, which is low in bus–based small–scale SMPs. This limits the performance of data intensive applications, like PCURE, especially for low dimensionality and consequently ineffective cache utilization. On the other hand, for data sets with high dimensional vectors (e.g. DS3), the negligible parallelization overheads and the effective cache utilization can result in super–linear speedups ($\simeq$ 4.5x) even on a HT–enabled dual–processor Intel Xeon system.

Apart from the performance speedup due to the OpenMP parallelization, our measurements demonstrate the quadratic execution time of the CURE algorithm with respect to the number of records and its linear execution time with respect to the dimensionality and the number of representative points. PCURE exhibits well defined and predictable runtime behavior, which allows us to adjust the problem input parameters (e.g. appropriate data set size) or the parallel execution environment (e.g. number of threads) appropriately.

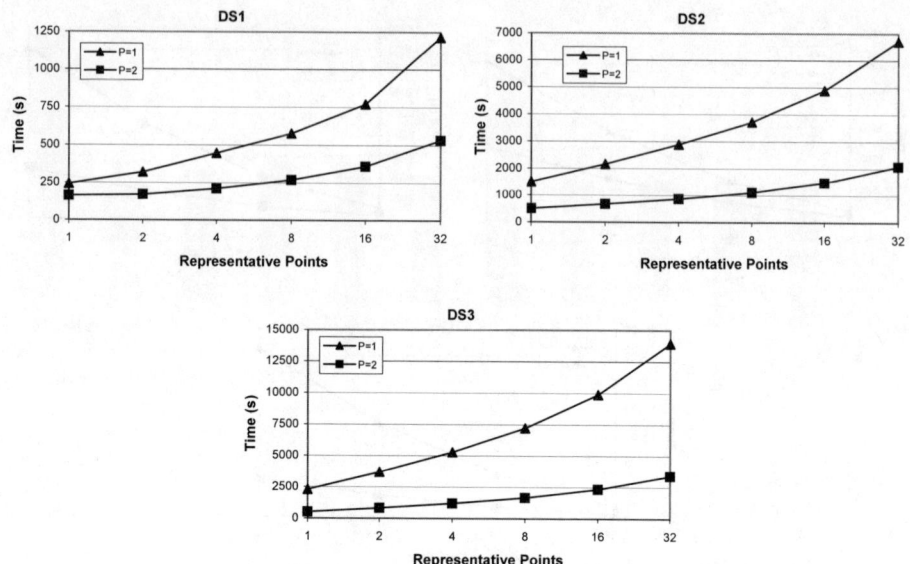

**Fig. 4.** Execution Time / Number of representatives

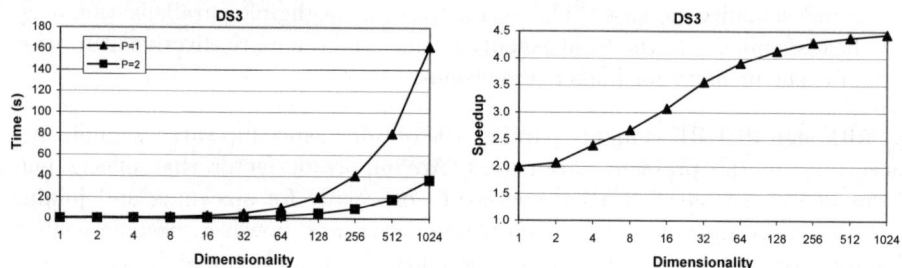

**Fig. 5.** Execution Time / Dimensionality

### 3.2 Nested Parallelism

In this section, we experimentally justify the necessity to exploit nested parallelism for achieving load balancing in the update phase of PCURE. Furthermore, we study the scalability of our parallel implementation on a large multiprocessor system. In both cases, we compute 1000 clusters for 50K records of our second data set (DS2), using the default input parameters. Moreover, the results refer only to the clustering phase of PCURE.

We evaluate every possible loop scheduling policy supported by OpenMP (static, dynamic and guided), trying to achieve the best performance using 4 threads on the 2 hyper–threaded processors for the case of single–level parallelism. Finally, we compare the measured speedups with those attained when nested parallelism has been enabled. Figure 6 depicts these speedups for the

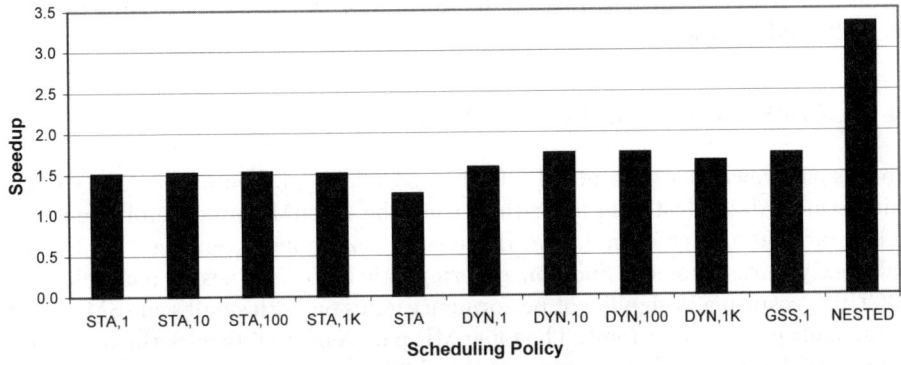

**Fig. 6.** Single Level vs. Nested Parallelism

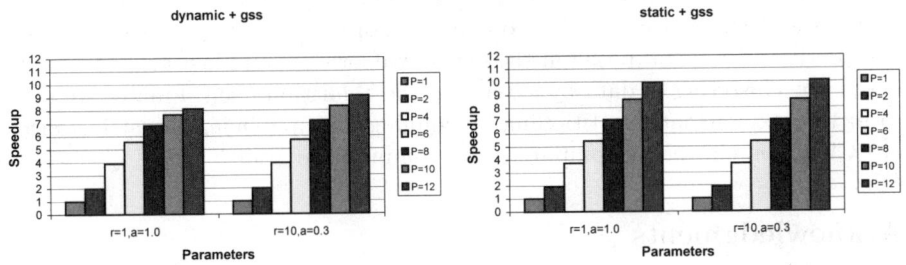

**Fig. 7.** Performance Scalability on the IBM SP2

update phase of the algorithm when a single level or two levels of parallelism are exploited. For the first case, the maximum speedup is achieved when the dynamic scheduling policy with chunk equal to 100 is used and it does not exceed 1.74x. On the contrary, when nested parallelism is enabled, the speedup for the update phase is significantly higher. The same performance behavior is observed regardless of the scheduling policies used for the two nested loops. The sequential execution time of the update phase was 3688 seconds.

Finally, Figure 7 shows the performance speedups of PCURE on up to 12 processors of a 16–way Power3 IBM SP2 system. This particular experiment provides also a demonstration of our portable and efficient support of nested parallelism. We include additional measurements for the pure centroid based version of PCURE (R=1, a=1.0). Furthermore, we run the same experiments using the static, instead of the dynamic, scheduling policy for the outer loop of the clustering phase. Our results show that PCURE scales efficiently, which is attributed to the effective exploitation of nested parallelism and the high performance memory subsystem of the underlying hardware. We also observe that the algorithm scales better when more representatives per cluster are used. When the number of processors increases, the static scheduling policy starts to outperform the dynamic one because of its better data locality since each processor is always assigned the same portion of the array that holds nearest neighbor information.

## 4 Conclusions and Future Work

In this paper, we have presented PCURE, a parallel implementation of the CURE hierarchical data clustering algorithm using the OpenMP programming model. The portable and efficient OpenMP runtime environment manages to handle efficiently the asymmetric and non–deterministic loop–level nested parallelism in PCURE, resulting in significant performance gains on both small scale SMPs and large multiprocessor systems. The OpenMP parallelization tackles the quadratic time complexity of the algorithm and thus high quality clustering can be applied to large data sets.

Future work includes execution of PCURE on distributed memory environments based on our OpenMP implementation on clusters of multiprocessors [6]. To reduce problem size, we plan to exploit additional techniques like sampling and partitioning, as proposed in the original algorithm, and fast grid or density based pre–clustering of data by extending the algorithm itself. Finally, we plan to further exploit data partitioning by running multiple independent instances of PCURE on clusters of computers and computation grids.

## Acknowledgments

We would like to thank CEPBA for providing us access to their IBM SP2 system and Dr. Han for the source code availability of CURE.

## References

1. Arlia, D., Coppola, M.: Experiments in Parallel Clustering using DBSCAN. In: Sakellariou, R., Keane, J.A., Gurd, J.R., Freeman, L. (eds.) Euro-Par 2001. LNCS, vol. 2150, pp. 326–331. Springer, Heidelberg (2001)
2. OpenMP Architecture Review Board. OpenMP specifications, Available at: http://www.openmp.org
3. Dimakopoulos, V.V., Tzoumas, X., Leontiadis, E.: A Portable Compiler for OpenMP v. 2.0. In: Proceedings of the 5th European Workshop on OpenMP (EWOMP 2003), Aachen, Germany (October 2003)
4. Guha, S., Rastogi, R., Shim, K.: CURE: An Efficient Clustering Algorithm for Large DataBases. In: Proceedings of the ACM SIGMOD International Conference on Management of Data (1998)
5. Hadjidoukas, P.E., Amsaleg, L.: Portable Support and Exploitation of Nested Parallelism in OpenMP. In: Proceedings of the 6th European Workshop on OpenMP (EWOMP 2004), Stockholm, Sweden (October 2004)
6. Hadjidoukas, P.E., Polychronopoulos, E.D., Papatheodorou, T.S.: A Modular OpenMP Implementation for Clusters of Multiprocessors. Journal of Parallel and Distributed Computing Practices (PDCP), Special Issue on OpenMP: Experiences, Implementations and Applications 2, 153–168 (2004)

7. Judd, D., McKinley, P., Jain, A.: Large-Scale Parallel Data Clustering. In: Proceedings of the International Conference on Pattern Recognition (1996)
8. Nagesh, H.S., Goil, S., Choudhary, A.: A Scalable Parallel Subspace Clustering Algorithm for Massive Data Sets. In: Proceedings of the International Conference on Parallel Processing (ICPP 2000) (2000)
9. Olson, C.F.: Parallel Algorithms for Hierarchical Clustering. Parallel Computing 21, 1313–1325 (1995)
10. Pizzuti, C., Talia, D.: P–AutoClass: Scalable Parallel Clustering for Mining Large Data Sets. IEEE Transactions on Knowledge and Data Engineering 15(3) (May 2003)
11. Stoffel, K., Belkoniene, A.: Parallel K–Means Clustering for Large Data Sets. In: Amestoy, P.R., Berger, P., Daydé, M., Duff, I.S., Frayssé, V., Giraud, L., Ruiz, D. (eds.) Euro-Par 1999. LNCS, vol. 1685, pp. 1451–1454. Springer, Heidelberg (1999)

# OpenMP and C++

Christian Terboven and Dieter an Mey

Center for Computing and Communication,
RWTH Aachen University, Aachen, Germany
{terboven,anmey}@rz.rwth-aachen.de

**Abstract.** In this paper we present our experiences parallelizing the C++ programs DROPS and FIRE with OpenMP. Though the OpenMP specification includes C++, several shortcomings and missing features can be noticed in both the current OpenMP compilers and the specification. We propose solutions of how to overcome these problems and formulate wishes for the future OpenMP specification 3.0.

## 1 Introduction

For the time being the FORTRAN programming language is still dominating the field of high performance computing, but in recent years the usage of C++ is catching up in the HPC domain as well. We parallelized two sophisticated C++ applications and experienced several shortcomings in current compilers and some missing features in the current OpenMP specification. However, the parallelization of the application codes was quite successful and we presented our proposed solutions in [11], [12] and [13]. Here we concentrate on how the object-oriented programming style can be used in the context of OpenMP and how to exploit C++ language features to improve scalability on ccNUMA-architectures.

The Navier-Stokes solver DROPS [2] is developed at the IGPM (Institut für Geometrie und Praktische Mathematik) at the RWTH Aachen University as part of the interdisciplinary project SFB 540 [3], where complicated flow phenomena are investigated. The object-oriented programming paradigm offers a high flexibility and elegance of the program code, facilitating development and investigation of numerical algorithms. Template programming techniques and the C++ Standard Template Library (STL) are heavily used. The DROPS code, which contains over 21000 lines of C++ code, achieves a speedup of 8 on an UltraSPARC-IV based system with 16 threads.

The Flexible Image Retrieval Engine (FIRE) has been developed at the Human Language Technology and Pattern Recognition Group of the RWTH Aachen University [4]. It is designed as a research system with flexibility in mind. That is, it is easily extensible and highly modular by using C++ language features, which in addition turned out to be quite helpful for the OpenMP parallelization. The FIRE system was successfully used in the ImageCLEF 2004 and 2005 content-based image retrieval evaluations [5], [6]. The FIRE code has been parallelized using nested OpenMP [1] and a nearly linear speedup can be achieved, thus reaching an efficiency of over 90% with 44 threads on an UltraSPARC-IV based system.

We evaluated the performance of our experiments on two architectures with different characteristics. The Sun Fire E6900 servers consist of 24 dual-core UltraSPARC-IV processors running at 1.2 GHz clock speed with a total of 96 GB of memory offering a rather flat memory model. The dual-core processors are treated as two independent processors by the Solaris 10 operating system. In addition, two different types of Sun Fire V40z machines were used. They have 4 AMD Opteron 875 dual-core processors and 16GB of RAM or 4 AMD 848 processors and 8GB of RAM, respectively. These processors have a clock speed of 2.2GHz. The dual-core machines are running the Solaris 10 operating system that treats the cores as individual processors and the single-core machines have Linux 2.6 as operating system. The Sun Fire V40z systems have a ccNUMA-architecture where data locality is very important. On the Solaris systems we used the Sun Studio 11 compiler suite, on the Linux systems the Intel 9.0 compiler suite.

For the experiments in subsection 2.1 we also tested the IBM C++ 8.0 compiler on an Power4+ based system and the PGI 6.1 and Microsoft 8.0 compilers, both on an AMD Opteron based system.

The remainder of this paper is organized as follows: in section 2 we present our approaches of combining object-oriented programming and OpenMP. We analyze current compilers with respect to their behavior with variables of class-type in OpenMP clauses and present methods to parallelize non-conforming loops and high-level codes. In section 3 we show how to improve the performance of using STL containers and other data types on ccNUMA-architectures. Section 4 contains our conclusion.

## 2 Parallelization and Object-Oriented Programming

In some cases we encountered an unexpected behavior when variables of class-type were used in OpenMP clauses. Therefore we created a set of small test cases to evaluate implementations regarding their support for privatization of variables of class-type. The results are presented in subsection 2.1. In C++ programs one often finds loops which do not conform to the requirements for parallelization using a loop-worksharing construct. We present four different approaches of parallelizing such loops in subsection 2.2 and compare the achieved performance. When parallelizing high-level object-oriented codes the user has some choices of where to put the parallelization constructs. For the PCG linear equation solver of DROPS we compare different levels in subsection 2.3.

### 2.1 Variables of Class-Type in OpenMP Clauses

To evaluate an implementation's behavior when privatizing variables of class-type we used a simple experimental approach using the class *Object1* as shown in program 1. We created a test program that instantiates a class of type *Object1*

**Program 1.** Test program to evaluate a compiler's behavior.

```
1   class Object1 {
2   public:
3     Object1() {
4       cout << "Object1::constructor" << endl; }
5     ~Object1() {
6       cout << "Object1::destructor" << endl; }
7     Object1(const Object1& o) {
8       cout << "Object1::copy constructor" << endl; }
9     Object1 &operator=(Object1& o) {
10      cout << "Object1::assignment operator" << endl;
11      return *this; }
12    void dummyfunc() const {
13      cout << "Object1::dummyfunc" << endl; }
14  };
15
16  int main(int argc, char* argv[]) {
17    Object1 o1;
18
19  #pragma omp parallel scoping-attribute(o1)
20    {
21      o1.dummyfunc();
22    }
23  } // end of main
```

and contains a parallel region and then experimented with different data scoping attributes for the instantiated variable.

We first looked at the *shared* scoping attribute. As the OpenMP specification [7] states in section 2.8.3.2 "All threads within a team access the same storage area for each shared object" we did not expect any other function calls to appear in the output than the calls to *dummyfunc*. This is true for all tested compilers.

Regarding the *private* scoping attribute the OpenMP specification states in section 2.8.3.3 "A new list item of the same type, with automatic storage duration, is allocated for the construct" and "The new list item is initialized, or has an undefined initial value, as if it had been locally declared without an initializer". As the C++ language is defined with a consistent memory model [8] the word 'undefined' is not related to to the state of the object but to the value, because when instantiating a variable of class-type a constructor has to be invoked. For our experiment setup we expect each thread to instantiate a new variable and therefore in addition to the calls to *dummyfunc* a pair of *constructor* and *destructor* calls should appear in the output for each thread. This is true for all tested compilers, except for the PGI compiler, where no constructor call appears.

For the *firstprivate* scoping attribute the OpenMP specification states in section 2.8.3.4 that "For class type, a copy constructor is invoked to perform the initialization". We therefore expect the *constructor* calls to be replaced by calls to the *copy constructor*. This is true for all tested compilers, except for the IBM and PGI compilers. For the IBM compiler we noticed an additional pair of *copy constructor* and *destructor*, for the PGI compiler the construction is missing again.

The *lastprivate* scoping attribute required a little modification, as it is only available for the loop-worksharing construct. We therefore introduced a loop over the number of threads with a *static* schedule and a chunksize of one. We expect one additional call to the *assignment operator*. This is true for all tested compilers, except the IBM and PGI compilers have the same discrepancy as above.

We are aware of the fact that more complex data types might change a compiler's behavior, but our application codes behave equal to our simplified test cases. For the scenarios discussed above all compilers except the PGI compiler can be used, but the following experiments will show severe problems in the current implementations. The Microsoft compiler is unable to run the next two experiments as it states that "Dynamic initialization of threadprivate symbols is currently not supported", therefore we will only look at the Intel, Sun and IBM compilers.

In section 2.8.2 the OpenMP specification states that "... each copy of a threadprivate object is initialized once, in the manner specified by the program, but at an unspecified point in the program prior to first reference to that copy". We declared the variable *o1* as *threadprivate* and removed the scoping attribute from the parallel region. According to the OpenMP specification, we expect the master's instance of the variable to be constructed before the parallel region and all other instances to be constructed just before the first parallel region. This is true for the Intel C++ compiler, but not for the Sun and IBM compilers, as shown in table 1.

For the *copyin* scoping attribute the OpenMP specification states in section 2.8.4.1 that "The copy is done after the team is formed and prior to the start of

**Table 1.** Compiler output with two threads for test cases with *threadprivate*

| Test case | Intel Output | Sun Output | IBM Output |
|---|---|---|---|
| threadprivate | 0: constructor | 0: constructor | 0: constructor |
| | 1: constructor | 0: copy constructor | 0: copy constructor |
| | 0: dummyfunc | 1: copy constructor | 1: copy constructor |
| | 1: dummyfunc | 0: dummyfunc | 0: dummyfunc |
| | 1: destructor | 1: dummyfunc | 1: dummyfunc |
| | 0: destructor | 0: destructor | 0: destructor |
| threadprivate + copyin | 0: constructor | 0: constructor | 0: constructor |
| | 1: constructor | 0: copy constructor | 1: copy constructor |
| | 0: assignment operator | 1: copy constructor | 1: assignment operator |
| | 1: assignment operator | 1: assignment operator | 0: dummyfunc |
| | 0: dummyfunc | 0: dummyfunc | 1: dummyfunc |
| | 1: dummyfunc | 1: dummyfunc | 1: destructor |
| | 1: destructor | 0: destructor | 0: destructor |
| | 0: destructor | | |

execution of the parallel region". We therefore expect a call to the *assignment operator* for each slave thread, but not for the master thread. The Intel C++ compiler does not differ substantially from the behavior we expected as it includes just an additional call to the *assignment operator*. While this does not change the expected result, it does not lead to optimal performance.

The obvious solution to overcome these problems is to use pointers as *threadprivate* variables which have to be set up manually prior to the first parallel region of a program.

In addition to these difficulties it should be mentioned that the privatization of class member variables is neither mentioned in the specification nor supported by the tested compilers. Instead of a clear error message we experienced sometimes confusing results like missing symbols during the linking stage, depending on the compiler. Our proposed workaround is to use a variable with the same name in the local scope which then can be privatized, though this might break the class interface under some circumstances.

### 2.2 Parallelization of Non-conforming Loops

In C and C++ programs one often finds loops that cannot be parallelized using the loop-worksharing construct because they do not conform to the OpenMP specification. In section 2.5.1 the OpenMP specification defines the requirements for the loop-worksharing construct.

The requirement "... signed integer variable ..." prohibits the parallelization of loops using *size_t* data type as the loop index variable. Changing the variable type to *long* allows the parallelization and should be feasable for most programs.

The restriction "... the corresponding for-loop must have the following canonical form: for(init-expr; var relational-op b; incr-expr) ..." prohibits the parallelization of loops using pointer arithmetic or iterators, though the number of loop iterations could be computed on entry to the loop. Examples of STL iterator style loops are the setup-routines in DROPS, which use iterators to run over custom data types building the stiffness matrices. We compare four approaches to parallelize iterator loops using the test code shown in program 2.

**Program 2.** Loop using C++ STL iterators.

```
1  list<CComputeItem> list1;
2  list<CComputeItem>::iterator it;
3  for (it = list1.begin(); it != list1.end(); it++) {
4      it->compute();
5  }
```

The first approach follows the idea to create a parallelizable loop by adding an additional loop in which the iterator pointers are stored in an array. Processing this array can be parallelized using the loop-worksharing construct, as shown in program 3.

**Program 3.** Creation of a parallelizable loop.

```
1    valarray<CComputeItem*> items(list1.size());
2    for (it = list1.begin(); it != list1.end(); it++) {
3      items[l] = &(*it);
4      l++;
5    }
6    #pragma omp parallel for default(shared)
7    for (long l = 0; l < list1.size(); l++) {
8      items[l]->compute();
9    }
```

The second approach uses Intel's Task-Queuing worksharing construct [9] which is an extension to OpenMP. Currently it is only available in the Intel compiler suite, but a proposal for the future OpenMP 3.0 specification is under discussion. For each value of the loop index variable the loop body is enqueued into a work queue, which is then processed in parallel by all threads, as shown in program 4.

**Program 4.** Parallel loop using Intel's Task-Queuing.

```
1    #pragma intel omp parallel taskq
2    {
3      for (it = list1.begin(); it != list1.end(); it++) {
4    #pragma intel omp task
5        {
6          it->compute();
7        }
8      } // end for
9    } // end omp parallel
```

The third approach places the loop into a parallel region, the loop body into a single-worksharing construct and the implicit barrier is omitted by specifying the *nowait* clause. This is shown in program 5.

The fourth approach is shown in program 6 and it has been proposed to us to be included in this experiment as well. The value of the incremented iterator is assigned in a critical region to a private iterator and the private iterator is then used to call the compute function.

To evaluate the performance of the different approaches and to decide which one to use in the parallelization of the application codes, we varied two parameters: DIM specifies the number of list entries and thereby the number of loop iterations. ITS specifies the amount of work to be done inside the *compute()* function and thereby the amount of work in the loop body. In figure 1 some results are shown for variations of DIM and ITS. The Intel compiler supports all four approaches and the performance is all the same for values of DIM and ITS

**Program 5.** Single-Nowait loop parallelization.

```
1  #pragma omp parallel private(it)
2  {
3    for (it = list1.begin(); it != list1.end(); it++) {
4  #pragma omp single nowait
5    {
6      it->compute();
7    }
8    } // end for
9  } // end omp parallel
```

**Program 6.** Critical region loop parallelization.

```
1  list<CComputeItem>::iterator priv_it;
2  #pragma omp parallel private(priv_it)
3  {
4    for (long l = 0; l < list1.size(); l++) {
5  #pragma omp critical
6    {
7      priv_it = it++;
8    }
9      priv_it->compute();
10   } // end for
11 } // end omp parallel
```

reflecting typical scenarios of our application. The Sun compiler only supports approaches one, three and four. For varying values of DIM and ITS the creation of a parallelizable loop clearly outperforms the other techniques as shown in the right part of figure 1. Therefore we preferred approach one in the parallel version of DROPS.

It should be mentioned that while approach one delivers the best performance in most cases, it also requires the most intrusive changes to the code. The Task-Queuing worksharing construct would provide a more elegant parallelization and could deliver the same performance as approach one, if it would provide a way to control the scheduling. Approaches two and four are expensive because of their synchronization overhead if the amount of work in the loop body is small and the number of loop iterations is high.

### 2.3 Parallelization of High-Level Codes

In C++ programs making extensive use of object-oriented programming, a large amount of computing time may be spent in member functions of variables of class-type. An example is shown in program 7 where a part of the PCG-type linear equation solver of DROPS is shown. The data types *Vec* and *Mat* represent vector or matrix implementations, respectively, and hide the implementation

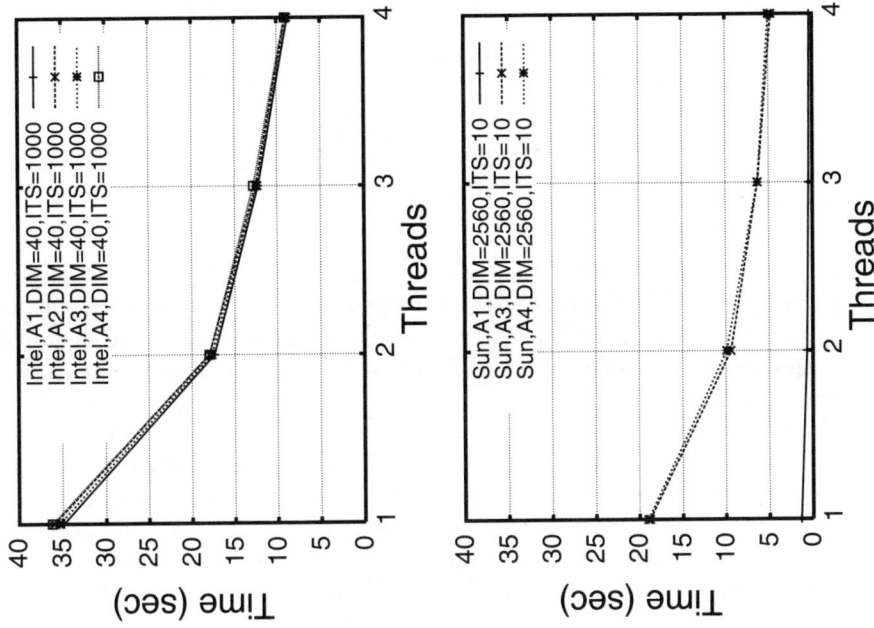

**Fig. 1.** Performance of parallelized iterator loops

from the user by providing an abstract interface. For example the expression in line 5 stores the result of the (sparse) matrix-vector-multiplication of matrix $A$ with vector $p$ in vector $q$.

**Program 7.** Single-Nowait loop parallelization.

```
1   PCG(const Mat& A, Vec& x,const Vec& b, ...) {
2     Vec p(n), z(n), q(n), r(n); Mat A(n, n);
3     [...]
4     for (int i = 1; i <= max_iter; ++i) {
5       q = A * p;
6       double alpha = rho / (p * q);
7       x += alpha * p;
8       r -= alpha * q;
9     [...]
```

To parallelize such codes the programmer has in general two choices:

1. *Internal parallelization*: a complete parallel region is embedded in the member function (e.g. *operator∗*). The advantage is that the parallelization is completely hidden by the interface and the interface does not need to be changed, though additional functionality to control the parallelization could

**Table 2.** Execution time (secs) of different parallelization levels

| Compiler | Parallelization | 1 | 2 | 4 | 8 |
|---|---|---|---|---|---|
| Intel | Internal | 219 | 108 | 65.5 | — |
| Intel | External | 216 | 107 | 65.0 | — |
| Sun | Internal | 375 | 185 | 112 | 42 |
| Sun | External | 320 | 161 | 97 | 37 |

be provided. The disadvantage results from the embedding of the complete parallel region. At the beginning of a parallel region a team of threads has to be created, which at the end of the parallel region has to be dissolved, both counting for some overhead. There is no chance to enlarge the parallel region or to reduce barriers.

2. *External parallelization*: the parallel region starts and ends outside of member functions (in the PCG code example it could span the whole loop body) and inside member functions orphaned worksharing constructs are used. This can reduce the overhead of thread creation and termination. The disadvantage is that the interface is changed implicitly, because the parallelized member functions may only be called by a serial program part or out of a parallel region, but not out of another worksharing construct. In many cases the compiler is unable to detect this calling situation and there is no way to find out (e.g. by an OpenMP runtime function) whether the member function is called in a worksharing construct.

We do not consider the absolute performance but only the scalability. In table 2 the runtime in seconds for the two approaches is shown. While the performance of the Intel compiler is all the same, it differs for the Sun compiler.

For performance reasons and from a library writer's point of view (e.g. C++ template) the external parallelization is in favour. The ability to check for calling situations out of a worksharing construct would broaden the applicability of that programming style.

## 3 ccNUMA-Architectures

C++ programs using STL data types like *std::valarray* to represent arrays for numerical computation can profit from additional functionality. The development might get simplified compared to native arrays on the one hand, but on the other hand it might leads to problems when used on ccNUMA-architectures. This is discussed in subsection 3.1. Subsection 3.2 illustrates how to affect memory locality for general data types.

### 3.1 STL Data Types

After instantiation all elements of a variable of type *std::valarray* are guaranteed to be initialized with zero. This leads to a disadvantage on ccNUMA-architectures, because the initialization with zero touches the data and thereby leads to a physical data (page) placement by the operating system. On a ccNUMA-architecture the data is placed in the memory of that CPU on which the current thread is running. If in the parallel part of the program all threads need to access one instance of the array, e.g. in a matrix-vector-multiplication, the speedup is limited.

A typical solution to such a problem is a parallel initialization with the same memory access pattern as in the computation, if the operating system supports the *fist-touch* memory placement policy. The problem with *std::valarray* is that the initialization is done inside the data type and a later parallel initialization would not lead to page migration. The same is true for *std::vector*. We considered several approaches to utilize ccNUMA-architectures with these data types.

A modification of *std::valarray* or *std::vector* is possible so that the initialization is done in parallel with a given memory access pattern. After memory allocation this memory is filled with zero in a loop which is parallelized using OpenMP. The problem of this approach is that it is limited to a given compiler, as typically every compiler provides its own STL implementation. Therefore this approach is not portable. We implemented it using the Intel C++ 9.0 compiler and it is shown in figure 2 as version *myvalarray*.

If *std::vector* is used instead of *std::valarray* a custom allocator can be specified. We implemented an allocator that uses *malloc()* and *free()* for memory allocation and initializes the memory with zero in a loop parallelized with OpenMP, whose schedule and chunksize parameters are specified as template parameters and therefore known at compile time. For the translation from *std::valarray* to *std::vector* one has to be careful to not experience a serial performance drop down, nevertheless this approach is portable. From a C++ programmer's perspective we found this approach the most elegant one. Its performance is shown in figure 2 as version *vector*.

The third approach is to use page migration functionality provided by the Solaris operating system. The *madvise()* function gives an advice to the virtual memory system for a given memory range. Specifying *MADV_ACCESS_LWP* advises to physically migrate the memory pages to the memory of that CPU which is accessing next. Again this approach is not portable when the Solaris operating system is not available. Its performance is shown in figure 2 as version *madvise*. We proposed a "next-touch" enhancement to the OpenMP specification that would provide functionality similar to *madvise()* independent from the programming language, compiler and operating system.

In figure 2 the speedup of the sparse matrix-vector-multiplication in DROPS for the original implementation based on *std::valarray* and the implementation using *std::vector* with a custom distributed allocator is shown.

## 3.2 Other Data Types

In many cases native pointers are used to dynamically manage data. We investigated a program, which uses pointers to dynamically address its main data structure, a two-dimensional grid. The grid is initialized in the serial program part and therefore the scalability on a ccNUMA-architecture is limited, because

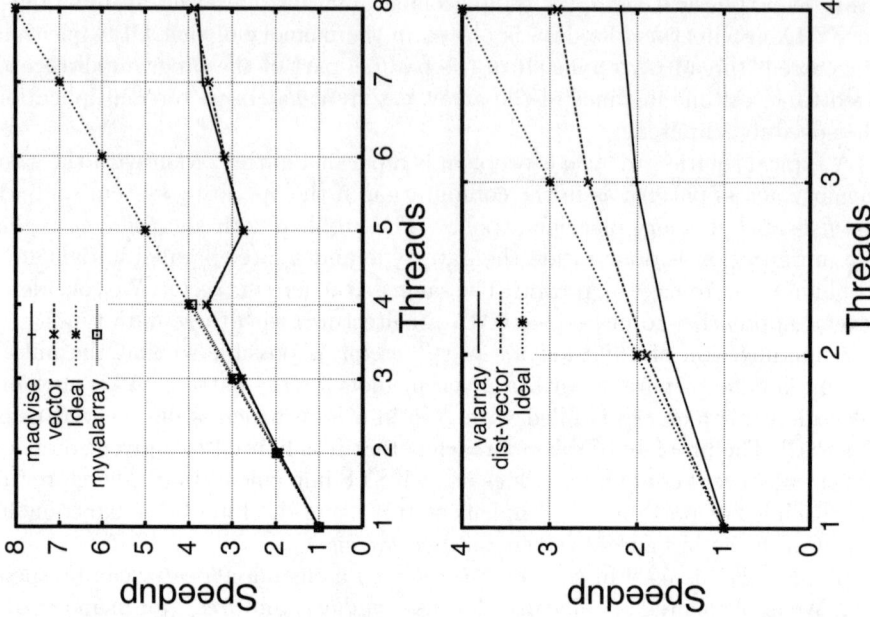

**Fig. 2.** Speedup of C++ STREAM-benchmark on SF V40z ccNUMA-architecture and speedup of DROPS sparse matrix-vector-multiplication

all threads have to access the master thread's memory where the data is allocated. Again it is possible to use the *madvise()* functionality to improve the situation on the cost of portability. To provide a portable solution based on C++ language features and OpenMP it is possible to initialize the pointer array using a parallel loop. Both approaches deliver the same performance.

To control data placement of general classes in C++ a mixin [10] can be used. Thereby it is possible to overwrite the *new* and *delete* operator for a given class without modifying its code. By using additional template parameters the same flexibility as with a custom allocator can be reached.

## 4 Conclusion

In this work we presented our experiences in parallelizing C++ applications with OpenMP. Though the current OpenMP specification includes C and C++, it does not address the special needs of a C++ programmer. The usage of OpenMP in C++ programs is not yet very widespread, nevertheless we successfully parallelized the two software packages DROPS and FIRE.

The primary problems we encountered were caused by incomplete or nonconforming implementations of the C++ compilers and the resulting errors may be hard to find for novice users. Missing features of the specification could be overcome in most cases using the presented approaches. Task-Queuing or additional runtime functions to inquire whether a code is called within a worksharing construct would make the C++ programmer's life easier.

In order to achieve high scalability on an ccNUMA-architecture data locality has to be considered. Since most operating systems except Solaris do not yet provide page migration functionality, the user has to manually optimize data distribution. The C++ programming language can address these issues elegantly and with minimal programming efforts.

## References

1. Terboven, C.: Shared-Memory Parallelisierung von C++ Programmen. Diploma thesis, RWTH Aachen University, Aachen, Germany (2006)
2. Groß, S., Peters, J., Reichelt, V., Reusken, A.: The DROPS package for numerical simulations of incompressible flows using parallel adaptive multigrid techniques, ftp://ftp.igpm.rwth-aachen.de/pub/reports/pdf/IGPM211_N.pdf
3. Reusken, A., Reichelt, V.: Multigrid Methods for the Numerical Simulation of Reactive Multiphase Fluid Flow Models (DROPS), http://www.sfb540.rwth-aachen.de/Projects/tpb4.php
4. Deselaers, T., Keysers, D., Ney, H.: Features for image retrieval – a quantitative comparison. In: Rasmussen, C.E., Bülthoff, H.H., Schölkopf, B., Giese, M.A. (eds.) DAGM 2004. LNCS, vol. 3175, pp. 228–236. Springer, Heidelberg (2004)
5. Clough, P., Müller, H., Sanderson, M.: The CLEF Cross Language Image Retrieval Track (ImageCLEF) 2004. In: Peters, C., Clough, P., Gonzalo, J., Jones, G.J.F., Kluck, M., Magnini, B. (eds.) CLEF 2004. LNCS, vol. 3491, pp. 597–613. Springer, Heidelberg (2005)
6. Clough, P., Müller, H., Deselaers, T., Grubinger, M., Lehmann, T., Jensen, J., Hersh, W.: The clef 2005 cross-language image retrieval track. In: Peters, C., Gey, F.C., Gonzalo, J., Müller, H., Jones, G.J.F., Kluck, M., Magnini, B., de Rijke, M., Giampiccolo, D. (eds.) CLEF 2005. LNCS, vol. 4022, pp. 535–557. Springer, Heidelberg (2006)
7. OpenMP Architecture Review Board: OpenMP Application Program Interface, v2.5 (2005)
8. Stroustup, B.: The C++ Programming Language. Addison-Wesley, Reading (2000)
9. Shah, S., Haab, G., Petersen, P., Throop, J.: Flexible Control Structures for Parallelism in OpenMP (1999)
10. Bracha, G., Cook, W.: Mixin-based inheritance. In: Proceedings of the Conference on Object-Oriented Programming: Systems, Languages and Applications (OOPSLA)
11. Terboven, C., Spiegel, A., an Mey, D., Gross, S., Reichelt, V.: Experiences with the OpenMP Parallelization of DROPS, a Navier-Stokes Solver written in C++. In: IWOMP 2005: Proceedings of the First International Workshop on OpenMP (2005)
12. Terboven, C., Spiegel, A., an Mey, D., Gross, S., Reichelt, V.: Parallelization of the C++ Navier-Stokes Solver DROPS with OpenMP. In: PARCO 2005: Proceedings of Parallel Computing (2005)
13. Terboven, C., Deselaers, T., Bischof, C., Ney, H.: Shared-Memory Parallelization for Content-based Image Retrieval. In: ECCV 2006: Workshop on Computation Intensive Methods for Computer Vision (to appear, 2006)

# Common Mistakes in OpenMP and How to Avoid Them
## A Collection of Best Practices

Michael Süß and Claudia Leopold

University of Kassel, Research Group Programming Languages / Methodologies,
Wilhelmshöher Allee 73, D-34121 Kassel, Germany
{msuess, leopold}@uni-kassel.de

**Abstract.** Few data are available on common mistakes made when using OpenMP. This paper presents a study on the programming errors observed in our courses on parallel programming during the last two years, along with numbers on which compilers and tools were able to spot them. The mistakes are explained and best practices for programmers are suggested to avoid them in the future. The best practices are presented in the form of an OpenMP checklist for novice programmers.

## 1 Introduction

One of the main design goals of OpenMP was to make parallel programming easier. Yet, there are still fallacies and pitfalls to be observed when novice programmers are trying to use the system. We have therefore conducted a study on a total of 85 students visiting our lecture on parallel programming, and observed the mistakes they made when asked to prepare assignments in OpenMP. The study is described in detail in Sect. 2.

We are concentrating on the most common mistakes from our study for the rest of this paper. They are briefly introduced in Tab. 1, along with a count of how many teams (consisting of two students each) have made the mistake each year. We have chosen to divide the programming mistakes into two categories:

1. *Correctness Mistakes:* all errors impacting the correctness of the program.
2. *Performance Mistakes:* all errors impacting the speed of the program. These lead to slower programs, but do not produce incorrect results.

Sect. 3 explains the mistakes in more detail. Also in this section, we propose possible ways and best practices for novice programmers to avoid these errors in the future. Sect. 4 reports on tests that we conducted on a variety of OpenMP-compilers to figure out, if any of the programming mistakes are spotted and/or possibly corrected by any of the available compilers. In Sect. 5, all suggestions made this far are condensed into an OpenMP programming checklist, along with others from our own programming experiences. Sect. 6 reviews related work, while Sect. 7 sums up our results.

**Table 1.** The list of frequently made mistakes when programming in OpenMP

| No. | Problem | 2004 | 2005 | Sum |
|---|---|---|---|---|
| | *Correctness Mistakes* | | | |
| 1. | Access to shared variables not protected | 8 | 10 | 18 |
| 2. | Use of locks without `flush` | 7 | 11 | 18 |
| 3. | Read of shared variable without `flush` | 5 | 10 | 15 |
| 4. | Forget to mark `private` variables as such | 6 | 5 | 11 |
| 5. | Use of `ordered` clause without `ordered` construct | 2 | 2 | 4 |
| 6. | Declare loop variable in `#pragma omp parallel for` as `shared` | 1 | 2 | 3 |
| 7. | Forget to put down `for` in `#pragma omp parallel for` | 2 | 0 | 2 |
| 8. | Try to change num. of thr. in parallel reg. after start of reg. | 0 | 2 | 2 |
| 9. | `omp_unset_lock()` called from non-owner thread | 2 | 0 | 2 |
| 10. | Attempt to change loop variable while in `#pragma omp for` | 0 | 2 | 2 |
| | *Performance Mistakes* | | | |
| 11. | Use of `critical` when `atomic` would be sufficient | 8 | 1 | 9 |
| 12. | Put too much work inside `critical` region | 2 | 4 | 6 |
| 13. | Use of orphaned construct outside parallel region | 2 | 2 | 4 |
| 14. | Use of unnecessary `flush` | 3 | 1 | 4 |
| 15. | Use of unnecessary `critical` | 2 | 0 | 2 |
| | Total Number of Groups | 26 | 17 | 43 |

## 2 Survey Methodology

We have evaluated two courses for this study. Both consisted of students on an undergraduate level. The first course took place in the winter term of 2004 / 2005, while the second one took place in the winter term of 2005 / 2006. The first course had 51 participants (26 groups of mostly two students), the second one had 33 participants (17 groups). The lecture consisted of an introduction to parallel computing and parallel algorithms in general, followed by a short introduction of about five hours on OpenMP. Afterwards, the students were asked to prepare programming assignments in teams of two people, which had to be defended before the authors. During these sessions (and afterwards in preparation for this paper), we analyzed the assignments for mistakes, and the ones having to do with OpenMP are presented in this paper.

The assignments consisted of small to medium-sized programs, among them:

– find the first $N$ prime numbers
– simulate the dining philosophers problem using multiple threads
– count the number of connected components in a graph
– write test cases for OpenMP directives / clauses / functions

A total of 231 student programs in C or C++ using OpenMP were taken into account and tested on a variety of compilers (e. g. from SUN, Intel, Portland

Group, IBM, as well as on the free OMPi compiler). Before we begin to evaluate the results, we want to add a word of warning: Of course, the programming errors presented here have a direct connection to the way we taught the lecture. Topics we talked about in detail will have led to fewer mistakes, while for other topics, the students had to rely on the specification. Moreover, mistakes that have been corrected by the students before submitting their solution are not taken into account here. For these reasons, please take the numbers presented in Tab. 1 as what they are - mere indications of programming errors that novice programmer might make.

## 3 Common Mistakes in OpenMP and Best Practices to Avoid Them

In this section, we will discuss the most frequently made mistakes observed during our study, as well as suggest possible solutions to make them occur less likely. There is one universal remark for instructors that we want to discuss beforehand: We based our lecture on assignments and personal feedback, and found this approach to be quite effective: As soon as we pointed out a mistake in the students programs during the exam, a group would rarely repeat it again. Only showing example programs in the lecture and pointing out possible problems did not have the same effect.

There are some mistakes, where we cannot think of any best practises to avoid the error. Therefore, we will just shortly sketch these at this point, while all other mistakes are discussed in their own section below (the number before the mistake is the same as in Tab. 1):

2. *Use of locks without flush*: Before version 2.5 of the OpenMP specification, lock operations did not include a `flush`. The compilers used by our students were not OpenMP 2.5 compliant, and therefore we had to mark a missing `flush` directive as a programming error.
5. *Use of ordered clause without ordered construct*: The mistake here is to put an `ordered` clause into a `for` worksharing construct, without specifying with a separate `ordered` clause inside the enclosed `for` loop, what is supposed to be carried out in order.
8. *Try to change number of threads in parallel region after start of region*: The number of threads carrying out a parallel region can only be changed before the start of the region. It is therefore a mistake to attempt to change this number from inside the region.
10. *Attempt to change loop variable while in #pragma omp for*: It is explicitly forbidden in the specification to change the loop variable from inside the loop.
11. *Use of critical when atomic would be sufficient*: There are special cases when synchronisation can be achieved with a simple `atomic` construct. Not using it in this case leads to potentially slower programs and is therefore a performance mistake.

13. *Use of orphaned construct outside parallel region*: When using a combined worksharing construct, sometimes our students would forget to put down the `parallel`, producing an orphaned construct. In other cases, the parallel region was forgotten altogether, leading e. g. to orphaned `critical` constructs.
14. *Use of unnecessary flush*: `flush` constructs are implicitly included in certain positions of the code by the compiler. Explicitly specifying a `flush` immediately before or after these positions is considered a performance mistake.
15. *Use of unnecessary critical*: The mistake here is to protect memory accesses with a `critical` construct, although they need no protection (e.g. on private variables or on other occasions, where only one thread is guaranteed to access the location).

## 3.1 Access to Shared Variables Not Protected

The most frequently made and most severe mistake during our study was to not avoid concurrent access to the same memory location. OpenMP provides several constructs for protecting critical regions, such as the `critical` construct, the `atomic` construct and locks. Although all three of these constructs were introduced during the lecture, many groups did not use them at all, or forgot to use them on occasions. When asked about it, most of them could explain what a critical region was for and how to use the constructs, yet to spot these regions in the code appears to be difficult for novice parallel programmers.

A way to make novice programmers aware of the issue is to use the available tools to diagnose OpenMP programs. For example, both the Intel Thread Checker and the Assure tool find concurrent accesses to a memory location.

## 3.2 Read of Shared Variable without Flush

The OpenMP memory model is a complicated beast. Whole sections in the OpenMP specification have been dedicated to it, as well as a whole paper written about it [1]. One of its complications is the error described here. Simply put, when reading a shared variable without flushing it first, it is not guaranteed to be up to date. Actually, the problem is even more complicated, as not only the reading thread has to flush the variable, but also any thread writing to it beforehand. Many students did not realize this and just read shared variables without any further consideration. On many common architectures this will not be a problem, because flushes are carried out frequently there. Quite often, the problem does not surface in real-world programs, because there are implicit flushes contained in many OpenMP constructs, as well.

In other cases, students simply avoided the problem by declaring shared variables as `volatile`, which puts an implicit `flush` before every read and after every write of any such variable. Of course, it also disables many compiler optimizations for this variable and therefore is often the inferior solution.

The proper solution, of course, is to make every OpenMP programmer aware of this problem, by clearly stating that every read to a shared variable must be preceded by a `flush`, except in very rare edge-cases not discussed here. This `flush` can be explicitly written down by the programmer, or it can be implicit in an OpenMP construct.

Version 2.5 of the OpenMP specification includes a new paragraph on the memory model. Whether or not this is enough to make novice programmers aware of this pitfall remains to be seen.

### 3.3 Forget to Mark Private Variables as Such

This programming error has come up surprisingly often in our study. It was simply forgotten to declare certain variables as private, although they were used in this way. The default sharing attribute rules will make the variable shared in this case.

Our first advice to C and C++ programmers to avoid this error in the future is to use the scoping rules of the language itself. C and C++ both allow variables to be declared inside a parallel region. These variables will be private (except in rare edge cases described in the specification, e.g. static variables), and it is therefore not necessary to explicitly mark them as such, avoiding the mistake altogether.

Our second advice to novice programmers is to use the `default(none)` clause. It will force each variable to be explicitly declared in a data-sharing attribute clause, or else the compiler will complain. We will not go as far as to suggest to make this the default behaviour, because it certainly saves the experienced programmer some time to not have to put down each and every shared variable in a shared clause. But on the other hand, it would certainly help novice programmers who probably do not even know about the `default` clause.

It might also help if the OpenMP compilers provided a switch for showing the data-sharing attributes for each variable at the beginning of the parallel region. This would enable programmers to check if all their variables are marked as intended. An external tool for checking OpenMP programs would be sufficient for this purpose as well.

Another solution to the problem is the use of autoscoping as proposed by Lin et al. [2]. According to this proposal, all data-sharing attributes are determined automatically, and therefore the compiler would correctly privatize the variables in question. The proposed functionality is available in the Sun Compiler 9 and newer.

Last but not least, the already mentioned tools can detect concurrent accesses to a shared variable. Since the wrongly declared variables fall into this category, these tools should throw a warning and alert the programmer that something is wrong.

### 3.4 Declare Loop Variable in #pragma omp Parallel for as Shared

This mistake shows a clear misunderstanding of the way the `for` worksharing construct works. The OpenMP specification states clearly that these variables are implicitly converted to private, and all the compilers we tested this on performed the conversion. The surprising fact here is that many compilers did the conversion silently, ignoring the shared declaration and not even throwing a warning. More warning messages from the compilers would certainly help here.

## 3.5 Forget to Put Down for in #pragma omp Parallel for

The mistake here is, to attempt to use the combined worksharing construct `#pragma omp parallel for`, but forget to put down the `for` in there. This will lead to every thread executing the whole loop, and not only parts of it as intended by the programmer.

In most cases, this mistake will lead to the mistake specified in Sect. 3.1, and therefore can be detected and avoided by using the tools specified there.

One way to avoid the mistake altogether is to specify the desired `schedule` clause, when using the `for` worksharing construct. This is a good idea for portability anyways, as the default `schedule` clause is implementation defined. It will also lead to the compiler detecting the mistake we have outlined here, as `#pragma omp parallel schedule(static)` is not allowed by the specification and yields compiler errors.

## 3.6 omp_unset_lock() Called from Non-owner Thread

The OpenMP-specification clearly states:

> The thread which sets the lock is then said to own the lock. A thread which owns a lock may unset that lock, returning it to the unlocked state. A thread may not set or unset a lock which is owned by another thread.
> ([3, p. 102])

Some of our students still made the mistake to try to unlock a lock from a non-owner thread. This will even work on most of the compilers we tested, but might lead to unspecified behaviour in the future.

To avoid this mistake, we have proposed to our students to use locks only when absolutely necessary. There are cases when they are needed (for example to lock parts of a variable-sized array), but most of the times, the `critical` construct provided by OpenMP will be sufficient and easier to use.

## 3.7 Put Too Much Work Inside Critical Region

This programming error is probably due to the lack of sensitivity for the cost of a critical region found in many novice programmers. The issue can be split into two subissues:

1. Put more code inside a critical region than necessary, thereby potentially blocking other threads longer than needed.
2. Go through the critical region more often than necessary, thereby paying the maintenance costs associated with such a region more often than needed.

The solution to the first case is obvious: The programmer needs to check if each and every line of code that is inside a critical region really needs to be there. Complicated function calls, for example, have no business being in there most of the time, and should be calculated beforehand if possible.

As an example for the second case, consider the following piece of code, which some of our students used to find the maximum value in an array:

```
#pragma omp parallel for
for (i = 0; i < N; ++i) {
  #pragma omp critical
  {
     if (arr[i] > max) max = arr[i];
  }
}
```

The critical region is clearly in the critical path in this version, and the cost for it therefore has to be paid $N$ times. Now consider this slightly improved version:

```
#pragma omp parallel for
for (i = 0; i < N; ++i) {
  #pragma omp flush (max)
  if (arr[i] > max) {
    #pragma omp critical
    {
       if (arr[i] > max) max = arr[i];
    }
  }
}
```

This version will be faster (at least on architectures, where the `flush` operation is significantly faster than a critical region), because the critical region is entered less often. Finally, consider this version:

```
#pragma omp parallel
{
  int priv_max;
  #pragma omp for
  for (i = 0; i < N; ++i) {
     if (arr[i] > priv_max)  priv_max = arr[i];
  }
  #pragma omp flush (max)
  if (priv_max > max) {
    #pragma omp critical
    {
       if (priv_max > max) max = priv_max;
    }
  }
}
```

This is essentially a reimplementation of a reduction using the `max` operator. We have to resort to reimplementing this reduction from scratch here, because reductions using the `max` operator are only defined in the Fortran version of OpenMP (which in itself is a fact that many of our students reported to have caused confusion). Nevertheless, it is possible to write programs this way, and by showing novice programmers techniques like the ones sketched above, they get more aware of performance issues.

## 4  Compilers and Tools

There are a multitude of different compilers for OpenMP available, and we wanted to know, if any of them were able to detect the programming errors

**Table 2.** How Compilers deal with the Problems

| No. | File | icc | pgcc | sun | guide | xlc | ompi | assure | itc |
|---|---|---|---|---|---|---|---|---|---|
| | *Correctness Mistakes* | | | | | | | | |
| 1. | access_shared | - | - | - | - | - | - | eE | eE |
| 2. | locks_flush | - | - | - | - | - | - | - | - |
| 3. | read_shared_var | - | - | - | - | - | - | - | (eE) |
| 4. | forget_private (=access_shared) | - | - | - | - | - | - | eE | eE |
| 5. | ordered_without_ordered | - | - | - | - | - | - | - | eW |
| 6. | shared_loop_var | cE | cC | cW+C | cE | cC | cC | cE | cE |
| 7. | forget_for | - | - | - | - | - | - | (eE) | (eE) |
| 8. | change_num_threads | - | - | - | - | - | - | - | - |
| 9. | unset_lock_diff_thread | - | - | - | - | - | - | - | - |
| 10. | change_loop_var | - | - | - | - | cW | - | - | - |
| | *Performance Mistakes* | | | | | | | | |
| 11. | crit_when_atomic | - | - | - | - | - | - | - | - |
| 12. | too_much_crit (no test!) | | | | | | | | |
| 13. | orphaned_const | - | - | rW | - | - | - | - | - |
| 14. | unnec_flush | - | - | - | - | - | - | - | - |
| 15. | unnec_crit | - | - | - | - | - | - | - | - |

sketched in Sect. 3. Therefore we have written a short testcase for each of the programming mistakes. Tab. 2 describes the results of our tests on different compilers.

The numbers in the first column are the same as in Tab. 1. The second column contains the names of our test programs. We could not think of a sound test for problem 12 (put too much work inside critical region), and therefore the results for this problem are omitted. Test program four is the same as test program one, and therefore the results are the same as well. The rest of the table depicts results for the following compilers (this list is not sorted by importance, nor in any way representative, but merely includes all the OpenMP-compilers we had access to):

- Intel Compiler 9.0 (icc)
- Portland Group Compiler 6.0 (pgcc)
- Sun Compiler 5.7 (sun)
- Guide component of the KAP/Pro Toolset C/C++ 4.0 (guide)
- IBM XL C/C++ Enterprise Edition 7.0 (xlc)
- OMPi Compiler 0.8.2 (ompi)
- Assure component of the KAP/Pro Toolset C/C++ 4.0 (assure)
- Intel Thread Checker 2.2 (itc)

The last two entries (assure and itc) are not compilers, but tools to help the programmer find mistakes in their OpenMP programs. As far as we know, Assure was superseded by the Intel Thread Checker and is no longer available, nevertheless it is still installed in many computing centers. We were not able to find any lint-like tools to check OpenMP programs in C, there are however solutions for Fortran available commercially.

The alphabetic codes used in the table are to be read as follows: the first (uncapitalized) letter is one of (c)ompiletime, (r)untime or (e)valuation time, and describes, when the mistake was spotted by the compiler. Only Assure and the Intel Thread Checker have an evaluation step after the actual program run. The second (capitalized) letter describes, what kind of reaction was generated by the compiler, and is one of the following: (W)arning, (E)rror or (C)onversion. Conversion in this context means that the mistake was fixed by the compiler without generating a warning. Conversion was done for problem six, where the compilers privatized the shared loop variable. W+C means, that the compiler generated a warning, but also fixed the problem at the same time. There is one last convention to describe in the alphabetic codes: When there are braces around the code, it means that a related problem was found by the program, which could be traced back to the actual mistake. An example: When the programmer forgets to put down `for` in a parallel worksharing construct (problem seven), it will lead to a data race. This race is detected by the Intel Thread Checker, and therefore the problem becomes obvious. All tests were performed with all warnings turned to the highest level for all compilers.

It is obvious from these numbers that most of the compilers observed are no big help in avoiding the problems described in this paper. Tools such as the Intel Thread Checker are more successful, but still it is most important that programmers avoid the mistakes in the first place. This paper and the programmers checklist presented in the next section are a step in this direction.

## 5  OpenMP Programmers Checklist

In this section, we summarize the advice given to novice programmers of OpenMP this far and rephrase it to fit into the easy to use format of a checklist. For this reason, the form of address is changed and the novice programmer is addressed directly. The checklist also contains other items, which we have accumulated during our own use of and experiences with OpenMP.

### General

- It is tempting to use fine grained parallelism with OpenMP (throwing in an occasional `#pragma omp parallel for` before loops). Unfortunately, this rarely leads to big performance gains, because of overhead such as thread creation and scheduling. You therefore have to search for potential for coarser-grained parallelism.
- Related to the point above, when you have nested loops, try to parallelize only the outer loop. Loop reordering techniques can sometimes help here.
- Use reduction where applicable. If the operation you need is not predefined, implement it yourself as shown in Sect. 3.7.

- Beware of nested parallelism, as many compilers still do not support it, and even if it is supported, nested parallelism may not give you any speed increases.
- When doing I/O (either to the screen or to a file), large time savings are possible by writing the information to a buffer first (this can sometimes even be done in parallel) and then pushing it to the device in one run.
- Test your programs with multiple compilers and all warnings turned on, because different compilers will find different mistakes.
- Use tools such as the Intel Thread Checker or Assure, which help you to detect programming errors and write better performing programs.

**Parallel Regions**

- If you want to specify the number of threads to carry out a parallel region, you must invoke `omp_set_num_threads()` *before* the start of that region (or use other means to specify the number of threads *before* entering the region).
- If you rely on the number of threads in a parallel region (e.g. for manual work distribution), make sure you actually get this number (by checking `omp_get_num_threads()` after entering the region). Sometimes, the runtime system will give you less threads, even when the dynamic adjustment of threads is off!
- Try to get rid of the `private` clause, and declare private variables at the beginning of the parallel region instead. Among other reasons, this makes your data-sharing attribute clauses more manageable.
- Use `default(none)`, because it makes you think about your data-sharing attribute clauses for all variables and avoids some errors.

**Worksharing Constructs**

- For each loop you parallelize, check whether or not every iteration of the loop has to do the same amount of work. If this is not the case, the static work schedule (which is often the default in compilers) might hurt your performance and you should consider dynamic or guided scheduling.
- Whatever kind of schedule you choose, explicitly specify it in the worksharing construct, as the default is implementation-defined!
- If you use `ordered`, remember that you always have to use both the `ordered` clause and the `ordered` construct.

**Synchronisation**

- If more than one thread accesses a variable and one of the accesses is a write, you must use synchronization, even if it is just a simple operation like $i = 1$. There are no guarantees by OpenMP on the results otherwise!
- Use `atomic` instead of `critical` if possible, because the compiler might be able to optimize out the `atomic`, while it can rarely do that for `critical`.
- Try to put as little code inside critical regions as possible. Complicated function calls, for example, can often be carried out beforehand.

- Try to avoid the costs associated with repeatedly calling critical regions, for instance by checking for a condition before entering the critical region.
- Only use locks when necessary and resort to the `critical` clause in all other cases. If you have to use locks, make sure to invoke `omp_set_lock()` and `omp_unset_lock()` from the same thread.
- Avoid nesting of critical regions, and if needed, beware of deadlocks.
- A critical region is usually the most expensive synchronisation construct (and takes about twice as much time to carry out as e. g. a barrier on many architectures), therefore start optimizing your programs accordingly — but keep in mind that these numbers only account for the time needed to actually perform the synchronisation, and not the time a thread has to wait on a barrier or before a critical region (which of course depends on various factors, among them the structure of your program or the scheduler).

**Memory Model**

- Beware of the OpenMP memory model. Even if you only read a shared variable, you have to flush it beforehand, except in very rare edge cases described in the specification.
- Be sure to remember that locking operations do not imply an implicit flush before OpenMP 2.5.

## 6 Related Work

We are not aware of any other studies regarding frequently made mistakes in OpenMP. Of course, in textbooks [4] and presentations teaching OpenMP, some warnings for mistakes are included along with techniques to increase performance, but most of the time, these are more about general pitfalls regarding parallel programming (like e. g. warnings to avoid deadlocks). There is one interesting resource to mention though: the blog of Yuan Lin [5], where he has started to describe frequently made errors with OpenMP. Interestingly, at the time of this writing, he has not touched any errors that we have described as well, which leads us to think that there are many more sources of errors hidden inside the OpenMP specification and that our OpenMP checklist is by no means all-embracing and complete.

## 7 Summary

In this paper, we have presented a study on frequently made mistakes with OpenMP. Students visiting the authors' courses on parallel programming have been observed for two terms to find out, which were their most frequent sources of errors. We presented 15 mistakes and recommendations for best practices to avoid them in the future. These best practices have been put into a checklist for novice programmers, along with some practices from the authors' own experiences. It has also been shown, that the OpenMP-compilers available today are not able to protect the programmer from making these mistakes.

## Acknowledgments

We are grateful to Björn Knafla for proofreading the paper and for his insightful comments. We thank the University Computing Centers at the RWTH Aachen, TU Darmstadt and University of Kassel for providing the computing facilities used to test our sample applications on different compilers and hardware. Last but not least, we thank the students of our courses, without whom it would have been impossible for us to look at OpenMP from a beginners perspective.

## References

1. Hoeflinger, J.P., de Supinski, B.R.: The OpenMP memory model. In: Proceedings of the First International Workshop on OpenMP - IWOMP 2005 (June 2005)
2. Lin, Y., Copty, N., Terboven, C., an Mey, D.: Automatic scoping of variables in parallel regions of an OpenMP program. In: Chapman, B.M. (ed.) WOMPAT 2004. LNCS, vol. 3349, Springer, Heidelberg (2005)
3. OpenMP Architecture Review Board: OpenMP specifications (2005), http://www.openmp.org/specs
4. Chandra, R., Dagum, L., Kohr, D.: Parallel Programming in OpenMP. Morgan Kaufmann Publishers, San Francisco (2000)
5. Lin, Y.: Reducing the complexity of parallel programming (2005), http://blogs.sun.com/roller/page/yuanlin

# Formal Specification of the OpenMP Memory Model

Greg Bronevetsky[1] and Bronis R. de Supinski[2]

[1] Department of Computer Science,
Cornell University,
Ithaca, NY 14850, USA
greg@bronevetsky.com,
[2] Center for Applied Scientific Computing,
Lawrence Livermore National Laboratory,
Livermore, CA 94551, USA
bronis@llnl.gov

**Abstract.** OpenMP [1] is an important API for shared memory programming, combining shared memory's potential for performance with a simple programming interface. Unfortunately, OpenMP lacks a critical tool for demonstrating whether programs are correct: a formal memory model. Instead, the current official definition of the OpenMP memory model (the OpenMP 2.5 specification [1]) is in terms of informal prose. As a result, it is impossible to verify OpenMP applications formally since the prose does not provide a formal consistency model that precisely describes how reads and writes on different threads interact.

This paper focuses on the formal verification of OpenMP programs through a proposed formal memory model that is derived from the existing prose model [1]. Our formalization provides a two-step process to verify whether an observed OpenMP execution is conformant. In addition to this formalization, our contributions include a discussion of ambiguities in the current prose-based memory model description. Although our formal model may not capture the current informal memory model perfectly, in part due to these ambiguities, our model reflects our understanding of the informal model's intent. We conclude with several examples that may indicate areas of the OpenMP memory model that need further refinement however it is specified. Our goal is to motivate the OpenMP community to adopt those refinements eventually, ideally through a formal model, in later OpenMP specifications.

## 1 Introduction

Modern systems are being increasingly built using multi-threaded architectures. These include systems with multiple processors on the same node and/or multiple cores on the same chip. Given the proximity of the processors/cores on such machines, they typically feature a single memory accessible to any processor. As such, these machines are most easily and effectively programmed in a multi-threaded shared memory style.

OpenMP [1] has emerged as a popular shared memory API because it combines the performance advantages of shared memory with an easy-to-use API. However, despite the relative simplicity of the API, OpenMP applications remain difficult to write. The difficulty arises from several inherent complexities of multi-threaded execution, including non-determinism, a large space of possible executions and a very relaxed memory consistency model. Thus, although OpenMP allows programmers to improve application performance significantly, this comes at a cost of significantly higher program complexity. This complexity makes OpenMP programs much more vulnerable to bugs than sequential programs and, thus, more expensive to debug. Ultimately, confidence in the correctness of the final application is reduced.

Formal verification is a family of techniques where a program or protocol is formalized into a mathematically well-defined form. Correctness is verified using a variety of techniques that range in their complexity and their correctness guarantees, from model checking to theorem proving [9]. While formal verification is generally too complex to apply to real-world applications, it is feasible for the basic algorithms on which real applications are based.

Existing work on formally verifying shared memory algorithms [8] requires us to represent the entire computational content of the algorithm formally, including algorithm logic and the details of the underlying system. In particular the underlying memory model must be formalized. While some formal memory models exist [7] [3], none exists for OpenMP. Instead, the official description of OpenMP's memory model (section 1.4 of version 2.5 of the OpenMP specification [1]) is written in detailed English, which is generally clear but not nearly precise enough for formal verification tasks. Similarly, while the OpenMP memory model was recently clarified further [6], this clarification is also informal.

This paper focuses on verification of OpenMP programs through a proposed formal memory model that we derived from the existing prose model [1]. Our formalization provides a two-step process to verify if an observed OpenMP execution is conformant. In addition to this formalization, our contributions include a discussion of ambiguities in the current prose-based memory model description. Although our formal model may not capture the current informal memory model perfectly, in part due to these ambiguities, our model reflects our understanding of the informal model's intent. We present several examples that demonstrate a need for further refinement of the OpenMP memory model however it is specified. Our goal is to motivate the OpenMP community eventually to adopt those refinements, ideally through a formal model, in later OpenMP specifications.

This paper is divided as follows. Section 2 provides an overview of the OpenMP memory model. Section 3 discusses aspects of that model that we find ambiguous (despite one of the authors having significant input into it). Section 4 outlines the formalization of this model. Section 5 defines the language of the operations used in the formal model. Sections 6 and 7 provide the details of the two phases used by the formal specification. Finally, section 8 provides several example programs and their outcomes under the formal model specified in this paper.

## 2  OpenMP Memory Model

The OpenMP memory model provides for two types of memory: shared and threadprivate. There is a single shared memory that is visible to reads and writes on all threads. Furthermore, each thread has its own threadprivate memory that is accessible to only the reads and writes on that thread. OpenMP's shared memory semantics are akin to but a little weaker than weak ordering [4]. While each thread may read from and write to data in shared memory, there is no guarantee that one thread can immediately observe a write by another thread. Thus, the value associated with a given read may not reflect all prior writes from other threads. Instead, each thread conceptually has a *temporary view* of shared memory and a `flush` operation limits the reordering of operations and synchronizes a thread's temporary view with shared memory.

Simple, intuitive concepts motivate the OpenMP memory model. In order to ensure that a read by thread j returns the value of a write by thread i, the program must provide synchronization that guarantees the following sequence of events:

1. Thread i writes to the variable
2. Thread i flushes the variable
3. Thread j flushes the variable
4. Thread j reads the variable

and no other writes to the variable are happening at the same time. Any behavior outside the above sequence can produce undefined read results and/or leave the variable's value in shared memory undefined. However, the OpenMP memory model is very complex with many potential pitfalls in practice despite the simplicity of the underlying concepts, as we will discuss.

A thread's temporary view can be its cache, registers or other devices that speed up memory operations by not forcing the processor to go to main memory for every shared access. Reads and writes to shared variables access the thread's temporary view of shared memory. If the thread reads a shared variable and the temporary view doesn't hold a value for this variable, the read goes directly to shared memory. If a thread writes to a shared variable, it only updates the thread's temporary view of that variable. However, the system is then free to non-deterministically push the value of the write from a thread's temporary view to shared memory at any time. Since there are no atomicity constraints (e.g., a 64-bit write may not be executed as a single operation), if two writes executed on two threads are not ordered via synchronization, the value of the variable in shared memory may become garbage and is thus undefined (until it is overwritten by some later write). Similarly, if a write to a variable and a read from the same variable are executed on different threads and are not related via appropriate flushes and synchronization, the value read is undefined.

In addition to uncertainty about when shared reads and writes will actually access shared memory, OpenMP allows the compiler and the hardware to execute application operations out of order relative to their order in the original source code (called "program order"). In particular, implementations are allowed to reorder shared operations that access different shared memory variables. It is not specified whether it is legal to reorder operations that do have data dependence (ex: A=B and B=1), although it is possible to imagine aggressive compiler transformations that may do that.

OpenMP's `flush` operation is the application's primary means of limiting the asynchrony of memory and the degree of out-of-order execution. A given `flush` operation applies to a list of shared variables and has two major effects:

- it synchronizes the thread's temporary view with shared memory for the variables in the list;
- it prevents reordering of the thread's operations on variables in the list.

The first effect ensures that any preceding writes to the list variables by the thread have completed in the shared memory before the `flush` completes. It also ensures that the first read that follows the `flush` to each of the list variables must come directly from shared memory. The second effect ensures that shared memory operations that accesses a variable in the `flush`'s variable list are executed in program order relative to the `flush`. Furthermore, all `flush` operations with overlapping variable lists must be executed in program order.

A program's `flush` operations also restrict the interleaving of operations by different threads. All threads must observe any two `flush` operations with overlapping variable lists in some sequential order. Thus, we can organize non-flush operations on different threads into a partial temporal order that in turn determines which writes are visible to which reads.

OpenMP provides several synchronization operations in addition to reads, writes and flushes. These include `locks`, `barriers`, `critical` sections, `ordered` sections and `atomic` updates. All of these operations are preceded and/or followed by implied `flush` operations that apply either to all variables or just the variable involved in the operation.

## 3 Ambiguities in the OpenMP Memory Model

Despite the precise prose that defines the OpenMP memory model, we had several questions as we formulated our formal memory model based on it. Some of the questions indicate ambiguities that should be resolved in future specifications. Other questions arise from discrepancies between the prose and our understanding of the intent of the OpenMP language committee. We present several of these questions in this section.

### 3.1 Dependence-Breaking Compilers

The OpenMP memory model clearly defines reordering restrictions with respect to flush operations. However, reordering restrictions for non-`flush` operations are much less clear. For example, most sequential compilers reorder operations that access different variables; does the memory model allow these? The memory model is definitely intended to allow them but only supports them with this sentence: "The flush operation restricts reordering of memory operations that an implementation might otherwise do." We read this to mean that the memory model imposes no other reordering restrictions. This would mean that compilers may reorder operations that access the same shared variable. In particular, they can reorder not only reads but also writes. In general, the compiler can reorder any accesses not separated by a `flush`, including conflicting accesses to the same variable, provided that it preserves the application's sequential semantics.

For example, in the sample code shown in Figure 1 the application's sequential semantics would be preserved if the two writes to B were exchanged, since in a single-threaded execution the write $B = A$ is guaranteed to assign 5 to $B$. However, if this code

```
if(threadNum==0) {
    Barrier
    A=20;
    Barrier
} else {
    A=5;
    Barrier
    Barrier
    B=5;
    B=A;
    print B;
}
```

Fig. 1.

were to be executed by two threads, the write $B = A$ would assign B to 20, rather than 5. As such, reordering these two writes, while apparently legal in OpenMP, can produce unexpected results. Since there exist apparently legal dependence-breaking compiler optimizations that violate the spirit of the OpenMP memory model, the OpenMP specification should include a clear statement about the validity of different types of variable access reordering.

### 3.2 Intra-thread Dependencies

The OpenMP memory model clearly states that a `flush` does not complete until the values of all preceding writes have been completed in shared memory. However, it is not clear if the OpenMP memory model enforces program order, i.e., processor consistency [5].

In Section 2, we presented the events required for a read by thread j to return the value written by thread i. If thread i writes another value between steps 1 and 2, what value should be read in step 4? The question is related to the reordering questions in the preceding section, but it is also different. If the first value is captured in the temporary view but not the second for some reason (for example, the writes are executed out of order), is it legal not to propagate the captured value? The memory model prose states otherwise: "the `flush` does not complete until the value of the variable has been written to the variable in memory." Simply put, the memory model does not address multiple writes to the same shared variable by the same thread between two `flush` operations. Ultimately, the question is: does OpenMP guarantee that writes by a given thread must be seen in program order by other threads as long as the appropriate `flushes` have been issued (i.e. writes, `flush`, `flush`, read)?

We can also ask about the impact of reads by thread i: suppose that thread i reads the variable between steps 1 and 2 and that value is different from what was written by the write in step 1 due to a write by some other thread. This scenario includes a race condition and the specification is clear that the variable's value becomes undefined. However, completing the write would now be inconsistent with program order. Does the race imply that the `flush` should not see the write from step 1 and the read in step 4 will get some other value? The specification provides little detail on how local state evolves so the issue is unclear.

### 3.3 Effect of Privatization

The memory model section, section 1.4, of the 2.5 specification [1] states that OpenMP has two types of memory: shared and threadprivate. The bulk of the section defines the semantics of the shared memory. It provides few details of the second type, which corresponds to threadprivate variables and to variables included in private clauses. The only issue discussed is the interaction with nested parallelism.

The memory model does not address any interactions between the two types. In particular, it does not discuss the impact on shared variables that are included in private clauses. However, section 2.8.3.3, which discusses the private clause, includes: "The value of the original list item is not defined upon entry to the region. The original list item must not be referenced within the region. The value of the original list item is not defined upon exit from the region." Including a shared variable in a private clause essentially writes the shared variable with an undefined value, an effect that is easily overlooked by someone trying to understand the OpenMP memory model. We understand that this effect is being reconsidered for the OpenMP 3.0 specification. However, our point here is that any interactions between the two types of memory should be included in the memory section. In the very least, a forward reference is needed.

### 3.4 Captured Writes

The OpenMP memory model states that "If a thread has captured the value of a write in its temporary view of a variable since its last `flush` of that variable, then when it executes another `flush` of the variable, the `flush` does not complete until the value of the variable has been written to the variable in memory." We find this ambiguous and believe others will also. What does it mean for a thread to capture a value of a write? Does this only refer to a write by the thread that executes the `flush`? We believe that to be the intent but the actual wording could refer to writes on other threads that have been read by the given thread. Our point is that English is a rich and complex language in general and the phrase "precise English" is an oxymoron. For this reason, a formal, mathematical model is needed.

## 4 Formal Specification

The following sections describe the OpenMP memory model in formal, mathematical language. This specification takes as input an application and a trace that shows how this application executed on top of some implementation of OpenMP (a trace is a tuple of lists of executed shared memory operations, one list for each thread, with the operations stored in the order in which they were executed on that thread, along with their results, if any). It then uses a set of rules to judge if the application could have generated the trace and if a valid interleaving of thread operations exists under the OpenMP memory model that results in the values read in the trace.

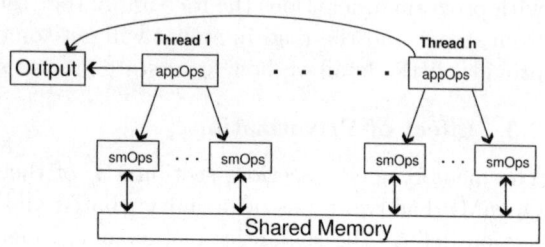

Fig. 2.

Our OpenMP formalization is an operational model (outlined in Figure 2). It defines a system state and valid transition rules for modifying the state. At a high level, this model defines the state of one or more application threads running on top of shared memory and transition rules for evaluating the next application operation on some thread. Applications are specified as lists of high-level operations such as $(var_A = var_B \otimes var_C)$ and $(While(var = val)\ bodyList)$, called "application operations" or "appOps". Each appOp is made up of one or more simpler operations such as $(Read\ var_A)$ or $(Write\ var_B\ val)$, called "shared memory operations" or "smOps". Every thread's state transition either:

- Evaluates the next smOp that makes up the thread's currently-executing appOp; or
- Moves to evaluation of the thread's next appOp in its remaining application source code.

The first action can change the shared memory state. The second action typically removes an appOp from the remaining application source code but can add appOps in the case of a while loop appOp that performs multiple loop iterations. A trace records each thread's view of a particular execution of the system. As such, it is a tuple of lists of smOps, one for each thread, (each list is some thread's "sub-trace"). Each sub-trace contains the smOps executed by its respective thread and any values they returned (e.g., the entry $(Read\ var \mapsto val)$ corresponds to a read of variable $var$ that returned the value $val$). Traces do not specify the interleaving of smOps from different threads.

We break our operational model into two sub-models, the Compiler Phase and the Runtime Phase, so that we can reason independently about different aspects of the memory model. The compiler phase evaluates each thread's source code independently from any other thread to verify that the application could have generated the list of smOps in each sub-trace. Its state consists of:
- a list of the current thread's remaining appOps;
- a list of smOps generated by that thread so far;
- the suffix of the thread's sub-trace that contains the yet unverified smOps.

During each state transition the compiler phase evaluates the next appOp, breaks it up into its constituent smOps (ex: the appOp $(var_A = var_B \otimes var_C)$ breaks up into ($Read\ var_B$), ($Read\ var_C$) and ($Write\ var_A$) smOps) and checks whether these smOps are contained in the sub-trace. Whenever an appOp uses values from shared memory (e.g., the value returned by a read), it looks them up in the sub-trace. The trace corresponds to the application's source code if the compiler phase independently verifies this for each sub-trace.

The runtime phase determines if the smOps in the individual threads' sub-traces correspond to each other. More specifically, it evaluates the threads' sub-traces in parallel to determine whether a conformant interleaving exists that results in the associated read values. It assumes that the smOps in the individual threads' sub-traces correspond to the application's source code. Therefore, its state consists of:

- the writes, atomic updates and flushes that each thread performed (one list per thread);
- a partial order that relates those smOps in time (used for determining the values that a read may return);
- the system's synchronization state: currently held locks, critical and ordered sections and the identities of threads that are currently blocked on a barrier;
- the smOps that remain to be evaluated for each thread (one list per thread).

During each state transition the runtime phase chooses a thread and evaluates its pending smOp. It may evaluate smOps out of order if this does not break their data dependences, (determined during the compiler phase). Evaluation of the read and atomic update smOps examines the values available to be read and verifies that the value returned by the read or atomic update in the trace could actually have been read during this interleaving. Every state transition also causes the state to change, including updating the synchronization state and adding new operations to the above partial order. Since the runtime phase is non-deterministic, the trace is self-consistent if the exists some interleaving of the different threads' smOps such that all reads and atomic updates performed by the formal model match their return values recorded in the trace.

Section 5 details the full language of appOps and smOps. Sections 6 and 7 provide more details on the mechanics of the compiler phase and runtime phase, respectively. Due to lack of space, we do not cover the full mathematical details of the formalism, which are available elsewhere [2]. Instead, we express them in a more verbal style here.

## 5 Language Specification

### 5.1 Application Operations

Our application language (specified in Table 1) models the major relevant features of C/Fortran and OpenMP. It contains basic computational and control flow operations as well as flushes and locks. Section number references refer to the OpenMP 2.5 specification [1]. The while loop primitive makes the application language Turing-complete in its use of shared memory operations. As mentioned, these operations are sufficient for our examples; the complete language covers the remaining OpenMP synchronization operations such as barriers and ordered sections [2].

## Table 1.

| | |
|---|---|
| $var_A = var_B \otimes var_C$ <br> • Represents any local computation performed by the application. <br> • $\otimes$ is a Turing-complete binary operation that does not use shared memory. <br> • $var_A$, $var_B$ and $var_C$ are shared variables. <br> • Corresponds to $(Read\ var_B)$, $(Read\ var_C)$ and $(Write\ var_A\ val)$ smOps. | $Lock\ lockVar$ <br> $Unlock\ lockVar$ <br> • Model the omp_set_lock and omp_unset_lock function calls [section 3.3]. <br> • $lockVar$ is a shared variable only accessed via $Lock$ and $Unlock$ operations. <br> • Correspond to a $BlockSynch$ smOp surrounded by $(Flush_{mm}\ allVars)$ smOps ($Lock$ and $Unlock$ correspond to different $BlockSynch$ smOps) |
| $Flush\ varList$ <br> • Models explicit flushes [sections 1.4.2 and 2.7.5]. <br> • $varList$ is a list of shared variables. <br> • An explicit flush operation with a list maps to $Flush\ varList$, where $varList$ is its variable list. <br> • An explicit flush operation without a list maps to $Flush\ allVarList$, where $allVarList$ contains all application shared variables. <br> • Corresponds to a single $Flush_{mm}$ smOp that applies to the same $varList$. | $While(var = testVal)\ bodyList$ <br> • A while loop control flow primitive. <br> • $var$ is a shared variable. <br> • $testVal$ is a value. <br> • $bodyList$ is a list of appOps. <br> • Corresponds to a single $(Read\ var)$ smOp. |
| | $Print\ var$ <br> • Outputs the value of a given shared variable to the user; primarily used in examples to reason about outcomes of application executions. <br> • $var$ is a shared variable. <br> • Corresponds to a single $(Read\ var)$ smOp. |
| $Atomic\ var \oplus =\ updVal$ <br> • Models the atomic update construct [section 2.7.4]. <br> • $\oplus$ may be one of the following operations: $+, *, -, /, \&, \hat{}, |, <<,$ or $>>$ (++ and -- are modeled via +=1 and -=1). <br> • $var$ is a shared variable. <br> • $updVal$ is a constant. <br> • Corresponds to an $Atomic_{mm}$ smOp surrounded by $(Flush_{mm}\ (var))$ smOps. | $End$ <br> • The last operation in the application's source code. <br> • Ensures each thread's sub-trace ends correctly. |

## 5.2 Shared Memory Operations

We use a very simple shared memory operation language that is sufficient for the functionality needs of the higher-level appOps. The smOps include reads, writes, atomic updates, flushes and blocking synchronizations (from which higher-level synchronizations are built) and are detailed in Table 2.

## 6 Compiler Phase

The compiler phase, diagrammed here, independently evaluates each thread of the application. It relates the application's source code to the smOps recorded in the thread's subtrace. The evaluation pass reads the appOps of the application source code in program order and unwraps its while loops as appropriate. In the process, it translates each appOp into its constituent smOp(s). These application smOps are looked up in the thread's subtrace during this evaluation process to verify that they actually do appear there. The values of all shared reads and atomic writes are also looked up in the trace.

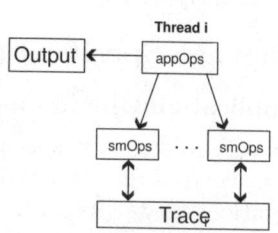

**Table 2.** Types of shared memory operations

| | |
|---|---|
| *Write var val*: writes *val* to variable *var*.<br>• *var* is a shared variable.<br>• *val* is a constant. | *BlockSynch blockF updF*:<br>generic blocking synchronization operation.<br>• Used to implement synchronization semantics of<br>higher-level operations such as locks and |
| *Read var* ↦ *val*: read of variable *var* returns *val*.<br>• *var* is a shared variable.<br>• *val* is a constant. | barriers.<br>• *blockF* is function.<br>○ Result depends on the formal system synchronization |
| $Atomic_{mm}$ *var* ⊕ = *updVal* ↦ *finalVal*:<br>atomically updates variable *var* to *finalVal*.<br>• *var* is a shared variable.<br>• *updVal* is a constant.<br>• Reads current value, *val*, of *var*.<br>• Computes *finalVal* = *val* ⊕ *updVal*.<br>• Writes *finalVal* to *var*.<br>• Actions are atomic: unsynchronized atomic updates<br>do not make the value of *var* indeterminate.<br>• Does not have any flush semantics<br>(unlike the *Atomic* appOp).<br>• ⊕ may be: +, *, −, /, &, ^, \|, <<, or >>. | state.<br>○ Returns False if the thread may continue executing<br>(i.e., is not blocked).<br>○ Returns True if the thread is blocked.<br>• *updF* is a function.<br>○ Result depends on the formal system<br>current synchronization state.<br>○ Returns the next synchronization state.<br>○ Applied only when *blockF* returns True.<br>○ Ensures the synchronization state reflects that the<br>thread has become unblocked. |
| $Flush_{mm}$ *varList*:<br>flushes this thread's temporary view<br>of variables in *varList*.<br>• *varList* is a list of shared variables.<br>• Updates thread's temporary view of those variables<br>with writes from other threads and vice versa.<br>• Provides flush semantics for explicit and<br>implicit flush operations. | • *blockF* and *updF* vary with each high-level<br>synchronization construct.<br>• The compiler phase (Section 6) defines<br>*blockF* and *updF*.<br>• The runtime phase (Section 7), where synchronization<br>state is defined, applies *blockF* and *updF*. |

This phase also defines a dependence order $\overrightarrow{DepO}$ on each thread's smOps, which the evaluation in the runtime phase must not violate. The remainder of this section defines the state and transition function of the compiler phase.

This phase's operational model is applied to the sub-trace corresponding to each thread. During each transition it evaluates the next appOp of the *app* list and verifies that its smOps occur in the sub-trace and have the appropriate step counter labels. The phase fails if it cannot verify those smOps. Whenever an appOp's evaluation depends on the outcome of a read, the read value is looked up in the trace and used in the appOp. For example, the while loop transition behaves differently depending on whether the value returned by its read is *testVal* or not.

The full trace is valid only if the above transition system independently passes each of its sub-traces. The Dependence Order $\overrightarrow{DepO}$ is preserved after this compiler pass for use in the runtime pass to ensure that whenever smOps are evaluated out of order, this new ordering does not violate their read-write dependences.

### 6.1 Compiler State

$[n, app, trace_{sub}, \overrightarrow{DepO}]$

- $n$: the number of smOps evaluated by this thread thus far. Initially $n = 0$.
- *app* : The list containing the appOps that remain to be evaluated by the thread. Initially, it is the original source code of the application.
- $trace_{sub}$ : The list containing the thread's sub-trace that is to be validated relative to application source code. The $m^{th}$ smOp generated on this thread is listed as $< smOp, m >$ (recall that the smOps in $trace_{sub}$ may have been

executed out of order, meaning that they may be listed out of program order). No two entries in $trace_{sub}$ have the same $m$ field.
- $\overrightarrow{DepO}$: The dependence order established so far between thread's smOps; initially the null relationship.

## 6.2 Compiler Transitions

The valid state transitions are shown in Table 3. One compiler transition exists for each appOp type. While loops have two transitions, one for the while loop performing an extra iteration and another for the while loop's termination. The transition used depends on the associated value of the loop variable, as described following the transitions. Whenever the partial order $\overrightarrow{DepO}$ is updated with new ordering relations, the new $\overrightarrow{DepO}$ is the transitive closure of the old $\overrightarrow{DepO}$ and the the new relations.

## 7 Runtime Phase

The first pass verifies that the smOps from each thread's sub-trace could have come from the given application. The second pass, the runtime phase, verifies that the values returned by reads and atomic updates would occur with some OpenMP conformant interleaving of the smOp traces. It evaluates the

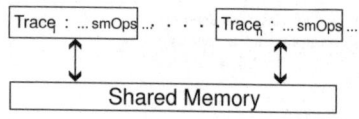

Fig. 3.

traces from all the threads in parallel, interleaving operations from different threads, as diagrammed in Figure 3. The transition system in Table 3 specifies this evaluation procedure. During each transition we choose some thread and evaluate the next smOp from this thread's sub-trace. We then check that the value returned for any *Read* or *Atomic* update could have been read under the OpenMP memory model. Conceptually, our runtime phase does not have a single shared memory. Instead, each write or atomic update simply becomes available to reads on its own thread and other threads the moment it is evaluated. Overall, this phase determines the trace is valid if at least one interleaving of thread operations agrees with the trace, since the procedure is non-deterministic. As discussed in Section 7.3, we consider an interleaving of smOps to agree with the trace if:

- it verifies the values returned by all reads and atomic updates; and
- either all smOps have been evaluated or the remaining smOps correspond to a deadlock.

### 7.1 Runtime State

The state of an application with $r$ threads is:
$\sigma, \overrightarrow{FlshO}; < t_1|subtrace_1, \overrightarrow{LclO_1} >, ...$
$..., < t_r|subtrace_r, \overrightarrow{LclO_r} >$

**Table 3.** Valid application state transitions

| Computation | Lock Acquire |
|---|---|
| **Current State:**<br>$[n, (\mathbf{var_A} = \mathbf{var_B} \otimes \mathbf{var_C}) :: app, trace_{sub}, \overrightarrow{DepO}]$<br>Next State: $[n+3, app, trace_{sub}, \overrightarrow{DepO}']$<br>and the following are true:<br>• $< Read\ var_B \mapsto val_B, n > \ \in\ trace_{sub}$<br>• $< Read\ var_C \mapsto val_C, n+1 > \ \in\ trace_{sub}$<br>• $< Write\ var_A\ (val_B \otimes val_C), n+2 > \ \in\ trace_{sub}$<br>• $\overrightarrow{DepO}'$ extends $\overrightarrow{DepO}$ as follows:<br>  ○ The write depends on the reads.<br>  ○ The read from $var_B$, the read from $var_C$ and the write to $var_A$, depend on the most recently evaluated writes or atomic updates to $var_A$, $var_B$ or $var_C$, respectively (if any).<br>  ○ All three smOps depend on the most recent read that was part of a while loop iteration test (i.e., they depend on control flow). | **Current State:**<br>$[n, (\mathbf{Lock\ lockVar}) :: app, trace_{sub}, \overrightarrow{DepO}]$<br>Next State: $[n+3, app, trace_{sub}, \overrightarrow{DepO}']$<br>and the following are true:<br>• $< Flush_{mm}\ allVars, n > \ \in\ trace_{sub}$<br>• $< BlockSynch\ lockBlock\ lockUpd, n+1 > \ \in\ trace_{sub}$<br>• $< Flush_{mm}\ allVars, n+2 > \ \in\ trace_{sub}$<br>• $\overrightarrow{DepO}'$ extends $\overrightarrow{DepO}$ as follows:<br>  ○ The smOps are ordered to place the lock acquisition between the two flushes.<br>  ○ The $Flush_{mm}$ smOps depend on all prior writes to or atomic updates of any variable and the lock acquire ($BlockSynch$) depends on the most recently evaluated acquire or release of $lockVar$.<br>  ○ All three smOps depend on the most recent read that was part of a while loop iteration test.<br>• $lockBlock$ is a function that returns $True$ (blocked) if $lockVar$ is currently held by some thread and $False$ otherwise.<br>• $lockUpd$ takes the current runtime state and returns one where $lockVar$ is recorded as being held. |
| **While Loop**<br>**Current State:**<br>$[n, (\mathbf{While}(var = testVal)\ bodyList) :: app,$<br>$\quad trace_{sub}, \overrightarrow{DepO}]$<br>Next State if $readVal = testVal$:<br>$[n+1, bodyList :: (While(var = testVal)\ bodyList) :: app,$<br>$\quad trace_{sub}, \overrightarrow{DepO}']$<br>Next State if $readVal \neq testVal$:<br>$[n+1, app, trace_{sub}, \overrightarrow{DepO}']$<br>and the following are true:<br>• $< Read\ var \mapsto readVal, n > \ \in\ trace_{sub}$<br>• $\overrightarrow{DepO}'$ extends $\overrightarrow{DepO}$ as follows:<br>  ○ The $Read$ of $var$ depends on the most recently evaluated write or atomic update of $var$ (if any).<br>  ○ The read depends on the most recent read that was part of a while loop iteration test. | **Lock Release**<br>**Current State:**<br>$[n, (\mathbf{Unock\ lockVar}) :: app, trace_{sub}, \overrightarrow{DepO}]$<br>Next State: $[n+3, app, trace_{sub}, \overrightarrow{DepO}']$<br>and the following are true:<br>• $< Flush_{mm}\ allVars, n > \ \in\ trace_{sub}$<br>• $< BlockSynch\ unlockBlock\ unlockUpd, n+1 > \ \in\ trace_{sub}$<br>• $< Flush_{mm}\ allVars, n+2 > \ \in\ trace_{sub}$<br>• $\overrightarrow{DepO}'$ extends $\overrightarrow{DepO}$ as follows:<br>  ○ The smOps are ordered to place the lock release between the two flushes.<br>  ○ The $Flush_{mm}$ smOps depend on all prior writes to or atomic updates of any variable and the lock release ($BlockSynch$) depends on the most recently evaluated acquire or release of $lockVar$.<br>  ○ All three smOps depend on the most recent read that was part of a while loop iteration test.<br>• $unlockBlock$ always returns False (not blocked)<br>• $unlockUpd$ updates the current runtime state s.t. $lockVar$ is recorded as being not held. |
| **Atomic Update**<br>**Current State:**<br>$[n, (\mathbf{Atomic\ var} \oplus = \mathbf{updVal}) :: app, trace_{sub}, \overrightarrow{DepO}]$<br>Next State: $[n+3, app, trace_{sub}, \overrightarrow{DepO}']$<br>and the following are true:<br>• $< Flush_{mm}\ (var), n > \ \in\ trace_{sub}$<br>• $< (Atomic_{mm}\ var \oplus = updVal \mapsto finalVal), n+1 > \ \in\ trace_{sub}$<br>• $< Flush_{mm}\ (var), n+2 > \ \in\ trace_{sub}$<br>• $\overrightarrow{DepO}'$ extends $\overrightarrow{DepO}$ as follows:<br>  ○ The smOps are ordered to place the atomic update between the two flushes.<br>  ○ The $Atomic_{mm}$ smOp depends on the most recently evaluated write to or atomic update of $var$.<br>  ○ The $Flush_{mm}$ smOps depend on all prior writes to or atomic updates of $var$.<br>  ○ All three smOps depend on the most recent read that was part of a while loop iteration test. | **Flush**<br>**Current State:**<br>$[n, (\mathbf{Flush\ varList}) :: app, trace_{sub}, \overrightarrow{DepO}]$<br>Next State: $[n+1, app, trace_{sub}, \overrightarrow{DepO}']$<br>and the following are true:<br>• $< Flush_{mm}\ varList, n > \ \in\ trace_{sub}$<br>• $\overrightarrow{DepO}'$ extends $\overrightarrow{DepO}$ as follows:<br>  ○ The $Flush_{mm}$ smOp depends on all previously evaluated writes to or atomic updates of variables in $varList$.<br>  ○ The $Flush_{mm}$ depends on the most recent read that was part of a while loop iteration test. |
| **Print**<br>**Current State:**<br>$[n, (\mathbf{Print\ var}) :: app, trace_{sub}, \overrightarrow{DepO}]$<br>Next State: $[n+1, app, trace_{sub}, \overrightarrow{DepO}']$<br>and the following are true:<br>• $< Read\ var \mapsto readVal, n > \ \in\ trace_{sub}$<br>• $\overrightarrow{DepO}'$ extends $\overrightarrow{DepO}$ as follows:<br>  ○ The read of $var$ depends on the most recently evaluated write or atomic update of $var$ (if any).<br>  ○ The read depends on the most recent read that was part of a while loop iteration test. | **End**<br>**Current State:**<br>$[n, (\mathbf{End}) :: app, trace_{sub}, \overrightarrow{DepO}]$<br>Next State: $[n+1, [], trace_{sub}, \overrightarrow{DepO}']$<br>and $\forall < smOp, m > \ \in\ trace_{sub}$, $m \leq n$<br>(the sub-trace has no more smOps). |

where:
- $\sigma$: The state of all synchronizations.
  - Contains one component for each type of synchronization in full model.
  - $\sigma.HeldLocks$: lock component (only component in abbreviated model)
    - Set of pairs $< lockVar, t_i >$, corresponding to lock variables $lockVar$ currently held by thread $t_i$.
    - Initially $= \emptyset$.
- $\overrightarrow{FlshO}$: The flush order established so far; initially, the null relationship.
- $subtrace_i$: The suffix of thread $t_i$'s sub-trace with its smOps yet to be evaluated; initially $t_i$'s full sub-trace.
- $\overrightarrow{LclO_i}$: Thread $t_i$'s local order established so far; initially, the null relationship.

The partial orders $\overrightarrow{FlshO}$ and $\overrightarrow{LclO_i}$ are defined on the events that happen on different threads. $\overrightarrow{FlshO}$ applies to events on all threads. $\overrightarrow{LclO_i}$ applies to events on thread $t_i$. How these two orders relate events determines the values returned by reads.

$\overrightarrow{LclO_i}$ is the program order of thread $t_i$ in our runtime pass, the order in which it evaluates $t_i$s operations. If event $E_1$ is evaluated on thread $t_i$ before event $E_2$ then we have $E_1 \overrightarrow{LclO_i} E_2$. For any event $E$ that happened on some thread $t_i$, we define "$\overrightarrow{LclO_i} \sqcup^i E$" to be an order that is identical to $\overrightarrow{LclO_i}$, except that event $E$ follows all events that have been completed on thread $t_i$.

$\overrightarrow{FlshO}$ is the global sequential flush order, defined by the relative times that different threads evaluate flushes. Let $E$ and $F$ be two events such that $F$ is a flush of the form $Flush_{mm}\ varList$. These two rules relate $E$ and $F$:
- If the *same* thread evaluates $E$ and $F$ and $E$ is a $(Read\ var)$, $(Write\ var)$ or $(Atomic_m m\ var \oplus=\ updVal)$ and $var \in varList$ then if $E$ was evaluated before $F$ then $E\ \overrightarrow{FlshO}\ F$, otherwise $F\ \overrightarrow{FlshO}\ E$.
- If $E$ is a flush of the form $Flush_{mm}\ varList2$ (on *any* thread) and $varList \cap varList2 \neq \emptyset$ then if $E$ was evaluated before $F$ then $E\ \overrightarrow{FlshO}\ F$, otherwise $F\ \overrightarrow{FlshO}\ E$.

The transitive closure of these rules defines $\overrightarrow{FlshO}$. For any event $E$ that happened on some thread $t_i$ we define "$\overrightarrow{FlshO} \sqcup^j_{var} E$" to be an order that is identical to $\overrightarrow{FlshO}$, except that event $E$ follows any flush operation evaluated on $t_j$ that has $var$ in its variable list. (note that $t_i$ may or may not be the same as $t_j$)

We use these orders in two key concepts: operation **races** and **eclipsing** operations. Two operations **race** if they are not related via $\overrightarrow{FlshO}$. A write or atomic update $WA_{ecl}$ on thread $t_i$ **eclipses** a write or atomic update $WA$ on thread $t_j$ from view by read $R$ on thread $t_k$ (all accessing the same variable) if $WA_{ecl}$ sits between $WA$ and $R$ under the order $\overrightarrow{FlshO} \cup \overrightarrow{LclO_i} \cup \overrightarrow{LclO_k}$. Similarly, a read $R_{ecl}$ on thread $t_i$ **eclipses** a write or atomic update $WA$ on thread $t_j$ from view by read $R$ on thread $t_k$ (all accessing the same variable) if $R_{ecl}$ sits between $WA$ and $R$ under the order $\overrightarrow{FlshO} \cup \overrightarrow{LclO_i} \cup \overrightarrow{LclO_k}$ and $R_{ecl}$ returns a value different from that written by $WA$.

## 7.2 Transition System

The runtime phase transition system contains one rule for each smOp. Each transition evaluates $s_i$, the first smOp in $subtrace_i$, provided that:
- No $s'_i$ previously evaluated on thread $t_i$ exists such that $s_i \; \overrightarrow{DepO} \; s'_i$;
- the return value in $subtrace_i$ is available for reading as defined below, if $s_i$ is a read or an atomic update;
- its $blockF$ function evaluates to false and its $updF$ function would update the synchronization state $\sigma$ to reflect $s_i$'s evaluation, if $s_i$ is a blocking synchronization operation.

If these conditions are not satisfied for thread $t_i$, its next smOp will not be evaluated until they are. The phase succeeds once $subtrace_i$ is empty on every thread $t_i$ or there is a deadlock, as discussed in Section 7.3; otherwise the phase backtracks to examine other interleavings. If no interleavings succeed, the phase fails and the trace demonstrates non-conformance.

The values available for reading in $subtrace_i$ depend on the established $\overrightarrow{FlshO}$ and $\overrightarrow{LclO}$ orders and the writes and atomic updates that the transition system has previously evaluated. Specifically, let $RA$ be a read or atomic update of variable $var$ on thread $t_i$. Let $pastWriteSet$ be the set of all un-eclipsed writes and atomic updates that precede $RA$ under $\overrightarrow{FlshO} \cup \overrightarrow{LclO_i}$ and let $presentRemoteWriteSet$ be the set of writes and atomic updates that race $RA$. Then a given value $val$ is `available for reading` by $RA$ if:
- $presentRemoteWriteSet$ contains any writes; or
- $presentRemoteWriteSet$ contains an atomic update the final value of which is $val$; or
- $pastWriteSet$ contains a pair of writes that race each other; or
- $pastWriteSet$ contains a write that wrote $val$ or an atomic update the final value of which is $val$; or
- $pastWriteSet$ is empty (i.e. $RA$ is not preceded by any writes to $var$ and thus got its value from uninitialized memory).

In other words, $val$ is available if it is the most recently written value to $var$ or if $var$ is uninitialized or racing writes exist to it (so $RA$ can return anything).

For any $s_i$, its transition rule:
- removes $s_i$ so $subtrace'_i = tail(subtrace_i)$ (recall that $s_i = head(subtrace_i)$);
- updates $\overrightarrow{FlshO}$ and $\overrightarrow{LclO_i}$ to include the ordering relationships between $E_{s_i}$, $s_i$'s evaluation event, and those of all previously evaluated smOps, as discussed above;
- updates synchronization state to $\sigma' = updF(\sigma)$ if $s_i$ is a $BlockSynch$ smOp.

Additional actions depend on the type of smOp, as detailed in Table 4.

## 7.3 Fairness and Deadlocks

The transition rules verify that a trace conforms with the OpenMP memory model if an interleaving of operations exists that agrees with the outcomes of the trace's smOps. An interleavings in which some smOp of some thread never executes is not sufficient since the phase will not validate that thread's sub-trace. Thus, our model has a basic fairness guarantee on valid traces that we now make explicit.

A trace is `Fair` if an interleaving of thread transitions exists such that no thread's current smOp is enabled for evaluation an infinite number of times without being evaluated. In particular, *BlockSynch* is only enabled in states where its *blockF* returns false, reads and atomic updates are enabled when their values are `available for reading` and writes and flushes are always enabled for execution. For finite traces this fairness condition guarantees that every smOp on every thread will eventually be evaluated unless there is a deadlock or the ordering of smOps on a thread's sub-trace violates the application's dependence order. For infinite traces it ensures no thread may be enabled for unblocking an infinite number of times without actually unblocking. In particular, if a thread is waiting to acquire a lock that periodically becomes available, it will eventually acquire it.

However, OpenMP does not guarantee deadlock freedom. A poorly written OpenMP program can contain a deadlock. Thus, our fairness guarantee also allows applications that deadlock. If the application reaches a point where every thread's next smOp is a *BlockSynch* whose *blockF* returns true, then the proposed interleaving deadlocks. Ordinarily, our transition system would reject the interleaving since each thread's last smOp (the *BlockSynch*) would not be validated against the trace. In order to allow (poorly written) applications that may deadlock, we explicitly accept deadlocked interleavings if every thread's last smOp is a *BlockSynch* for which *blockF* returns true.

A situation similar to deadlocks can occur when the sub-traces of one or more threads violate the dependence order established during the compiler phase. The problem is that the next smOp on such threads will never be evaluated since its evaluation would follow the evaluation of an smOp that should have preceded it according to the dependence order. Such traces are illegal and are rejected by the above model.

## 8 Examples

In the examples below we use the following shorthand:
- $var_A = const$ corresponds to $var_A = var_{const} + var_{zero}$ where $var_{const}$ and $var_{zero}$ are variables that are initialized to $const$ and 0 and never modified.
- *Barrier* corresponds to a barrier synchronization (not explicitly defined due to lack of space) and a $Flush_{mm}$ of all variables.

### 8.1 Uninitialized Read

Figure 4 contains an example code where the read on thread 0 may return any value. The reason is that if the read executes before the write, its *pastWriteSet* will be empty. Therefore, the read may return any value since the value would

**Table 4.** Valid shared memory state transitions

| Blocking synchronization | Read |
|---|---|
| Current State: <br> $\sigma, \overline{FlshO}; ..., < t_i | < \textbf{BlockSynch blockF updF}, m >::$ <br> $subtrace_i, \overrightarrow{LclO_i} >, ...$ <br> Next State: <br> $\sigma', \overline{FlshO}; ..., < t_i | subtrace_i, \overrightarrow{LclO_i}' >, ...$ <br> and the following are true: <br> • The function $blockF(\sigma)$ returns $False$, meaning that this thread does not need to block. <br> • $\sigma' = updF(\sigma)$, meaning that that synchronization state is transformed to reflect the fact that thread $t_i$ is unblocked. <br> • $\overline{FlshO}' = \overline{FlshO} \sqcup_{var}^i E_{s_i}$ for all variables $var$. <br> • $\overline{LclO_i}' = \overline{LclO_i} \sqcup^i E_{s_i}$. | Current State: <br> $\sigma, \overline{FlshO}; ..., < t_i | < \textbf{Read var} \mapsto \textbf{readVal}, m >::$ <br> $subtrace_i, \overrightarrow{LclO_i} >, ...$ <br> Next State: <br> $\sigma', \overline{FlshO}; ..., < t_i | subtrace_i, \overrightarrow{LclO_i}' >, ...$ <br> and the following are true: <br> • The value $readValue$ is available for reading. <br> • $\overline{FlshO}' = \overline{FlshO} \sqcup_{var}^i E_{s_i}$. <br> • $\overline{LclO_i}' = \overline{LclO_i} \sqcup^i E_{s_i}$. |
| **Atomic Update** | **Write** |
| Current State: <br> $\sigma, \overline{FlshO}; ...,$ <br> $< t_i | < \textbf{Atomic}_{mm} \textbf{ var} \oplus = \textbf{updVal} \mapsto \textbf{finalVal}, m >::$ <br> $subtrace_i, \overrightarrow{LclO_i} >, ...$ <br> Next State: <br> $\sigma', \overline{FlshO}; ..., < t_i | subtrace_i, \overrightarrow{LclO_i}' >, ...$ <br> and the following are true: <br> • $\overline{FlshO}' = \overline{FlshO} \sqcup_{var}^i E_{s_i}$. <br> • $\overline{LclO_i}' = \overline{LclO_i} \sqcup^i E_{s_i}$. | Current State: <br> $\sigma, \overline{FlshO}; ...,$ <br> $< t_i | < \textbf{Write var val}, m >::$ <br> $subtrace_i, \overrightarrow{LclO_i} >, ...$ <br> Next State: <br> $\sigma', \overline{FlshO}; ..., < t_i | subtrace_i, \overrightarrow{LclO_i}' >, ...$ <br> and the following are true: <br> • $\overline{FlshO}' = \overline{FlshO} \sqcup_{var}^i E_{s_i}$. <br> • $\overline{LclO_i}' = \overline{LclO_i} \sqcup^i E_{s_i}$. |
| | **Flush** |
| | Current State: <br> $\sigma, \overline{FlshO}; ...,$ <br> $< t_i | < \textbf{Flush}_{mm} \textbf{ varList}, m >::$ <br> $subtrace_i, \overrightarrow{LclO_i} >, ...$ <br> Next State: <br> $\sigma', \overline{FlshO}'; ..., < t_i | subtrace_i, \overrightarrow{LclO_i}' >, ...$ <br> and the following are true: <br> • $\overline{FlshO}' = \overline{FlshO} \sqcup_{var}^j E_{s_i}$ for all variables $var$ and threads $t_j$. <br> • $\overline{LclO_i}' = \overline{LclO_i} \sqcup^i E_{s_i}$. |

| Thread 0 | Thread 1 |
|---|---|
| Flush | var=1 |
| print var | Flush |

**Fig. 4.** Uninitialized read example

| Thread 0 | Thread 1 |
|---|---|
| var=0 | Barrier |
| Barrier | var=1 |
| Flush | Flush |
| print var | |

**Fig. 5.** Initialized read example

come from uninitialized memory. In order to avoid such uninitialized reads we can transform this program into the one in Figure 5.

In the modified program the barrier ensures that thread 0's read must follow some write to $var$, meaning that its $pastWriteSet$ cannot be empty. In future examples, whenever we make a statement about variables' initial value, we mean that the example's operations were preceded by a barrier, which was itself preceded by writes that initialized those variables. Equivalently, we could assume that the initialization occurs prior to the first parallel construct; we construct our examples with existing threads for notational simplicity.

## 8.2 Example A.2

The example in Figure 6 comes directly from example A.2 from the OpenMP 2.5 specification [1], converted from the original C/C++ and Fortran into our simplified language. Figure 7 shows a typical operation interleaving of this code (All other interleavings produce the same results).

Initially, $x = 2$

| Thread 0 | Thread 1 |
|---|---|
| x=5 | print(x) |
| Barrier | Barrier |
| print(x) | print(x) |

**Fig. 6.** Example A.2

| Thread 0 | Thread 1 |
|---|---|
| Write flag 2 | |
| Barrier | Barrier |
| Write x 5 | |
| | Read $x \mapsto$ ??? (print x) |
| Barrier | Barrier |
| Read $x \mapsto 5$ (print x) | |
| | Read $x \mapsto 5$ (print x) |

**Fig. 7.** Sample execution

This interleaving features three reads. The first read is evaluated on thread 1 before the barriers. As such, in any possible interleaving it must race the write to $x$ on thread 0. Since the write is in the first read's *presentRemoteWriteSet*, the read may return any value, regardless of $x$'s initial value. The two other reads are in a different situation. The barriers force them to follow the write in any interleaving. Because of the $Flush_{mm}$ inside each barrier, both reads follow the write on thread 0 in $\overrightarrow{FlshO}$. As such, the write is in their *pastWriteSet*. With no other available writes, this means that both reads must return 5, the value written by thread 0. Our formalism is consistent with the explanation of example A.2 [1].

## 8.3 Faulty Spinlock

Figure 8 shows a basic spinlock. At first it appears that this program will print a finite sequence of 0's, followed by a 1. However, despite the abundance of flushes there is a race between the write on thread 0 and the reads on thread 1. The smOp interleaving that reveals this race is shown in Figure 9.

The problem here is that the reads on thread 1 may happen before the flush on thread 0. Thus, the values read by these reads are unspecified, meaning that the values printed may be garbage. Fortunately, our fairness assumption guarantees the flush on thread 0 will eventually be evaluated. Another iteration of the while loop on thread 1 will produce a flush call, which will cause thread 0's write to precede subsequent reads on thread 1 under $\overrightarrow{FlshO} \uplus \overrightarrow{LclO_1}$. This in turn causes them to read 1, terminating the while loop.

While this seems to be a contrived example, suppose that we have a shared memory implementation where 64-bit writes are broken up into multiple 16-bit messages and the write on thread 0 actually writes some large 64-bit value. In this case the reads on thread 1 may read $flag$ while it is only partially updated

| Initially, $flag = 0$ | |
|---|---|
| Thread 0 | Thread 1 |
| flag=1 | Flush |
| Flush | while(flag=0){ |
| | print(flag) |
| | Flush |
| | } |
| | print(flag) |

**Fig. 8.** Example of a faulty spinlock

| Thread 0 | Thread 1 |
|---|---|
| $Write\ flag\ 0$ | |
| Barrier | Barrier |
| $Write\ flag\ 1$ | |
| | $Flush_{mm}\ allVars$ |
| | $Read\ flag \mapsto ???$ (while) |
| | $Read\ flag \mapsto ???$ (print) |
| | ... |
| $Flush_{mm}\ allVars$ | |
| | $Flush_{mm}\ allVars$ |
| | $Read\ flag \mapsto 1$ (while) |
| | $Read\ flag \mapsto 1$ (print) |

**Fig. 9.** Sample faulty spinlock interleaving

| Initially, $flag = 1$ | |
|---|---|
| Thread 0 | Thread 1 |
| Atomic flag+=1 | Flush |
| | while(flag=0){ |
| | print(flag) |
| | Flush |
| | } |
| | print(flag) |

**Fig. 10.** Correct Spinlock

with only some of the 16-bit messages, causing the prints to output garbage. Indeed, the only way to prevent this situation is to ensure that the write to the flag is atomic, something that only the `atomic` construct can provide.

Given this new knowledge we can augment the program above to use an atomic update, as shown in Figure 10. In this case the above interleaving produces the expected behavior since even when the reads on thread 1 race with the atomic update on thread 0 (i.e. the atomic update is in their $presentRemoteWriteSet$), they do not get garbage values but rather either 0 or 1. (atomic update appOps contain their own $Flush_{mm}$ smOps)

## 8.4 Flush-Free Spinlock

The example in Figure 11 is the same as the one above except that the flushes have been removed. This program must either print a sequence of zero of more 0's, followed by a 1 or an infinite sequence of 0's. To understand why this is, lets examine the smOp interleaving shown in Figure 12.

Before thread 0 executes the atomic update, the fact that reads on thread 1 have empty $presentRemoteWriteSets$ and $pastWriteSets$ that contain only

Initially, $flag = 0$

| Thread 0 | Thread 1 |
|---|---|
| Atomic flag+=1 | while(flag=0){ |
| | print(flag) |
| | } |
| | print(flag) |

**Fig. 11.** Flush-free spinlock example

| Thread 0 | Thread 1 |
|---|---|
| $Write\ flag\ 0$ [*] | |
| Barrier | Barrier |
| | $Read\ flag \mapsto 0$ (while) |
| | $Read\ flag \mapsto 0$ (print) |
| | ... |
| $Flush_{mm}\ (flag)$ | |
| $Atomic_{mm}\ flag+ = 1 \mapsto 1$ | |
| $Flush_{mm}\ (flag)$ | |
| | ... |
| | $Read\ flag \mapsto 0$ (while) |
| | $Read\ flag \mapsto 0$ (print) |
| | ... |
| | $Read\ flag \mapsto 1$ (print) [**] |
| | $Read\ flag \mapsto 1$ (while) |
| | $Read\ flag \mapsto 1$ (print) |

**Fig. 12.** Sample flush-free spinlock interleaving

the initialization write [*], causes them to return 0. When thread 0's atomic update does occur, thread 1 may not update its temporary view - ever. The atomic update is in the *presentRemoteWriteSet* of its reads. Thus, the value may never be observed by thread 0, which can iterate its loop forever, printing out 0's. In the trace above, the view is eventually updated and some read [**] returns 1. Therefore, all subsequent reads of $flag$ on thread 1 must also read 1 because read [**] eclipses write [*] under order $\overrightarrow{FlshO} \cup LclO_0 \cup \overrightarrow{LclO_1}$.

This example portrays an important lesson. Although fairness is an important condition and critical for avoiding infinite loops, it does not prevent them. Programs without appropriate flushes may still loop infinitely because a thread's temporary view may not be updated.

### 8.5 Multi-thread Writer Race

The example in Figure 13 shows the effect of a race between writes. Suppose that the above application has smOp interleaving as in Figure 14. Before threads 0 and 1 do their flushes, the reads on thread 2 are racing with the writes on threads 0 and 1 under the order $\overrightarrow{FlshO} \cup \overrightarrow{LclO_2}$. This is still true after thread 0 performs its flush since the reads on thread 2 are still racing with thread 1's write. The problem persists even after thread 1's flush. At this point both writes are in the past of all subsequent reads on thread 2 according to $\overrightarrow{FlshO} \cup \overrightarrow{LclO_2}$. However, the two writes are not related to each other under $\overrightarrow{FlshO} \cup \overrightarrow{LclO_2}$, meaning that they race. This means that the third read on thread 2 may also return an unspecified value.

In reality, this example can happen in the aforementioned implementation where 64-bit writes are broken up into 16-bit messages and no filtering is done

Initially, $flag = 0$

| Thread 0 | Thread 1 | Thread 2 |
|---|---|---|
| flag=1 | flag=42 | Flush |
| Flush | Flush | print(flag) |
| | | Flush |
| | | print(flag) |
| | | Flush |
| | | print(flag) |

**Fig. 13.** Multi-thread writer race example

| Thread 0 | Thread 1 | Thread 2 |
|---|---|---|
| Write flag 0 | | |
| Barrier | Barrier | Barrier |
| Write flag 1 | | |
| | Write flag 42 | |
| | | $Flush_{mm}\ allVars$ |
| | | Read $flag \mapsto$ ??? (print) |
| | $Flush_{mm}\ allVars$ | |
| | | $Flush_{mm}\ allVars$ |
| | | Read $flag \mapsto$ ??? (print) |
| | $Flush_{mm}\ allVars$ | |
| | | $Flush_{mm}\ allVars$ |
| | | Read $flag \mapsto$ ??? (print) |

**Fig. 14.** Sample multi-thread writer race interleaving

to tell which 16-bit message comes from which 64-bit write. Since the writes on threads 0 and 1 are unrelated by any synchronization, their individual messages may arrive in memory in arbitrary order, causing the resulting stored value to contain pieces from both writes.

## 8.6 Writes from Same Thread

The example in Figure 15 shows how writes on one thread that were placed in a given order by the program's source code will be seen to occur in this order by any reads on other threads that have ordered themselves correctly relative to the writes (via flushes). However, in the absence of proper ordering, anything can happen.

Initially, $flag = 0$

| Thread 0 | Thread 1 |
|---|---|
| flag=1 | Flush |
| flag=2 | print(flag) |
| Flush | |

**Fig. 15.** Example of a writes from the same thread

| Thread 0 | Thread 1 |
|---|---|
| Write flag 0 | |
| Barrier | Barrier |
| Write flag 1 [*] | |
| Write flag 2 [**] | |
| $Flush_{mm}\ allVars$ | |
| | $Flush_{mm}\ allVars$ |
| | Read $flag \mapsto 2$ (print) |

**Fig. 16.** Properly ordered interleaving

Figure 16 shows a properly ordered trace. Thread 0 goes first, issues both writes and performs a flush. Note that since both writes were to $flag$, they were related via $\overrightarrow{DepO}$ and had to be evaluated in that order. Furthermore, when

| Thread 0 | Thread 1 |
|---|---|
| $Write\ flag\ 0$ | |
| Barrier | Barrier |
| | $Flush_{mm}\ allVars$ |
| $Write\ flag\ 1\ [*]$ | |
| $Write\ flag\ 2\ [**]$ | |
| $Flush_{mm}\ allVars$ | |
| | $Read\ flag \mapsto ???$ (print) |

**Fig. 17.** Uordered interleaving

the read on thread 1 was evaluated, both writes precede it according to order $\overrightarrow{FlshO\ \cup\ LclO_1\ \cup\ LclO_2}$ and write [**] follows write [*] under to the same ordering. As a result, the write [*] is eclipsed by write [**] under the definition of $WriteEclipse(flag, R, Write\ [*], W\ [**], \overrightarrow{FlshO\ \cup\ LclO_1\ \cup\ LclO_2})$. Thus, the read only has write [**] in its past, no writes in its present and therefore returns 2.

Figure 17 shows what happens when the read is not properly ordered relative to the writes. In this case both writes are in the read's present since they are not ordered relative to the read via $\overline{FlshO}$. Thus, the read may return any value. Indeed, any later read is also free to return any value until thread 1 calls a $Flush_{mm}$, placing the two writes on thread 0 into the past under order $\overrightarrow{FlshO\ \cup\ LclO_0\ \cup\ LclO_1})$.

### 8.7 Atomic Updates Racing with Reads

Figure 18 shows a code example where atomic updates to a given variable may not be seen in a linear order to a reader thread that has not performed the appropriate flushes. This behavior is shown in Figure 19. In this trace the reads on thread 1 are preceded by the initialization write on thread 0 and two atomic updates on thread 1. Thus, the first read [*] has the initialization write in its *pastWriteSet* and the two atomic updates in its *presentRemoteWriteSet*. Therefore, the read is free to return any of the three available values: 0, 1 or 2. In this trace it returns 2.

Now examine the other reads. Although they do follow read [*], the absence of flushes on thread 1 means that under the ordering $\overrightarrow{FlshO\ \cup\ LclO_0\ \cup\ LclO_1}$ read [*] does not eclipse any of the writes or atomic updates on thread 0. As such, their *pastWriteSets* and *presentRemoteWriteSets* are identical to those of read [*] and so they are free to return any of the same values: 0, 1 or 2.

## 9 Conclusion

The OpenMP 2.5 specification includes a section that details the OpenMP memory model [1]. This section significantly improves previous specifications – the

Initially, $flag = 0$

| Thread 0 | Thread 1 |
|---|---|
| Atomic flag+=1 | print flag |
| Atomic flag+=2 | print flag |
| | print flag |

**Fig. 18.** Atomic values racing with reads example

| Thread 0 | Thread 1 |
|---|---|
| $Write\ flag\ 0$ | |
| Barrier | Barrier |
| $Flush_{mm}\ (flag)$ | |
| $Atomic_{mm}\ flag\ +=\ 1) \mapsto 1$ | |
| $Flush_{mm}\ (flag)$ | |
| $Flush_{mm}\ (flag)$ | |
| $Atomic_{mm}\ flag\ +=\ 1) \mapsto 2$ | |
| $Flush_{mm}\ (flag)$ | |
| | $Read\ flag \mapsto 2\ (print()\ [*]$ |
| | $Read\ flag \mapsto 1\ (print)$ |
| | $Read\ flag \mapsto 0\ (print)$ |

**Fig. 19.** Sample interleaving for the Atomic Updates Racing with Reads example

previous C/C++ specifications did not address the issue directly at all. Instead, users and implementers had to synthesize a model as best they could from several disparate sections. However, the memory model is still described in informal prose, which lacks precision by definition.

This paper presents a formal OpenMP memory model, derived from the model in the current specification. We tried to faithfully adhere to that prose description. However, as we have discussed, it has several ambiguities, which we resolve in our formal model by relying on our understanding of the intent of the language committee. Our operational model supports the verification of the conformance of OpenMP implementations. It consists of two phases: a compiler phase that extracts the constituent operations of the application and a runtime phase that verifies that a compliant execution could produce the values that appear in the trace. We have applied this model to several examples. Overall, our work demonstrates the need for the OpenMP community to adopt further refinements of the OpenMP memory model. Ideally those changes will lead to a formal model in later OpenMP specifications.

## References

1. OpenMP Architecture Review Board. OpenMP application program interface, version 2.5
2. Bronevetsky, G., de Supinski, B.: Fully formal specification of the OpenMP memory model. Cornell Computer Science (in preparation) (2005)
3. Collier, W.W.: Reasoning About Parallel Architectures (1992)
4. Scheurich, C., Dubois, M., Briggs, F.: Memory access buffering in multiprocessors. In: Proceedings of the 13th Annual International Symposium on Computer Architecture (ISCA), pp. 434–442 (1986)
5. Goodman, J.R.: Cache consistency and sequential consistency. Technical Report 61, SCI Committee (1989)
6. Hoeflinger, J., de Supinski, B.: The openmp memory model. In: International Workshop on OpenMP (IWOMP) (2005)

7. Pugh., W., Manson., J., Adve, S.V.: The java memory model. In: Symposium on Principles of Programming Languages (POPL ) (2005)
8. Matthews, J., Tasiran, S., Tuttle, M., Joshi, R., Lamport, L., Yu, Y.: Checking cache-coherence protocols with tla+. Formal Methods in System Design 22(2), 125–131 (2003)
9. Robinson, A., Voronkov, A. (eds.): Handbook of Automated Reasoning (2000)

# Applications II

# Performance and Programmability Comparison Between OpenMP and MPI Implementations of a Molecular Modeling Application*

Russell Brown[1] and Ilya Sharapov[2]

[1] Sun Microsystems, Inc., 15 Network Circle, Menlo Park, CA 94025, USA
russ.brown@sun.com
[2] Sun Microsystems, Inc., 16 Network Circle, Menlo Park, CA 94025, USA
ilya.sharapov@sun.com

**Abstract.** Important components of molecular modeling applications are estimation and minimization of the internal energy of a molecule. For macromolecules such as proteins and amino acids, energy estimation is performed using empirical equations known as force fields. Over the past several decades, much effort has been directed towards improving the accuracy of these equations, and the resulting increased accuracy has come at the expense of greater computational complexity. For example, the interactions between a protein and surrounding water molecules have been modeled with improved accuracy using the generalized Born solvation model, which increases the computational complexity to $O\left(n^3\right)$.

Fortunately, many force-field calculations are amenable to parallel execution. This paper describes the steps that were required to transform the Born calculation from a serial program into a parallel program suitable for parallel execution in both the OpenMP and MPI environments. Measurements of the parallel performance on a symmetric multiprocessor reveal that the Born calculation scales well for up to 144 processors, and that programmability and performance are better for the OpenMP implementation than for the MPI implementation.

## 1 Introduction

Molecular modeling is one of the most demanding areas of scientific computing today. Although the high computational requirements of molecular simulation can produce long computation times, parallel execution may used to increase the size of molecules that can be analyzed in manageable time. Several software applications exist for molecular modeling and estimation of the internal energy of molecules using classical equations known as *force fields* [4]. An open-source application related to the well-known AMBER [18] software is Nucleic Acid Builder or NAB[1] [15], which we use as the basis for our analysis in this work.

---

* This material is based upon work supported by DARPA under Contract No. NBCH3039002.
[1] http://www.scripps.edu/case

A typical force field includes several energy terms, one of which models the interactions between a biomolecule and the solvent, or surrounding water molecules. This energy is known as the Born free energy of solvation. Using a method known as the generalized Born [3,8,17] approximation, the electrostatic contribution to this energy is computed as the sum of pairwise interactions:

$$E_{\text{Born}} = -\frac{1}{2} \sum_i \sum_{j>i} q_i q_j \left[ 1 - \frac{e^{-\kappa \sqrt{d_{ij}^2 + R_i R_j e^{\frac{-d_{ij}^2}{4 R_i R_j}}}}}{\epsilon_w} \right] \quad (1)$$

In this equation, $d_{ij}$ represents the distance between atoms $i$ and $j$, $\epsilon_w$ represents the dielectric constant of water, and $\kappa$ represents a Debye-Huckel screening constant [16]. $R_i$ represents the *effective Born radius* that is a measure of the amount by which the atom $i$ is screened from the solvent by all of the surrounding atoms $k$. The effective Born radius is calculated as:

$$R_i^{-1} = \frac{1}{\rho_i} + \sum_{k \neq i} f(d_{ik}, \rho_i, \rho_k) \quad (2)$$

In this equation, $\rho_i$ and $\rho_k$ represent the intrinsic radii of atoms $i$ and $k$, and $f()$ is a smooth function of the interatomic distance and the intrinsic radii [11].

Because of the presence of $R_i$ and $R_j$ in equation (1), computation of the Born free energy and its first and second derivatives can involve considerable complexity. However, it is possible to reduce this computational complexity via precomputation, the details of which are beyond the scope of this paper and are reported elsewhere [6]. This precomputation produces two vectors, $R$ and $A$, each of length $n$ (where $n$ is the number of atoms). These vectors are used to produce four matrices: **N** (of size $n$ by $n$) and **F**, **G** and **D** (each of size $3n$ by $n$). Products of these matrices are used to form the Hessian matrix **H** (of size $3n$ by $3n$) that contains the second derivatives. This precomputation reduces the computational complexity of the Born free energy and its first derivatives to $O(n^2)$, as well as reducing the complexity of the second derivatives to $O(n^3)$.

For molecular modeling applications, it is desirable to estimate the internal energy of a molecule and to minimize that energy. Minimization may be performed using the iterative Newton-Raphson method [4] that is calculated as:

$$x_1 = x_0 - \mathbf{H}^{-1}(\mathbf{x_0}) \nabla \mathbf{E}(\mathbf{x_0}) \quad (3)$$

In the above equation, $x_0$ represents the initial Cartesian coordinates of the atomic nuclei prior to a step of Newton-Raphson iteration, $x_1$ represents the Cartesian coordinates after the step of Newton-Raphson iteration, $\nabla E(x_0)$ represents the gradient vector of first derivatives that are calculated from the initial Cartesian coordinates, and $\mathbf{H}^{-1}(\mathbf{x_0})$ represents the inverse of the Hessian matrix of second derivatives that are calculated from the initial Cartesian coordinates. In practice, inversion of the Hessian matrix is avoided by solving the linear system via techniques such as Cholesky factorization.

## 2 Implementation

As shown in Figure 1, each iteration of Newton-Raphson minimization is subdivided into four phases of computation: (1) calculation of the non-Born energy terms and their derivatives; (2) calculation of the Born free energy, its first derivatives, and the matrices **N**, **F**, **G** and **D**; (3) matrix multiplication to produce the Hessian matrix **H** of second derivatives; and (4) solution of the linear system via Cholesky factorization.

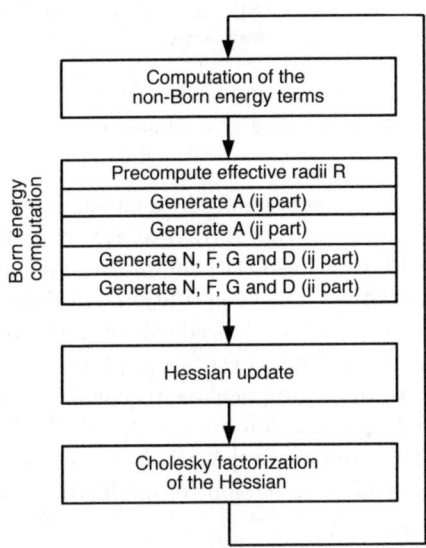

**Fig. 1.** Phases 1-4 of Newton-Raphson. The $ij$ and $ji$ parts are computed separately.

We can visualize the phases of an iteration by monitoring the changes in low-level activity in the system. Figure 2 gives an example of the low-level activity that illustrates the change in the *cycle per instruction* (CPI) measurements for different phases of execution of the MPI version of NAB. The CPI measures the efficiency of the CPU; low CPI values indicate good performance. Superscalar processors are capable of executing multiple instructions in one cycle and therefore CPI values can be less than one for carefully tuned sections of code. In our experiments we measured the CPI and other low-level statistics, such as cache miss rates, using UltraSPARC® processor on-chip hardware counters [9].

In Phase 1 we observe fairly poor processor efficiency, which can be explained by non-contiguous memory references that occur in the computation of the non-Born energy terms. Relatively little computation is performed during this phase, which doesn't significantly impact the overall performance. In Phase 2 we can see five distinct regions corresponding to the five Born energy computations outlined in Figure 1. Phase 3 updates the Hessian matrix via three matrix-matrix

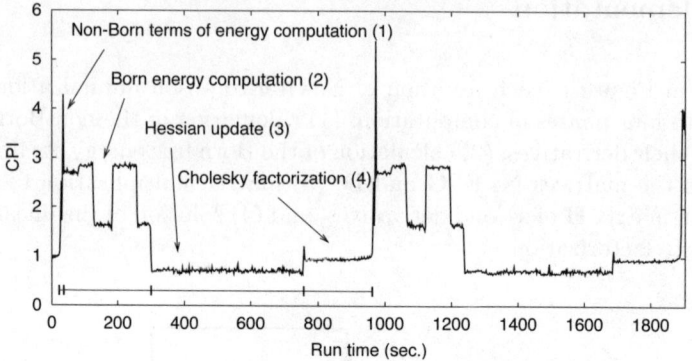

**Fig. 2.** CPI for Phases 1-4 of Newton-Raphson minimization, using 5-way parallelism for Phase 2 and 25-way parallelism for Phases 3 and 4

multiplications. Phase 4 performs Cholesky factorization. Phases 3 and 4 are executed using subroutines from the ScaLAPACK[2] scientific library [5].

The computation in Phase 1 has $O(n)$ complexity and contributes only minimally to the total computation. The computational in Phases 3 and 4 has $O(n^3)$ complexity and is performed by parallelized subroutines from a scientific library. The computation in Phase 2 exhibits $O(n^2)$ complexity and must be parallelized in order that the total computation achieve reasonable scalability [1]. We have parallelized the Phase 2 computation via an OpenMP[3] [7] implementation, as well as via an MPI[4] [10] implementation. Prior studies have reported comparisons between OpenMP and MPI versions of applications software [2,12,13,14].

### 2.1 OpenMP

The computation of Phase 2 of Newton-Raphson minimization involves several summations such as $\sum_i \sum_{j>i}$ or $\sum_i \sum_{j \neq i}$, each of which implies a loop nest having $i$ as the outer loop index, and $j$ as the inner loop index. Each loop nest updates a vector and a matrix, as shown for the shared vector $A$ and the shared matrix $\mathbf{N}$ in C code fragment (4) that is parallelized via OpenMP by adding a `#pragma omp parallel for` directive to the serial code.

Code fragment (4) exhibits a race condition for the update of $A[j]$ and $\mathbf{N}[j][j]$. Because the $j$ loop index is not partitioned amongst the OpenMP threads, all of the threads can potentially update $A[j]$ and $\mathbf{N}[j][j]$. There is no guarantee that these updates will occur atomically, and therefore the threads may overwrite one another's updates. This race condition is removed by splitting the loop nest to create two loop nests. The first loop nest uses $i$ as the outer index, and $j$ as the inner index. It updates $A[i]$, $\mathbf{N}[i][i]$ and $\mathbf{N}[i][j]$ as shown in code fragment (5).

---

[2] http://www.netlib.org/scalapack
[3] http://www.openmp.org
[4] http://www.mpi-forum.org

```
#pragma omp parallel for private(j)
for (i = 0; i < n; i++) {
  for (j = i+1; j < n; j++) {
    A[i] += f1(i, j);
    A[j] += f2(i, j); // Incorrect!
    N[i][i] += f3(i, j);
    N[i][j] += f4(i, j);
    N[j][j] += f5(i, j); // Incorrect!
    N[j][i] += f6(i, j);
  }
}
```
(4)

```
#pragma omp parallel for private(j)
for (i = 0; i < n; i++) {
  for (j = i+1; j < n; j++) {
    A[i] += f1(i, j);
    N[i][i] += f3(i, j);
    N[i][j] += f4(i, j);
  }
}
```
(5)

The second loop nest uses $j$ as the outer index, and $i$ as the inner index. It updates $A[j]$, $N[j][j]$ and $N[j][i]$ as shown in code fragment (6).

```
#pragma omp parallel for private(i)
for (j = 0; j < n; j++) {
  for (i = 0; i < j; i++) {
    A[j] += f2(i, j);
    N[j][j] += f5(i, j);
    N[j][i] += f6(i, j);
  }
}
```
(6)

For either loop nest, a given value of $i$ is combined with the same values of $j$. However, a race condition will exist unless $A[i]$ and $N[i][i]$ are updated in the first loop nest, and unless $A[j]$ and $N[j][j]$ are updated in the second loop nest. In contrast, $N[i][j]$ and $N[j][i]$ may be updated in either loop nest. No race condition can exist for these updates because each update involves both $i$ and $j$ indices that guarantee partitioning of the matrix elements amongst the threads.

However, by updating $N[i][i]$ and $N[i][j]$ in the first loop nest, and by updating $N[j][j]$ and $N[j][i]$ in the second loop nest, the matrix elements are partitioned amongst the OpenMP threads by matrix row. Groups of r contiguous rows may be partitioned amongst the threads by adding a `schedule(static, r)` clause to the `#pragma omp parallel for` directive. This partitioning is known as *row cyclic* partitioning, and promotes locality of memory access by each OpenMP thread for a matrix that is allocated in row-major order.

Instead of using the schedule(static, r) clause, we map the OpenMP threads to matrix rows explicitly, as shown for MPI in code fragment (7), in order to maintain compatibility between the OpenMP and MPI implementations of NAB. This approach is an example of the SPMD OpenMP programming style that can improve the performance of OpenMP applications when used in conjunction with code tuning techniques [13]. A performance analysis tool such as the Sun$^{\text{TM}}$ Performance Analyzer[5] is essential to the code tuning process.

## 2.2 MPI

The parallelization of Phases 2 to 4 of Newton-Raphson minimization is more complex for MPI than for OpenMP. The increased complexity is due principally to the ScaLAPACK library that is used with the MPI implementation of NAB. The ScaLAPACK library does not support global, shared vectors and matrices; instead, it supports vectors and matrices that are distributed across all of the MPI processes. Under this distributed paradigm, each process has exclusive access to a unique subset of the global vector or matrix elements, which we will call the *sub-vector* or *sub-matrix*. Each process initializes its sub-vector or submatrix, and then the ScaLAPACK subroutines distribute computation such as matrix multiplication and Cholesky factorization across all of the processes.

Before a vector or matrix can be processed by a ScaLAPACK subroutine, it must be distributed onto a *process grid*. The process grid is a group of MPI processes that are placed on a rectangular grid of nprow rows by npcol columns. Each process has unique row and column coordinates myrow and mycol that indicate the location of the process on the grid. The matrix elements are mapped onto the process grid in a *block cyclic* manner [5] wherein a matrix of $m$ rows by $n$ columns is subdivided into blocks of mb rows by nb columns of contiguous matrix elements. Block cyclic mapping is the two-dimensional analog of the one-dimensional row cyclic mapping that was described for OpenMP, and is used by ScaLAPACK to achieve reasonable load balancing across the MPI processes.

Each MPI process has exclusive access to a unique sub-vector or sub-matrix. Thus, the $i$ and $j$ loop indices of a loop nest must be restricted to only those values that are required to update the accessible sub-vector or sub-matrix elements. The loop nest must be split because a process that can access $\mathbf{N}[i][j]$ cannot necessarily access $\mathbf{N}[j][i]$. Code fragment (7) satisfies these constraints and accomplishes block cyclic partitioning of the matrix elements amongst the MPI processes. Each process asserts a unique subset of the matrix $i$ and $j$ indices, which is selected by the myrow and mycol coordinates of the process.

Code fragment (7) is incorrect for several reasons. First, although the first loop nest can access $\mathbf{N}[i][j]$, it cannot necessarily access $\mathbf{N}[i][i]$. Similarly, although the second loop nest can access $\mathbf{N}[j][i]$, it cannot necessarily access $\mathbf{N}[j][j]$. These access restrictions arise because the diagonal elements of the matrix $\mathbf{N}$ are not guaranteed to belong to the process that calculates updates to those elements. A solution to this problem is for each process to maintain a private copy of

---

[5] http://developers.sun.com/sunstudio

```
for (i = 0; i < n; i++) {
  if ( (i/mb)%nprow != myrow ) continue;
  for (j = i+1; j < n; j++) {
    if ( (j/nb)%npcol != mycol ) continue;
    A[i] += f1(i, j); // Incorrect!
    N[i][i] += f3(i, j); // Incorrect!
    N[i][j] += f4(i, j);
  }
}

for (j = 0; j < n; j++) {
  if ( (j/mb)%nprow != myrow ) continue;
  for (i = 0; i < j; i++) {
    if ( (i/nb)%npcol != mycol ) continue;
    A[j] += f2(i, j); // Incorrect!
    N[j][j] += f5(i, j); // Incorrect!
    N[j][i] += f6(i, j);
  }
}
```
(7)

the diagonal elements. When all of the processes have finished updating their private copies of the diagonal elements, these private copies are combined and the result is rebroadcast to each process. The MPI_Allreduce function is used for this combine and rebroadcast operation.

The second problem with code fragment (7) arises due to a ScaLAPACK convention that requires that a distributed vector such as the vector $A$ exist only in column zero of the process grid. Hence, only a process that exists in column zero of the grid possesses a sub-vector of the vector $A$. A particular process that calculates updates to the vector $A$ may not lie in column zero of the grid and therefore may not be able to access the elements of the sub-vector that it needs to update. This problem is very similar to the first problem discussed above, and it has a similar solution. Each process must maintain a private copy of the entire vector $A$. When all of the processes have finished updating their private copies of the vector $A$, these private copies are combined and the result is rebroadcast to each process via the MPI_Allreduce function.

The third problem with code fragment (7) is that the matrix elements $N[i][j]$ and $N[j][i]$ are not accessed as elements of the global matrix $N$, but rather as elements of a sub-matrix that is owned by a particular process. The sub-matrix is not addressed using the global $[i][j]$ or $[j][i]$ address directly. Instead, the global address is converted to an offset into the sub-matrix. This global-to-local address mapping involves four integer divisions and four modulus operations for access to *each* element of the matrix. Fortunately, the divisors are mb, nb, nprow and npcol, which are constants for a given matrix, and the dividends are $i$ and $j$ whose values lie in the range $0 \leq i < n$. These features of the divisors and dividends permit precomputation of all of the division and modulus operations. The results of the precomputation are stored in eight lookup tables, each of length $n$. Because a matrix requires $O(n^2)$ memory, these tables represent only a small fraction of the size of a typical matrix, and hence offer the possibility of accelerated computation at the expense of minimal additional storage.

## 3 Programmability

Creating OpenMP and MPI versions of the same application using different approaches to parallelization allowed us to compare the effort required to implement both versions. Parallelization of Phase 2 required the splitting of nested loops for both versions. Once this splitting was completed, the OpenMP version was straightforward. However, creating the MPI version required substantial additional effort (required by ScaLAPACK) to map global matrices onto a two-dimensional process grid and to modify one-dimensional row cyclic partitioning to obtain two-dimensional block cyclic partitioning of the matrix elements.

We have estimated the relative complexity of the two versions of NAB by counting the non-comment source code lines that are related to the Newton-Raphson minimization and to the calculation of the Born energy and its derivatives. Three categories of source code lines were counted: (1) source code lines that are required for serial execution, (2) source code lines that are required to modify the serial code for parallel execution by OpenMP, and (3) source code lines that are required to modify the serial code for parallel execution by MPI. The serial line count is 1643. The OpenMP line count is 180. The MPI line count is 962. These line counts reveal that adaptation of the serial code for MPI produced significantly (*i.e.*, a factor of five) more source code than adaptation of the serial code for OpenMP. This finding suggests that the programmer's productivity may be higher when an application that relies on linear algebra is parallelized using OpenMP instead of MPI.

## 4 Performance and Scalability

In order to obtain an accurate comparison between the OpenMP and MPI implementations of the code, we performed all of the measurements using the same server: a Sun Fire™ E25K server with 72 dual-core UltraSPARC IV processors. We have compared the performance of the two implementations for up to 144 OpenMP threads and MPI processes using the 1AKD, 1AFS and 1AMO molecular models from the RCSB Protein Data Bank[6] that comprise 6,370, 10,350 and 19,030 atoms (including hydrogen atoms), respectively.

We start by comparing different versions of the MPI implementation of NAB. Figure 2 (see the Implementation section) shows the profile of our baseline MPI implementation which does not use the table lookup optimization for address mapping. The division and modulus operations that are used in address mapping between the global matrix and sub-matrices are relatively inefficient and inhibit pipelined execution, thereby prolonging Phase 2. Moreover, for this version the parallelization of Phase 2 is implemented in a row cyclic manner similar to

---

[6] http://www.rcsb.org/pdb

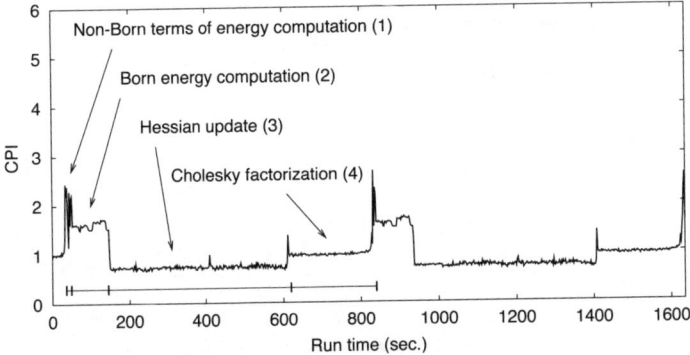

**Fig. 3.** CPI for MPI implementation with table lookup optimization, and using 5-way parallelism for Phase 2 and 25-way parallelization for Phases 3-4

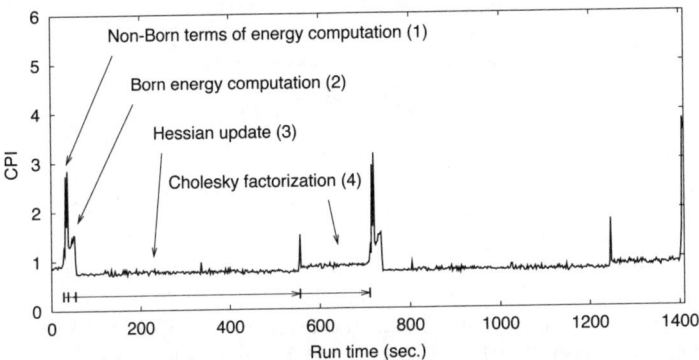

**Fig. 4.** CPI for the MPI implementation using 25-way parallelism for Phases 2-4

the approach that was discussed for OpenMP. For a given row of the process grid, all of the columns perform redundant computation. Therefore, when this baseline implementation executes using 25 processes on a 5 by 5 process grid, the degree of parallelization is limited to 5. The Phase 2 execution time is 269 seconds.

Figure 3 shows the execution profile for the improved version of the code in which the address mapping is implemented with lookup tables. Because these tables are small they are stored entirely within the processor caches, which facilitate table lookup and reduce the Phase 2 execution time to 105 seconds.

Figure 4 shows the execution profile for full parallelization of Phase 2. Full parallelization is achieved by having all process columns of a particular process row perform unique (instead of redundant) computation, thus increasing the degree of parallelization from 5 to 25 for a 5 by 5 process grid. The execution time for Phase 2 is 20 seconds, and thus this phase is no longer a barrier to the scalabiliity of the overall computation, as required by Amdahl's law [1].

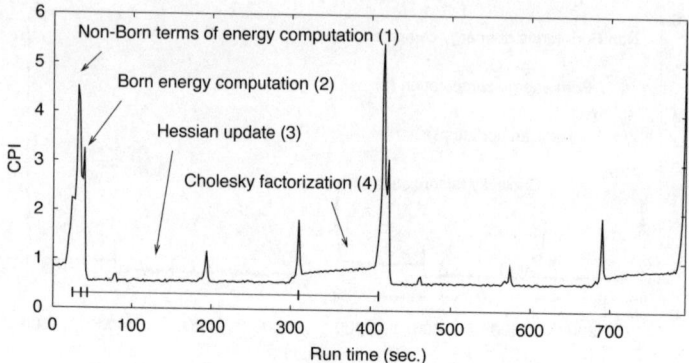

**Fig. 5.** CPI for the OpenMP implementation using 25-way parallelism for Phases 2-4

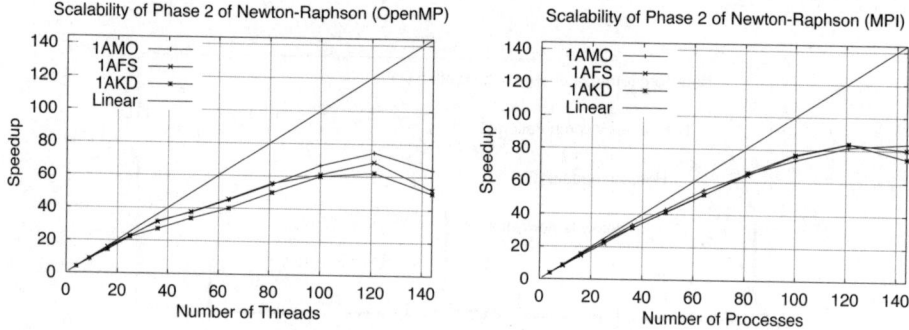

**Fig. 6.** Scalability of Phase 2 of Newton-Raphson for the 1AKD, 1AFS and 1AMO models, using a 144-core Sun Fire E25K server. For both plots, the scalability at 144 OpenMP threads or MPI processes increases in the order 1AKD, 1AFS and 1AMO. The "Linear" plot represents perfect scalability.

Figure 5 shows the execution profile for the OpenMP implementation of NAB that fully parallelizes Phase 2 of the Newton-Raphson minimization. The execution time for Phase 2 is 10 seconds, compared to 20 seconds for the fastest MPI implementation. This improvement can be attributed to the direct access to global matrix elements by OpenMP, which avoids the address mapping required by ScaLAPACK for distributed sub-matrix access in the MPI implementation.

In addition to measuring the relative speeds of the OpenMP and MPI implementations of NAB, we have examined the relative scalabilities of these implementations. Figure 6 shows that Phase 2 scales better for MPI than for OpenMP. This observation has been reported for other parallel applications [12]. In an attempt to understand the disparity between the scalabilities of the OpenMP and MPI implementations of NAB, we used the Sun Performance Analyzer to measure the level 2 cache miss rates during Phase 2 of the computation. We obtained these measurements for the 1AMO model and 4, 9, 16, 25 and 36 OpenMP threads or MPI processes executed on a Sun Fire E6900 server with

24 dual-core UltraSPARC IV processors. The cache miss rates per instruction were 0.0034 for OpenMP and 0.0012 for MPI, measured for between 4 and 36 OpenMP threads or MPI processes. Hence, relative to MPI we observed a nearly three-fold increase in the cache miss rate for OpenMP.

## 5 Conclusions

This article is a case study of parallelizing a molecular modeling application. We have demonstrated that energy minimization computations can be implemented in a highly-scalable way and can utilize up to 144 processors efficiently. In our experiments, the programmability and performance of the OpenMP version were superior to those of the MPI version, but the scalability of the MPI version was superior to that of the OpenMP version. Two other studies [2,14] have found the performance of OpenMP to be inferior to, or at best equal to that of MPI. One other study has found that the SPMD OpenMP programming style combined with careful code tuning can result in better performance for OpenMP [13]. In our experiments, it is likely that OpenMP performs better than MPI not only due to our use of the SPMD OpenMP programming style, but also due to the array indexing overhead imposed by ScaLAPACK in the MPI version.

We observe that the MPI version can execute on clusters of processors, in addition to symmetric multiprocessors. Clusters may have better nominal price/performance characteristics than large symmetric multiprocessors, although typically clusters have slower interconnects. A performance comparison between symmetric multiprocessors and clusters is beyond the scope of this work.

## Acknowledgments

We thank David Case, Guy Delamarter, Gabriele Jost, Daryl Madura, Eugene Loh and Ruud van der Pas for helpful comments.

## Trademark Legend

Sun, Sun Microsystems, UltraSPARC and Sun Fire are trademarks or registered trademarks of Sun Microsystems, Inc. in the United States and other countries.

## References

1. Amdahl, G.M.: Validity of the single-processor approach to achieving large scale computing capabilities. In: AFIPS Conference Proceedings, Reston, VA, pp. 483–485. AFIPS Press (1967)
2. Armstrong, B., Kim, S.W., Eigenmann, R.: Quantifying differences between OpenMP and MPI using a large-scale application suite. In: Valero, M., Joe, K., Kitsuregawa, M., Tanaka, H. (eds.) ISHPC 2000. LNCS, vol. 1940, Springer, Heidelberg (2000)

3. Bashford, D., Case, D.: Generalized born models of macromolecular solvation effects. Ann. Rev. Phys. Chem 51, 129 (2000)
4. Berkert, U., Allinger, N.: Molecular mechanics. ACS Monograph 177, American Chemical Society (1982)
5. Blackford, L., Choi, J., Cleary, A., D'Azevedo, E., Demmel, J., Dhillon, I., Dongara, J., Hammarling, S., Henry, G., Petitet, A., Stanley, K., Walker, D., Whaley, R.C.: Scalapack Users' Guide. Society for Industrial and Applied Math (1977)
6. Brown, R., Case, D.: Second derivatives in generalized born theory. J. Comput Chem 27, 1662–1675 (2006)
7. Chandra, R., Menon, R., Dagum, L., Kohr, D., Maydan, D., McDonald, J.: Parallel Programming in OpenMP. Morgan Kaufmann, San Francisco (2000)
8. Feig, M., Im, W., Brooks, C.J.: Implicit solvation based on generalized born theory in different dielectric environments. J. Chem. Phys. 120, 903–911 (2004)
9. Garg, R.P., Sharapov, I.: Techniques for Optimizing Applications: High Performance Computing. Prentice-Hall, Englewood Cliffs (2001)
10. Gropp, W., Lusk, E., Skjellum, A.: Using MPI: Portable Parallel Programming with the Message-Passing Interface, 2nd edn. MIT Press, Cambridge (1999)
11. Hawkins, G., Cramer, C., Truhlar, D.: Parametrized models of aqueous free energies of solvation based on pairwise descreening of solute atomic charges from a dielectric medium. J. Phys. Chem 100, 19824–19839 (1996)
12. Jin, H., Frumkin, M., Yan, J.: The OpenMP implementation of the NAS parallel benchmarks and its performance. NASA Ames Research Center, editor, Technical Report NAS-99-01 (1999)
13. Krawezik, G., Cappello, F.: Performance comparison of MPI and three OpenMP programming styles on shared memory multiprocessors. In: SPAA 2003: Proceedings of the Fifteenth Annual ACM Symposium on Parallel Algorithms, pp. 118–127. ACM Press, New York (2003)
14. Luecke, G.R., Lin, W.H.: Scalability and performance of OpenMP and MPI on a 128-processor SGI Origin 2000. Concurrency and Computation: Practice and Experience 13(10), 905–928 (2001)
15. Macke, T.: NAB, a language for molecular manipulation. PhD thesis, the Scripps Research Institute (1996)
16. Srinivasan, J., Trevathan, M., Beroza, P., Case, D.: Application of a pairwise generalized born model to proteins and nucleic acids: inclusion of salt effects. Theor. Chem. Acc. 101, 426–434 (1999)
17. Still, W., Tempczyk, A., Hawley, R., Hendrickson, T.: Semianalytical treatment of solvation for molecular mechanics and dynamics. J. Am. Chem. Soc 112, 6127–6129 (1990)
18. Weiner, P., Kollman, P.: AMBER: Assisted model building with energy refinement. A general program for modeling molecules and their interactions. J. Comp. Chem 2, 287–303 (1981)

# OpenMP Implementation of SPICE3 Circuit Simulator

Tien-Hsiung Weng[1], Ruey-Kuen Perng[1], and Barbara Chapman[2]

[1] Department of Computers Science and Information Engineering, Providence University
{thweng, rkperng}@pu.edu.tw
[2] Department of Computer Science, University of Houston
chapman@cs.uh.edu

**Abstract.** In this paper, we describe our experience of creating an OpenMP implementation of the SPICE3 circuit simulator program. The aim of this work is to present a case study showing the development of a shared memory parallel code with minimum effort and minimal code modification. We present our implementation and discuss the results of the case study in terms of what future compiler tools may be needed to help OpenMP application developers with similar porting goals. Our experiments, based on SRAM model simulation running on a SunFire V880 UltraSPARC-III 750 MHz with 4 CPUs, are promising.

## 1 Introduction

SPICE3 is a general purpose circuit simulation program for DC, transient, linear AC, pole-zero, sensitivity, and noise analyses developed by UC Berkeley [1][2] and written in C. Several commercial codes are based on SPICE. It is used to simulate circuits for various applications from switching power supplies to SRAM cells and sense amplifiers. Doing so requires the simultaneous solution of a number of equations that capture the behavior of electrical/electronic circuits. The number of equations can be quite large for a modern electronic circuit with transistor counts from several hundred thousands to millions, and thus the simulation of circuits has become complex and quite time-consuming. Thus, a shared memory parallel program version is needed to achieve cost-effective performance.

Circuit simulator programs have been parallelized using Pthreads [3]. Although good performance has been achieved, Pthreads provides a low-level and cumbersome programming model that is particularly useful for task parallelism. It requires major code rewriting and thus a major porting effort. Moreover, the resulting code is difficult to maintain in view of the many calls to Pthreads library routines and explicit coding of parallelism.

OpenMP [4] is an industry standard for shared memory parallel programming agreed on by a consortium of software and hardware vendors. It consists of a collection of compiler directives, library routines, and environment variables that can be easily inserted into a sequential program to create a portable program that will run in parallel on shared-memory architectures. It is considerably easier for a non-expert programmer to develop a parallel application under OpenMP than under either Pthreads or the de facto message passing standard MPI. OpenMP also permits the incremental development of parallel code. Thus it is not surprising that OpenMP has quickly become widely accepted for shared-memory parallel programming.

For many scientific applications, especially numerical codes written in Fortran, parallelization is chiefly a matter of distributing the computation in loops that modify large arrays. Thus parallelization via OpenMP is simply a matter of inserting directives to indicate parallel regions and loops, and specifying which variables are shared or private with few modifications of the original source code needed. Unfortunately, this is not the case for other applications, particularly if they are written in C/C++. Challenges arise in the OpenMP implementation of C codes with dynamic linked-list data structures such as the SPICE3 circuit simulator, but also encountered in agent-based model simulations [5], such as simulation of the immune response to a pathogen, financial markets applications, and more. In [5], Massaioli at. al. discuss three techniques for realizing pointer-chasing loops in OpenMP: 1) By explicit decomposition of the lists into approximately equal-sized chunks, storing pointers to these chunks in an array, and then adding *omp for* worksharing directives. With this approach, the list decomposition is difficult to parallelize. 2) By adding the *omp taskq* directive, which is available only in the Intel KSR KAP/Pro compiler, but which is not (yet) part of the official standard and thus introduces a portability problem. 3) By adding an *omp single nowait* clause to independently parallelizable linked-list loops. If the size of the system being simulated is sufficiently large, this may scale well.

Our aim in this work was to realize an OpenMP implementation of the SPICE3 circuit simulator with as few modifications to the sequential program as possible. Our approach relies on performing loop transformations and then adding OpenMP directives to the resulting loops. We discuss both possibilities for improving the OpenMP SPICE3 parallel program and developing it with minimum efforts using task queue without modification of the original program. Then, we present the result of our evaluation of this parallel version of SPICE3 on SunFire SPARC-III platforms and give our conclusions and future plans.

## 2 OpenMP Implementation

In this section, we give an overview of the original sequential SPICE3 program simulator. Next, in Section 2.2 we present our OpenMP implementation of SPICE3. We describe the steps taken to create the OpenMP program. In section 2.3, we discuss the possibilities for and challenges of further improvement of the parallel program. We also discuss our development of parallel SPICE3 implementation with minimum efforts and without modification to the original program using Intel *omp taskq* in Section 2.4.

### 2.1 Brief Overview of Sequential SPICE3

For modern electronic circuit design, the transient circuit simulation is the most frequently used simulation in SPICE3. Figure 1 shows the basic configuration of the SPICE3 transient simulation algorithm. There is a pre-processing portion which parses the circuit netlist and generates the appropriate data structures. Then, the matrix representing the circuit is created and the data structures related to the matrix are set up. Actual transient analysis occurs next. For each time point in the transient analysis, the model calculations for each device, such as MOSFET, resistor, or capacitor device, are performed. The electrical parameters such as conductance and current for each instance instantiated from the corresponding device are computed and put into the matrix elements. After the device and instance calculations, all elements in the matrix for the linear system in transient analysis are ready for the sparse matrix solver in SPICE3. Then the matrix calculations for the linear system, such as the LU decomposition and forward/backward elimination in each iteration, are carried out. This process will continue until the final transient time is reached. Finally, the simulation results are output.

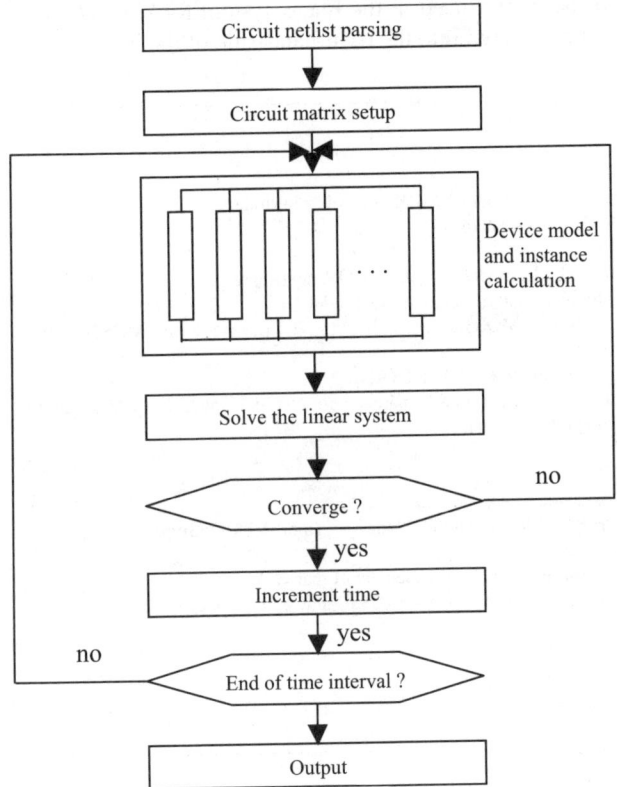

**Fig. 1.** Basic configuration of SPICE3 Simulator

## 2.2 Transforming Sequential Application to OpenMP

In order to reduce the effort in developing parallel code, we first try to compile the original code with the auto-parallelization option of the Sun compiler switched on to find loops that may be a good candidate for potential parallelization. Unfortunately, few loops are parallelizable, and their computational workload is very light. The most time consuming workload loops are the matrix calculation and model and instance calculation and these are not recognized as being parallelizable.

In this work, we focus exclusively on parallelizing this model and instance calculation part, shown in Figure 1. We refer to it as the device loading routine, because all the model parameters related to the device, and the parameters for the instantiations of the device are computed and loaded into the corresponding matrix elements. There are many devices, such as MOSFET, resistor, capacitor, diode, and bipolar transistor, supported by SPICE3. For each device, SPICE3 provides at least one model for the instances corresponding to this device used in the circuit simulated. For example MOS3 is one of the models for the instances of MOSFET device. The parameters such as the conductance and current are calculated according to the model equations built into the device loading routines. The conductance calculated will contribute to the elements of the matrix used in the linear system for simulation, while the calculated current will be entered into the right-hand-side of the linear system.

```
int MOS3load(inModel,ckt)
   GENmodel *inModel;
   register CKTcircuit *ckt;
{  register MOS3model *model = (MOS3model *) inModel;
   register MOS3instance *here;
   ........
   for( ; model != NULL; model = model->MOS3nextModel ) {
      /* loop through all the instances of the model */
      for (here = model->MOS3instances; here != NULL; here=here->MOS3nextInstance) {
         ........
         if ( ckt->CKTmode & MODETRAN ) {
            error = NIintegrate(ckt,&geq,&ceq,here->MOS3capbd, here->MOS3qbd);
            if(error) return(error);
         }
         ........
         // Right hand side of Ax = b
         *(ckt->CKTrhs + here->MOS3gNode) -=  (model->MOS3type * (ceqgs + ceqgb + ceqgd));
         ....
         //  Sum of contributions for the element of matrix A
         *(here->MOS3DdPtr) += (here->MOS3drainConductance);
         ........
   }} /* end of for loop */
   return(OK);
}
```

**Fig. 2.** Compact Sequential code of MOS3load of SPICE3

In this paper, we use a SRAM circuit as an example to demonstrate the SPICE3 simulation in its OpenMP implementation. The SRAM circuit consists of many instances of the MOSFET device with MOS3 model. Therefore the time-consuming part of the original sequential routine was the *MOS3load* function, which is the device-loading routine in SPICE3. It contains a nested pointer loop traversing an orthogonal linked-list. The actual size of the source code of the loop is approximately 1.3K LOC (Line of Code) and Figure 2 reproduces the compact example code. The size of iteration is depending on the size of the circuit, the number of devices such as transistor, capacitor, etc. simulated may vary widely. Currently, it is not possible to parallelize a pointer-chasing loop by just directly adding an OpenMP directive in a portable manner: the *omp taskq* directive available in the Intel compiler is not a standard feature. Since there is a *return* statement within this loop, it has multiple exits, and cannot be parallelized without modification in any case. Likewise, we cannot employ an *omp parallel for reduction* to obtain the sum of the values for each element of the linked list.

```
int MOS3load(inModel,ckt)
    GENmodel *inModel;
    register CKTcircuit *ckt;
{   ......
    register MOS3model *model = (MOS3model *) inModel;
    register MOS3instance *here;

    ......
    MOS3instance **MOS3instanceArray;
    MOS3instanceCount = model->MOS3instanceCount;
    MOS3instanceArray = model->MOS3instanceArray;
#pragma omp parallel default(none) shared(ckt, CONSTKoverQ,MOS3instanceCount,MOS3instanceArray)
 #pragma omp for private(vt,Check, SenCond,EffectiveLength,DrainSatCur, SourceSatCur, \
     GateSourceOverlapCap,GateDrainOverlapCap,GateBulkOverlapCap,Beta,OxideCap,vgs,vds,vbs, \
     vbd,vgb,vgd,xfact,vgdo,delvbs,delvbd,delvgs,delvds,delvgd,cbhat,cdhat,tempv, \
     cdrain,capgs,capgd,capgb,von,evbs,evbd,vdsat,cdreq,xrev,xnrm, ceqbd,ceqbs,ceqgb, \
     ceq,geq,vgs1,vgd1,vgb1,arg,sarg,sargsw,error,gcgs,ceqgs,gcgd,ceqgd,gcgb,model,here)
    for( i = 0; i < MOS3instanceCount; i++) {
        here = MOS3instanceArray[i];
        model = here->MOS3modPtr;
        ......
        #pragma omp critical(lockA)
        {  // Right hand side of Ax = b
           *(ckt->CKTrhs + here->MOS3gNode) -= (model->MOS3type * (ceqgs + ceqgb + ceqgd));
           ......
           // Sum of contributions for the element of matrix A
           *(here->MOS3DdPtr) += (here->MOS3drainConductance);
           *(here->MOS3GgPtr) += ((gcgd+gcgs+gcgb));
           ......
        } /* end critical */
    } /* end of for loop */
    return(OK);
} /* end of MOS3load() */
```

**Fig. 3.** OpenMP implementation of *MOS3load* in SPICE3

There is a straightforward way to parallelize the sequential nested loop shown in Figure 2. First, at the level of the circuit matrix setup of Figure 1, we introduce a data structure to store the address of each linked-list element of an instance in an array of pointers, *MOS3instanceArray[i]*, as well as to keep track of the total number of elements in the lists in a variable *model->MOS3instanceCount*. Second, we perform loop coalescing to reduce the number and nesting level of loops, as well as to generate loops with larger loop iteration count. The result is shown in Figure 3.

This loop now involves an array of pointers and integer index instead of pointers. Finally, we may now directly add a *parallel omp for* directive to the loop since the loop iterations are independent except for first, shared pointers that point to the variables that are used to update the right hand side of the linear system, $Ax = b$, and second, shared pointers that point to the elements of matrix $A$, which is used to sum the contributions for those elements. The *omp critical* synchronization directive is used to resolve this conflict, as shown in Figure 3. We have chosen to present the large number of private variables required, given the lack of a *default(private)* clause; we believe that such a clause would be beneficial for such codes. In our case, there are 56 private variables to be declared manually and given the presence of pointer variables and aliasing, automatic scoping of variables in parallel regions (as proposed in [6]) would be highly desirable but may be difficult for the compiler to perform.

Other model device-loading functions such as *CAPload* (capacitor load), *DIOload* (diode load), *VSRCload* (voltage-source load), and many more as shown in Figure 4, have a very similar program structure as *MOS3load*, so they can be parallelized the same way.

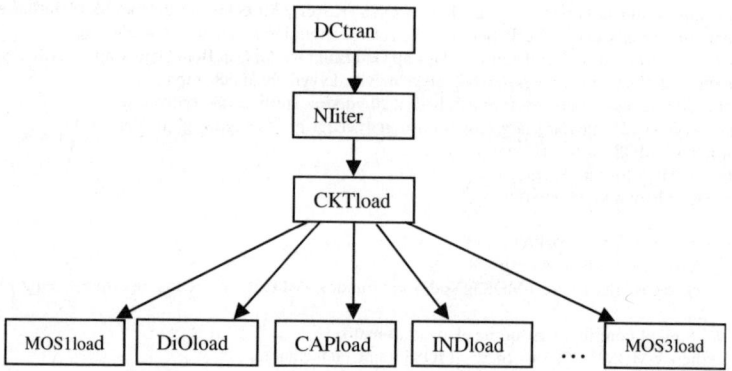

**Fig. 4.** Structure of a partial callgraph for a transient simulation of SPICE3 simulator program

The shortcoming of parallelizing this code by creating the parallel region within the *MOS3load* routine is that the routine is called many times, thereby involving considerable fork-join overheads in addition to the cost of the synchronization. Barrier and critical section overheads are, however, unavoidable.

### 2.3 Possible Improvements

The most important issue is to reduce the significant fork-join overhead incurred, since *MOS3load* is invoked several hundreds to thousands of times by other functions. The total number of calls to *MOS3load* depends on the number of time points and the nonlinear iteration counts at each time in the simulation. As the number of threads increases, fork-join overheads increase significantly. To improve this situation, we need to move the parallel region to include the calling functions. Unfortunately, it is non-trivial to do so. To explain the difficulties, we first manually created an incomplete or partial callgraph for a transient simulation of SPICE3 simulator program that shows invocations of our *MOS3load* function in Figure 4. This callgraph is built from dynamically bound calls resulting from variable pointer assignment mechanism. It is a quite time consuming task and it would be preferable to have a tool to assist in doing this, in particular to help a novice developer gain an understanding of the code.

```
........
for (i=0;i<DEVmaxnum;i++) {
    if ( (((*DEVices[i]).DEVload != NULL) && (ckt->CKThead[i] != NULL) ){
        error = (*((*DEVices[i]).DEVload))(ckt->CKThead[i],ckt);
        if (ckt->CKTnoncon)
}}
.....
```

**Fig. 5.** Dynamically bound call from *CKTload* routine

Suppose we are able to move the *omp parallel* from the *MOS3load* function to *CKTload*. There will be no improvement because there is only one dynamically bound call from the *CKTload* routine; as shown in Figure 5, this is realized by the value of the pointer *(*((*DEVices[i]).DEVload))(ckt->CKThead[i],ckt)*, which can point to *CAPload()*, *MOS2load()*, *MOS3load()*, *INDload()*, and/or many other device loading functions corresponding to calling relationships shown in Figure 4. It can be improved by moving it to the function *NIiter*. From inside *for(;;)* loop in *NIiter* function, there is call to the *CKTload* function; this loop iterates until the convergence criterion is met. Further, there is a call to *NIiter* from inside the *while(1)* loop of the *DCtran* function of Figure 4; these calls continue until the final transient time is reached. In other words, the call from *DCtran* to *NIiter* represents the outer loop and the call from *NIiter* to *CTKload* represents the inner loop of Figure 1. The parallelization of the code becomes more tedious, however.

### 2.4 Implementing Parallel SPICE3 without Modification

In this section, we discuss the OpenMP implementation of SPICE3 using Intel *omp taskq* without any modification made to the original program. Our implementation with few modifications discuss in Section 2.2 has been parallelized to perform device instances calculation per thread. To parallelize using *Intel omp taskq* at this level is not possible without more modifications efforts, because there are several branch statements such as *return* and especially the *continue* at the inner loop of *MOS3load* function as shown Figure 6, which is not allowed by the compiler.

In order to parallelize it without any modification, we have to parallelize at a coarser grain that is at the level of device instances of a model (each model consists of many device instances) per task. This is shown in Figure 7. With this level of granularity (model per thread), the load imbalance problem can occur, since each model may consists of different number of device instances, but then the OpenMP directives can be inserted directly to original program.

```
int MOS3load(inModel,ckt)
{ .................
  #pragma omp parallel shared(ckt, CONSTKoverQ,inModel)
    for( ; model != NULL; model = model->MOS3nextModel ) {
    #pragma intel omp taskq private(.....)
      for (here = model->MOS3instances; here != NULL; here=here->MOS3nextInstance) {
        #pragma intel omp task
        { .....
          if (SendCond) continue;
          .....
          if (here->MOS3senPertFlag == OFF) continue;
          .....
        } /* end of omp task */
    }} /* end of for loop */
  return(OK);
}
```

Fig. 6. Device instance level of granularity

```
int MOS3load(inModel,ckt)
{
#pragma omp parallel default(none) shared(ckt, CONSTKoverQ,inModel)
  #pragma intel omp taskq private(vt,Check, SenCond,EffectiveLength, \
                                  DrainSatCur,SourceSatCur, ........... ,model,here)
  for(model=inModel; model!=NULL;model=model->MOS3nextModel) {
  #pragma intel omp task
    for(here=model->MOS3instances; here!=NULL; here=here->MOS3nextInstance){
      ...
      #pragma omp critical(lockA)
      {  // Right hand side of Ax = b
         // Sum of contributions for the element of matrix A
      } /* end critical */
  }}
  return(OK);
}
```

Fig. 7. All instances of a model level of granularity

The above source code of Figure 7 has been successfully compiled by Intel OpenMP compiler on Linux Itanium of 4-CPUs machine. We include this experimental result in the next section.

## 3 Experiments

Our experimental results are based on an 8K SRAM model simulation compiled with SUN compiler and run on a SunFire V880 Ultra SPARC-III 750 MHz with 4 CPUs and 4G memory. In this simulation, *MOS3load* is invoked 780 times; this numbers depends on the number of time points simulated and the nonlinear iteration counts of each time. The increase of this number does not affect the scalability of the performance. On the other hand, there are 61,584 total instances, which represent the size (number of iterations) of the coalesced loop iterations in *MOS3load*. This in turn depends on the number of device instances involved in the circuit simulation. Figures 8 and 9 show that it performs well up to three processors for entire SPICE3 program (note that the rest of SPICE3 is not parallelized here).

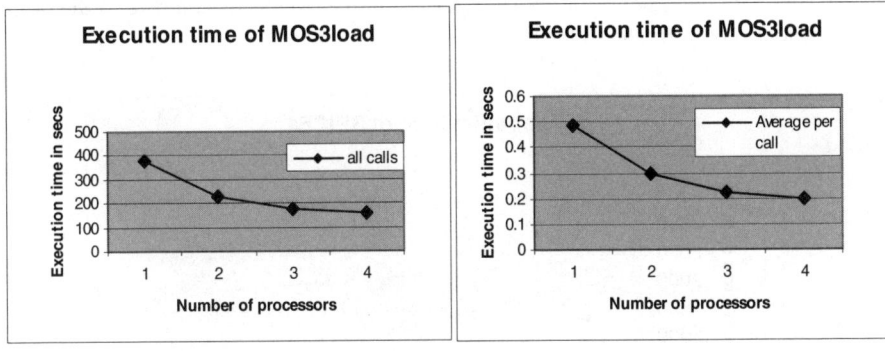

**Fig. 8.** The performance of OpenMP MOS3load function

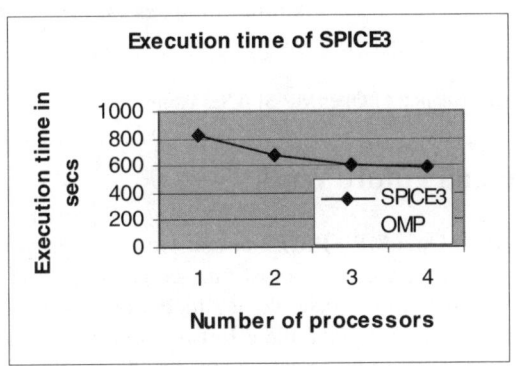

**Fig. 9.** The performance of OpenMP SPICE3

Other than the device model and instance calculation, the sparse matrix computation in SPICE3 is fairly time consuming. We are currently studying the parallelization of a public domain linear algebra sparse matrix package implemented in SPICE3 by Kundert [7]. It is not part of this paper. The code uses an orthogonal linked list data structure to store the sparse matrix. The matrix computation technique is based on the direct method with LU decomposition. With four processors, the fork-join overhead costs more than the execution time of each call to *MOS3load*, since its average execution time is only about 0.2 seconds. To scale well, we would need to simulate a larger number of device instances.

We further perform the experiment with Intel taskq as shown in Figure 7 and compare it with our approach as in Figure 3. They both were compiled with Intel icc compiler version 9.0 under the option –O2 –openmp and run on a 4-processors IBM eServer xSeries 380 Itanium 733 MHz with 16GB memory. Again, we only parallelize the MOS3load part and not the rest of SPICE3, but this time we simulate based on a 16K SRAM model simulation, which has a larger number of device instances. The result in Figure 10 shows that both perform well for entire SPICE3 program if matrix calculation part is also parallelized. But, with the taskq, it incurs significant runtime overheads compare to the original sequential code without –open mp compilation option label as 'seq'.

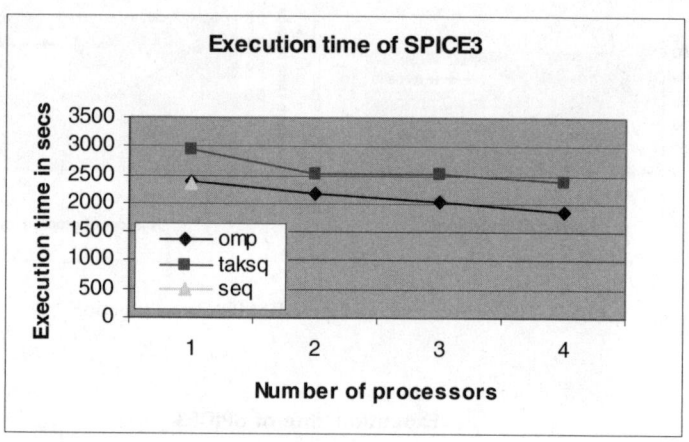

**Fig. 10.** The performance of OpenMP SPICE3 using Intel taskq runing on Itanium

## 4 Conclusions and Future Work

We have developed an OpenMP SPICE3 circuit simulator program. The matrix and model device calculations are the two most time-consuming parts of the computation. We present our implementation of the device model and instance calculation part of SPICE3. Our goals were to minimize the effort required and the amount of modification of the original program. Our experimental results are promising in this respect, despite the data structures used. We discussed possible improvements; however, they do require more programming effort. We also explained the need for a compiler tool that provides the novice user with a precise callgraph, even in the presence of dynamically bound calls (in section 2.3). We are continuing our work by creating an OpenMP version of the sparse matrix calculation (Sparse matrix package in SPICE3) that is also a time-consuming part of SPICE3.

## Acknowledgements

We are grateful to our colleagues in the Sun Center of Excellence in Geosciences at the University of Houston for their provision of a platform that enabled use to perform the experiments reported on in this paper.

# References

1. Nagel, L.W.: SPICE2 - A Computer program to simulate semiconductor circuits, University of California, Berkeley, ERL. Memo ERL-M520 (May 1975)
2. Quarles, T.L.: Analysis of Performance and Convergence Issues for Circuit Simulation, University of California, Berkeley, ERL. Memo ERL-M89 (April 1989)
3. Lee, P.M., Ito, S., Hashimoto, T., Sato, J., Touma, T., Yokomizo, G.: A Parallel and Accelerated Circuit Simulator with Precise Accuracy. In: Procedings of the 15th International Conference on VLSI Design (2002)
4. OpenMP Architecture Review Board, Fortran 2.0 and C/C++ 1.0 Specifications, At: http://www.openmp.org
5. Massaioli, F., Castiglione, F., Bernaschi, M.: OpenMP parallelization of agent-based models. Journal of Parallel computing (submitted, 2005)
6. Lin, Y., Terboven, C., an Mey, D., Copty, N.: Automatic Scoping of Variables in Parallel Regions of an OpenMP Program. In: Chapman, B.M. (ed.) WOMPAT 2004. LNCS, vol. 3349, pp. 83–97. Springer, Heidelberg (2005)
7. Kundert, K.: Sparse Matrix Techniques. In: Ruehli, A. (ed.) Circuit Analysis, Simulation and Design, North-Holland, Amsterdam (1986)

# Automatic Generation of Parallel Code for Hessian Computations

H. Martin Bücker, Arno Rasch, and Andre Vehreschild

RWTH Aachen University, Institute for Scientific Computing,
Seffenter Weg 23, D-52074 Aachen, Germany
{buecker,rasch,vehreschild}@sc.rwth-aachen.de

**Abstract.** Given a program to compute some function, automatic differentiation can be used to mechanically generate another program capable of evaluating first- and higher-order derivatives of that function. A new strategy for the computation of Hessians by automatic differentiation is proposed where the generated code is automatically parallelized using OpenMP. The approach is applied to compute second-order derivatives of an atmospheric reference model and performance results on a Sun Fire E6900 system are reported.

## 1 Introduction

First- and second-order derivatives are required in various areas of scientific computing, for instance in algorithms for nonlinear optimization and nonlinear equations [21,17,20,28,27] or optimal experimental design [2]. Sometimes these derivatives are easy to calculate by hand. However, in a growing number of cases arising from real-world applications in science and engineering, the underlying functions are represented by large programs written in C, C++, Fortran or MATLAB and are too complicated. That is, it is no longer reasonable to expect the user to provide code to compute the corresponding Jacobians or Hessians by hand. Instead, one is often relying on numerical approximations by divided differences. While this approach based on numerical differentiation is easy to implement by calling the program multiple times with perturbed input values, its significant drawback is the presence of truncation errors. Fortunately, automatic differentiation remedies this issue by transforming a given computer program to a new program capable of evaluating (higher-order) derivatives without truncation error. Compared to the original program, the number of floating point operations of the corresponding program generated by this technique is increased, sometimes significantly for higher-order derivatives. Therefore, there is a need for parallelism in derivative computations [13,3,6,7,9,14,15,25,26,19,29].

In this article, we propose a novel strategy for automatically parallelizing Hessian computations using OpenMP which is implemented in ADIFOR [5,18], a software tool for automatic differentiation of Fortran 77 programs. The feasibility of this approach is demonstrated by an application to the atmospheric reference model MSIS–86 [24] predicting temperature and concentration profiles of species in the Earth's atmosphere above 120 km.

The structure of this note is as follows. In Sect. 2, the technology of automatic differentiation is briefly sketched. The new strategy to automatically parallelize the computation of Hessians is introduced in Sect. 3. This strategy is applied to the atmospheric reference model MSIS–86 in Sect. 4 where the parallel performance of the approach is reported.

## 2 Automatic Differentiation

Automatic Differentiation (AD) is a technology for automatically augmenting computer programs with statements for the computation of derivatives. The basic idea behind AD is that any computer program $P$ performs a—potentially very long—sequence of elementary mathematical operations like binary addition or multiplication, or intrinsic functions, of which the derivatives are known. For each elementary mathematical operation occurring in $P$, the AD technology generates a corresponding derivative computation. Combining the elementary derivative operations according to the chain rule yields a new program $P^{AD}$ that is capable of not only computing the original function implemented by $P$, but also derivatives of selected outputs, called *dependent variables*, with respect to certain input parameters, referred to as *independent variables*.

In the so-called *forward mode* of automatic differentiation the derivatives are computed along with the original function. For example, for a statement $c = f(a, b)$, where $f$ denotes an elementary binary operation, the derivative of $c$ can be computed by

$$\nabla c = \frac{\partial c}{\partial a}\nabla a + \frac{\partial c}{\partial b}\nabla b \,. \tag{1}$$

It is assumed that the derivatives of the elementary operation $f$ are known, and the gradients $\nabla a$ and $\nabla b$ are computed along with the values $a$ and $b$. The size of the gradients corresponds to the number of directional derivatives propagated through the code and may in general be greater than one, say $n$. Therefore the computation in (1) may actually involve a loop iterating over the $n$ entries of the gradients. Thus, the AD-generated program $P^{AD}$ needs $O(n)$ times more operations than the original program $P$.

There is a number of software tools available, implementing the AD technology for various languages such as Fortran, C, C++, or MATLAB. For a detailed list of tools visit the web portal of automatic differentiation, www.autodiff.org. A thorough introduction into the theory of AD is given in [30,22]; applications of AD in different numerous areas are contained in [23,4,16,10].

AD can also be employed to compute higher-order derivatives. For the statement $c = f(a, b)$, the Hessian of $c$ can be computed by

$$\nabla^2 c = \frac{\partial c}{\partial a}\nabla^2 a + \frac{\partial c}{\partial b}\nabla^2 b + \frac{\partial^2 c}{\partial a^2}(\nabla a \cdot \nabla a^T) + \frac{\partial^2 c}{\partial b^2}(\nabla b \cdot \nabla b^T)$$
$$+ \frac{\partial^2 c}{\partial a \partial b}(\nabla a \cdot \nabla b^T + \nabla b \cdot \nabla a^T) \,. \tag{2}$$

This way, the computation of the Hessian $\nabla^2 c$ can be performed by $O(n^2)$ times the number of operations of the original program $P$, where $n$ is the number of independent variables. In practice, however, one would save half of the operations and storage by exploiting the symmetry of the Hessians. For example, in [1], the Hessians are stored in the LAPACK packed symmetric scheme.

## 3 Automatically Parallelizing Hessian Computations

The execution time and memory requirement of the *differentiated* program computing first- and second-order derivatives along with the original function values increases by an order of $n^2$, relative to the original program. Especially for large $n$ the execution time of the differentiated program dramatically increases, and, if the program is executed multiple times, e.g., within an optimization framework requiring Hessian evaluations at different points, this overhead in CPU time is even multiplied. To overcome this situation we suggest the parallel execution of the differentiated code.

The idea is based on the fact that, for large $n$, almost all the computational work is derivative computation, and the type of the operations for the derivative computations are always similar, e.g., vector linear combinations for first-order derivatives, which can be easily parallelized using appropriate OpenMP directives. In [11,12] two strategies for parallelizing the computation of first-order derivatives with OpenMP have been proposed. Both strategies can be implemented in software tools for AD, making such an AD tool capable of generating parallel differentiated code.

In this work, we specifically extend the strategy presented in [12] to the parallel computation of Hessians, and report on a recent implementation using the ADIFOR 3.0 [18] automatic differentiation software and the language-independent Hessian module presented in [1] which is interfacing with ADIFOR and ADIC [8].

Recognizing that the loops iterating over the Hessians have always the same length, namely $n(n+1)/2$, if symmetry is exploited, and that each entry in the Hessian can be computed independently, we suggest the following strategy for automatically parallelizing AD-generated code with OpenMP:

1. The whole differentiated code is executed in parallel, i.e., the call to the differentiated subroutine is performed within a parallel region. In case a driver routine for the differentiated subroutine is generated, as provided by ADIFOR 3.0, the corresponding OpenMP directives can be automatically inserted in this driver without changing the calling sequence of the driver routine. This way, the user can call the driver just like in the serial case.
2. All program variables holding second-order derivatives are *shared*. For the second-order derivatives occurring in the lexical extent of the parallel region that has been created in the previous step, this could be achieved by using the OpenMP *shared* clause. All remaining variables holding second-order derivatives, i.e., Hessian variables occurring outside the lexical extent of the parallel region, need to be static in order to make them *shared*. In Fortran,

this is achieved by explicitly adding the `save` attribute to these Hessian variables.

In the present implementation, all second-order derivatives are stored in one-dimensional arrays of length $n(n + 1)/2$, representing Hessians in LAPACK packed symmetric storage format. When a Hessian array is updated, like, e.g., $\nabla^2 c$ in (2), the work is shared by the available threads. Since the total size of the Hessian arrays is known in advance, the portion of work delegated to each thread can be determined in advance and kept fixed for all the loops involving Hessian computations. When entering the parallel region, we divide the work on the Hessian arrays such that $p$ threads are assigned disjoint portions of size approximately $n(n+1)/(2p)$. More precisely, for each thread, we compute a pair of indices (`LB`,`UB`) specifying the lower and upper bound of the chunk of the Hessian assigned to this thread. These indices are *private* to each thread. In the current implementation, we also store a pair of indices (`my_i`,`my_j`) indicating the row and column of the Hessian entry that corresponds to the element number `LB` in the one-dimensional array representation in LAPACK packed symmetric storage scheme. These indices are used to pick the correct values from the gradient vectors when computing the outer products element-wise for the elements `LB` to `UB` in the one-dimensional array representation of the Hessian.

3. The computations related to the original function are performed redundantly by each thread. Hence, all variables from the original code must be private. If the variables are in the scope of the parallel region generated in the first step, additional *private* or *firstprivate* clauses for these variables need to be inserted. Intermediate variables occurring outside the lexical scope are *private* by default, except for static variables. These need to be made private by using the *threadprivate* directive. In the present implementation, inactive static variables, those static variables that do not require derivative information, are not recognized and need a *threadprivate* directive explicitly added by the user.

4. Computations of the first-order derivatives are also performed redundantly. The rules for the variables containing gradients are the same as for the variables related to the original function, described in the previous step. The reason for redundant computation of the first-order derivatives is that, for nonlinear operations, the computation of the Hessian requires the full gradients of the arguments for each entry of the Hessian. Hence, additional barriers would be needed before every nonlinear operation when parallelizing the first-order derivative computation. However, we suggest redundant evaluation of gradients in order to save those barriers. In fact, the method presented in this note has the clear advantage that it does not create any barriers within the differentiated code.

As an example, we demonstrate the computation of first- and second-order derivatives of a binary multiplication using the parallelization strategy described above. For a binary multiplication $c = a \cdot b$, the terms of (2) involving $\partial^2 c/\partial a^2$

```fortran
      SUBROUTINE ad_fh_fmulad(nmax,qmax,n,q,c,g_c,h_c,a,g_a,h_a,b,g_b,
     +     h_b)

      INTEGER i, j, k, n, q, nmax, qmax
      DOUBLE PRECISION c, a, b, g_c(nmax), g_a(nmax), g_b(nmax),
     +     h_c(qmax), h_a(qmax), h_b(qmax)

      INTEGER LB, UB, my_i, my_j
      COMMON /adomp/ LB, UB, my_i, my_j
c$omp threadprivate (/adomp/)
      i = my_i
      j = my_j
c-- Hessian computation
      DO k = LB, UB
         h_c(k) = b*h_a(k) + a*h_b(k) + g_a(i)*g_b(j) + g_a(j)*g_b(i)
         IF (i.eq.j) then
            i = i+1
            j = 1
         ELSE
            j = j+1
         ENDIF
      ENDDO
c-- gradient computation
      DO i = 1, n
         g_c(i) = b*g_a(i) + a*g_b(i)
      ENDDO
      END
```

**Fig. 1.** Fortran code for a parallel update of the Hessian h_c for a binary multiplication with scalar arguments a and b. The Hessians of a and b are denoted by h_a and h_b, respectively. The corresponding gradients are denoted by g_a and g_b. The gradient of c is updated in a redundant fashion.

and $\partial^2 c/\partial b^2$ vanish. So, the gradient $\nabla c$ and Hessian $\nabla^2 c$ can be computed by

$$\nabla c = b \cdot \nabla a + a \cdot \nabla b \quad \text{and}$$
$$\nabla^2 c = b \cdot \nabla^2 a + a \cdot \nabla^2 b + \nabla a \cdot \nabla b^T + \nabla b \cdot \nabla a^T \ .$$

The corresponding Fortran code with OpenMP directives is given in Figure 1.

The example presented in this figure performs a parallel update of the Hessian (returned in the one-dimensional array h_c) for a binary multiplication with scalar arguments a and b. The Hessians of a and b are denoted by h_a and h_b, respectively. The corresponding gradients are denoted by g_a and g_b. The gradient vector of c is also updated, but in a redundant fashion. Practically, the code for the derivative computations is encapsulated in subroutines which are provided in a library. As described before, each thread has a private pair of integers, (LB,UB), specifying lower and upper bounds of the chunk it is assigned to. The parallel computation of the outer products is performed using temporary integer variables (i,j) selecting the correct elements from the gradient vectors. Another private pair of integers, (my_i,my_j), is needed to initialize the temporary variables (i,j). The values for LB, UB, my_i, and my_j are computed once, and passed to the library routines via a *threadprivate* common block. Alternatively, these values could be passed via the argument list.

## 4 Parallel Second-Order Derivatives for MSIS–86

Atmospheric reference models predicting temperature and concentration profiles of species in the upper atmosphere were first developed in the early sixties based on theoretical considerations and satellite drag data. A prominent example of an atmospheric reference model is a suite of models known as the Mass Spectrometer Incoherent Scatter (MSIS) models [24] developed at NASA's Goddard Space Flight Center.

In a preparatory step, the MSIS–86 code was slightly modified to strictly conform to the Fortran 77 standard. All variables are initialized before their first use. In addition, all single precision variables and constants are promoted to double-precision in order to avoid under- or overflow in the derivative computations which tend to have a larger dynamic range of values than computations occurring in the original function.

In the following experiments, we consider the function

$$\varrho = f(\delta, \lambda, \mathbf{p})$$

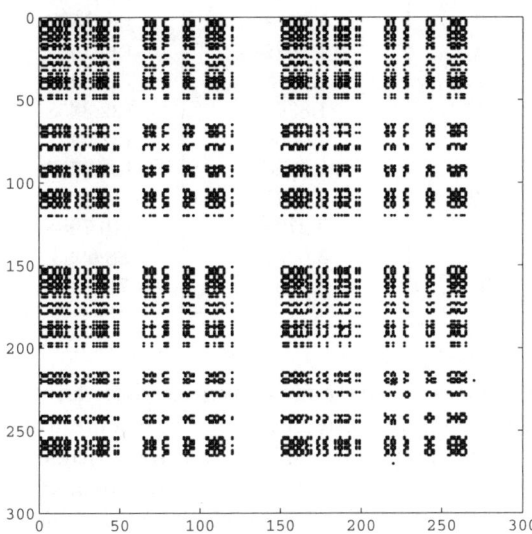

**Fig. 2.** The sparsity pattern of the Hessian $\partial^2 \varrho / \partial \mathbf{p}^2$

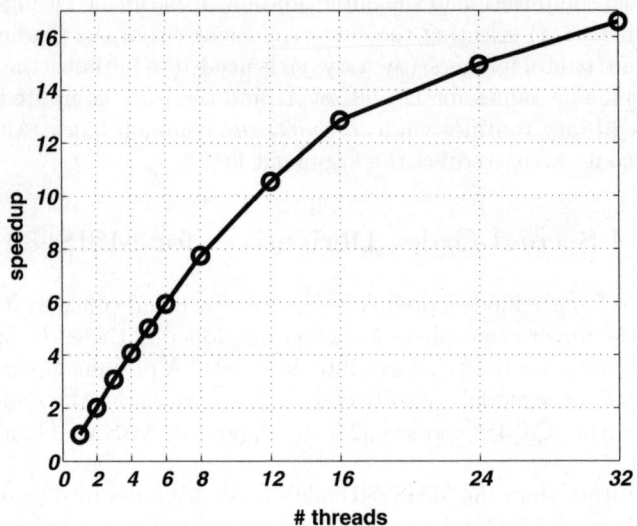

**Fig. 3.** Speedup of the parallel Hessian computation for the MSIS–86 model

**Table 1.** Approximate memory bandwidth consumed by the application running with various numbers of threads, and achieved Mflops rate per thread

| # threads | memory bandwidth [GB/s] | Mflops/thread |
|---|---|---|
| 1 | 0.5944 | 129.7795 |
| 2 | 0.9678 | 131.3873 |
| 3 | 1.3082 | 134.1513 |
| 4 | 1.5915 | 133.5491 |
| 5 | 1.8365 | 132.3858 |
| 6 | 2.0469 | 131.0441 |
| 8 | 2.4253 | 129.3738 |
| 12 | 2.8741 | 119.2794 |
| 16 | 3.2792 | 110.7988 |
| 24 | 3.6324 | 88.6993 |
| 32 | 3.8884 | 75.9986 |

computing the total mass density $\varrho$ for a given geodetic latitude $\delta$, longitude $\lambda$, and 300 scalar parameters **p** representing measurements from several rockets, satellites and incoherent scatter radars. Note that the altitude and local apparent solar time are kept constant. The function $f$ is evaluated on an equidistant $10 \times 10$ grid varying $\delta$ and $\lambda$. Selecting $\varrho$ and **p** as dependent and independent variable, respectively, we automatically generate OpenMP-parallelized derivative code for computing $\partial \varrho / \partial \mathbf{p}$ and $\partial^2 \varrho / \partial \mathbf{p}^2$, in addition to $\varrho$. The sparsity pattern of the Hessian $\partial^2 \varrho / \partial \mathbf{p}^2$ for fixed parameters $\delta$ and $\lambda$ is displayed in Figure 2.

Performance experiments have been conducted on a Sun Fire E6900 system, equipped with 24 UltraSparc IV dual core processors running at 1.2 GHz clock speed, and 96 GByte of main memory. The speedup for 1 to 32 threads is given in Figure 3 where the speedup is related to the parallel version with a single thread.

For a small number of threads, the speedup is almost linear. For larger numbers of threads, only a moderate increase of the speedup can be observed. This is likely caused by the dramatic increase of memory traffic generated by large numbers of threads. In the Sun Fire E6900 system, there are 6 CPU boards containing 4 processors each. Since the majority of the data is located on one board, the limiting resource is the memory bandwidth between board and backplane, which is 4.8 GB/s on the Sun Fire E6900. The approximate memory bandwidth consumed by the application and the Mflops rate per thread is summarized in Table 1. In particular, for large number of threads, the memory bandwidth consumed by the application is quite close to the theoretical maximum of 4.8 GB/s. Nevertheless, the result is remarkable since speedup is achieved in a fully automatic way that requires no interaction with the user.

## 5 Concluding Remarks

Given a serial source code, a set of techniques referred to as automatic differentiation can be used to generate code for computing gradients and Hessians. Depending on the size of the Hessians, this code may consume considerably more execution time and memory, compared to the original program. Therefore, we suggest to use OpenMP to automatically parallelize the computation of the Hessians, which is by far the most expensive task. The key idea of our new approach is the fact that the computation of each element of a Hessian can be performed independently by relatively simple loops which are easy to parallelize. The total size of the Hessian is typically large leading to large data structures and loops involving many iterations. Hence, the shared-memory model is particularly well-suited for this parallelization approach. Since the evaluation of the function computed by the original code is tightly interleaved with the first- and second-order derivative computation, we perform a redundant evaluation of the original function and its gradients, enabling parallel processing without synchronization. The new implementation of the parallelization strategy using ADIFOR 3.0 [18] and the Hessian module [1] is able to generate ready-to-use parallel code for computing first- and second-order derivatives.

Finally, we report on an application of the proposed strategy to the MSIS–86 atmospheric model leading to an augmented model capable of evaluating second-order derivatives in parallel. Performance results have shown the feasibility of the new approach. We stress that this approach is fully automatic, which, integrated in automatic differentiation tools, allows the user to accurately compute large Hessians in an efficient way.

## Acknowledgments

The authors are grateful to Mike Fagan of Rice University for sharing his insight into the ADIFOR 3.0 system. This work was stimulated by Marc Kalkuhl and Wolfgang Wiechert of the Department of Simulation, University of Siegen, Germany, by their interest in high precision satellite orbit simulations. We also thank Samuel Sarholz, Alexander Spiegel, and Christian Terboven for valuable discussions on the performance analysis. This research is partially supported by the Deutsche Forschungsgemeinschaft (DFG) within SFB 401 "Modulation of flow and fluid–structure interaction at airplane wings," RWTH Aachen University, Germany.

## References

1. Abate, J., Bischof, C., Carle, A., Roh, L.: Algorithms and design for a second-order automatic differentiation module. In: Proceedings of the 1997 International Symposium on Symbolic and Algebraic Computing (ISSAC 1997), pp. 149–155. ACM Press, New York (1997)
2. Atkinson, A.C., Donev, A.N.: Optimum Experimental Designs. Oxford Statistical Science Series, vol. 8. Oxford University Press, Oxford (1992)
3. Benary, J.: Parallelism in the reverse mode. In: Berz et al. [4], pp. 137–148
4. Berz, M., Bischof, C., Corliss, G., Griewank, A. (eds.): Computational Differentiation: Techniques, Applications, and Tools. SIAM, Philadelphia (1996)
5. Bischof, C., Carle, A., Khademi, P., Mauer, A.: ADIFOR 2.0: Automatic Differentiation of Fortran 77 Programs. IEEE Computational Science & Engineering 3(3), 18–32 (1996)
6. Bischof, C., Griewank, A., Juedes, D.: Exploiting parallelism in automatic differentiation. In: Houstis, E., Muraoka, Y. (eds.) Proceedings of the 1991 International Conference on Supercomputing, pp. 146–153. ACM Press, New York (1991)
7. Bischof, C.H.: Issues in parallel automatic differentiation. In: Griewank, A., Corliss, G. (eds.) Automatic Differentiation of Algorithms, pp. 100–113. SIAM, Philadelphia (1991)
8. Bischof, C.H., Roh, L., Mauer, A.: ADIC — An extensible automatic differentiation tool for ANSI-C. Software: Practice and Experience 27(12), 1427–1456 (1997)
9. Bücker, H.M., Buschelman, K.R., Hovland, P.D.: A Matrix-Matrix Multiplication Approach to the Automatic Differentiation and Parallelization of Straight-Line Codes. In: Brinkschulte, U., Großpietsch, K.-E., Hochberger, C., Mayr, E.W. (eds.) Workshop Proceedings of the International Conference on Architecture of Computing Systems ARCS 2002, Germany, April 8–12, 2002, pp. 203–210. VDE Verlag (2002)
10. Bücker, H.M., Corliss, G.F., Hovland, P.D., Naumann, U., Norris, B. (eds.): Automatic Differentiation: Applications, Theory, and Implementations. Lecture Notes in Computational Science and Engineering, vol. 50. Springer, Heidelberg (2005)
11. Bücker, H.M., Lang, B., an Mey, D., Bischof, C.H.: Bringing Together Automatic Differentiation and OpenMP. In: Proceedings of the 15th ACM International Conference on Supercomputing, Sorrento, Italy, June 17–21, 2001, pp. 246–251. ACM Press, New York (2001)

12. Bücker, H.M., Lang, B., Rasch, A., Bischof, C.H., an Mey, D.: Explicit Loop Scheduling in OpenMP for Parallel Automatic Differentiation. In: Almhana, J.N., Bhavsar, V.C. (eds.) Proceedings of the 16th Annual International Symposium on High Performance Computing Systems and Applications, Moncton, NB, Canada, June 16–19, 2002, pp. 121–126. IEEE Computer Society, Los Alamitos (2002)
13. Bischof, C.H., Hovland, P.D.: Automatic Differentiation: Parallel Computation. In: Floudas, C.A., Pardalos, P.M. (eds.) Encyclopedia of Optimization, vol. I, pp. 102–108. Kluwer Academic Publishers, Dordrecht, The Netherlands (2001)
14. Carle, A.: Automatic Differentiation. In: Dongarra, J., Foster, I., Fox, G., Gropp, W., Kennnedy, K., Torczon, L., White, A. (eds.) Sourcebook of Parallel Computing, pp. 701–719. Morgan Kaufmann, San Francisco, CA (2003)
15. Carle, A., Fagan, M.: Automatically Differentiating MPI-1 Datatypes: The Complete Story. In: Corliss et al. [16], pp. 215–222
16. Corliss, G., Faure, C., Griewank, A., Hascoët, L., Naumann, U.: Automatic Differentiation of Algorithms: From Simulation to Optimization. Springer, New York (2002)
17. Dennis Jr., J.E., Schnabel, R.B.: Numerical Methods for Unconstrained Optimization and Nonlinear Equations. Prentice-Hall, Englewood Cliffs (1983)
18. Fagan, M., Carle, A.: Adifor 3.0 overview. Technical Report CAAM–TR00–03, Rice University, Department of Computational and Applied Mathematics (2000)
19. Fischer, H.: Automatic differentiation: Parallel computation of function, gradient and Hessian matrix. Parallel Computing 13, 101–110 (1990)
20. Fletcher, R.: Practical Methods of Optimization, 2nd edn. John Wiley & Sons, New York (1987)
21. Gill, P.E., Murray, W., Wright, M.H.: Practical Optimization. Academic Press, New York (1981)
22. Griewank, A.: Evaluating Derivatives: Principles and Techniques of Algorithmic Differentiation. SIAM, Philadelphia (2000)
23. Griewank, A., Corliss, G.: Automatic Differentiation of Algorithms. SIAM, Philadelphia (1991)
24. Hedin, A.E.: MSIS-86 thermospheric model. Journal of Geophysical Research 92(A5), 4649–4662 (1987)
25. Hovland, P.: Automatic Differentiation of Parallel Programs. PhD thesis, University of Illinois at Urbana-Champaign, Urbana, IL, USA (1997)
26. Hovland, P.D., Bischof, C.H.: Automatic differentiation of message-passing parallel programs. In: Proceedings of the First Merged International Parallel Processing Symposium and Symposium on Parallel and Distributed Processing, Orlando, FL, March 30–April 3, 1998, pp. 98–104. IEEE Computer Society Press, Los Alamitos (1998)
27. Kelley, C.T.: Iterative Methods for Optimization. SIAM, Philadelphia (1999)
28. Nocedal, J., Wright, S.J.: Numerical Optimization. Springer, New York (1999)
29. Heimbach, P., Hill, C., Giering, R.: An efficient exact adjoint of the parallel MIT general circulation model, generated via automatic differentiation. Future Generation Computer Systems 21(8), 1356–1371 (2005)
30. Rall, L.B.: Automatic Differentiation. LNCS, vol. 120. Springer, Heidelberg (1981)

# Geographical Locality and Dynamic Data Migration for OpenMP Implementations of Adaptive PDE Solvers

Markus Nordén, Henrik Löf, Jarmo Rantakokko, and Sverker Holmgren

Uppsala University, Department of Information Technology
Box 337, 751 05 Uppsala, Sweden
markus.norden@it.uu.se

**Abstract.** On cc-NUMA multi-processors, the non-uniformity of main memory latencies motivates the need for co-location of threads and data. We call this special form of data locality, *geographical locality*. In this article, we study the performance of a parallel PDE solver with adaptive mesh refinement. The solver is parallelized using OpenMP and the adaptive mesh refinement makes dynamic load balancing necessary. Due to the dynamically changing memory access pattern caused by the runtime adaption, it is a challenging task to achieve a high degree of geographical locality.

The main conclusions of the study are: (1) that geographical locality is very important for the performance of the solver, (2) that the performance can be improved significantly using dynamic page migration of misplaced data, (3) that a migrate-on-next-touch directive works well whereas the first-touch strategy is less advantageous for programs exhibiting a dynamically changing memory access patterns, and (4) that the overhead for such migration is low compared to the total execution time.

## 1 Introduction

Today, most parallel solvers for large-scale PDE applications are implemented using a local address space programming model such as MPI. During the last decade there has also been an intensified interest in using shared address space programming models like OpenMP for these type of applications. A main reason is that an increasing number of applications require the use of adaptive mesh refinement (AMR), and in this case the work and data need to be dynamically repartitioned at runtime to get good parallel performance. Using a local address space model, an extensive programming effort is needed to develop parallel PDE solver implementations that include such mechanisms. Using a shared address space model, the programming effort for producing a working parallel code can be reduced significantly. Another driving force for the use of shared address space models is the recent development in computer architecture; Emerging computer systems are built using multi-threaded and/or multi-core processors, future standard computational nodes will comprise an increasing number of threads that share a single address space. Codes using a programming model like OpenMP can then be transparently and easily used on different size systems, ranging from laptops with a single multi-threaded CPU to large shared memory systems with many such CPUs.

Most larger shared memory computers are built from nodes (chips) with one or several processors (cores), forming a cache-coherent non-uniform memory architecture (cc-NUMA). In a NUMA system, the latency for a main memory access depends on whether data is accessed at a local memory location or at a remote location. One characteristic property of this type of computer system is the *NUMA-ratio*, which is defined as the quotient of the remote and local access times. The non-uniform memory access time leads to that the *geographical locality* of data potentially affects the application performance. Here, optimal geographical locality corresponds to that the data is distributed over the nodes in a way that matches with the thread accesses in the best possible way. Good geographical locality can be achieved by carefully selecting the node where data is allocated at initiation, and/or by introducing some form of dynamic migration of data between the nodes during execution [1,2,3,4].

A main reason for the complexity of local address space implementations of AMR PDE solvers is that the programmer *must* explicitly control and modify the partitioning of work and data during execution. If suitable algorithms for partitioning and load balancing are used and the migration of data is efficiently implemented, a local address space implementation will regularly exhibit good parallel performance. In a shared address space model, the native work sharing constructs and transparent communication result in that it is much less demanding to develop a working parallel code. However, the aspects of work partitioning and load balancing must normally still be considered to obtain robust and competitive parallel performance. In programming models like OpenMP, the work partitioning and load balancing can easily be performed using the same well-developed and efficient algorithms as for local address space models, resulting in that the potential for good parallel efficiency is retained. Data distribution is not considered in OpenMP, and poor geographical locality could possibly lead to deteriorated performance. In this paper, we study the implementation of a structured adaptive mesh refinement (SAMR) PDE solver and attempt to answer the following questions:

- How large is the impact of geographical locality on the performance?
- Can the performance be improved through dynamic migration of misplaced data?
- How large is the migration overhead?

The rest of the paper is organized as follows: In Section 2 we describe existing parallel SAMR solvers and techniques used for distribution of work and data. In Section 3 we consider the model PDE application that is solved, in Section 4 we introduce the NUMA computer system used, and in Section 5 we give some details about the implementation and experimental setup. In Section 6 we present performance results, and in Section 7 we conclude.

## 2 Parallel SAMR Solvers

Most existing parallel implementations of large-scale SAMR PDE solvers [5], e.g., AMROC [6], PARAMESH [7], GrACE [8], and SAMRAI [9] exploit a local address space model implemented using MPI. The parallelization is based on grid blocks and each MPI process is responsible for the computations corresponding to one or more blocks or parts of blocks. For easier balancing of the computational work, large blocks are often first split into smaller blocks. All blocks are then assigned to the processors with a load balancing algorithm. In the computations, a small amount of communication is always needed for interpolating grid function values between some blocks in different processes.

Both patch-based and domain-based approaches for dynamic work partitioning and load balancing in local address space SAMR solvers have been developed. In patch-based methods the blocks at one level are partitioned over all processes while in the domain-based the computational domain is partitioned and the partitions are projected to the different grid levels. Hybrid versions combining patch-based and domain based methods have also been considered [10]. Furthermore, the algorithms can be categorized into scratch-remap and diffusion type. Using a scratch-remap strategy, a new partitioning is computed without considering previous partitionings, but the new partitioning is re-mapped according to previous data distribution in order to minimize data migration. In diffusion algorithms the partitioning is computed with a previous partitioning as a starting point. Scratch-remap strategies tries to optimize load balance and communication while the main objective in diffusion algorithms is to minimize data migration with load balance and communication as secondary objectives [11]. So far, there is no single algorithm that performs best for all types of applications or not even for all states of a particular application, see e.g. [10]. General dynamic load balancing algorithms for SAMR solvers remains an open field of research. In hierarchical AMR methods a common choice is to use space filling curves for clustering blocks into partitions [6,12,7,8,9], using Morton or Hilbert ordering. Space filling curves are fast and offer both locality within and between levels in the grid hierarchy. For flat unstructured AMR methods, graph partitioning methods are more common [13,11,14] and have better locality properties than the space filling curves. For flat, SAMR applications and multi-block grids the graph partitioning algorithms are also preferable [15,16]. For the experiments performed in this paper, it is important to minimize the data migration in addition to achieve a good load balance and minimize communication. A diffusion partitioning algorithm then becomes the natural choice, and we have chosen to us an algorithm of this type from the Jostle package [14].

In [17], a shared memory parallelization of a SAMR solver is presented. The code is parallelized using OpenMP and the experiments are performed on an SGI Origin system. A reasonable amount of geographical data locality is achieved by using SGI Origin's first touch data placement policy, i.e., data is allocated in the local memory of the thread first touching a grid block. In [12] a shared address space parallelization using POSIX-multi-threading is discussed. Here, explicit localization of data is implemented by using private memory in the threads for storing the blocks, i.e., each thread has access to the grid hierarchy but stores only its blocks in private data structures in local memory. Moreover, to guarantee geographical locality the threads are explicitly bind to single CPU's.

In [18] different programming models using MPI, OpenMP and hybrid MPI-OpenMP for parallelizing a SAMR solver are compared on a Sun Fire 15K, which is also a NUMA system. Parallelization is performed both at block level and at loop level. It is shown that the coarse grain block level parallelization with MPI gives the best performance as long as the number of blocks is large enough for a good load balance, otherwise a mixed MPI-OpenMP model is better due to better distribution of the work. The standard OpenMP implementation suffers from poor geographical data locality and does not perform as well as the corresponding MPI implementations.

## 3 The PDE Solver

As a representative model problem we solve the advection equation

$$u_t = u_x + u_y$$

with periodic boundary conditions on a square. The initial solution is a Gaussian pulse. As time evolves the pulse moves diagonally out through one of the corners of the domain and comes back in from the opposite corner without changing shape. The PDE is discretized by a second-order accurate finite difference method in space and the classical fourth order Runge-Kutta method in time. We use a structured cartesian grid and divide the domain into a fixed user-defined number of blocks. As a simple error estimate in the adaption criterion we use the maximum value of the solution in a block. In a real-life application, a more sophisticated error estimate e.g. based on applying the spatial difference operator on a coarse and a fine block discretization would be used [19]. However, this would not affect the parallel performance much, and the conclusions drawn from the experiments presented later will not change. If the error estimate of a block exceeds a threshold, the resolution of the grid is refined with a factor two in the entire block. On the contrary, if the error is small enough, the grid in the block is coarsened with a factor two.

The code is written in Fortran 90 and parallelized using OpenMP. The parallelization is coarse grained over entire blocks, i.e. each thread is responsible for a set of blocks. The blocks have two layers of ghost cells which are updated by reading data from the neighboring blocks. When the grid resolution changes in any of the blocks, the entire grid block structure is repartitioned using the Jostle diffusion algorithm and the work partitioning between the threads is changed accordingly.

Before the main time-evolution loop starts, the solution is initialized. This is done in parallel, according to an initial partitioning that was defined when the grid was created. After this the error is estimated and the grid adapted if necessary. This procedure, initialization and adaption, is repeated until the error estimates are satisfied in all blocks. Thereafter, the grid is repartitioned and the main computations starts. The computational kernel of our SAMR application is presented in pseudocode in Figure 1. In the code, the procedure Diff() performs the necessary interpolation between grid blocks and applies the spatial difference operator for all blocks in the grid. In the experiments presented later, we perform

```
1    do t=1,Nt
2      if (t mod adaptInterval=0) then
3        Estimate error per block.
4        Adapt blocks with inappropriate resolution.
5        Repartition the grid.
6        Migrate blocks (if migration is activated).
7      end if
8      F1=Diff(u);
9      F2=Diff(u+k/2*F1)
10     F3=Diff(u+k/2*F2)
11     F4=Diff(u+k*F3)
12     u=u+k/6*F1+k/3*F2+k/3*F3+k/6*F4
13   end do
```

**Fig. 1.** Pseudocode for the computational kernel of our SAMR application

a total of 20000 time steps. Adaption, partitioning and migration (if active) is performed every AdaptInterval time step, where we use AdaptInterval=20. We use a discretization with 16x16 blocks, and the adaption criterion results in three different block sizes: 100x100, 200x200 and 400x400. When a block is refined or coarsened new memory is allocated and the old block is discarded. At a typical iteration the resident working set was about 350 MB. We define the load balance $\gamma$ by

$$\gamma = \frac{\max_i p_i}{\frac{1}{n}\sum_{i=0}^{n} p_i}, \qquad (1)$$

where $p_i$ is the amount of work in partition $i$ and $n$ the number of partitions. Using $n = 4$ and sampling the load balance $\gamma_j$ after each partitioning, we got an arithmetical mean of 1.09 with a standard deviation 0.12. Hence, the diffusion type partitioner used gives a reasonably good load balance.

## 4 The NUMA System

All experiments presented below were performed on a Sun Fire 15000 system, where a dedicated domain consisting of four nodes was used. Each node contains four 900 MHz UltraSPARC-IIICu CPUs and 4 GByte of local memory, and each CPU has an off-chip 8MB L2 cache. Within a node, the access time to local main memory is uniform. The nodes are connected via a crossbar interconnect, forming a cc-NUMA system with NUMA-ratio approximately 2.0.

The codes was compiled with the Sun STUDIO 11 compiler using the flags `-fast -xarch=v8plusa -xchip=ultra3cu`, and the experiments were performed using the 4/04 release of Solaris 9. When an application starts the Solaris scheduler assigns each thread a home node (called *locality group* or *lgroup* in Solaris terminology). Although threads are allowed to execute on any node the scheduler tries to keep the threads to their home node. By default, memory is allocated according to a first-touch strategy which, in an ideal case, means that memory will be allocated to the home node to create good geographical locality.

In Solaris, dynamic migration of pages between nodes can be performed using a directive with a migrate-on-next-touch semantic using the `madvise(3C)` library call [20]. The directive tags pages for migration and the kernel resets the address translation for these pages. Since the TLB is handled by software in Solaris, dirty translations needs to be invalidated by a TLB shoot down procedure for all CPUs that have executed the address space. After the shot down the pages have no physical address associated with them. When a thread accesses one of these pages a minor page fault occurs and the contents of the page is migrated, i.e. physically copied, to a new page allocated in the node where the faulting thread executes. If the new page is physically allocated to the node where the contents resides, there is a fast-path, no data is actually copied. The overhead of the migration can be divided into two parts: the overhead from TLB shoot down and the cost of copying data. The shoot down overhead is dependent on how many pages are shot down and for how many CPU:s. Due to kernel consistency issues this procedure needs to be serialized using global locking, see Teller [21].

A migrate-on-next-touch directive is also available on the Compaq Alpha Server GS-series [22]. On SGI Origin-systems [23], dynamic page migration is also available. However, it is implemented using access counters, and no migrate-on-next-touch feature is available. Instead, HPF-style explicit directives for data distribution can be inserted in the code. Tikir et al [24] showed that a migrate-on-next-touch directive can be used to create a transparent data distribution engine based on hardware access counts. Also, Spiegel et. al [25] showed how to use the migrate-on-next-touch call to speed up an hybrid CFD solver.

## 5 Experimental Methodology and Setup

In the SAMR application studied here, the data distribution corresponding to the initial first-touch allocation will not be optimal since we need to maintain a good load balance. The partitioner will in many cases assign blocks where some or all of them were initially allocated on a node different from the home node of a given thread. Also, in our implementation, when a block is refined or coarsened, new memory is allocated and the old block is discarded. As the adaption phase precedes the partitioner, new blocks might be allocated by the first-touch strategy to a remote node depending on the outcome of the partitioner.

To increase geographical locality we can use page migration to migrate the data of each partition to the home locality group (node) of the corresponding thread. The simplest strategy would be to migrate all blocks after each partitioning. However, if the partitioner has a notion of locality, such as a diffusion partitioner, the number of blocks that change partition will be lower than the total amount of blocks in each partition. As a consequence we keep track of inter-partition block movements and only migrate the blocks that change partition after each partitioning step. Since the migration is driven by page faults we need to be careful how we touch the data. Each block has a layer of ghost cells to simplify the interpolation between blocks of different levels of refinement. The ghost cells are normally accessed first in a communication step which means that these pages will be migrated to a neighboring thread if a migrate-on-next-touch directive is used. To avoid this behavior we added a sequence of code where the thread touches the entire block, including the ghost cells directly after the migrate-on-next-touch call.

## 5.1 Experimental Setups

To investigate the performance impact of geographical locality we have performed experiments where the application was executed on using four threads on the following four configurations

**UMA** All threads confined to the same node. Migration not active.
**NUMA** One thread per node. Migration not active.
**NUMA-MIG** One thread per node. Migrate data belonging to blocks that are transfered to another node. Force immediate migration by touching pages.

To make sure that threads stay in their home nodes we used the SUNW_MP_PROCBIND environment variable to bind each thread to a specific CPU. We also kept the system unloaded apart from the application studied. In the UMA case, all accesses will be local. However, there is a risk that the performance of the code will be inhibited by the limited bandwidth provided by a single node. In the NUMA cases the aggregate bandwidth to main memory is four times higher. We align data to page boundaries by interposing the Fortran 90/95 allocate() routine. Since the SAMR application allocates new blocks in parallel we used the mtmalloc allocator. This allocator is part of Solaris and it is much more scalable than the standard allocator. We mapped all allocations to the valloc() routine of mtmalloc. This will result in that the smallest possible block of data is a memory page. The memory waste was found to be very low. In total, the application allocated 241540 8kB pages which is close to 2 GB of data in 7636 calls to allocate. The waste due to alignment was about 40 MB.

To quantify the effect of geographical locality we measure the number of remote accesses generating from the CPUs using the UltraSPARC-IIICu hardware counters. We define the number of remote accesses as the difference between the total amount of local L2-cache accesses (EC_miss_local) and the total amount of L2-cache misses (EC_misses). The hardware counter data was sampled using the Sun Performance Analyzer. To reduce the file size of the hardware counter sampling only 4000 times steps of the 20000 were executed. We believe that the basic miss ratio characteristics can still be observed using only a subset of the iterations. Furthermore, the Solaris kernel (kstat) provides counters for the amount of pages migrated to and from a node and the Solaris tool trapstat was used to sample the amount of time spent handling address translations.

## 6 Results

Table 1 shows both total execution time and hardware counter data from the three setups. We can see from Table 1 that the NUMA case runs slower than

**Table 1.** Execution time measurements and hardware counter data from the three different experimental setups

|  | UMA | NUMA | NUMA-MIG |
|---|---|---|---|
| Total Execution Time | 4.09 h | 6.64 h | 3.99 h |
| L2 Miss Ratio | 4.3% | 3.9% | 4.2% |
| L2 Remote Ratio | 0.2% | 62.9% | 8.1% |

the UMA case and the NUMA-MIG case. The number of remote accesses is also much higher for this case compared to the UMA and NUMA-MIG cases which shows that the NUMA case exhibits a low degree of geographical locality. It is also clear that the effect of page migration is large since the amount of remote accesses for the NUMA-MIG case is much lower compared to the NUMA case. Remember that we can not completely remove all remote accesses since the usage of ghost cells will result in a small amount of communication.

Figure 2 shows the entire execution for the UMA, NUMA and NUMA-MIG cases. To be able to compare performance we have aligned the graphs vertically and all three graphs have the same scale. It is clear that the impact of geographical locality is significant even though the NUMA-ratio of the SF15K is only about two. By comparing the execution time of UMA and NUMA-MIG we see that we can increase the geographical locality using a migrate-on-next-touch directive. Surprisingly, the execution time for the NUMA-MIG case is lower than the UMA case. This can be explained by the fact that in the UMA case all CPU:s of the node will be used resulting in a very high memory pressure on that node. In the NUMA cases each thread will have the entire node for itself resulting in a higher aggregate bandwidth for the application to use.

Table 2 shows migration statistics for the application in the NUMA-MIG case. At a typical iteration the resident working set was about 350 MB which corresponds to 44800 8kB pages. Migration was triggered every 20:th time steps which gives a total of 1000 times. If all the data migrate at each migration this would correspond to a total traffic of 44.8 million pages. Comparing this figure to the total amount of pages migrated (2.21 million) we can conclude that a small fraction (5%) of all the pages are migrated. Assuming that the migrations are evenly distributed over time, only 2213 pages (18.1 MB) are migrated at each

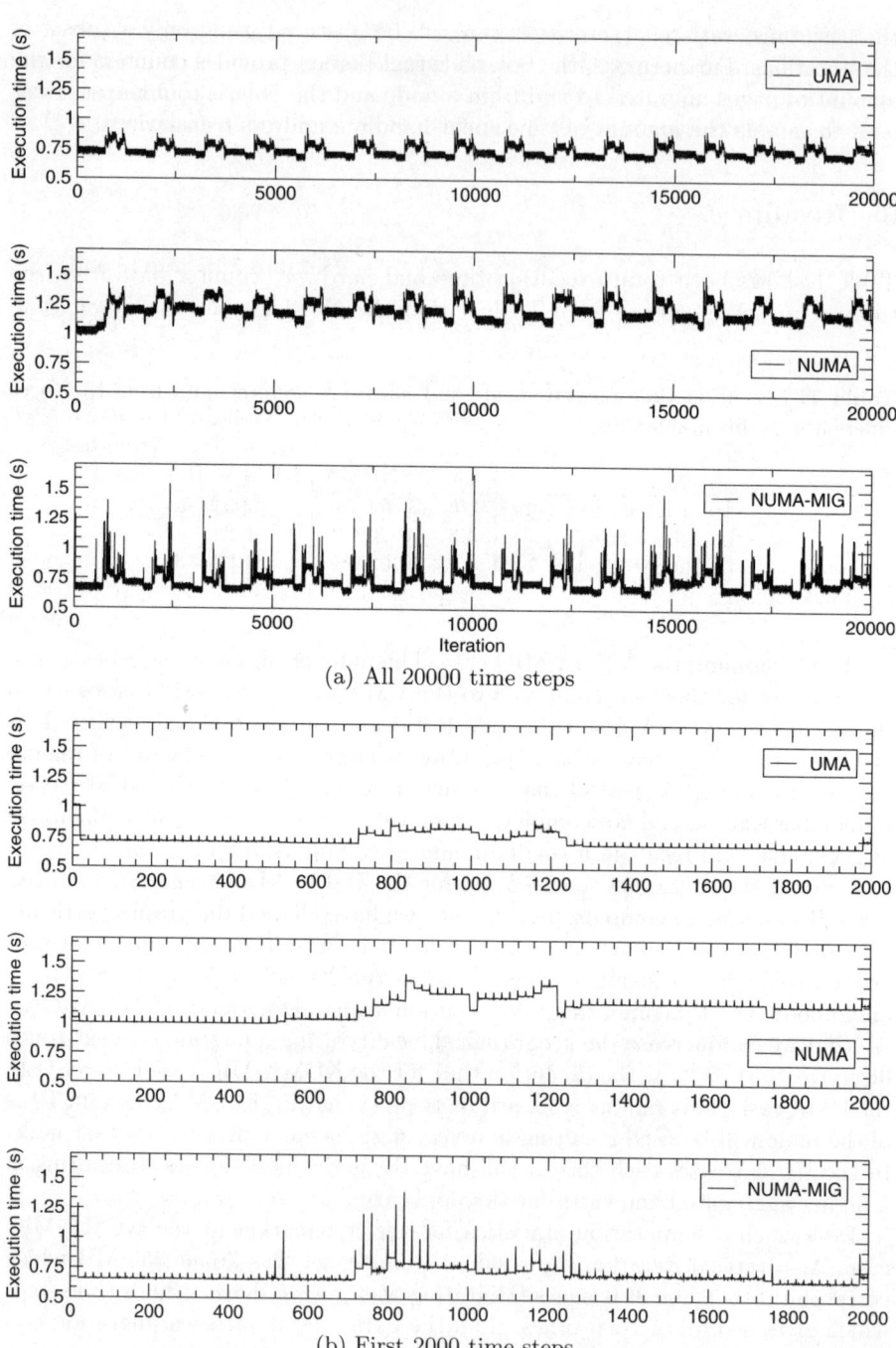

**Fig. 2.** The impact of geographical locality on performance. Migration, adaption and partitioning is triggered every 20:th time step.

**Table 2.** Migration statistics for the NUMA-MIG case collected using Solaris kernel statistics (kstat). Column 4 shows the net data flow from a node where the thread executed. A negative value indicates that more pages were migrated from the node. Column 5 shows the sum of columns 2 and 3 ie the total amount of migrations for one node. The total amount of migrated pages for all nodes was 2212844 (16.88 GB).

|          | Migrated To | Migrated From | Total Net Flow | Total Traffic   |
|----------|-------------|---------------|----------------|-----------------|
| Thread 0 | 346479      | 356903        | -10424         | 703382 (2.9 GB) |
| Thread 1 | 318407      | 319417        | -1010          | 637824 (5.4 GB) |
| Thread 2 | 249414      | 243203        | 6211           | 492617 (3.8 GB) |
| Thread 3 | 192122      | 186899        | 5223           | 379021 (3.4 GB) |

migration. Hence, we conclude that the amount of data migrated is fairly low. This together with the fact that the NUMA-MIG case executes faster than the UMA case indicates that the overheads of migration are low for the experiments performed. Using the trapstat tool we found that the solver (all cases) spends at most 1.0% of its time (2.4 mins) in the Solaris page fault trap handler. This fact further supports the conclusion that the overheads from migration are low.

## 7 Conclusions

In this paper we have investigated the impact of geographical locality for an adaptive PDE solver. This application has a dynamic access pattern which implies that a system needs to support some kind of runtime data distribution to minimize the effects of geographical locality. Our results show that the impact of geographical locality is large even though the NUMA-ratio of the system used is only two. We also show that we can significantly improve geographical locality and overall performance using a library call with a migrate-on-next-touch semantic.

The overheads of migration was found to be low which can be attributed to two facts. First, our experiments were performed using only four threads. The overheads from page migration will probably increase with the number of nodes and CPUs. Second, for SAMR to be efficient the refined area of the mesh needs to be rather small. This indicates that the amount of data that needs to be migrated will be low. The refinement patterns of AMR solvers often vary a lot depending on the physics of the problem studied. If data migration is to be used in a more general setting the frequency of invoking a migrate-on-next-touch call must be tuned to match the refinement patterns of the studied problem. If large amounts of data needs to be migrated we may have to reduce the number of migrate calls to amortize the overhead over several time steps. However, for the model problem studied in this article, using a diffusion type partitioner resulted in fairly low amounts of data migrations.

We believe that a call or directive with a migrate-on-next-touch semantic can be a useful addition to an architecture-independent language like OpenMP. Since such a directive is invoked by the programmer we do not have to spend system resources monitoring geographical locality were thread-data affinity is not critical for performance. Furthermore if a system support a transparent mechanism for increasing geographical locality, a migrate-on-next-touch directive could serve as a useful hint to the system.

## References

1. Wilson, K.M., Aglietti, B.B.: Dynamic page placement to improve locality in CC-NUMA multiprocessors for TPC-C. In: Proceedings of the 2001 ACM/IEEE conference on Supercomputing, pp. 33–33. ACM Press, New York (2001)
2. Corbalan, J., Martorell, X., Labarta, J.: Evaluation of the memory page migration influence in the system performance: the case of the SGI O2000. In: Proceedings of the 17th annual international conference on Supercomputing, pp. 121–129. ACM Press, New York (2003)
3. Holmgren, S., Nordén, M., Rantakokko, J., Wallin, D.: Performance of PDE Solvers on a Self-Optimizing NUMA Architecture. Parallel Algorithms and Applications 17(4), 285–299 (2002)
4. Bull, J.M., Johnson, C.: Data Distribution, Migration and Replication on a cc-NUMA Architecture. In: Proceedings of the Fourth European Workshop on OpenMP (2002), http://www.caspur.it/ewomp2002/
5. Rendleman, C.A.: Parallelization of structured, hiearchical adaptive mesh refinement algorithms. Computing and Visualization in Science 3, 147–157 (2000)
6. Deiterding, R.: Construction and application of an amr algorithm for distributed memory computers. In: Adaptive Mesh Refinement – Theory and Applications, Proc. of the Chicago Workshop on Adaptive Mesh Refinement Methods, pp. 361–372. Springer, Heidelberg (2003)
7. MacNeice, P.: Paramesh: A parallel adaptive mesh refinement community toolkit. Computer physics communications 126, 330–354 (2000)
8. Parashar, M., Browne, J.: System engineering for high performance computing software: The hdda/dagh infrastructure for implementation of parallel structured adaptive mesh refinement. In: IMA Volume on Structured Adaptive Mesh Refinement (SAMR) Grid Methods, pp. 1–18 (2000)
9. Wissink, A.M., Hornung, R.D., Kohn, S.R., Smith, S.S., Elliott, N.: Large scale parallel structured amr calculations using the samrai framework. In: Proceedings of SC 2001 (2001)
10. Steensland, J.: Efficient partitioning of structured dynamic grid hierarchies. Doctoral thesis, Scientific computing, Department of Information Technology, University of Uppsala, Uppsala dissertations from the faculty of science and technology 44 (2002)
11. Schloegel, K., Karypis, G., Kumar, V.: A unified algorithm for load-balancing adaptive scientific simulations. In: Proceedings Supercomputing 2000 (2000)
12. Dreher, J., Grauer, R.: Racoon: A parallel mesh-adaptive framework for hyperbolic conservation laws. Parallel Computing 31, 913–932 (2005)
13. Maerten, B.: Drama: A library for parallel dynamic load balancing of finite element applications. In: Amestoy, P.R., Berger, P., Daydé, M., Duff, I.S., Frayssé, V., Giraud, L., Ruiz, D. (eds.) Euro-Par 1999. LNCS, vol. 1685, pp. 313–316. Springer, Heidelberg (1999)
14. Walshaw, C., Cross, M., Everett, M.: Parallel dynamic graph partitioning for adaptive unstructured meshes. Parallel Distributed Computing 47(2), 102–108 (1997)

15. Rantakokko, J.: Partitioning strategies for structured multiblock grids. Parallel Computing 26, 1661–1680 (2000)
16. Steensland, J., Söderberg, S., Thuné, M.: A comparison of partitioning schemes for blockwise parallel samr algorithms. In: Sørevik, T., Manne, F., Moe, R., Gebremedhin, A.H. (eds.) PARA 2000. LNCS, vol. 1947, pp. 160–169. Springer, Heidelberg (2001)
17. Balsara, D., Norton, C.: Highly parallel structured adaptive mesh refinement using parallel language-based approaches. Parallel Computing 27, 37–70 (2001)
18. Rantakokko, J.: Comparison of parallelization models for structured adaptive mesh refinement. In: Danelutto, M., Vanneschi, M., Laforenza, D. (eds.) Euro-Par 2004. LNCS, vol. 3149, pp. 615–623. Springer, Heidelberg (2004)
19. Ferm, L., Lötsetdt, P.: Space-time adaptive solutions of first order pdes. Journal of Scientific Computing 26(1), 83–110 (2006)
20. Sun Microsystems: Solaris Memory Placement Optimization and Sun Fire servers (2003), http://www.sun.com/servers/wp/docs/mpo_v7_CUSTOMER.pdf
21. Teller, P.J.: Tranlation-Lookaside Buffer Consistency. Computer 23(6), 26–36 (1990)
22. Bircsak, J., Craig, P., Crowell, R., Cvetanovic, Z., Harris, J., Nelson, C.A., Offner, C.D.: Extending OpenMP for NUMA machines. Scientific Programming 8, 163–181 (2000)
23. Laudon, J., Lenoski, D.: The SGI Origin: a ccNUMA highly scalable server. In: Proceedings of the 24th annual international symposium on Computer architecture, pp. 241–251. ACM Press, New York (1997)
24. Tikir, M.M., Hollingsworth, J.K.: Using Hardware Counters to Automatically Improve Memory Performance. In: SC 2004: Proceedings of the 2004 ACM/IEEE conference on Supercomputing, Washington, DC, USA, p. 46. IEEE Computer Society, Los Alamitos (2004)
25. Spiegel, A., an Mey, D.: Hybrid Parallelization with Dynamic Thread Balancing on a ccNUMA System. In: Brorson, M. (ed.) Proceedings of the 6th European Workshop on OpenMP, Royal Institute of Technology (KTH), Sweden, pp. 77–81 (2004)

# Proposed Extensions to OpenMP

# A Comparison of Task Pool Variants in OpenMP and a Proposal for a Solution to the Busy Waiting Problem

Alexander Wirz, Michael Süß, and Claudia Leopold

University of Kassel, Research Group Programming Languages / Methodologies,
Wilhelmshöher Allee 73, D-34121 Kassel, Germany
`wirz@student.uni-kassel.de`, {`msuess`, `leopold`}`@uni-kassel.de`

**Abstract.** Irregular algorithms are difficult to parallelize using existing OpenMP constructs. This paper concentrates on algorithms that deploy task pools, i.e., data structures for dynamic load balancing. We present several task pool variants that we have implemented in OpenMP, and compare their performance. Due to the lack of a mechanism in OpenMP to put a thread to sleep, we had to use busy waiting in our implementations. To eliminate this need, we suggest an extension to OpenMP that allows to put a thread to sleep on demand.

## 1 Introduction

OpenMP [1] provides powerful constructs to parallelize regular programs, i.e., programs that execute a similar set of operations on different elements of a regular data structure such as an array. Irregular applications, in contrast, are difficult to parallelize using the existing OpenMP constructs. For irregular applications, the units of work can usually not be distributed statically among a fixed number of threads, because they are created dynamically at runtime and their number depends on the given input. Moreover, it is often not possible to predict the amount of work to be done in a unit for any particular input data.

One approach to achieve dynamic load balancing is the use of task pools. A task pool is a data structure that stores dynamically created work units (tasks) to support distribution to a certain number of threads. Section 2 gives an overview about some task pool variants that we have implemented in OpenMP, and presents the results of our runtime experiments with three irregular applications using task pools: Quicksort, Labyrinth-Search and Sparse Cholesky Factorization. Performance numbers gathered with the workqueuing model proposed by Shah et al. [2] are included for comparison in this section as well.

One problem we have been confronted with during the implementation of our task pools is the lack of a suitable mechanism in OpenMP to put a thread to sleep while waiting for a condition to become true. The programmer therefore has to resort to busy waiting, which may be wasteful on the available computing resources. Fig. 1 sketches a solution to this problem by proposing an extension to OpenMP, which is spelled out in Sect. 3. Sect. 4 surveys related work, and Sect. 5 summarizes our results.

- **# pragma omp yield**:
    release the processor so that another thread can run on it
- **#pragma omp sleepuntil (scalar_expression)**:
    sleep until scalar_expression becomes true

**Fig. 1.** Scheduling in a nutshell

## 2 Task Pools

Task pools are used to achieve dynamic load balancing in irregular applications. A task pool stores tasks that are created dynamically at runtime. It also provides a set of operations that allow threads to insert and extract tasks concurrently in a threadsafe manner. The remainder of this section is organized as follows: Sect. 2.1 introduces the high level interface for the programmer used by all our task pool variants. In Sect. 2.2, the different task pool variants are described, while Sect. 2.3 highlights the most severe implementation problem we had with all variants: lack of a suitable mechanism to put threads to sleep while waiting for a condition to become true. Finally, in Sect. 2.4, we introduce three example applications that are used in Sect. 2.5 to assess the performance of the different task pool variants: Quicksort, Labyrinth-Search and Sparse Cholesky Factorization.

### 2.1 Application Programming Interface

All implemented task pools use the same application programming interface. This API provides functions to initialize and destroy the task pool structure, as well as to insert and extract tasks concurrently. Listing 1.1 shows an example of the relevant part of an OpenMP program that uses our API.

```
1   task_data_t *task_data;
2   tpool_t *pool = tpool_init(num_threads, sizeof(task_data_t));
3   task_data = generate_initial_task();
4   tpool_put(pool, 0, task_data);
5   #pragma omp parallel shared(pool)
6   {
7     task_data_t *my_task_data;
8     int me = omp_get_thread_num();
9     while(TPOOL_EMPTY != tpool_get(pool, me, &my_task_data))
10        do_work(my_task_data);
11  }
12  tpool_destroy(pool);
```

**Listing 1.1.** OpenMP program using task pools

First, a task pool must be initialized by using the *tpool_init()* function. This function must only be called once, and only by a single thread. Afterwards, the task pool can be used to store (*tpool_put()*) and extract (*tpool_get()*) tasks. The latter function blocks until it either successfully extracts a task from the pool, or discovers that the task pool is empty and all threads using the pool are idle.

Finally, function *tpool_destroy()* frees the memory used by the task pool. All task pool variants and test applications were implemented in C.

## 2.2 Variants of Task Pools

We implemented several variants of task pools. Some of them (*sq1*, *sdq1* and *dq8*) were ported to OpenMP from existing POSIX threads and Java implementations described by Korch and Rauber [3]. Others (*dq9* and *dq9-1*) have been developed by the authors as enhancements of the *dq8* variant. The remainder of this section explains the variants.

**Central Task Queue:** The simplest way to design a task pool, called *sq1*, is to use a single shared task queue. Each thread is allowed to access this queue with functions *tpool_put()* and *tpool_get()*. We used OpenMP lock variables to ensure that only one thread can access the task queue at a time. The variant has the drawback that when two or more threads are trying to access the task pool simultaneously, they have to wait for each other. Therefore, the task pool can become a bottleneck for applications that use a large number of threads or access the pool frequently. Nevertheless, this variant offers good load balancing capabilities and performs well for applications that create only few tasks or access the task pool rarely.

**Combined Central and Distributed Task Queues:** To reduce waiting times caused by access conflicts, the task pool variant *sdq1* uses distributed task queues. It manages a private task queue for each thread and permits only the owner thread to access the queue. Therefore no synchronization operations are needed for the private queues. An extra central queue is maintained for load balancing. Whenever a private queue is empty, the owner thread tries to fetch a task from the central queue. To ensure the exchange of tasks among the threads, the size of the private queues is limited. If a thread tries to enqueue a new task and discovers that its private queue is full, it will move the new task to the central queue.

**Distributed Queues with Dynamic Task Stealing:** In contrast to *sdq1*, the task pool variants *dq8*, *dq9* and *dq9-1* use multiple shared queues to reduce the possibility of access conflicts: each thread has its own private and its own shared queue. If a thread runs out of tasks in its private queue, it will take a task from its shared queue. If the shared queue is also empty, the thread will try to steal a task from the shared queue of another thread, and then return it from *tpool_get()*.

Although the task pool variants *dq8*, *dq9* and *dq9-1* are conceptually similar, they use different strategies for filling the shared queues. Like *sdq1*, *dq8* uses private queues with a limited size. If a private task queue is full, the new tasks are moved to the shared queue.

Unlike *dq8*, variants *dq9* and *dq9-1* adjust the size of the private queues dynamically, based on the state of the shared queue. The size of a private queue in *dq9* and *dq9-1* is not limited to a certain value. The private queues of these task pool variants can contain an arbitrary number of tasks. The reason is that *dq9* and *dq9-1* try to keep most tasks in the private queues to reduce the number of operations on shared queues. A thread will move a task into its shared queue in *tpool_put()* only if the shared queue is running empty. If both the private and the shared queue are empty, a new task will be inserted into the shared queue in *dq9*, but into the private queue in *dq9-1*.

Another difference between *dq9-1* and *dq9* is the point in time, when task stealing is started. While *dq8* and *dq9* do not attempt to steal tasks before a private queue is empty, *dq9-1* initiates task stealing as soon as the number of tasks in the private queue drops below a predefined threshold value. This is done to prevent the private queue from running empty.

All of these different variants are summarized in Tab. 1.

**Table 1.** Comparison of implemented task pool variants (Q. stands for Queue, the number in braces stands for the actual number of tasks used in our tests)

| Name | Num. Q. | Num. Shared Q. | Size of Priv. Q. | Task-stealing Time |
|---|---|---|---|---|
| sq1 | 1 | 1 | - | - |
| sdq1 | $num\_threads + 1$ | 1 | limited (2) | - |
| dq8 | $num\_threads * 2$ | $num\_threads$ | limited (2) | priv. queue empty |
| dq9 | $num\_threads * 2$ | $num\_threads$ | unlimited | priv. queue empty |
| dq9-1 | $num\_threads * 2$ | $num\_threads$ | unlimited | priv. queue low (2) |

### 2.3 Implementation Problem: Busy Waiting

The implementation of the task pools sketched in the previous section was relatively straightforward with OpenMP, but we encountered a problem for which OpenMP does not provide an adequate solution: Each time a thread tries to extract a task but detects an empty task pool, it has to wait until another thread inserts a new task. Korch and Rauber [3] solved the problem for their implementations with condition variables in POSIX threads, and the *wait()*-*notify()* mechanism in Java, respectively. Unfortunately, there is no mechanism in OpenMP to put a thread to sleep until an event occurs or a condition becomes true. In our task pool implementations, we therefore had to fall back on busy waiting, which results in unnecessary idle cycles. For this reason, Sect. 3 suggests OpenMP extensions to solve the problem. These simple extensions can be used to avoid busy waiting in a task pool, and are also helpful in other contexts.

### 2.4 Benchmarks

To compare the performance of our task pool variants, we have implemented three irregular applications: Quicksort, Labyrinth-Search and Sparse Cholesky

Factorization. Quicksort is a popular sorting algorithm, initially invented and described by Hoare [4]. The Labyrinth-Search application finds the shortest path through a labyrinth using the breadth-first search algorithm. To ensure that all labyrinth cells with the same distance from the entry cell are visited before any other cells are processed, we use two task pools. The tasks in the first pool correspond to cells with distance $d$ from the entry cell. The second task pool is used to collect tasks (cells) with distance $d + 1$.

Cholesky Factorization is an algorithm to solve systems of linear equations $Ax = b$. It exploits the fact that a symmetric positive definite matrix $A$ can be decomposed into $A = LL^T$, where $L$ is a lower triangular matrix with positive diagonal elements. Using this decomposition, the original equation can be solved more efficiently. Information on Cholesky Factorization can be found, for instance, in the book by George and Liu [5]. To test our task pool variants, we have implemented only the most expensive part of Cholesky Factorization: numerical factorization. Numerical factorization computes the nonzero elements of the result matrix $L$. We have implemented a so-called right-looking factorization scheme. Each task computes one column of the result matrix, dividing all elements of this column by the square root of its diagonal. Then, all columns which depend on the recently computed column are updated by adding a multiple of the computed column to them.

## 2.5 Results

Performance measurements were carried out on an AMD Opteron 848 class computer with four processors at 2.2 GHz, and on a Sun Fire E6900 with 24 dual-core Ultra Sparc IV processors at 1.2 GHz. On the AMD system, a maximum of four threads was used, while on the Sun system, a maximum of eight threads was used. Although more threads would have been possible on the latter machine, eight processors is the maximum number that this machine supports without encountering NUMA-effects (as it consists of multiple mainboards with 4 dual-core processors each). On the AMD system, the benchmarks were compiled with the Intel C++ Compiler 9.0 using options -O2 and -openmp. On the Sun Fire E6900, the Guide compiler with options -fast --backend -xchip=ultra3cu --backend -xcache=64/32/4:8192/512/2 --backend -xarch=v8plusb was used. We have not used the native SUN compiler, because it does not support the workqueuing extension (see next paragraph).

For comparison, we implemented Quicksort and the Cholesky factorization using Intel's proposed workqueuing model. It was first introduced by Shah et al. [2], as an integrated approach to achieve dynamic load balancing for irregular applications. Those authors suggest an OpenMP extension which allows to split the work into units (tasks) that are distributed dynamically to the threads of a program using a task queue. Since both the Intel C++ and the Guide compilers already support the workqueueing model, we implemented two benchmarks using this proposed OpenMP extension on the same set of machines. Unfortunately, we could not implement the Labyrinth-Search algorithm with this model, because we did not find a way to use two different queues and ensure that all tasks from

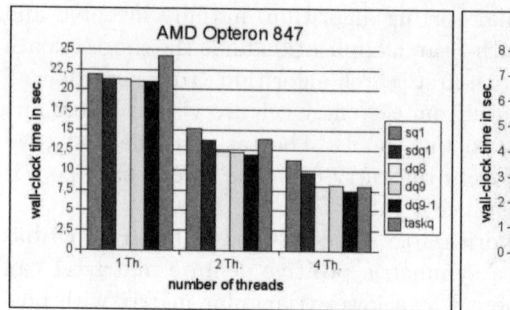 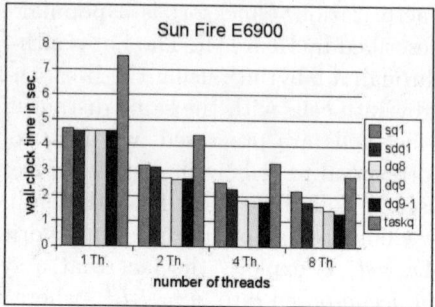

**Fig. 2.** Wall–clock times for Quicksort in seconds. Each time shown is the average of three runs.

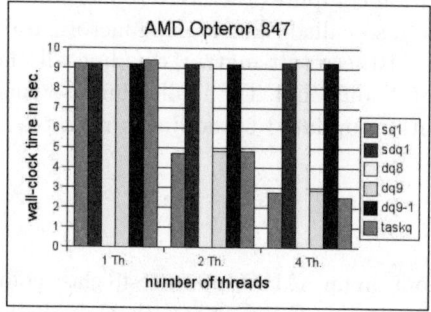 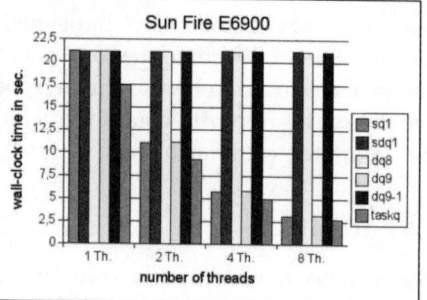

**Fig. 3.** Wall–clock times for the Cholesky factorization. Each time shown is the average of three runs.

one queue are executed before the program starts to execute tasks from the second queue.

Fig. 2 shows the wall-clock times in seconds for the Quicksort benchmark application with different task pool variants and the Intel taskq implementation. We used an array with 100.000.000 elements as input data on the AMD Opteron system and an array with 10.000.000 elements on the Sun Fire E6900. The results for the Cholesky factorization are shown in Fig. 3. For the Cholesky factorization, a 500x500 matrix was used as input. Fig. 4 shows the results for the Labyrinth-Search benchmark application with different task pool variants.

Our experiments indicate that the performance of different task pool variants depends on the type of application. Quicksort and Labyrinth-Search, which create a large number of tasks, achieve better performance using task pools with distributed task queues. Cholesky factorization, in contrast, generates only a few tasks, and therefore good load balancing is crucial. The use of private queues turns out to be a drawback in this case, because all tasks remain in the private queues and the idle threads have no chance to fetch them. The performance of *dq9* is good, though, because this variant makes the distribution of tasks among

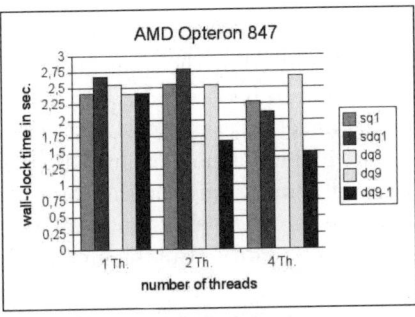 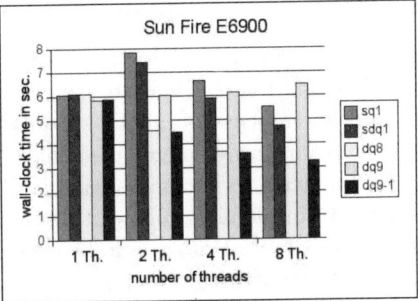

**Fig. 4.** Wall–clock times for Labyrinth-Search. Each time shown is the average of three runs.

the queues dependent on the number of tasks in the pool. If there are only a few tasks in the pool (shared queues are empty and at least one thread is idle), a new task will be inserted into a shared queue (and not, like e.g. for *dq9-1* into a private queue). If there are enough tasks in the pool, however, *dq9* will insert a new task into a private queue to avoid synchronization operations. Using this technique, *dq9* achieves much better performance than the other task pool variants with distributed queues.

Fig. 3 shows that the only task pool variant that uses one central queue (*sq1*) achieves the best performance for Cholesky Factorization. The reason is the good load balancing offered by *sq1*: all tasks are kept in one central queue, where all threads can access them. Due to the small number of tasks generated by the algorithm, a central queue does not slow down the program because the application accesses the task pool only rarely.

The bottom line from our experiments is that there is no clear winning taskpool implementation. It depends on the application, which task pool variant is suited best.

As can be seen, the performance of the task pools implemented inside the two compilers using Intels taskq is comparable to (and in some cases even better than) our implementations for the Cholesky example. When many tasks are generated and stored in the pools (as is the case for Quicksort), our optimized task pools are able to outperform the Intel implementations, though.

## 3 Solving the Problem of Busy Waiting

As has already been stated in Sect. 2.3, there is a problem regarding busy waiting and OpenMP. The problem is shortly rehashed on a broader scale in Sect. 3.1. Afterwards, Sect. 3.2 specifies our proposed solution, and Sect. 3.3 gives our reasons for the design. Finally, in Sect. 3.4, the specification is applied to our examples, and some ways to use the new functionality are shown. A reference implementation of the suggested changes to the OpenMP functionality can be found in a special release of the OMPi Compiler [6] that is available from the authors on request.

## 3.1 Problem Description

The problem of busy waiting has already been discussed by the same authors [7]. It manifests if a thread has to wait for a condition to become true before it can continue. In the case of our task pools, for instance, function *tpool_get()* is supposed to return an element from the pool, but if there is no element left, it has to wait for work to become available. The most sparing way for the computing resources to implement this waiting is to put the thread to sleep until the condition becomes true. Unfortunately, there is no functionality available in OpenMP to support the waiting, though.

As a valid workaround, the programmer may poll a condition repeatedly, thereby wasting processor time. This approach is known as *busy waiting*. To give another example, busy waiting is also required for pipelined algorithms, where a stage has to wait until a previous stage has completed its work. Busy waiting is best avoided, especially when other threads are waiting for the processor to become available, or when power consumption is an issue, e.g. in embedded systems.

Novice OpenMP programmers may resort to using locks to solve the problem. In their approach, the waiting thread tries to set an already set lock, and is put on hold as a result. As soon as work is available, a different thread will unset the lock, thereby enabling the waiting thread to continue. Although this approach often works, it is not compliant with the OpenMP specification, because the lock is unset by a different thread than the owner thread, which leads to unspecified behaviour. Furthermore, there is no guarantee that a thread waiting on a lock is put to sleep at all (busy waiting is also allowed), and therefore this approach is even more flawed.

The problem described above cannot be solved in OpenMP satisfactory as of now, since there are no directives for scheduling available. Therefore, Sect. 3.2 suggests a possible addition to the OpenMP specification that makes the suggested workarounds (busy waiting or non-compliant use of locks) obsolete. The problem has already been noticed by Lu et al. [8], who suggested the introduction of condition variables (as found in POSIX threads) in 1998. Our solution tries to combine the power of condition variables with the ease of use of OpenMP.

Let us make one more fact perfectly clear: The newly proposed functionality is not useful for the common case in computing centers today, where one processor is exclusivly available for each thread. It is intended for the more general case that multiple threads are competing for the available processors. With the advent of multi-core CPUs in common desktop systems and the expected shift to multi-threaded applications, we soon expect this case to be the dominant one.

## 3.2 Specification

We suggest two new directives:

#### #pragma omp yield

Similar to the POSIX function sched_yield (), this function tells the scheduler to pick a new thread to run on the current processor. If no new thread is available, it returns immediately. The directive provides a simple way to pass on knowledge on what is important and what not from the programmer to the runtime system and operating system scheduler. As a second new directive, we propose:

**#pragma omp sleepuntil (scalar_expression)**
This directive puts the current thread to sleep until the specified scalar expression becomes true (non-zero). The expression is occasionally tested by the runtime system in the background. Before each test, a flush is carried out automatically, to keep the temporary view of memory consistent with memory. An implementation of the directive is not required to wake up the sleeping thread immediately after the expression becomes true, nor does it have to wake it up if the expression becomes true and becomes false again shortly afterwards. Not all threads waiting on the same expression have to wake up at the same time either. It is unspecified, how many times any side-effects in the evaluation of the scalar expression occur.

### 3.3 Rationale

The *yield* directive is inspired by its POSIX counterpart, *sched_yield()*. It offers an easy to use way to influence the scheduling policies of the operating system. This can be important when computing resources are sparse and the programmer wants to optimize program throughput. An example of this would be calling the *yield* directive at the end of every pipeline step in a pipelined application, to get values through the pipeline as fast as possible.

We know of no scheduling primitive in any other parallel programming system that is as powerful and easy to use as the proposed *sleepuntil*. Thanks to the OpenMP memory model (and its ability to read variables without locking them using only *flush*, as compared to e. g. POSIX Threads), this directive is as powerful as condition variables, yet it lacks their difficult usage. The directive can be emulated by wasting time in a loop, but this would be busy waiting and wasteful to the available computing resources, as outlined in Sect. 3.1.

The proposed changes are fully backwards compatible to the existing OpenMP specification, since no behaviour of existing OpenMP functionality is altered in any way.

### 3.4 Application

We have emphasized in Sect. 3.1 that there is no opportunity for a parallel algorithm using taskpools to wait for a new element out of an otherwise empty pool, without constantly polling the pool. There are two approaches to solve this problem with our newly proposed directives. The first one calls the *yield* directive whenever there is no work in the pool, which will put the thread to sleep if another thread is waiting for a processor to become available. Chances are, that a different thread will produce work for the taskpool. If there is no other thread from the same application, a context switch may occur and a different application will run on the processor, allowing for a higher throughput on the machine. Finally, if there is no other thread waiting for the processor, the call to the *yield* directive will just return and no harm is done. A second possible solution is the following:

```
#pragma omp sleepuntil (!tpool_is_empty (pool))
```

This solution offers a more fine-grained control over when the thread is supposed to wake up again, as the thread will sleep until something has been put into the taskpool and not just an unspecified amount of time as with the *yield* solution. After wake-up, it is still necessary to check if the taskpool is not empty again, as no locking of any sort is involved here. The thread might have been woken up at a time when the pool was not empty, but when it tries to actually get a task from the pool, a different thread might have already popped the task.

It is difficult to measure the impact of the proposed directives, as they are most useful on fully loaded systems. We have therefore overloaded a system by starting our benchmark applications with 32 threads using the most simple taskpool sq1. The results are shown in Fig. 5.

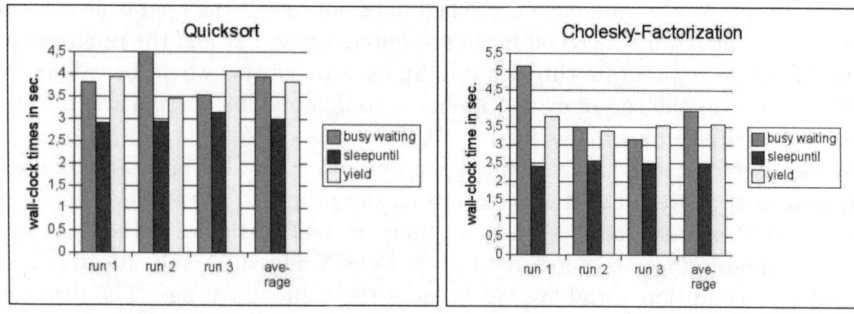

**Fig. 5.** The impact of the proposed directives on a fully loaded system (Sun Fire E6900 with 32 Threads running on 8 processors), measured wall-clock times in seconds over multiple runs with sq1

A different use case for both new directives is testing. When testing OpenMP compilers or performing tests for OpenMP programs, it is often useful to *force* the scheduler into certain timing behaviours that could not be tested otherwise (e.g. stalling one thread, while all other members of the team go ahead and run into a barrier). This is not possible with the present OpenMP specification (except with busy waiting again), and can be very useful to test for hard to catch errors. An example to stall execution of one thread for 100 milliseconds is shown below:

```
double now = omp_get_wtime();     /* save current time into now */
#pragma omp sleepuntil (omp_get_wtime() >= now + 0.1)
```

## 4 Related Work

This paper is an indirect follow-up paper to our own work on task-pools [7]. Some less advanced task pool variants were presented there, along with a first mention of the problem of busy waiting. The present paper includes more advanced task pool variants, two new example algorithms, and a proposal to solve the problem of busy waiting.

A detailed analysis of several task pool implementations with pthreads and Java threads can be found in the article of Korch and Rauber [3]. They conclude, that a combination of private and a public queues for each thread works best for their three benchmark applications.

An OpenMP extension that could help to deal with irregular problems, the workqueuing model, has been suggested by Shah et al. [2], and performance measurements for this extension have already been discussed in Sect. 2.5.

Another approach was proposed by Balart et al. [9]. They suggest to relax the specifications of the sections directive allowing a section to be instantiated multiple times. Additionaly they suggest to execute code outside of any section by a single thread. Each time this thread detects a section instance, it will insert this section into an internal queue. The section instances inserted into the queue are executed by a team of threads.

## 5 Concluding Remarks and Perspectives

Efficient parallelization of irregular algorithms is an ambitious goal that often can be tackled with task pools. We have presented several variants of task pools along with their implementation in OpenMP. To assess the performance of the variants, we have implemented three irregular algorithms: Quicksort, Labyrinth-Search and Cholesky Factorization. Results show that the correct selection of a task pool variant has a significant impact on the performance of an application. There is no universally best variant, but the suitability depends on the pattern of accesses to the task pool. Applications that generate many tasks and access the task pool frequently benefit from the usage of distributed private queues. Applications that access the task pool infrequently, in contrast, need good load balancing, and therefore gain more profit from a central shared task queue.

The second contribution of this paper has been a proposal to the OpenMP language committee. We suggested two directives: *yield* and *sleepuntil*. Both enable the programmer to influence the scheduling process, and to put threads to sleep on demand. By using these directives, the need for busy waiting is eliminated.

A reference implementation of the extended OpenMP functionality can be found in a special release of the OMPi Compiler [6] that is available from the authors on request. In the future, we plan to explore more applications with OpenMP, trying to find ways to improve the specification in the process. Our progress will be visible in the UKOMP project [10]. The project will serve as our testing ground for new functionality we discover to be useful, and also enables other developers to give feedback on how they like our changes.

## Acknowledgments

We are grateful to Björn Knafla for proofreading the paper and for his insightful comments. We thank the University Computing Centers at RWTH Aachen, TU Darmstadt and University of Kassel for providing the computing facilities used to test our sample applications on different compilers and hardware.

## References

1. OpenMP Architecture Review Board: OpenMP specifications (2005), http://www.openmp.org/specs
2. Shah, S., Haab, G., Petersen, P., Throop, J.: Flexible control structures for parallelism in OpenMP. In: Proceedings of the First European Workshop on OpenMP - EWOMP (1999)
3. Korch, M., Rauber, T.: A comparison of task pools for dynamic load balancing of irregular algorithms. Concurrency and Computation: Practice and Experience 16(1), 1–47 (2004)
4. Hoare, C.: Quicksort. The Computer Journal 5, 10–15 (1962)
5. George, A., Liu, J.W.H.: Computer Solution of Large Sparse Positive-Definite Systems. Prentice-Hall, Englewood Cliffs (1981)
6. Dimakopoulos, V.V., Georgopoulos, A., Leontiadis, E., Tzoumas, G.: OMPi compiler homepage (2003), http://www.cs.uoi.gr/~ompi/
7. Süß, M., Leopold, C.: A user's experience with parallel sorting and OpenMP. In: Proceedings of the Sixth European Workshop on OpenMP - EWOMP 2004 (2004)
8. Lu, H., Hu, C., Zwaenepoel, W.: OpenMP on networks of workstations. In: Proc. of Supercomputing 1998 (1998)
9. Balart, J., Duran, A., Gonzàlez, M., Martorell, X., Ayguadé, E., Labarta, J.: Nanos mercurium: a research compiler for OpenMP. In: Proceedings of the European Workshop on OpenMP (2004)
10. Süß, M.: University of Kassel OpenMP – UKOMP homepage (2005), http://www.plm.eecs.uni-kassel.de/plm/index.php?id=ukomp

# A Proposal for OpenMP for Java

Michael Klemm, Ronald Veldema, Matthias Bezold, and Michael Philippsen

University of Erlangen-Nuremberg, Computer Science Department 2
Martensstr. 3, 91058 Erlangen, Germany
{klemm,veldema,philippsen}@cs.fau.de
mail@msbezold.de

**Abstract.** The current OpenMP 2.5 specification does not include a binding for Java. However, Java is a wide-spread programming language that is even used for HPC programming. We propose an adaptation of OpenMP to Java by retrofitting the basic OpenMP directives to Java and further propose some new concepts to make OpenMP fit into Java's language philosophy.

We discuss how Java's memory model matches OpenMP's memory model and how the OpenMP bindings for Java and C++ differ. We also suggest how to achieve flexibility of an OpenMP implementation by allowing both Java threads (*java.lang.Thread*) and Java tasks (*java.util.concurrent.FutureTask*) as an underlying means of parallelization.

Support for object-orientation is added to allow OpenMP to better suit the Java programming model. For example, we suggest a parallel for-each loop over Java collections, OO-based reductions, and object-cloning semantics to adapt data-sharing clauses to Java. Also, we suggest a minimal runtime library to allow object-pooling to circumvent any implicit synchronization involved in object allocations.

Finally, we present some performance numbers for a reference implementation in a research compiler.

## 1 Introduction

Java becomes more and more pervasive in the programming landscape with numerous high-performance (and free) implementations (Sun's and IBM's JVMs, Jackal [21], etc.). For many applications Java's performance is on-par with other programming languages (including C++ and Fortran) [3,12].[1] Hence, Java becomes suitable for high-performance computing as well. Some advantages of Java over, for example, C/C++ and Fortran are its higher productivity, its safety features, and its large standard library. Finally, because of its high level of pervasiveness, many prospective HPC programmers are already experienced in Java but have no thorough knowledge of C/C++ or Fortran.

Java currently has two ways to allow parallel execution of programs. First, the Remote Method Invocation (RMI) facility allows distributed execution of a program that spans several nodes. Second, shared-memory parallelism is achieved by means of the Java Threading API. We argue that using Java's threading model directly results in two unforeseen problems. First, it is a cumbersome and error-prone task to transform a

---

[1] Java performance often depends on the programming style used.

sequential program into a parallel one by manually creating threads and implementing explicit data exchange by hand. Second, compiler analysis is made exponentially harder as the parallelism is hidden from it. For example, static data-race detection tools (for Java) such as ESC/Java [8] currently have to resort to complex model checking techniques to search for all possible interleavings of object usages (and thread creations). An OpenMP-style programming allows for easier static data-race detection.

With Java 1.5, the *java.util.concurrent* package provides more support for multithreaded programming, i.e. it contains a large set of thread-aware collections and threading support mechanisms. For example, it contains code for forms of light-weight threads (tasks) as well as scalable thread-safe blocking queues and concurrent hashtables. The collections from *java.util.concurrent* are universally useful and applicable to both standard Java and the proposal of this paper, Java-OpenMP. However, the new task subsystem enlarges the landscape of expressible parallelism. With Java-OpenMP it is easier to choose, even automatically, from the available types of parallelisms.

Finally, because Java has no support for fast thread-local storage, OpenMP's data-sharing clauses (*shared*, *private*, etc.) fulfill a need in the Java environment. Java only allows one to simulate true thread-local storage by putting variables into the thread class, a reference to which can be acquired by calling the *currentThread()* method at any time.

Combining the above observations we propose to adapt the C++ OpenMP specification to fit Java and to retrofit any Java features onto it. We present the basic set of OpenMP directives and their suggested semantics under the proposed Java-OpenMP model. We further propose a series of extensions to the basic model so that OpenMP is a better match to Java's object-oriented programming model.

## 2 Related Work

Various commercial and production quality compilers for C/C++ and Fortran, e.g. the Intel compiler suite [14] or the compilers of the Portland Group [15], can compile OpenMP programs to native code.

With OdinMP/CCp [5] for C/C++ and Omni OpenMP [20] as well as OpenUH [16] for both C/C++ and Fortran there also are open-source implementations of the OpenMP specification. These are source-to-source compilers that transform OpenMP programs to equivalent programs that make use of a native threading API. Source-to-source compilers rely on another compiler to generate the executable code. Only for the Itanium architecture, OpenUH uses the Open64 [13] compiler back-end to directly emit a native executable.

At present, we are not aware of any standardized Java binding of the OpenMP specification. An earlier paper on the JOMP project [6] dealt with a subset of OpenMP for Java. The JOMP compiler follows the source-to-source approach and transforms a JOMP program into an equivalent Java program, which uses the Java Threading API for parallelism. In JOMP (as in our proposal), special Java line comments are treated as OpenMP directives. The proposal we present here follows the OpenMP specification as closely and as completely as possible and suggests new features that allow a tight integration of OpenMP into the Java language.

## 3 Differences to the C++ OpenMP Specification

The semantics of most of the OpenMP features can be applied from the C/C++ OpenMP standard without any changes. This includes: *//#omp for*, *//#omp section*, *//#omp sections*, *//#omp single*, *//#omp master*, *//#omp barrier*, *//#omp flush*, and others. Whereas we discuss only the necessary differences between Java-OpenMP and C++-OpenMP, we reason where we feel that Java's design philosophy is better served by adapting the C++-OpenMP definitions.

### 3.1 Pragmas and Conditional Compilation

Although pragmas and conditional compilation are essential elements of the OpenMP standard, Java's language specification does neither provide the pragma concept nor conditional compilation. Although one might use a C-style pre-processor on Java source code, such scenarios are discouraged as the mapping between the unprocessed code and the pre-processed code is potentially lost.

With Java 1.5 the concept of *annotations* was added as an alternative. Annotations are type-safe meta-statements that the Java compiler, the JVM, and other tools are able to recognize. For example, annotations can be used to describe RMI stubs and skeletons so that they can be automatically created. Unfortunately, annotations cannot be used for Java-OpenMP for several reasons. First, they cannot be used at the level of Java statements or blocks, but only at the level of declarations (e. g. at class level, method level, etc.). Second, an annotation must always have a well-defined annotation type that in turn must be defined in some Java package visible to the Java compiler. Thus, a non-OpenMP compiler would have to provide dummy annotations of the same annotation types as an OpenMP-compliant compiler does. This is clearly not acceptable.

The only reasonable way to implement pragmas in an OpenMP binding for Java is not to modify the language itself but instead to add a special kind of Java line comment for OpenMP directives:

//#omp <directive-name> <clauses> newline

A non-OpenMP compiler treats such a directive as a regular Java comment and ignores it. The OpenMP-compliant compiler recognizes the line comment as an OpenMP directive. This approach has also been used by the JOMP compiler [6].

Similarly, we propose an adoption of the conditional compilation sentinels that are introduced by the OpenMP specification for Fortran. A Java line comment starting with //# is recognized as a conditional compilation sentinel. Again, the non-OpenMP compiler safely ignores the rest of the line while the Java-OpenMP compiler parses and compiles the statement as desired.

### 3.2 Extensions

This section proposes how OpenMP's constructs can be applied or extended to an object-oriented context.

```
package omp;
interface CustomCloner<T> {
    T clone(T other);
}
```

**Fig. 1.** Interface for creating custom copies

```
Object a = ...; int b = ...; CustomCloner cloner = ...;
//#omp ... firstprivate(a:cloner, b)
```

**Fig. 2.** Example for custom object copying

**Data-sharing Attributes**

The argument passing semantics used in the C++ binding for the data-sharing clauses are inappropriate for a Java binding. In the C++ binding the behavior depends on the data type. For example, if a shared $x$ is a pointer to an object, the pointer is manipulated. But if $x$ is of an object type, the copy constructor of the object is invoked. Hence, the C++ semantics of a *firstprivate* pointer would imply that only the reference is privatized whereas the object is still shared among the individual threads. A *firstprivate* object would result in a private copy.

Since Java does not have the concept of a copy constructor as C++ does and since it only has references to objects[2], a programmer cannot choose whether an object is to be copied upon privatization or not.

The semantics of OpenMP's data-sharing clauses have to be adapted to overcome this limitation of Java. We propose the following solution: in case of a *shared* object reference, the reference is copied, whereas *firstprivate* and *lastprivate* always create copies of the object by means of Java's *Object.clone()* method.

Moreover, an additional parameter may be added to the *firstprivate* and *lastprivate* clauses to be able to specify a custom behavior for copying objects. That way, one can for example specify a deep clone instead of a shallow one or use objects of classes which do not implement the *Cloneable* interface. We introduce a special interface (*CustomCloner*, see Fig. 1) which offers a method *clone* to clone an object in a customized manner. Objects of this type can be placed right after each variable in the *firstprivate* and *lastprivate* clauses, separated by a colon (see Fig. 2).

**Parallelization of Iterations over Collections**

Most Java programs make heavy use of Java's collection API and for-each loops. For example, a molecular dynamics application written in Java will probably be programmed with collections of molecule objects rather than flat arrays of doubles. Without any extensions of OpenMP such Java programs cannot be parallelized by means of OpenMP's work-sharing clauses.

We therefore propose the constructs *//#omp iterator* and *//#omp parallel iterator* (similar to *//#omp for* and *#omp parallel for*) to allow iterations over container objects to be parallelized (see Fig. 3). When these constructs are used, the iterator which is

---

[2] A Java reference is similar to a C++ pointer but pointer arithmetic is not allowed.

```
LinkedList c = new LinkedList();
c.add("this");
c.add("is");
c.add("a");
c.add("demo");

//#omp parallel iterator
for (String s : c)
    System.out.println("s");
```

**Fig. 3.** Parallel iteration over a Java collection

```
package omp;
interface Reducer<T> {
    T reduce(T a, T b);
    T init();
}
```

**Fig. 4.** Interface for object-oriented reductions

purely sequential in regular Java is replaced by a parallel execution of the loop. Conceptually, during compilation, the code of Fig. 3 is transformed by first calling *c.toArray()*, which flattens the collection into an array. The loop itself is then rewritten into a traditional *for* loop with an ordinary loop counter instead of the iterator. The resulting loop is then parallelized by the *for* work-sharing construct. The same clauses as for the *for* construct are allowed. Like *'for'*, *'iterator'* has an additional convenience form called *'parallel iterator'*, which is then transformed into a parallel region containing an *'iterator'* region.

In the proposed form, the *iterator* construct is a specialization of the more generic *taskq* construct that the Intel OpenMP compiler already provides [1]. With *taskq* the programmer introduces a program region that encloses a set of *task* environments. Upon entrance into the *taskq* region, the encountering thread appends each *task* region to a work queue while it executes the enclosing structured block. All other worker threads then dequeue tasks one by one and process the code contained in the *task* region. If the enclosing *taskq* contains a loop, the loop is executed sequentially by the encountering thread.

Because C and C++ lack a standardized iterator interface, a C/C++ binding has to provide such a generic construct to allow to parallelize iterator loops over collections. For example, in C++ iterators are mostly expressed by classes that overload the '++' and '==' operators. In contrast, Java provides for-each loops and the standardized *Iterator* interface. Hence, the proposed *iterator* construct offers sufficient expressiveness and is a straightforward extension of concepts known to Java programmers.

**Object-Oriented Reductions**

Currently, OpenMP only defines reductions for primitive data types and for certain operators. However, in Java's object-oriented design philosophy most data is expressed as objects for which no reduction is allowed. Current practice forces the programmer to explicitly implement reduction algorithms.

```
class BigIntegerReducer implements omp.Reducer<BigInteger> {
    public BigInteger reduce(BigInteger a, BigInteger b) {
        return a.add(b);
    }
    public BigInteger init() { return new BigInteger("0"); }
}

BigInteger value = ...; BigIntegerReducer reducer = new
BigIntegerReducer();
//#omp parallel for reduction(reducer:value)
{
    ...
}
```

**Fig. 5.** Example of object-oriented reductions

We therefore propose an interface (see Fig. 4) that lets the programmer use objects in a reduction clause. The interface *Reducer* offers a method *reduce*, which reduces two objects, and a method *init*, which is used to initialize the values inside the region. That way, the class definition of the objects, which should be reduced, does not need to be changed or enhanced and existing classes can be used by creating a new *Reducer* implementation. Different kinds of reductions can be implemented for every class by creating different classes implementing the *Reducer* interface. Fig. 5 shows an example of an object-oriented reduction, which defines a custom reduction for the BigInteger class from the java.math package.

## 4 Specific Aspects of a Java Binding

This section discusses the most important aspects of a Java-OpenMP binding. A potential OpenMP Java binding needs to respect the semantics and restrictions that are imposed by the Java Language Specification [10].

### 4.1 Java-OpenMP Memory Model

Java's memory model [17] and OpenMP's memory model [18,11] are conceptually very close. Both Java's and OpenMP's memory models propose a relaxed-consistency model that supports the idea of a global memory where individual threads can temporarily fetch data from and then place it into a thread-local cache. At specific points in a thread's life-time, the thread flushes its cache by copying modified data back into the global memory.

OpenMP specifies that this 'flush' operation occurs when a flush or barrier construct is reached. Roughly, Java's memory model specifies that the same occurs when a synchronization statement is entered or left. It is therefore only natural to reuse Java's flushing behavior to implement the Java-OpenMP's *flush* construct. For a naive implementation it would be sufficient to use an empty synchronized block—such as *synchronized(this)* {}—for any *flush* operation.

### 4.2 Interaction Between Parallel Regions and Java Threads

Whereas C and C++ do not offer language constructs for multi-threading, Java's language specification provides explicit threading support. Thus, a Java-OpenMP specification needs to take the coexistence of its parallel regions and the standard system thread package into account. It needs to state what Java-thread functionality is available to the programmer and what is disallowed. Limitations to the programmer need to be independent of the way parallel regions are mapped by the compiler. For example, it is possible to map a Java-OpenMP parallel region either to a Java-thread or to a 'task' from the *java.util.concurrent* package (from here on called Java-task). Java-threads are fully preemptive threads which are scheduled by the JVM while Java-tasks can be used as user-level cooperative threads. Java-tasks are then multiplexed upon a number of Java threads. The Java-OpenMP specification must therefore define sufficient (but not too severe) restrictions on the usage of Java's threading features so that no harmful interactions with the transformation of parallel regions may result. The following list of restrictions serves this purpose. Please note that these restrictions also hold when *java.util.concurrent* is used for manual parallelization of a Java program. Thus, the restrictions are no artifical limitations that stem solely from the OpenMP Java binding. However, for Java-OpenMP the restrictions only apply within parallel regions. Outside, i.e. in regular Java code, all of Java's features may of course still be used as usual.

- The programmer cannot use *Thread.currentThread()* to differentiate between regular Java threads and the threads that execute a parallel region. For a Java lightweight task *Thread.currentThread()* will report incorrect values as the light-weight tasks are multiplexed over a set of regular Java threads.
  *Thread.currentThread()* is mostly used to implement some form of thread-local storage. This is no longer necessary when OpenMP's data-sharing primitives are available.
- The programmer may not use *synchronized()* for synchronization in parallel regions. Instead the various OpenMP thread synchronization constructs must be used. The reason for this is again the flexibility of the Java-OpenMP mapping to different threading models. If a parallel region is executed by tasks that are multiplexed onto several Java threads, *synchronized()* will not behave as desired.
- The programmer may not use *Object.wait()* because it may cause deadlocks of the threads and tasks used internally to implement parallel regions. For example, consider the case where two Java-tasks are multiplexed upon one Java-thread. A *wait()* inside one Java-task will block the entire Java-thread and therefore also the co-scheduled other Java-task(s).
- However, *Object.notify()* may be used as it is guaranteed to not block. This allows a parallel region to notify threads spawned by non-OpenMP parts of the application. For example, in an application where data is visualized by means of the Java Swing toolkit, even from within a Java-OpenMP parallel section, the application may need to notify the GUI threads that new data is ready for visualization.

## 4.3 Exception Management

The OpenMP specification disallows to prematurely leave a parallel region. This includes jumping out of a region using *break* statements or the like. The same applies to exceptions thrown inside a parallel region. For the C++ specification exception handling is not an issue because exceptions rarely occur in most C++ programs.

In contrast, Java makes heavy use of exceptions. Exceptions are not only visible at the programming level, but are also used for handling system events and errors (e. g. *NullPointerException, ArrayIndexOutOfBoundsException*, etc.). These system-level events may be thrown by many of the byte-code instructions of a JVM. Hence, every OpenMP-conforming implementation for Java needs to take special care when dealing with exceptions that are thrown inside parallel regions.

Unfortunately, it is not obvious how to appropriately react to an exceptional event. First, it is not an option to terminate the exception-throwing thread as it will no longer reach subsequent barriers, thus resulting in a deadlock of other threads in the team. Second, it is not permitted to prematurely proceed to the next barrier as the reason of the exception might still be affecting the program, again resulting in undefined behavior.

The desired behavior is that other threads of a team should abort as soon as possible if one thread has encountered an exception. For performance reasons, it is not acceptable that threads continuously actively poll for exceptions that might have occurred on other threads. For this purpose we propose the following semantics:

- If one thread throws an exception, it sets a cancellation flag for all other threads, registers its exception at the thread team, and then proceeds to terminate.
- All (other) threads check for the occurrence of an exception at cancellation points. Cancellation points are the start and the end of *#omp barrier, #omp critical, #omp parallel*, and work-sharing clauses.
- When a thread reaches a cancellation point and finds that another thread has requested cancellation, the thread itself is interrupted by an exception. This ensures that the thread's stack frames are unwound and execution of the thread is terminated.
- The exception is re-thrown by the master thread after all threads have terminated execution of the parallel region. If more than one exception was thrown, the master thread randomly selects an exception and re-throws it.

Similar semantics have also been proposed for the JCilk [7] parallel programming language. Although JCilk does not define cancellation points, it allows to terminate asynchronous computations. Such computations are aborted as soon as possible whenever a exception occurs therein. If multiple exceptions are thrown in asynchronous computations, then JCilk randomly selects one to be handled by the corresponding `catch` block.

## 5 Runtime Environment

### 5.1 Runtime Support Library

The runtime library functions as proposed by the OpenMP specification can directly be adopted to Java-OpenMP. Java, however, does not support globally scoped functions as

```
package omp;
public class Omp {
    // runtime support functions
    public static int getThreadNum() { ... }
    public static int getNumThreads() { ... }
    ...
    // object pooling functions
    public static Object objectPoolGet(Class cls) { ... }
    public static Object objectPoolGet(Class cls,
                                        int elts) { ... }
    public static void objectPoolPut(Object obj) { ... }
}
```

**Fig. 6.** Java runtime support functions and additional functions for object pooling

C++ does. Hence, the runtime library is to be implemented as a set of static methods in a class called *omp.Omp* (see Fig. 6). With Java 1.5, the qualified class name can be omitted as the Java-OpenMP compiler should automatically add a static import statement during compilation.

### 5.2 Object-Pooling Support

Because of Java's highly object-oriented programming style, most Java programs allocate lots of objects. To create arrays and objects, the *new* operator accesses the global heap. Many JVM implementations are synchronizing this access to guarantee that only one thread can perform a *new* operation or can garbage collect at the same time. Hence, Java programs that allocate lots of objects are less parallel than desired. An obvious solution to this JVM limitation is the usage of a per-thread object-pool.

This proposal does not restrict OpenMP's implementation to Java threads but also, for example, allows Java tasks for implementing parallel regions (see Section 4.2). However, to create an efficient thread-pool, and to use the correct and most efficient form of thread-local data, the thread-pool implementation needs to know if a given Java-OpenMP implementation is based on Java-threads or Java-tasks. Hence, Java-OpenMP must provide an API (see Fig. 6) that supplies the programmer with an efficient object-pool abstraction.

The *objectPoolPut* method allows a programmer to store an object into the pool for later retrieval. If *objectPoolGet* is invoked and no object of the specified type is available, a new object is allocated from the global heap. The *objectPoolGet* method with two parameters is used to create array objects of a specific size.

### 5.3 Environment Variables

OpenMP 2.5 specifies a set of shell-level environment variables that can be used to control program behavior at runtime. We propose that these variables should be made available to an application by means of Java's properties mechanism since this is the preferred way of passing environment settings to a Java application.

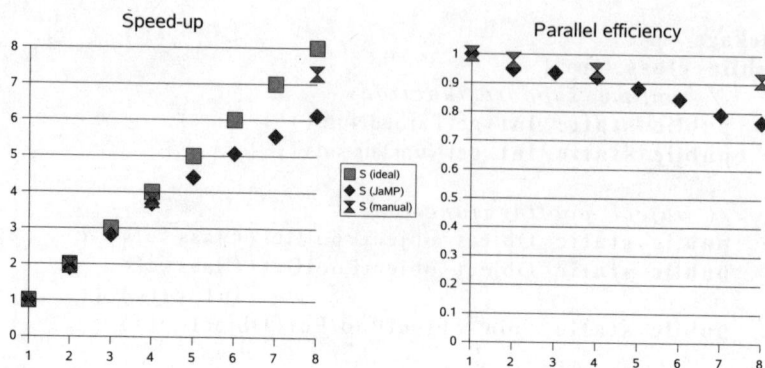

**Fig. 7.** Speed-up and parallel efficiency of the Lattice-Boltzmann Method on upto 8 CPUs of the AMD cluster using the DSM of Jackal

## 6 A Reference Implementation of Java-OpenMP in Jackal

Jackal [21] is a high-performance implementation of Java that provides a Distributed Shared Memory (DSM) both on Shared-Memory Processor (SMP) machines and clusters with fast interconnects.

Our Java-OpenMP implementation in Jackal is in part compiler-based and in part Java-based. The compiler detects the directives and transforms them into an internal representation. This representation is carried right into the backend for optimization purposes so that Java-OpenMP-level analyses and optimizations can be performed there. The reference implementation is thus not implemented as a pre-processor but rather as part of the compiler. A compiler-based implementation has a number of advantages over a pre-processor. First, the code generated by a pre-processor needs to obey the rules of the target language which restricts the possible transformations that can be applied. Second, code reuse is increased as other compiler front-ends can share the same compiler intermediate code level infrastructure.

The Java-part of the reference implementation contains the runtime support functions specified by Java-OpenMP. It also contains additional support for data-sharing, management of the thread teams, and exception handling.

Our current Java-OpenMP infrastructure implements the basic OpenMP directives as discussed in Section 3 except for the potential extensions to the data-sharing and reduction clauses as outlined in Section 3.2.

### 6.1 Performance of a Java-OpenMP Version of the Lattice-Boltzmann Method

The Lattice-Boltzmann Method (LBM) [22] is used to simulate fluids using cellular automata. Space and time are discretized and normalized. In our case, LBM operates on a 2D domain divided into cells. Each cell holds a finite number of states called *distribution functions*. In one time step the whole set of states is updated synchronously by deterministic, uniform update rules. The evolution of the state of a given cell depends only on its neighboring cells.

The kernel was parallelized in a straightforward manner by placing Java OpenMP comments in the Java source code. The domain is decomposed along the $y$-axis, that is, the outermost loop is distributed over the worker threads. A similar scheme is used for the MPI parallelized version in [19]. Fig. 7 shows the speed-up and the parallel efficiency achieved by both the Java-OpenMP parallelized LBM and the MPI parallelized LBM.

For 8 nodes, Java-OpenMP achieves a speed-up of about 6.1. This is about 83 % of the speed-up achieved by the manually optimized LBM kernel that is written in C using MPI for communication. The lower scalability is due to the DSM environment that strongly depends on the communication layer (in our case Ethernet). Jackal currently does not include message aggregation. Instead it transfers each cell individually whereas the MPI version transfers whole partitions, i.e. each process requests all the cells from neighboring processes with one MPI request.

## 7 Summary

In this paper, we have sketched a number of extensions and adaptations to the OpenMP C++ specification that are necessary to adapt it to Java's object-oriented programming model. A syntax for OpenMP directives was proposed that allows a non-OpenMP compiler to safely ignore OpenMP directives. We have shown differences of Java and C++ that require changes to the OpenMP specification. For example, OpenMP's data-sharing clauses need to be adapted to Java's notion of object references. Parallel iteration over the items of a Java collection requires a different syntax and semantics. We discussed the Java and OpenMP memory model and described how OpenMP parallel regions can interact with regular Java threads. We further proposed semantics for managing exceptions in the context of parallel regions. Runtime support is provided by means of a class that not only contains the functions that are specified by the OpenMP standard, but in addition offers object pooling. A reference implementation of the Java-OpenMP support in the Jackal DSM system demonstrates Java-OpenMP's viability.

## Future Work

First, the performance evaluation of the Java-OpenMP prototype should be based upon standardized and well-known benchmarks. A Java-OpenMP version of the NAS Parallel Benchmarks (NPB) [4,9] is in the works. The benchmark suite has to be backported from multi-threaded Java to sequential Java. Then, the OpenMP pragmas have to be inserted to re-parallelize the individual NPB benchmarks. Afterwards, the benchmarks suite is to be evaluated with large-scale clusters and SMPs.

Second, as soon as the upcoming OpenMP 3.0 draft is committed as a standard, the current prototype implementation has to be extended to follow the OpenMP 3.0 standard. If the upcoming standard also contains specifications of exception handling, user-level threads, and parallel execution of iterator loops, the new semantics have to be retrofitted onto the Java programming language as well.

Finally, we plan to port the Java-OpenMP binding into the Jikes RVM [2]. With the port to the Jikes RVM, we will show that a Java-OpenMP binding is not only feasible for our DSM environment but also usable on an SMP architecture.

## References

1. Intel C++ Compiler for Linux Systems User's Guide, Document Number 253254–031 (2004)
2. Alpern, B., Attanasio, C.R., Barton, J.J., Burke, M.G., Cheng, P., Choi, J.-D., Cocchi, A., Fink, S.J., Grove, D., Hind, M., Hummel, S.F., Lieber, D., Litvinov, V., Mergen, M.F., Ngo, T., Russell, J.R., Sarkar, V., Serrano, M.J., Shepherd, J.C., Smith, S.E., Sreedhar, V.C., Srinivasan, H., Whaley, J.: The Jalapeño Virtual Machine. IBM System Journal 29(1), 211–238 (2000)
3. Artigas, P.V., Gupta, M., Midkiff, S.P., Moreira, J.E.: High Performance Numerical Computing in Java: Language and Compiler Issues. In: Proc. of the 12th Intl. Workshop on Languages and Compilers for Parallel Computing, San Diego, CA, USA, pp. 1–17 (1999)
4. Bailey, D.H., Barszcz, E., Barton, J.T., Browning, D.S., Carter, R.L., Dagum, D., Fatoohi, R.A., Frederickson, P.O., Lasinski, T.A., Schreiber, R.S., Simon, H.D., Venkatakrishnan, V., Weeratunga, S.K.: The NAS Parallel Benchmarks. The Intl. Journal of Supercomputer Applications 5(3), 63–73 (1991)
5. Brunschen, C., Brorsson, M.: OdinMP/CCp - a Portable Implementation of OpenMP for C. Concurrency: Practice and Experience 12(12), 1193–1203 (2000)
6. Bull, J.M., Kambites, M.E.: JOMP—an OpenMP-like Interface for Java. In: Proc. of the ACM 2000 Conf. on Java Grande, San Francisco, CA, USA, pp. 44–53 (2000)
7. Danaher, J.S., Lee, I.A., Leiserson, C.E.: Programming with Exceptions in JCilk. The Journal of Science of Computer Programming (to appear, 2006)
8. Flanagan, C., Rustan, K., Leino, M., Lillibridge, M., Nelson, G., Saxe, J.B., Stata, R.: Extended static checking for java. In: Proc. of the ACM SIGPLAN 2002 Conf. on Programming Language Design and Implementation, Berlin, Germany, pp. 234–245 (2002)
9. Frumkin, M., Schultz, M., Jin, H., Yan, J.: Implementation of NAS Parallel Benchmarks in Java. Technical Report NAS-02-009, Ames Research Center, Moffett Field, CA, USA (2002)
10. Gosling, J., Joy, B., Steele, G., Bracha, G.: The Java Language Specification, 3rd edn. Addison-Wesley, Upper Saddle River, NJ, USA (2005)
11. Hoeflinger, J.P., de Supinski, B.R.: The OpenMP Memory Model. In: Proc. of the 1st Intl. Workshop on OpenMP (IWOMP 2005), Eugene, OR, USA (May 2005)
12. http://kano.net/javabench
13. http://open64.sourceforge.net
14. http://www.intel.com/cd/software/products/asmo-na/eng/compilers/index.htm
15. http://www.pgroup.com/products/cdkindex.htm
16. Liao, C., Hernandez, O., Chapman, B., Chen, W., Zheng, W.: OpenUH: An Optimizing, Portable OpenMP Compiler. In: Proc. of the 12th Workshop on Compiler for Parallel Computers, A Coruna, Spain, January 2006, pp. 356–370 (2006)
17. Manson, J., Pugh, W., Adve, S.V.: The Java Memory Model. In: Proc. of the 32nd ACM SIGPLAN-SIGACT Symp. on Principles of Programming Languages, Long Beach, CA, USA, pp. 378–391 (2005)
18. OpenMP C and C++ Application Program Interface, Version 2.0 (March 2002)
19. Pohl, T., Thürey, N., Deserno, F., Rüde, U., Lammers, P., Wellein, G., Zeiser, T.: Performance Evaluation of Parallel Large-Scale Lattice Boltzmann Applications on Three Supercomputing Architectures. In: Proc. of the IEEE/ACM Supercomputing Conf. SC 2004, Pittsburgh, PA, USA, August 2004, pp. 21–33 (2004)

20. Sato, M., Satoh, S., Kusano, K., Tanaka, Y.: Design of OpenMP compiler for an SMP cluster. In: Proc. of the 1st European Workshop on OpenMP, Lund, Sweden, September 1999, pp. 32–39 (1999)
21. Veldema, R., Hofman, R.F.H., Bhoedjang, R.A.F., Bal, H.E.: Runtime optimizations for a Java DSM implementation. In: 2001 joint ACM-ISCOPE Conf. on Java Grande, Palo Alto, CA, USA, June 2001, pp. 153–162 (2001)
22. Wolf-Gladrow, D.A.: Lattice-Gas Cellular Automata and Lattice Boltzmann Models. Lecture Notes in Mathematics, vol. 1725. Springer, Heidelberg (2000)

# A Proposal for Error Handling in OpenMP

Alejandro Duran, Roger Ferrer, Juan José Costa, Marc Gonzàlez,
Xavier Martorell, Eduard Ayguadé, and Jesús Labarta

Barcelona Supercomputing Center (BSC)
Departament d'Arquitectura de Computadors
Universitat Politècnica de Catalunya
Jordi Girona, 1-3, Barcelona, Spain
{aduran,rferrer,jcosta,marc,xavim,eduard,jesus}@ac.upc.edu

**Abstract.** OpenMP has been focused in performance applied to numerical applications, but when we try to move this focus to other kind of applications, like Web servers, we detect one important lack. In these applications, performance is important, but reliability is even more important, and OpenMP does not have any recovery mechanism. In this paper we present a novel proposal to address this lack.

In order to add error handling to OpenMP we propose some extensions to the current OpenMP specification. A directive and a clause are proposed, defining an scope for the error handling (where the error can occur) and specifying a behaviour for handling the specific errors.

Some examples of use are presented, and we present also an evaluation showing the impact of this proposal in OpenMP applications. We show that this impact is low enough to consider the proposal worthwhile for OpenMP.

## 1 Introduction and Motivation

OpenMP has become one of the most widespread programming models within the scientific domain for SMP machines. The language has proved to be a reliable paradigm throughout a great variety of numerical codes, and its success can be explained by two main reasons: the language simplicity and the fact that the model is under a continuous revision by both the industry and the academia. After the first OpenMP specification, there have been many proposals for the improvement of the language. All of them well based on experimental works that have led to changes in the language. Some have been included in the language specification (e.g the nested parallelism support or the definition of the workshare construct) and some others are currently under consideration (e.g. task queues or autoscoping of variables).

Currently, the OpenMP community is engaged in an open discussion; whether the OpenMP programming can be moved to the non numerical domain. There have been experimental works where OpenMP has been adopted for the parallelization of applications like Web servers [1] or even in games [2]. Traditionally, these applications have been parallelized using threading techniques, through hand coded transformations that degrade the programming style. Besides, it is quite common that these applications are originally provided with recovery mechanisms for specific runtime events. For instance, Web servers are programmed to have particular responses when connection descriptors are exhausted, a timeout expires, or, when in a critical situation, a memory allocation operation fails. Clearly, each case is not having the same consequences, so the error management should be different depending on the situation.

For these and other environments, performance is not always the main issue, while reliability is. An application crash might not be acceptable to happen, or at least, when a fall down is about to happen, the application is provided with escape mechanisms allowing safety actions to be taken always under a controlled behaviour.

It is pointless to run in parallel an application that has been specifically shielded to particular events, while the runtime implementation is open to situations that may crash the execution of the application. Numerical codes with considerable execution times are equally sensible to internal runtime fails. After days of execution, an application crash is not acceptable without any chance to react to it.

Currently, the OpenMP specification lacks any support to report an error in its code to the application. So, an application has no chance to react to errors that happen inside the OpenMP transformations and not in the application itself. This paper presents a set of new OpenMP directives and clauses with the aim of decoupling the specification and execution of the parallelism from the handling of OpenMP errors. The proposal defines a set of mechanisms to specify explicit error handling in the application, as well as runtime error recovery. The structure of this paper is as follows: Section 3 describes our proposed extensions to OpenMP. Section 4 discusses some implementation details of our prototype of the proposal. In section 5 we evaluated which is the overhead of the error handling code. Section 2 presents other works related to ours. And finally, in section 6 we discuss the conclusions of our work.

## 2 Related Work

Gatlin[3] pointed out one of the lacks of OpenMP that restrict its application fields. As long as OpenMP does not include explicit support for error recovery or even detection, it is going to be limited to the scientific domain. Although an explicit proposal is not presented, Gatlin explore three main lines for error detection and recovery: exception based mechanisms, call-back mechanisms and explicit error code mechanisms. Exception based mechanisms are inspired in the try/catch constructs in C++. Call-back mechanisms are introduced through the definition of a new clause to directives where to indicate the function containing the call-back code. Call-back mechanisms offer the advantage of keeping away from the site of the computation the code responsible for the error recovery. Finally, error based code introduces a new variable type (e.g: OMPerror) and a new clause where to supply an error variable (similar in use to the posix

ERRNO variable). This approach is quite general, but forces the programmer to add the error handling code to the computational parallel code. The proposal in this paper introduces similar mechanism to the error based and the call-back approaches.

Callback mechanisms have been applied in a wide range of different domains, but always under a common aim, that is, a proper response to particular runtime events. It is quite easy to find different applications for these mechanisms. From operating system implementation [4,5], to error reporting within a parser implementation [6], callback mechanisms are not a novelty. The proposal in this paper introduces the callback mechanisms in the OpenMP programming model.

Error recovery mechanisms have been extensively studied, especially for distributed memory systems [7]. Checkpointing has been the main strategy for recovering an application from a fail-stop fault. For message-passing-based applications, the most common technique is based on hand coded barrier instrumentation. The barrier synchronization mechanism is modified and checkpoint code is introduced [8]. The case for shared memory systems has not been studied as extensively, but again, checkpointing has been the principal approach. Within threading environments, Dieter et al. [9] have proposed the implementation of the checkpoint support inside the thread library internals.

Although not specifically related to checkpoint recovery, the proposal in this paper allows for decoupling such mechanisms from the parallelization process. In front of classical approaches based on hand coded and/or runtime mechanisms, the proposal in this paper provides the programmer with specific constructs to embed recovery mechanisms. Beyond checkpointing strategies, the proposal is generic enough to consider any kind of recovery mechanism.

## 3 Proposed Extension

### 3.1 Basic Concepts

**Sources of Error.** OpenMP errors can come from two sources. First, errors that come from the code that the compiler inserts to transform the serial application following the user directives. As this code is transparent to the user, there is no way for the user to code a response to anomalous situations in the execution of this inserted code.

The second source of error comes from the use of OpenMP intrinsics. Currently, although an error can occur when the user specifies an intrinsic there is no way to know whether the operation failed or not. This is different from the previous case in that the code is explicitly inserted by the programmer.

Our proposal covers both cases to provide a complete error management for OpenMP.

**Error Classification.** From the perspective of the application it would be interesting to be able to identify which kind of error has occurred when an anomalous situation arises. We propose two orthogonal classifications. First, it would be interesting to know the kind of the error that has occurred (e.g. memory exhaustion, invalid argument, ... ). Second, it would be interesting to classify errors based on the predicted impact they will have on the execution of the application (i.e. what chances the application has to finish being a valid execution).

For that, two new types would be introduced: omp_error_type and omp_error_severity. Table 1 shows a possible list of proposed error types. This list does not intend to be complete but just serve as an example. Table 2 lists a possible range for the severity of the errors. The decision on which level of severity corresponds to an error is left to each implementation. Table 2 gives a few examples from our implementation. Note that both, the type and the severity of the error, have no relationship at all between them (i.e. two errors may have the same type and have a different severity).

**Table 1.** omp_error_type possible values

| Constant | Meaning |
| --- | --- |
| OMP_ERR_INV_ARGUMENT | One or more arguments to a directive or intrinsic are invalid. |
| OMP_ERR_OP_NOT_IMPLEMENTED | The requested operation is not supported by the implementation. |
| OMP_ERR_NOT_ENOUGH_MEMORY | Memory could not be allocated to complete an operation. |
| OMP_ERR_NOT_ENOUGH_THREADS | Not all requested threads could be created. |
| OMP_ERR_UNKNOWN | None of the previous errors |

**Table 2.** omp_error_severity possible values

| Constant | Meaning | Examples |
| --- | --- | --- |
| OMP_ERR_MILD | The error will not hinder a correct execution | Invalid number of threads specified |
| OMP_ERR_MEDIUM | The error may potentially alter the specified behavior but it will probably still be a correct execution | Nested parallelism not supported |
| OMP_ERR_SEVERE | Unless corrected the error will result in undefined behavior | Not all threads could start a workshare |
| OMP_ERR_FATAL | If the application continues it will either have incorrect results or none at all (i.e. crash) | A barrier synchronization failed |

**Error Handling.** When an error occurs some action needs to be executed. We suggest to define a set of available actions and a way for the user to specify more complex actions using callbacks. Table 3 summarizes the different proposed set of available actions to execute when an error occurs.

As shown in table 3 the user is given the option to specify its own function which will be invoked on an error condition. The user can specify as well a list of expressions. These expressions will be evaluated just before the callback is executed and passed to it as parameters of the callback function. We propose to additionally have an implicit argument to the callback that will be a pointer to a structure containing information about the anomalous situation. The exact content of this structure would be implementation dependent. This additional

**Table 3.** Possible responses to an error

| Type | Meaning | Comments |
|---|---|---|
| OMP_ABORT | The execution is aborted | |
| OMP_IGNORE | The failed operation is ignored | Execution is continued on best effort. |
| OMP_RETRY | The failed operation is retried | Has an optional argument specifying the maximum number of retries. |
| OMP_SKIP | Skip if possible the offending code | Execution is continued on best effort. |
| OMP_SERIALIZE | Execute if possible the offending code with just one thread | |
| User callback | An user function is called to decide what to do | The callback may have arguments specified by the user |

argument could later be used to query about the nature of the error using a new set of intrinsics (see section 3.4). Figure 1 shows the proposed prototypes for the callback functions for both C/C++ and Fortran.

```
C/C++ prototype:
omp_action_t callback ( omp_err_info_t *info, ... );

Fortran prototype:
integer callback ( ctx )
type(omp_err_info)::ctx
```

**Fig. 1.** User callbacks prototypes

## 3.2 The ONERROR Clause

Our proposal introduces a new clause to all OpenMP directives, the ONERROR clause. This clause specifies how a given error (or set of errors) needs to be handled in the dynamic extent of the directive where the clause is specified.

```
onerror([err_list:]action[, arg_list])
```

**Fig. 2.** ONERROR clause syntax

```
C/C++ Syntax:
#pragma omp context [onerror(...)]
    statement

Fortran Syntax:
c$omp context [onerror(...)]
    statements
c$omp end context
```

**Fig. 3.** CONTEXT directive syntax

Figure 2 shows the proposed syntax for the ONERROR clause. *Action* specifies what action to perform when an error covered by the clause arises. Its value must be one from Table 3. *Err_list* is an optional comma separated list of error severities. It specifies to which errors (i.e. only those of a severity listed in the clause) the given action applies. If no error severity is specified the clause is applied to any error. The optional *arg_list* can be used when the specified *action* is a user callback. Then it can be comma separated list of expressions that will be passed to the callback function on its invocation. Additionally, some other actions can also have optional arguments (e.g. in OMP_RETRY the maximum number of retries can be specified).

When no ONERROR directive has been specified the implementation will decide how to handle errors.

## 3.3 The CONTEXT Directive

The use of the ONERROR clause can not handle intrinsics outside the scope of a directive and it can not specify the same error handling properties for multiple directives. To solve that, our proposal introduces a new OpenMP directive, called CONTEXT.

The CONTEXT directive allows to define an execution context to which different properties could be attached. This proposal only defines the error handling property but others could be defined (e.g. scheduling, number of threads, ...). The properties of the CONTEXT directive are activated when entering the dynamic extent of the directive and they are deactivated upon exit. Inside the dynamic extent this properties are applied to all the code (including subfunctions) unless they are overridden by a nested CONTEXT directive or an specific clause in another directive (e.g. ONERROR).

Figure 3 shows the proposed syntax for this directive. The CONTEXT directive can be followed by one or more ONERROR clauses.

We would like to note that a non-CONTEXT directive with an ONERROR clause can be also be seen as a compound of that directive nested inside a CONTEXT directive with the ONERROR clause.

Another possible idea would be to have the CONTEXT directive applied to a ?le. This way component wise policies could be de?ned without having to modify each function. It could be particularly useful for encapsulated components such as libraries.

## 3.4 Error Support Intrinsics

Additionally, our proposal defines a number of intrinsics that the programmer can use in the callback function to inquire about the different aspects of the risen error. These are:

**omp_error_severity_t omp_error_get_severity(omp_err_info_t \*)** -
    Returns the severity assigned by the runtime to the error.
**omp_error_type_t omp_error_get_type (omp_err_inf_t \*)** -
    Returns the type of the error.
**int omp_error_get_str (omp_err_info_t \*, char \*buf, int n)** -
    Returns a human readable description of the severity and type of the error.

**int omp_error_get_source_file (omp_err_info_t \*, char \*buffer, int n)** -
Returns the filename where the code that arose error is.
**int omp_error_get_source_line (omp_err_info_t \*)** -
Returns the line in the code where the error has arisen.

### 3.5 Examples

In this section we show a few examples using the proposed constructions.

Figure 4 shows a simple code that defines in the top level function that all errors that occur must be ignored. Note that in this simple way it can be defined a default error handling policy for all the application.

Figure 5 shows how the CONTEXT directive may include OpenMP intrinsics and directives that will share the error handling properties. Also, it shows how multiple ONERROR clauses can be used to specify different behaviors depending on the error severity.

```
1   int main ()
2   {
3   #pragma omp context\
4           onerror(OMP_IGNORE)
5       my_code();
6   }
```

```
1   void f(int num_threads) {
2   #pragma omp context onerror(OMP_ERR_MILD:OMP_IGNORE)\
3           onerror(OMP_ERR_SEVERE,OMP_ERR_FATAL:OMP_ABORT)
4   {
5       omp_set_num_threads(num_threads);
6   #pragma omp parallel
7   {
8       /* parallel code */
9   }
10  }
11  }
```

**Fig. 4.** Example defining an error policy for the application

**Fig. 5.** Example with multiple ONERROR clauses

Figure 6 shows a possible use of the ONERROR clause to save all computed data when a serious error occurs. The *savedata* callback will be called before the application aborts its execution allowing the application to save the data computed so far.

Figure 7 shows a more complex example of a server-like application. In this case if there is any error when starting the processing of requests the PARALLEL directive is aborted and it will be tried again in the next iteration of the server. If while processing a request there is any error (e.g. memory exhaustion ) the *process_error* callback will be called. The callback will close the related request and abort the SINGLE execution. In this way, while some requests may get lost when errors arise the server will be able to continue processing new requests.

## 4 Implementation

We have implemented partially the error support described in the previous section. The modifications have been developed in the NANOS environment[10]:

```
1   omp_error_action_t savedata (error)
2   omp_err_info_t *error;
3   {
4       /* save computed data */
5       return OMP_ABORT;
6   }
7
8   void f ( )
9   {
10  #pragma parallel do \
11          onerror(OMP_ERR_SEVERE,\
12          OMP_ERR_FATAL: savedata)
13  {
14      /* parallel code */
15  }
16  }
```

**Fig. 6.** Example where data is saved before abortion

```
1   omp_error_action_t process_error (error,request)
2   omp_err_info_t *error;
3   Request *request;
4   {
5       close_connection(request);
6       dequeue(request);
7       return OMP_SKIP;
8   }
9
10  void process_requests()
11  {
12  #pragma omp parallel onerror(OMP_SKIP)
13          while (request=next_in_queue(&ready_queue)){
14  #pragma omp single nowait \
15          onerror(process_error,request)
16          process_request(request);
17      }
18  }
```

**Fig. 7.** Example a server-like request processing aware of OpenMP errors

in its runtime and in the code generated by the Mercurium compiler[11]. The runtime has been extended to add several services for error handling. The compiler has been modified so the generated code makes use of this services. So far, only the PARALLEL, PARALLEL DO and DO constructs are supported in our prototype. But adding support for most of the other constructs should be straightforward.

This section considers some non obvious issues of the implementation that has to support error handling.

### 4.1 Dynamic Extent of the Error Policies

Since *ONERROR* prolongs its semantics along the dynamic extent of the application, error handling code is always needed, even if no *ONERROR* clauses were specified for an OpenMP construct. Additionally, the runtime has to be able to get the current error handling procedure despite the exact routine that triggers the error.

This issue has been solved saving the error context in the thread descriptor. It will be updated every time a new *ONERROR* clause is specified. To implement the dynamic extend semantics we store the previous error context in the thread stack so it can be restored later upon exit of the dynamic extent of the *ONERROR* clause.

### 4.2 Callback Argument Evaluation

When a callback is specified for an error severity level it can be given several arguments. These arguments must be evaluated when the error is detected and passed to the callback when it is invoked. The problem here is how to evaluate these arguments, since they may belong to a lexical scope not accessible in the point where the error is triggered.

In order to be able to evaluate the expressions involved in the arguments, the compiler stores in the thread stack a callback descriptor that saves references to all the variables appearing in the callback arguments. A reference to this descriptor is saved in the error context (already saved in the thread descriptor).

When an error is detected, and before the callback is invoked, an evaluation function is called. This evaluation function is generated by the compiler and it uses the callback descriptor to evaluate the argument expressions. These evaluated expressions are then passed to the user specified callback when it is called.

### 4.3 Additional Barriers

Some of the error actions require that the threads do not continue until all of them have accomplished correctly some step (e.g. for the PARALLEL construct under OMP_SKIP all threads must be created correctly before the work can start). This means that some additional barriers are needed, even if no error happens, to comply with the user specified behavior in those cases.

In our prototype we protect these barriers with a condition that ensures that each barrier is really necessary for the current error handling semantics. These avoids unnecessary overhead under most situations.

## 5 Evaluation

To evaluate the runtime error recovery support proposal for OpenMP we have used two approaches. The first one has consisted on testing a slightly modified version of the 2.0 EPCC OpenMP Microbenchmarks [12]. The second one has been running NAS 3.0 benchmarks [13].

### 5.1 Evaluation Purpose

The purpose of this evaluation is to see how the error recovery support impacts the performance of the application. Run applications did not have any error along its execution. We were only measuring the overhead of their ability to deal with possible runtime errors (even if they do not happen).

### 5.2 Common Environment

The evaluation was run in a dedicated machine with 16-way 375Mhz Power3 and 4 Gb of memory running AIX 5.2. The native compiler in this environment has been XL Fortran 95 8.1.1. The OpenMP runtime has been the modified NANOS with error recovery support described in section 4.

## 5.3 Tested Scenarios

As seen in implementation section, barriers may be required for proper implementation of the *ONERROR* semantic specified by the programmer. Since they are not always needed, for the purpose of the evaluation, two scenarios have been considered. The first one assumes no additional barriers will be needed while the other assumes that barriers are always needed. In this way, we will see the overhead for the best and the worst case scenario of the error support.

In order to observe the overhead, the execution time of error recovery enabled applications has been compared against a runtime implementation without this support.

## 5.4 EPCC Microbenchmarks

EPCC Microbenchmarks provide a set of several microbenchmarks intended to evaluate performance of OpenMP implementations. In this evaluation, only syncbench has been considered. It measures how long takes the runtime when entering and then leaving a PARALLEL, PARALLEL DO or DO constructs. This measure is performed several times in an outer loop.

Initial experimentation showed that difference between the two scenarios was not appreciable. To magnify the overhead we modified the microbenchmark outer loop to perform 100 iterations instead of 20. This fact already gives us an idea on how low is the impact of the added code (including additional barriers).

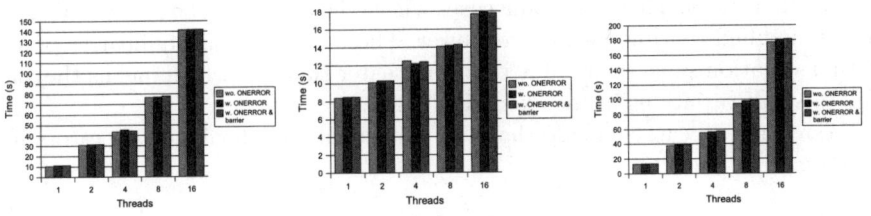

(a) for the PARALLEL directive  (b) for the DO workshare  (c) for the PARALLEL DO directive

**Fig. 8.** Execution times of the EPCC microbencnmarks

Results are depicted in figure 8. For every construction (PARALLEL, DO and PARALLEL DO) they show the execution time in the two aforementioned error semantics (i.e. that need a barrier and that they don't). As a reference, execution time for a runtime with no error recovery support is also showed.

As can be seen in the figure, in any of the two error semantics there is a significative increase in the execution time for the tested OpenMP constructions.

## 5.5 NAS Benchmarks

NAS 3.0 benchmark is a suite of numeric applications available in Fortran intended to test OpenMP implementations. In this analysis we have considered only the class A data set.

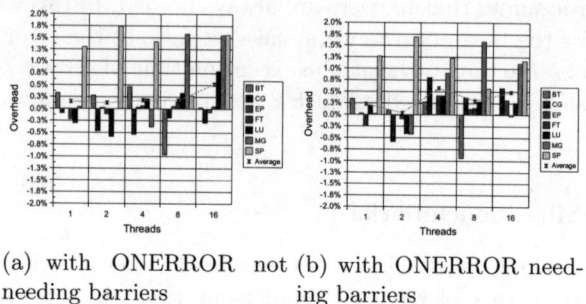

(a) with ONERROR not needing barriers

(b) with ONERROR needing barriers

**Fig. 9.** Overhead of the ONERROR support on the NAS benchmarks (class A)

Like in the EPCC microbenchmarks, the two different semantics described previously have been evaluated. Figure 9 shows the overhead of these scenarios. The overhead has been computed against an execution that had no error support. We can see that the average overhead is around 0.5% which is not a significant value. Even in the worst case (for the SP benchmark) the overhead is never greater than 1.8%. There are some cases where the overhead is negative meaning the unmodified runtime performed slower. These cases are not indicative of any unusual situation since their value is never greater than 1% which means they are due to small variations in the execution times.

So overall, we can conclude that the additional code for handling errors has no noticeable impact.

## 6 Conclusions

In this paper, we proposed an extension to the current OpenMP standard. This extension would allow applications to specify an error handling policy for the OpenMP constructions. Thus applications could increase its reliability by not only being able to react to errors in the user code but also being able to take actions when an error occurs in an OpenMP construct.

The proposal presents a new ONERROR clause for all OpenMP directives that allows to define such a policy. It also presents a new directive called CONTEXT that defines an stackable context of properties for each thread. Error policies can be attached to this context as well, allowing the definition of a common policy for multiple directives and even OpenMp intrinsics.

Because of the CONTEXT directive all OpenMP code inserted by the compiler must be able to detect an error and support any of the different error policies (including some additional barriers under some error semantics). Different experiments have been presented that show this extra code represents a negligible overhead when no error occurs in an application. This means that the inclusion of this support would be worthwhile as it can help increase the reliability of the applications. While at the same time, it will not hurt the performance of those that choose not to use it.

While this proposal may not fully cover all aspects of error handling in OpenMP we think it is a good start for its discussion.

## Acknowledgements

This research has been supported by the Ministry of Science and Technology of Spain under contract TIN2004-07739-C02-01.

## References

1. Balart, J., Duran, A., Gonzàlez, M., Ayguadé, E., Martorell, X., Labarta, J.: Experiences parallelizing a web server with openmp. In: First International Workshop on OpenMP (May 2005)
2. Isensee, P.: Utilizing Multicore Processors with OpenMP. In: Game Programming Gems, vol. 6, Charles River Media (2006)
3. Su Gatlin, K.: Openmp 3.0 feature: Error detection capability. In: Panel at First International Workshop on OpenMP (May 2005), http://www.nic.uoregon.edu/iwomp2005/Talks/gatlin-panel.pdf
4. Huang, Y., Chung, P.E., Kintala, C., Liang, D., Wang, C.: Nt-swift: Software-implemented fault tolerance for windows-nt. In: Proceedings of the 1998 USENIX WindowsNT Symposium (1998)
5. Beazley, D.M.: An embedded error recovery and debugging mechanism for scripting language extensions. In: Proceedings of the USENIX 2001 Annual Technical Conference (June 2001)
6. Donelly, C., Stallman, R.: The Bison Manual: Using the YACC-Compatible Parser Generator. Gnu Press (2004)
7. Elnozahy, M., Alvisi, L., Wang, Y.M., Johnson, D.B.: A survey of rollback-recovery protocols in message passing systems. Technical report, School of Computer Science, Carnegie Mellon University (October 1996)
8. Bronevetsky, G., Marques, D., Pingali, K., Stodghill, P.: Automated application-level checkpointing of mpi programs. In: Proceedings of Principles and Practice of Parallel Programming (PPoPP), June 2003, pp. 84–94 (2003)
9. Dieter, W., Lumpp Jr., J.: A user-level checkpointing library for posix threads programs. In: Proceedings of 1999 Symposium on Fault-Tolerant Computing Systems (FTCS) (June 1999)
10. Nanos project, http://www.cepba.upc.edu/nanos/
11. Balart, J., Duran, A., Gonzàlez, M., Martorell, X., Ayguadé, E., Labarta, J.: Nanos mercurium: a research compiler for openmp. In: Proceedings of the European Workshop on OpenMP 2004 (October 2004)

12. Bull, J.M.: Measuring synchronization and scheduling overheads in openmp. In: First European Workshop on OpenMP (September 1999)
13. Bailey, D.H., Barszcz, E., Barton, J.T., Browning, D.S., Carter, R.L., Dagum, D., Fatoohi, R.A., Frederickson, P.O., Lasinski, T.A., Schreiber, R.S., Simon, H.D., Venkatakrishnan, V., Weeratunga, S.K.: The NAS Parallel Benchmarks. The International Journal of Supercomputer Applications 5(3), 63–73 (1991)

# Extending the OpenMP Standard for Thread Mapping and Grouping

Guansong Zhang

**Abstract.** In this paper, we are exploring the idea of improving the OpenMP 2.5 standard to facilitate parallel programming on emerging architectures. This includes mapping threads to particular processors, improving the load balance among processors by work distribution, and supporting nested parallelism inherited from applications. We will demonstrate the performance gains of thread mapping with our experimental implementation, and propose new concepts in the standard to try addressing the issue in a broader sense.

**Keywords:** OpenMP, thread mapping, thread grouping, nested parallelism, SPMD programming.

## 1 Introduction

As several research papers pointed out, OpenMP [1] is due to an update [2], [3]. The set of features in the existing specification of the API provides essential functionality that were mostly selected from previous shared memory parallel APIs. Recently, the shared memory architectures have interesting developments, from multi-core processors to hyperthread and SMT, from accelerate boards to CELL broadband Engine[4]. The emerging architectures force us to explore new features for the OpenMP standard.

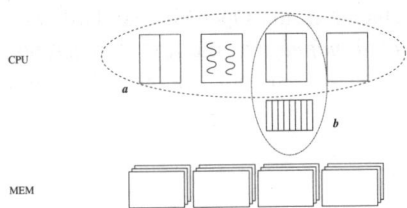

**Fig. 1.** Emerging architectures

Figure 1 is an example of an extreme case of the latest development of shared memory architectures. The picture includes dual core CPUs, hyperthreads or SMT processors. Although the system is not modeled on any real machine [1], it can illustrate the potential problems a programmer may face in the real world.

---
[1] Circle *a* and *b* may represent two different computers, or they are combined together to create an even more complicated structure. This is purely imaginary. We used the figure just to illustrate the complexity of the problem.

Programming on such complicated systems is going to require difficult compromise. The following items are elements that an ideal solution framework should address, quoted directly from OpenMP language committee discussion:

- *Modularity*: Modern software engineering makes heavy use of modular software. OpenMP's existing model, however, does not support compartmentalization of parallelism. For example, MPI defines a communicator. This can be passed to a library and the library is then restricted to the constraints defined by that communicator. Furthermore, the library developer can rename the communicator so the interactions between components of the library are not exposed outside the library.
- *Multi-level machines*: Shared memory machines will become increasingly hierarchical. The combination of SMT and NUMA all come together to create a nightmare for the existing "flat earth" model of OpenMP. The scalar OMP_NUM_THREADS has to expand to encompass a multi-level abstraction of some kind.
- *Mapping OpenMP threads onto processors*: Different machines will have different hierarchies of processors. A program must be able to query the system about the processor hierarchy and then adapt to it by controlling how the OpenMP threads map onto processors.
- *Worksharing between subteams*: To broaden the range of applications appropriate for OpenMP, we need to extend OpenMP so programmers can specify worksharing restricted to subteams.

This is a very ambitious goal. We are not sure if such a solution exist to fit all the requirements. Even if it does exist, the impact could be so large as to make all the previous OpenMP programs obsolete; or make the implementation of the standard so difficult that the drawbacks outweight the advantage it brings to the standard.

In this paper we will describe our attempt to solve this problem. It may not be the answer we are looking for, a lot of things need to be further improved.

To avoid a dramatic change in the OpenMP standard, we propose an incremental approach, to consider *thread mapping* as the first step and address the bigger issue in the second step. We will discuss these topics in the following sections.

## 2 Thread Mapping

The objective of *thread mapping* or *thread binding* is to define a way to associate threads to physical processors. Once mapped, the thread will stay on the processor during the execution without being moved from one processor to another by the operating system. This is also referred to as *thread affinity*.

A typical situation that needs thread mapping is shown in Figure 2, we have two dual core processors, each core is capable of running two threads simultaneously. So the system can provide up to eight *physical threads* to an OpenMP application.

In this system, suppose each processor core has its own L2 cache, and each CPU core has its own L1 cache.

If an OpenMP application running on such system decides to use only four OpenMP threads, then there are at least two mapping options one may choose from,

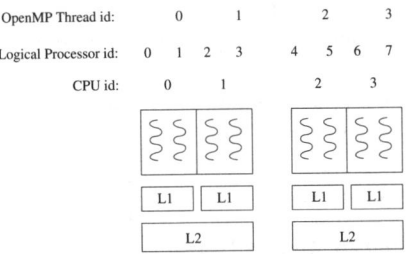

**Fig. 2.** Thread mapping

- let each CPU core have one OpenMP thread, so every processor core is fully utilized. Or
- let each CPU core have two OpenMP threads, leaving two CPUs idle. In this way threads can share more data in cache among the adjacent ones.

It is hard to predict which mapping option can achieve better performance at an abstract level. Depending on the hardware implementation and the program itself, the answer may be different. So it will be desirable to let users define the mapping to get better performance.

### 2.1 Logical Processors

Before we define the actual mapping, it is important to think from an application programmer's point of view, what kind of abstract view the hardware system should look like.

A parallel program itself may be complicated enough that a regular user may not want to know anything more than the number of processors offered in a system. This is partially reflected in the current OpenMP standard. In [1], an API function `omp_get_num_procs` returns the number of the processors available to the program at the time the routine is called. Most of the cases, it will be used to define the initial value for the internal control variable *nthreads-var*, which controls how many threads are going to be created when a parallel region is encountered later on.

The environment variable `OMP_NUM_THREADS` and the API function `omp_set_num_threads` provide ways to change this control variable. Yet in the current standard, there is no clear relationship between entities represented by this control variable and the number returned from the function `omp_get_num_procs`, especially when the two values are different.

The thread binding proposal we present here is to establish the affinity relations between the *OpenMP threads* controlled by the *nthreads-var* variable and the number of processors queried by `omp_get_num_procs`. In our proposal, we consider that the number returned in the function `omp_get_num_procs` is only for a group of *logical processors*. In Figure 2, this number can either be 4 or 8 depending whether the operating system made the extra threads available. It is in contrast to the *physical processors*, which may be only 4 traditionally.

As an initial step, we will organize the logical processor group as a linear array, giving each of them an id number, starting from 0, 1, and up to the total number of the processors minus one.

Then the thread binding problem becomes the problem of having $m$ OpenMP threads, and $n$ logical processor, how to assign those threads to the processors, especially when $m \neq n$.

## 2.2 The Mapping Definition and Its Effect

As explained above, we define the logical processors as a linear array.

We use two environment variables to specify the first processor number that the master thread binds to; and we specify the next processor number the second thread will bind on with a "stride" on the processor array.

For example, the following pair of environment variables

```
OMP_PROC_START=1; OMP_PROC_STRIDE=2
```

will give us the mapping drawn in Figure 2, where thread 0 is on logical processor 1, thread 1 is on logical processor 3, and so on. We will use a *round-robin* fashion to assign the other threads.

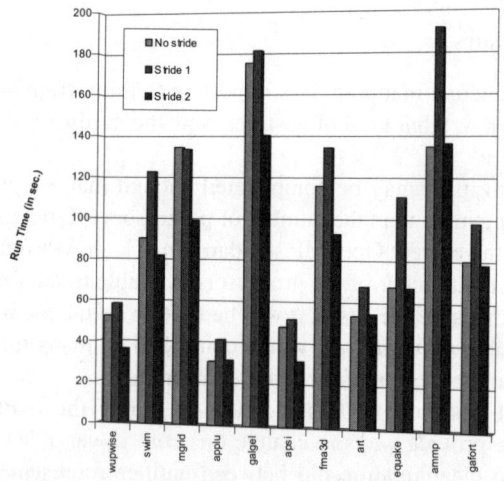

**Fig. 3.** SPEC OMP performance

We used SPEC OpenMP benchmark to demonstrate the performance impact of thread mapping. In Figure 3, a 64 processor core POWER5 machine is used to measure the OpenMP suite. The SMT option of POWER5 was on. So each processor is capable of running two threads. We set OMP_NUM_THREADS to be 64, and compared the following situations

- No processor mapping specification.
- Stride as one by OMP_PROC_START=0; OMP_PROC_STRIDE=1, and
- Stride as two by OMP_PROC_START=0; OMP_PROC_STRIDE=2

The rest of the execution environment, including compilation flags, are all the same. From the figure, one can see that stride two setting has the best overall performance numbers; Stride one setting is the worst: No stride setting is in between. And some of the difference can be as big as 40%, such as in equak, apsi and wupwise.

## 2.3 More About the Mapping

In the mapping scheme above, all the processors and threads are kept in a simple linear relation. More complicated mapping may be achieved though other structures. For example, we can define a arbitrary mapping as an integer array. We are not sure that such level of complexity is useful. We will see that more complicated mapping actually can be achieved in the next section when we extend the logical processor array to a *processor group*.

We also need to point out that the mapping we defined here is only the one from the OpenMP threads to the logical processors. We did not define how logical processors mapped to physical threads. It is left as an implementation defined issue.

# 3 Processor Group

In the previous section, all the logical processors are considered as a linear array. This is a simplification which will hide all the architecture structure of a real machine. In this section, we will extend the view to represent more complex architectures. To support this, we need to introduce the following concepts in the OpenMP APIs.

- *Hierarchical* level: This represents a programmer's view of the hierarchy level that the hardware support. For example, a 4-way Pentium processor with hyper thread enabled can be viewed as either one group of eight processors on the same level, an 8 element linear array; Or two levels of processor groups, with four processor groups at the top level and two processors in each group at the second level.
- *Processor group*: A handle to an opaque data structure which represents a group of processors. It will be a new internal type in the language extension. We will use a temporary name omp_procs to represent this type. (A previous example in the language is a variable of omp_lock type.) As in HPF[5], a variable of this type can be used with Fortran array syntax.
- omp run on *construct*: This will specify on which processor group OpenMP threads will continue to execute and form a new group. When a program is running, the sequential part will be executed on a master thread in the *master group*. When a parallel region is encountered, the group members will execute in parallel

and share the workshares bound to the region[2]. The mapping method introduced in the previous section will still work here for distributing the threads among the processors[3].

To express the three concepts in the OpenMP API, the following extensions are suggested[4]

```
const omp_procs * omp_get_procs(void);
```

This function will return a runtime variable, which has the data type as a pointer to omp_procs. This is the address of our special handle pointing to the underlining machine where the program will run. By default, all the program was executed as if it was run on this processor group. The const modifier indicating that a user can not change this variable.

The processor group provides a way to encapsulate detailed system informations for an application. We will use some annotated pseudo code as examples to further illustrate these ideas.

### 3.1 Group Operation

Suppose we have a machine setup as Figure 1 *a*, and we assume for certain reasons the single core chips are faster.

These functions

```
const omp_procs & processor = * omp_get_procs();
const omp_procs & processor = * omp_get_running_procs();
```

at the beginning of the program will both give us a variable named processor to represent the logical view of a processor group. We can define a member function to get the number of members of the group

```
const int omp_procs::get_num_members(void) const;
```

We use the [ ] operator to overload the function to get member element

```
const omp_procs::get_member(int index) const;
```

Then we may have,

```
const omp_procs g0=processor[0];
const omp_procs g1=processor[1];
const omp_procs g2=processor[2];
const omp_procs g3=processor[3];
```

---

[2] More discussion is needed for the situation that another group is specified on the second level of parallel region. Possible issues may include thread migrating and thread stealing.

[3] We may decide later whether to allow different mapping for different processor groups through multiple *internal control variables*

[4] The function names are all tentative, and can be all different in different language bindings. And they may not be limited to these.

Suppose the processors were organized as two levels of hierarchy. So,

```
// g0 and g2 have two processors each
Assert(g0.get_num_members()==2 && g2.get_num_members()==2);
// g1 and g3 only have one processor each
Assert(g1.get_num_members()==1 && g3.get_num_members()==1);

// omp_procs is a hierarchy,
// it won't be 6 if queried at this level
Assert(get_omp_procs()->get_num_members()==4);
```

Again, we get members as,

```
const omp_procs g4=g1[0]; const omp_procs g5=g1[1];
const omp_procs g6=g3[0]; const omp_procs g7=g3[1];
```

We like to let the two faster processors each having two logical processors,

```
// Create new threads on g1 and g3
omp_procs * gptr0=new(g1) omp_procs[2];
omp_procs * gptr1=new(g3) omp_procs[2];

// Let operator [] work with a processor group address,
// same as ptr->get_member(int index)
const omp_procs g8=gptr0[0];
const omp_procs g9=gptr0[1];

const omp_procs g10=gptr1[0];
const omp_procs g11=gptr1[1];
```

We organize them as a new flat array of processors.

```
// This uses c++ array constructor
const omp_procs g12[]={g8,g9,g4,g5,g10,g11,g6,g7};
```

Now we can write code like this,

```
// On all the processors we grouped previously
#pragma omp parallel on g12[:]
{
  // all the old omp code should be here
  ...
}
```

A triplet [::] with optional *start:end:stride* is used here to get a "section" of the array as a group, the same as used in a Fortran array.

We can also write,

```
#pragma omp parallel on g12
{
    // On all the odd numbered processor
    #pragma omp run on g12[1::2]
    {
        ...
    }
}
```

For C++ completeness, we should do the following at the end of the program,

```
delete[] gptr0; delete[] gptr1;
```

In Fortran and C, we do not have objects, we can only access an address of an object. Besides, we do not have the operator overloading, the corresponding concepts have to be implemented through functions[5]. Suppose we have a user defined function can get the address of g12[:] defined in the previous code. And we want to start a new parallel region on all the odd numbered processors in Fortran,

```
            USE OMP_LIB   ! or INCLUDE "omp_lib.h"
            PARAMETER (N=10)
            INTEGER (OMP_PROCS_KIND) :: GROUP, NEWGROUP
            INTEGER (OMP_PROCS_KIND), DIMENSION(N) :: GROUPARRAY

!           User defined function, which calls a C routine.
            INTEGER (OMP_PROCS_KIND) GET_THE_C_GROUP_HANDLE
            EXTERNAL GET_THE_C_GROUP_HANDLE

            CALL OMP_INIT_PROCS(GROUP)
!           Call the user defined function,
!           to get the group handle in the previous C code.
            GROUP = GET_THE_C_GROUP_HANDLE()

!           In F77, there is no allocatable array.
            IF (OMP_PROCS_GET_NUM_MEMBERS(GROUP) .GT. N) STOP

!           Use DO loop if we don't like an extra function name
            CALL OMP_INIT_PROCS_ARRAY(GROUPARRAY)
!           We know the group is g12, it is a flat 8 element array
!           Get the immediate members of the top level.
            CALL OMP_PROCS_GET_MEMBERS(GROUP, GROUPARRAY)

            CALL OMP_INIT_PROCS(NEWGROUP)
!           Lets use F90 array syntax,
            CALL
```

---

[5] To make it clear, OMP_PROCS_ is used as the prefix for all these type of functions. In addition, like omp_lock, we may need omp_init_procs and omp_destroy_procs.

```
        OMP_PROCS_SET_MEMBERS(NEWGROUP,GROUPARRAY(1::2))\\
!OMP$ PARALLEL ON NEWGROUP
        ...
!OMP$ END PARALLEL
        CALL OMP_DESTROY_PROCS(GROUP)
        CALL OMP_DESTROY_PROCS_ARRAY(GROUPARRAY)
        CALL OMP_DESTROY_PROCS(NEWGROUP)
```

We need to use explicit functions to group an array of processors as a simple handle and convert it back. (This is done through the [] operator in the previous C++ coding.) In fact, in C/C++ and Fortran, if we define the on clause working on both scalar and array type of omp_procs variables, the previous snips can be further simplified, as some of the conversions are not needed.

## 3.2 Programming Examples

We use a simple task to show a "real" program. Suppose we will calculate the sum of 100 numbers held in array a[] as a reduction. (This is not for performance, just to illustrate ideas.) And we assume that the machine is configured as two levels of hierarchy.

We can use the processor group denoted by the previous array g12 to compute this with 8 threads,

```
#pragma omp parallel on g12
{
  #pragma omp for reduction(+:s)
  for (int i = 0; i < 100; i++) s+=a[i];
}
```

Alternatively, we can do

```
// get a flat processor array
#pragma omp parallel on (* processor.all())
{
  #pragma omp for reduction(+:s)
  for (int i = 0; i < 100; i++) s+=a[i];
}
```

Here omp_procs * omp_procs::all(void) const is a member function of the processor group, which returns an address of a processor group consist of all the lowest level processors available.

The differences of the two snips is that the first one will use the same number of threads as the available processors in group g12, i.e., 8 threads; While the second segment only uses 6 threads, since we did not create new processors explicitly, the function will return all the 6 physical processors available.

If we reconfigure the system, and let the two faster CPUs each offering two logical processors, as we specified that how to map local processors to physical processors is still an implementation defined issue, then the second writing of the program should have the same effects as the first one before the reconfiguration[6].

Even though the processor group given by the system is hierarchical, all the previous examples are organizing processors as a flat array, now we will check ways to use nested levels explicitly. If a user want to take advantage of the architecture levels directly, (or organize a processor array as a hierarchy suitable to his nested parallel application) he may write the following code in Fortran,

```fortran
      INTEGER (OMP_PROCS_KIND) :: GROUP

      ! Get the processor hierarchy
      CALL OMP_INIT_PROCS(GROUP)
      GROUP = OMP_GET_PROCS()

!OMP$ PARALLEL ON (GROUP) REDUCTION(+:S)
      ID = OMP_GET_THREAD_NUM()
      MYSTART = 1 + 25*ID  ! 25 should be calculated
      MYEND = MYSTART + 25 - 1

      !This is a real hardware, mapping rules are needed
      IF ((ID .EQ. 1) .OR. (ID .EQ 3)) THEN
         ! We are on the two faster single core CPUs
!OMP$    PARALLEL DO NUM_THREADS(2) REDUCTION(+:S)
         DO I = MYSTART, MYEND
            S = S + A(I)
         ENDDO
!OMP$    END PARALLEL DO
      ELSE
         ! We are on the two processors with 2 CPU cores
!OMP$    PARALLEL DO REDUCTION(+:S)
         DO I = MYSTART, MYEND
            S = S + A(I)
         ENDDO
!OMP$    END PARALLEL DO
      ENDIF
!OMP$ END PARALLEL

      CALL OMP_DESTROY_PROCS(GROUP)
```

If more member functions as query function are defined for the processor group, a user may get complete information from the handle. For example, communication routines can be customized directly by a user.

```c
    void my_function(omp_procs * g) {
      // more query functions on *g
      ...
      #pragma omp parallel on *g
      ...
    }
```

---

[6] So with the concept of the processor group, a well configured system hierarchy can hide all the machine details from regular users. The rather confusing new operator in the previous example should be regarded only as an advanced feature.

## 4 Summary

In this paper, we first gave out an simple mapping mechanism for binding threads to particular processors. We showed that even with this simple scheme we can improve performance numbers for real benchmark suites.

We further extended the concept of logical processors to a processor group, which addressed the OpenMP programming issues raised earlier. The concepts we presented here are still in draft. A lot of issues, including syntax definition, still need to be refined.

We can summarize the main features of the framework as following,

- *Structured SPMD programming*: SPMD programming is the most common techniques used in parallel programming, especially for data parallel programming. As more processors are available in a real parallel system, and more applications begin to looking for speed up exploiting parallel processing, the traditional way of partition computation with data distribution alone can not fulfill users' need. Specifically work distribution or task distribution is needed among processors. Instead of just using the processor ID number as a conditional guard to distribute work, our programming model is based on a well designed structure — processor group. We call this programming style *structured SPMD* programming.
- *Backward compatibility*: When we introduce the concept of the processing group, we try to consider the backward compatibility. All the previous OpenMP codes will still be legal with out any change. They are running on the *default* processor group provided by the system.
- *Information encapsulation*: We hide most of the detailed information of a physical machine by the logical processor group. We believe this will help user to write portable code without too much machine specific information. In fact, we think most of the time the flat array machine configuration plus thread mapping are good enough for regular users to write efficient code.
- *System configurations*: Although we did not specify the details of system configuration here, it is possible to develop a resource manager that allows a physical system to be configured differently for different applications or partially available to particular applications. It is also possible to configure a system to have multiple groups with different attributes, so an application may target to a heterogeneous architecture.
- *Integrated solution*: The proposal we presented here actually addressed all the issues listed by the OpenMP language committee discussion group as in section 1. Currently we are not sure whether subgrouping threads will be merged in the future OpenMP programming, but we believe that providing a parallel context for subroutines and library functions with the concept of processor group is useful for the OpenMP programming model.
- *Incremental implementation*: There is no real runtime implementation to support the proposal in this paper yet. Some of the earlier work on distributed memory system can be traced through [6]. We think the framework can be implemented with incremental steps. In the first step, the main structure omp_procs and its supporting functions will be added as C++ library functions. Most of the concepts in the proposal will be expressed as function calls. In the second step, we can improve the language syntax, to bind the programming model to Fortran and C/C++. In

the third step, more compiler analysis will be used to improve the efficiency of the code. We hope that most of the operation overhead inside the runtime system can be optimized away or moved out of the hot spot with code motion. As those functions should not have any side effects on user variables.

Parallel programming was never an easy task, and new machine architectures posed even more challenges in it. Unfortunately the extra concepts in the language we present here do not make any of this simpler. Yet they provide a set of tools for users to have more controls over the system.

As we implemented the full features of the OpenMP 2.5 APIs, We are well aware of the possible overhead these extra concepts may bring to the standard. It is not the intention of this paper to discuss whether these kinds of complications are necessary. Rather we hope the paper can serve as a discussion base to see if this is the direction of the future OpenMP. It will be up to the users to decide which way OpenMP development should go.

## 5 Trademarks and Copyright

IBM, POWER are trademarks or registered trademarks of International Business Machines Corporation in the United States, other countries, or both.

Other company, product and service names may be trademarks or service marks of others.[7]

© Copyright International Business Machines Corporation, 2006. All rights reserved.

## References

1. OpenMP Architecture Review Board. Openmp application program interface version 2.5 (2005), http://www.openmp.org
2. Gonzalez, M., Oliver, J., Martorell, X., Ayguade, E., Labarta, J., Navarro, N.: OpenMP Extensions forThread Groups and Their Run-time Support. In: Midkiff, S.P., Moreira, J.E., Gupta, M., Chatterjee, S., Ferrante, J., Prins, J.F., Pugh, B., Tseng, C.-W. (eds.) LCPC 2000. LNCS, vol. 2017, Springer, Heidelberg (2001)
3. Chapman, B.M., Huang, L., Jin, H., Jost, G., de Supinski, B.R.: Support for flexibility and user control of worksharing in openmp (2005), http://www.nas.nasa.gov/News/Techreports/2005/PDF/nas-05-015.pdf
4. Cell broadband engine resource center (2005), http://www-128.ibm.com/developerworks/power/cell
5. Koelbel, C.H., Loveman, D.B., Schreiber, R.S.: The High Performance Fortran Handbook. MIT Press, Cambridge (1993)
6. Zhang, G., Carpenter, B., Fox, G., Li, X., Wen, Y.: Structured SPMD programming (1998), http://grids.ucs.indiana.edu/ptliupages/projects/HPJava/reports/structuredSPMD/javad.pdf

---

[7] The opinions expressed in this paper are those of the authors and not necessarily of IBM.

# Author Index

Amsaleg, Laurent  289
an Mey, Dieter  95, 217, 255, 300
Antonopoulos, Christos D.  133
Armstrong, Brian  24
Ayguadé, Eduard  191, 422

Bae, Hansang  24
Balart, Jairo  191
Berry, Michael W.  107
Bezold, Matthias  409
Bronevetsky, Greg  324
Brown, Russell  349
Brunst, Holger  5
Bücker, H. Martin  372

Chapman, Barbara  178, 267, 361
Chun, Huang  51
Costa, Juan José  422
Curtis-Maury, Matthew  133

de Supinski, Bronis R.  167, 324
del Cuvillo, Juan  230
Ding, Xiaoning  133
Dongarra, Jack  267
Duran, Alejandro  191, 422

Eigenmann, Rudolf  24

Ferrer, Roger  422
Fürlinger, Karl  15

Gao, Guang R.  230
Gerndt, Michael  15
Gonzàlez, Marc  191, 422
Gross, Louis J.  107
Gross, Sven  95

Ha, Soonhoi  242
Hadjidoukas, Panagiotis E.  289
Hasegawa, Hidehiko  153
Hernandez, Oscar  267
Hoeflinger, Jay P.  167
Holmgren, Sverker  382
Hörschler, Ingolf  217
Huang, Lei  178

Ierotheou, Constantinos  217

Jeun, Woo-Chul  242
Johnson, Steve  217
Jost, Gabriele  255

Kajiyama, Tamito  153
Karl, Wolfgang  65
Karlsson, Sven  78
Kee, Yang-Suk  242
Klemm, Michael  409
Ko, Walden  207
Kotakemori, Hisashi  153

Labarta, Jesús  191, 422
Leggett, Peter  217
Leopold, Claudia  312, 397
Liao, Chunhua  178
Lin, Yuan  36
Liu, Zhenying  178
Löf, Henrik  382

Malony, Allen D.  119, 279
Martorell, Xavier  191, 422
Mazurov, Oleg  255
Mohr, Bernd  5, 267
Moore, Shirley  267
Morris, Alan  279
Müller, Matthias S.  145

Nikolopoulos, Dimitrios S.  133
Nishida, Akira  153
Nordén, Markus  382
Nukada, Akira  153

Pan, Zhelong  24
Perng, Ruey-Kuen  361
Philippsen, Michael  409
Polychronopoulos, Constantine D.  207

Rantakokko, Jarmo  382
Rasch, Arno  372
Reichelt, Volker  95

Salman, Adnan 119
Sharapov, Ilya 349
Shende, Sameer S. 279
Song, Fengguang 267
Spiegel, Alexander 95, 217
Süß, Michael 312, 397
Suda, Reiji 153

Tao, Jie 65
Terboven, Christian 95, 300
Trinitis, Carsten 65
Turovets, Sergei 119

Vehreschild, Andre 372
Veldema, Ronald 409
Volkov, Vasily 119

Wang, Dali 107
Weng, Tien-Hsiung 361
Wirz, Alexander 397
Wolf, Felix 267

Xuejun, Yang 51

Zhang, Guansong 435
Zhu, Weirong 230

Printing: Mercedes-Druck, Berlin
Binding: Stein+Lehmann, Berlin

# Lecture Notes in Computer Science

Sublibrary 1: Theoretical Computer Science and General Issues

For information about Vols. 1– 4703
please contact your bookseller or Springer

Vol. 5038: C.C. McGeoch (Ed.), Experimental Algorithms. X, 363 pages. 2008.

Vol. 5036: S. Wu, L.T. Yang, T.L. Xu (Eds.), Advances in Grid and Pervasive Computing. XV, 518 pages. 2008.

Vol. 5018: M. Grohe, R. Niedermeier (Eds.), Parameterized and Exact Computation. X, 227 pages. 2008.

Vol. 5015: L. Perron, M.A. Trick (Eds.), Integration of AI and OR Techniques in Constraint Programming for Combinatorial Optimization Problems. XII, 393 pages. 2008.

Vol. 5011: A.J. van der Poorten, A. Stein (Eds.), Algorithmic Number Theory. IX, 455 pages. 2008.

Vol. 5010: E.A. Hirsch, A.A. Razborov, A. Semenov, A. Slissenko (Eds.), Computer Science – Theory and Applications. XIII, 411 pages. 2008.

Vol. 5008: A. Gasteratos, M. Vincze, J.K. Tsotsos (Eds.), Computer Vision Systems. XV, 560 pages. 2008.

Vol. 5004: R. Eigenmann, B.R. de Supinski (Eds.), OpenMP in a New Era of Parallelism. X, 191 pages. 2008.

Vol. 4996: H. Kleine Büning, X. Zhao (Eds.), Theory and Applications of Satisfiability Testing – SAT 2008. X, 305 pages. 2008.

Vol. 4988: R. Berghammer, B. Möller, G. Struth (Eds.), Relations and Kleene Algebra in Computer Science. X, 397 pages. 2008.

Vol. 4981: M. Egerstedt, B. Mishra (Eds.), Hybrid Systems: Computation and Control. XV, 680 pages. 2008.

Vol. 4978: M. Agrawal, D. Du, Z. Duan, A. Li (Eds.), Theory and Applications of Models of Computation. XII, 598 pages. 2008.

Vol. 4975: F. Chen, B. Jüttler (Eds.), Advances in Geometric Modeling and Processing. XV, 606 pages. 2008.

Vol. 4974: M. Giacobini, A. Brabazon, S. Cagnoni, G.A. Di Caro, R. Drechsler, A. Ekárt, A.I. Esparcia-Alcázar, M. Farooq, A. Fink, J. McCormack, M. O'Neill, J. Romero, F. Rothlauf, G. Squillero, A.Ş. Uyar, S. Yang (Eds.), Applications of Evolutionary Computing. XXV, 701 pages. 2008.

Vol. 4973: E. Marchiori, J.H. Moore (Eds.), Evolutionary Computation, Machine Learning and Data Mining in Bioinformatics. X, 213 pages. 2008.

Vol. 4972: J. van Hemert, C. Cotta (Eds.), Evolutionary Computation in Combinatorial Optimization. XII, 289 pages. 2008.

Vol. 4971: M. O'Neill, L. Vanneschi, S. Gustafson, A.I. Esparcia Alcázar, I. De Falco, A. Della Cioppa, E. Tarantino (Eds.), Genetic Programming. XI, 375 pages. 2008.

Vol. 4967: R. Wyrzykowski, J. Dongarra, K. Karczewski, J. Wasniewski (Eds.), Parallel Processing and Applied Mathematics. XXIII, 1414 pages. 2008.

Vol. 4963: C.R. Ramakrishnan, J. Rehof (Eds.), Tools and Algorithms for the Construction and Analysis of Systems. XVI, 518 pages. 2008.

Vol. 4962: R. Amadio (Ed.), Foundations of Software Science and Computational Structures. XV, 505 pages. 2008.

Vol. 4961: J.L. Fiadeiro, P. Inverardi (Eds.), Fundamental Approaches to Software Engineering. XIII, 430 pages. 2008.

Vol. 4960: S. Drossopoulou (Ed.), Programming Languages and Systems. XIII, 399 pages. 2008.

Vol. 4959: L. Hendren (Ed.), Compiler Construction. XII, 307 pages. 2008.

Vol. 4957: E.S. Laber, C. Bornstein, L.T. Nogueira, L. Faria (Eds.), LATIN 2008: Theoretical Informatics. XVII, 794 pages. 2008.

Vol. 4943: R. Woods, K. Compton, C. Bouganis, P.C. Diniz (Eds.), Reconfigurable Computing: Architectures, Tools and Applications. XIV, 344 pages. 2008.

Vol. 4942: E. Frachtenberg, U. Schwiegelshohn (Eds.), Job Scheduling Strategies for Parallel Processing. VII, 189 pages. 2008.

Vol. 4941: M. Miculan, I. Scagnetto, F. Honsell (Eds.), Types for Proofs and Programs. VII, 203 pages. 2008.

Vol. 4934: U. Brinkschulte, T. Ungerer, C. Hochberger, R.G. Spallek (Eds.), Architecture of Computing Systems – ARCS 2008. XI, 287 pages. 2008.

Vol. 4927: C. Kaklamanis, M. Skutella (Eds.), Approximation and Online Algorithms. X, 289 pages. 2008.

Vol. 4926: N. Monmarché, E.-G. Talbi, P. Collet, M. Schoenauer, E. Lutton (Eds.), Artificial Evolution. XIII, 327 pages. 2008.

Vol. 4921: S.-i. Nakano, M.. S. Rahman (Eds.), WALCOM: Algorithms and Computation. XII, 241 pages. 2008.

Vol. 4919: A. Gelbukh (Ed.), Computational Linguistics and Intelligent Text Processing. XVIII, 666 pages. 2008.

Vol. 4917: P. Stenström, M. Dubois, M. Katevenis, R. Gupta, T. Ungerer (Eds.), High Performance Embedded Architectures and Compilers. XIII, 400 pages. 2008.

Vol. 4915: A. King (Ed.), Logic-Based Program Synthesis and Transformation. X, 219 pages. 2008.

Vol. 4912: G. Barthe, C. Fournet (Eds.), Trustworthy Global Computing. XI, 401 pages. 2008.

Vol. 4910: V. Geffert, J. Karhumäki, A. Bertoni, B. Preneel, P. Návrat, M. Bieliková (Eds.), SOFSEM 2008: Theory and Practice of Computer Science. XV, 792 pages. 2008.

Vol. 4905: F. Logozzo, D.A. Peled, L.D. Zuck (Eds.), Verification, Model Checking, and Abstract Interpretation. X, 325 pages. 2008.

Vol. 4904: S. Rao, M. Chatterjee, P. Jayanti, C.S.R. Murthy, S.K. Saha (Eds.), Distributed Computing and Networking. XVIII, 588 pages. 2007.

Vol. 4878: E. Tovar, P. Tsigas, H. Fouchal (Eds.), Principles of Distributed Systems. XIII, 457 pages. 2007.

Vol. 4875: S.-H. Hong, T. Nishizeki, W. Quan (Eds.), Graph Drawing. XIII, 402 pages. 2008.

Vol. 4873: S. Aluru, M. Parashar, R. Badrinath, V.K. Prasanna (Eds.), High Performance Computing – HiPC 2007. XXIV, 663 pages. 2007.

Vol. 4863: A. Bonato, F.R.K. Chung (Eds.), Algorithms and Models for the Web-Graph. X, 217 pages. 2007.

Vol. 4860: G. Eleftherakis, P. Kefalas, G. Păun, G. Rozenberg, A. Salomaa (Eds.), Membrane Computing. IX, 453 pages. 2007.

Vol. 4855: V. Arvind, S. Prasad (Eds.), FSTTCS 2007: Foundations of Software Technology and Theoretical Computer Science. XIV, 558 pages. 2007.

Vol. 4854: L. Bougé, M. Forsell, J.L. Träff, A. Streit, W. Ziegler, M. Alexander, S. Childs (Eds.), Euro-Par 2007 Workshops: Parallel Processing. XVII, 236 pages. 2008.

Vol. 4851: S. Boztaş, H.-F.(F.) Lu (Eds.), Applied Algebra, Algebraic Algorithms and Error-Correcting Codes. XII, 368 pages. 2007.

Vol. 4848: M.H. Garzon, H. Yan (Eds.), DNA Computing. XI, 292 pages. 2008.

Vol. 4847: M. Xu, Y. Zhan, J. Cao, Y. Liu (Eds.), Advanced Parallel Processing Technologies. XIX, 767 pages. 2007.

Vol. 4846: I. Cervesato (Ed.), Advances in Computer Science – ASIAN 2007. XI, 313 pages. 2007.

Vol. 4838: T. Masuzawa, S. Tixeuil (Eds.), Stabilization, Safety, and Security of Distributed Systems. XIII, 409 pages. 2007.

Vol. 4835: T. Tokuyama (Ed.), Algorithms and Computation. XVII, 929 pages. 2007.

Vol. 4818: I. Lirkov, S. Margenov, J. Waśniewski (Eds.), Large-Scale Scientific Computing. XIV, 755 pages. 2008.

Vol. 4800: A. Avron, N. Dershowitz, A. Rabinovich (Eds.), Pillars of Computer Science. XXI, 683 pages. 2008.

Vol. 4783: J. Holub, J. Žďárek (Eds.), Implementation and Application of Automata. XIII, 324 pages. 2007.

Vol. 4782: R. Perrott, B. Chapman, J. Subhlok, R.F. de Mello, L.T. Yang (Eds.), High Performance Computing and Communications. XIX, 823 pages. 2007.

Vol. 4771: T. Bartz-Beielstein, M.J. Blesa Aguilera, C. Blum, B. Naujoks, A. Roli, G. Rudolph, M. Sampels (Eds.), Hybrid Metaheuristics. X, 202 pages. 2007.

Vol. 4770: V.G. Ganzha, E.W. Mayr, E.V. Vorozhtsov (Eds.), Computer Algebra in Scientific Computing. XIII, 460 pages. 2007.

Vol. 4769: A. Brandstädt, D. Kratsch, H. Müller (Eds.), Graph-Theoretic Concepts in Computer Science. XIII, 341 pages. 2007.

Vol. 4763: J.-F. Raskin, P.S. Thiagarajan (Eds.), Formal Modeling and Analysis of Timed Systems. X, 369 pages. 2007.

Vol. 4759: J. Labarta, K. Joe, T. Sato (Eds.), High-Performance Computing. XV, 524 pages. 2008.

Vol. 4750: M.L. Gavrilova, C.J.K. Tan (Eds.), Transactions on Computational Science I. XI, 181 pages. 2008.

Vol. 4746: A. Bondavalli, F. Brasileiro, S. Rajsbaum (Eds.), Dependable Computing. XV, 239 pages. 2007.

Vol. 4743: P. Thulasiraman, X. He, T.L. Xu, M.K. Denko, R.K. Thulasiram, L.T. Yang (Eds.), Frontiers of High Performance Computing and Networking ISPA 2007 Workshops. XXIX, 536 pages. 2007.

Vol. 4742: I. Stojmenovic, R.K. Thulasiram, L.T. Yang, W. Jia, M. Guo, R.F. de Mello (Eds.), Parallel and Distributed Processing and Applications. XX, 995 pages. 2007.

Vol. 4739: R. Moreno Díaz, F. Pichler, A. Quesada Arencibia (Eds.), Computer Aided Systems Theory – EUROCAST 2007. XIX, 1233 pages. 2007.

Vol. 4736: S. Winter, M. Duckham, L. Kulik, B. Kuipers (Eds.), Spatial Information Theory. XV, 455 pages. 2007.

Vol. 4732: K. Schneider, J. Brandt (Eds.), Theorem Proving in Higher Order Logics. IX, 401 pages. 2007.

Vol. 4731: A. Pelc (Ed.), Distributed Computing. XVI, 510 pages. 2007.

Vol. 4728: S. Bozapalidis, G. Rahonis (Eds.), Algebraic Informatics. VIII, 291 pages. 2007.

Vol. 4726: N. Ziviani, R. Baeza-Yates (Eds.), String Processing and Information Retrieval. XII, 311 pages. 2007.

Vol. 4719: R. Backhouse, J. Gibbons, R. Hinze, J. Jeuring (Eds.), Datatype-Generic Programming. XI, 369 pages. 2007.

Vol. 4711: C.B. Jones, Z. Liu, J. Woodcock (Eds.), Theoretical Aspects of Computing – ICTAC 2007. XI, 483 pages. 2007.

Vol. 4710: C.W. George, Z. Liu, J. Woodcock (Eds.), Domain Modeling and the Duration Calculus. XI, 237 pages. 2007.

Vol. 4708: L. Kučera, A. Kučera (Eds.), Mathematical Foundations of Computer Science 2007. XVIII, 764 pages. 2007.

Vol. 4707: O. Gervasi, M.L. Gavrilova (Eds.), Computational Science and Its Applications – ICCSA 2007, Part III. XXIV, 1205 pages. 2007.

Vol. 4706: O. Gervasi, M.L. Gavrilova (Eds.), Computational Science and Its Applications – ICCSA 2007, Part II. XXIII, 1129 pages. 2007.

Vol. 4705: O. Gervasi, M.L. Gavrilova (Eds.), Computational Science and Its Applications – ICCSA 2007, Part I. XLIV, 1169 pages. 2007.